CRC SERIES IN MEDICINAL CHEMISTRY

Series Editor-in-Chief

Matthew Verderame
Professor of Medicinal Chemistry
Head, Department of Chemical and
Physical Sciences
Albany College of Pharmacy
Albany, New York

Handbook of Chemotherapeutic Agents (Volumes I and II)

Handbook of CNS Agents and Local Anesthetics

Handbook of Autonomic Drugs and Autacoids

Handbook of Cardiovascular and Anti-Inflammatory Agents

Handbook of Hormones, Vitamins, and Radiopaques

CRC Handbook of CNS Agents and Local Anesthetics

Editor
Matthew Verderame
Professor of Medicinal Chemistry
Head, Department of Chemical and Physical Sciences
Albany College of Pharmacy
Union University
Albany, New York

CRC Press, Inc.
Boca Raton, Florida

Library of Congress Cataloging-in-Publication Data
Main entry under title:

Handbook of CNS agents and local anesthetics.

Includes bibliographies and index.
1. Neuropharmacology. 2. Central nervous system--
Effect of drugs on. I. Verderame, Matthew.
[DNLM: 1. Anesthetics, Local. 2. Central Nervous
System Agents. QV 76.5 H236]
RM315.H344 1986 617'.966 85-22390
ISBN 0-8493-3288-5

Direct all inquiries to CRC Press, Inc., 2000 Corporate Blvd., N.W., Boca Raton, Florida, 33431.

International Standard Book Number 0-8493-3288-5
Library of Congress Card Number 85-22390
Printed in the United States

DEDICATION

To my typist and companion and the mother of our four children — Fran.

PREFACE

The *CRC Handbook of CNS Agents and Local Anesthetics*, is a companion to five other handbooks in this pharmaceutically related set published by CRC Press. The titles of the other five follow: *Handbook of Chemotherapeutic Agents* (Volumes I and II), *CRC Handbook of Anti-Inflammatory and Cardiovascular Agents, Handbook of Autonomic Drugs and Autacoids*, and *Handbook of Hormones, Vitamins, and Radiopaques*.

The six handbooks represent a systematic collection of selective chemical and pharmacological reference data on drugs drawn from the major categories of therapeutically important agents. For the principal classes of drugs, or for individual drugs which do not belong to a particular class, the following topics were addressed: mechanism of action, structure-activity relationships, pharmacokinetics, uses and dosage, toxicity (including drug interactions), and physical-chemical and other data. Drug metabolism data were heavily emphasized in the pharmacokinetics sections. The term "toxicity" was used broadly to also include side effects. Generally included in the physical-chemical and other data sections was such information on the drug as the chemical name, formula, melting point, pKa value(s), solubility and stability data, physical description, salt forms, and any other property deemed worthy of inclusion.

Other topics generally considered included historical aspects leading to or associated with the development of the class of drugs, possible causes of the condition or disease state for which the drugs are used, life cycles of organisms against which the chemotherapeutic agents are used, testing methods employed in the screening of potentially useful compounds, and review articles covering the fields. Chemical syntheses were not emphasized. However, this topic received significant attention in a few sections. Figures were used appropriately to illustrate a point or concept, especially in the mechanism of action sections.

In this series of reviews, emphasis was placed on the most useful drugs associated with each physiological class or disease entity. The monograph format was used extensively, not only to denote the most useful members, but also to provide a close examination of these agents. Comparative pharmacological or therapeutic data on the principal drugs within the classes provide additional, practical information.

Monographs of drugs with useful multiple functions are appropriately placed in more than one section and cross-referenced. Though necessitating some repetition of data, these monographs generally complement each other and maintain a continuity of thought and application.

Newer and/or experimental drugs have also been included to round out the coverage and to present current knowledge and possible future trends.

This set of reference books should have appeal and utility for the practitioners and students of medicinal chemistry, pharmacology, pharmacy, and medicine. It should also be useful to researchers in the pharmaceutical industry and in other allied health professions.

A special debt of gratitude is extended to the many contributors who labored long and diligently to prepare their manuscripts for publication. Also, for their counsel and cooperation, the dedicated assistance of the Advisory Board is acknowledged with sincere appreciation.

The editorial assistance of the CRC staff and, in particular, the coordinating efforts of Ms. Amy Skallerup must not go unmentioned.

Finally, it is our hope that the reader will find these first edition handbooks interesting, useful, and valuable. Comments for improvement will be gratefully received.

Matthew Verderame, Ph.D.
Editor
Albany, New York

THE EDITOR

Matthew Verderame, Ph.D., is Professor of Medicinal Chemistry and Head of the Department of Chemical and Physical Sciences at the Albany College of Pharmacy, Union University. Dr. Verderame received his undergraduate education at the University of Connecticut, where he earned the Bachelor's degree in Chemistry in 1950. At the University of Wisconsin, he received an M.S. degree in Pharmacy in 1952 and the Ph.D. degree in 1955 in Pharmaceutical Chemistry. In 1954, he joined the faculty at the Albany College of Pharmacy as an Assistant Professor and moved through the ranks to his present position. Dr. Verderame's education was enhanced by his attendance of NSF-sponsored institutes at the University of North Carolina in 1962, Michigan State University in 1964, Duke University in 1969, and State University of New York in Albany in 1974. From 1955 through 1968, Dr. Verderame received consecutive year-long research grants from the Sterling-Winthrop Research Institute. In 1968 and again in 1973, he was cited for his contributions to teaching by receiving the Lederle Award. Courses which he regularly teaches include Medicinal Chemistry, Qualitative Organic Analysis, and Drug Analysis. The principal areas of synthetic research of Dr. Verderame include the preparation of various derivatives of cysteine, methionine, and piperazine as possible anti-infective agents. He also synthesized organo tin compounds and ether derivatives of cyclic hydrocarbons as possible anti-infective agents. He has authored several publications and has refereed technical papers. Dr. Verderame is a member of the American Chemical Society, Rho Chi, Sigma Xi, Phi Sigma, Rho Pi Phi, the American Association for the Advancement of Science, the New York Academy of Sciences, the American Association of University Professors, and Phi Lambda Upsilon.

CONTRIBUTORS

H. Jack Adams
Associate Director of Clinical Research
Astra Pharmaceutical Products, Inc.
Westborough, Massachusetts

John D. Alvin
Associate Professor of Pharmacology
School of Pharmacy
University of Pittsburgh
Associate for Toxicology
Pharmakon, Inc.
Pittsburgh, Pennsylvania

Norman R. Boisse
Associate Professor of Pharmacology and Physiology
College of Pharmacy and Allied Health Professions
Northeastern University
Boston, Massachusetts

Michael A. Eissenstat
Group Leader, Medicinal Chemistry
Sterling-Winthrop Research Institute
Rensselaer, New York

John M. Essery
Licensing Manager
Organon, Inc.
West Orange, New Jersey

Dean R. Haubrich
Director of Pharmacology
Sterling-Winthrop Research Institute
Rensselaer, New York

Dennis J. Hlasta
Group Leader
Department of Medicinal Chemistry
Sterling-Winthrop Research Institute
Rensselaer, New York

Beng T. Ho
Professor of Neuropharmacology
Departments of Neuro-Oncology and Chemotherapy Research
Division of Medicine
The University of Texas System Cancer Center
M. D. Anderson Hospital and Tumor Institute
Houston, Texas

Ronald L. Jacoby
Professor of Pharmaceutical Chemistry
School of Pharmacy
Ferris State College
Big Rapids, Michigan

Ronald J. Kotun
Corporate Toxicologist
Comprehensive Safety Compliance, Inc.
Associate for Toxicology
Pharmakon, Inc.
Pittsburgh, Pennsylvania

Patricia M. Kralik
Senior Research Assistant
Department of Neurology
Baylor College of Medicine
Houston, Texas

Daniel Luttinger
Senior Research Biologist
Department of Pharmacology
Sterling-Winthrop Research Institute
Rensselaer, New York

William F. Michne
Section Head, Medicinal Chemistry
Sterling-Winthrop Research Institute
Rensselaer, New York

John L. Neumeyer
Distinguished Professor of Medicinal Chemistry and Chemistry
College of Pharmacy and Allied Health Professions
Northeastern University
Boston, Massachusetts

David E. Nichols
Professor of Medicinal Chemistry and Pharmacology
Department of Medicinal Chemistry and Pharmacognosy
School of Pharmacy and Pharmacal Sciences
Purdue University
West Lafayette, Indiana

Robert A. Ronfield
Manager, Drug Metabolism Department
Pfizer, Inc.
Groton, Connecticut

Bertil Takman
Internal Consultant
Astra Pharmaceutical Products, Inc.
Worcester, Massachusetts

Paul A. Tenthorey
Senior Research Scientist
Pharmaceutical Research and Development Division
Bristol-Myers Company
Evansville, Indiana

Julius A. Vida
Vice President—Licensing
Science and Technology Group
Bristol-Myers Company
New York, New York

TABLE OF CONTENTS

GENERAL ANESTHETICS*

Ronald L. Jacoby

INTRODUCTION

General anesthetics are drugs that produce a reversible depression of the CNS, with a resulting loss of perception of all sensations, including pain. This state of insensibility is often accompanied by unconsciousness and is known as "anesthesia." A less clinical but more physiologic definition would be that anesthetics are agents capable of reversibly blocking neural conduction without any significant effect on resting membrane potential.[1] These are very broad definitions as they must be, to accommodate the diverse compounds, such as xenon, ether, and thiopental, that produce anesthesia.

The use of chemicals other than opium or alcohol, to relieve surgical pain had its beginning in 1844 when Horace Wells, a dentist, introduced nitrous oxide into dentistry. This was followed by William Morton's famous demonstration of ether in the "ether dome" at Massachusetts General Hospital in 1846. Chloroform was introduced as an anesthetic by a Scottish obstetrician, James Simpson, in 1847. Yet, surgery did not become commonplace until aseptic techniques were developed some 20 years later. The interesting history of surgical anesthesia and the life stories of the "inventors" of anesthesia are retold elsewhere.[2-4]

Of these three early anesthetics, only nitrous oxide is commonly used today. It is most often used in combination with other anesthetics since it lacks sufficient potency to be used alone for surgical anesthesia. Diethyl ether has fallen into disuse due to its flammability, prolonged induction and emergence time, and frequent side effects: postanesthetic nausea and vomiting. Chloroform is hepatotoxic, nephrotoxic, and often induces cardiac arrhythmias. The newer anesthetics of today still demonstrate these same disadvantages, although less frequently. Many other gases and volatile liquids have been used as anesthetics and they are listed in Table 1 along with their dates of introduction. Also listed in Table 1 are the i.v. anesthetics that are currently in use. Although these i.v. agents are convenient to use, side effects have limited their use to the production of rapid induction of surgical anesthesia and to their use in special surgical procedures.

Of the volatile anesthetics listed in Table 1 only nitrous oxide, halothane, methoxyflurane, enflurane, and isoflurane are extensively used today. All have disadvantages associated with their use, thus justifying the continued search for the ideal anesthetic. That ideal volatile anesthetic has been described as having the following characteristics: (1) be nonflammable, (2) be inexpensive, (3) require uncomplicated equipment for administration, (4) provide adequate muscular relaxation, (5) produce rapid and uncomplicated induction and emergence, (6) have no effect on myocardium or respiration at anesthetic doses, (7) be chemically and metabolically stable, (8) be sufficiently potent to permit adequate oxygen supply, and (9) have a wide margin of safety.[5]

Fixed or i.v. anesthetics by their very nature possess several of these characteristics. They are nonflammable, easy to administer, inexpensive, and their administration does not displace a part of the patient's oxygen supply. Nonetheless, thiopental sodium, the standard of comparison among i.v. agents, when used as the sole anesthetic agent, can produce serious circulatory, respiratory, and nervous system depression.

Signs and Stages of Anesthesia

Subjecting a patient to general anesthesia is a life-threatening procedure since it requires

* Submitted for publication July 13, 1982.

Table 1
GENERAL ANESTHETICS

Name	Date of introduction
Volatile Anesthetics	
Nitrous oxide U.S.P.[a]	1844
Ether U.S.P.[a] (diethyl ether)	1846
Chloroform N.F.[a]	1847
Ethyl chloride U.S.P.[a]	1848
Ethylene	1923
Vinyl ether U.S.P.[a]	1933
Ethyl vinyl ether	1947
Cyclopropane U.S.P.[a]	1934
Trichloroethylene	1934
Fluroxene	1956
Halothane U.S.P.[a] (Fluothane)	1956
Methoxyflurane U.S.P.[a] (Penthrane)	1960
Enflurane U.S.P.[a] (Ethrane)	1972
Isoflurane (Forane)	1974
I.V. Anesthetics	
Thiopental sodium for injection, U.S.P.[a]	1935
Thiamylal sodium for injection, U.S.P.[a]	1952
Methohexital sodium for injection, U.S.P.[a]	1960
Ketamine hydrochloride injection, U.S.P.[a]	1970

[a] U.S.P. XX or N.F. XV, 1980.

paralysis of function of the CNS. Anesthetics are among the most dangerous drugs used today in that their therapeutic ratios range from two to four. Anesthesiologists have tried to accurately describe the physiologic changes that occur as anesthesia deepens to help determine the level of anesthesia in patients. Best known is the system of Guedel, devised in 1920, to describe the progressive deepening of anesthesia produced by diethyl ether. Guedel divided general anesthesia into four stages:

- **Stage I. Analgesia** — This stage begins with the administration of anesthetic and extends to loss of consciousness. A mild depression of cortical centers allows minor surgical procedures to be performed. A few anesthetics, like nitrous oxide, produce significant analgesia during this stage making them well suited for use in dentistry.
- **Stage II. Delirium** — This stage lasts from the loss of consciousness to the onset of surgical anesthesia. Depression of cerebral cortex and reticular formation results in excitement of motor activity and involuntary activity. This stage is marked by erratic respiration, uncoordinated muscular movements, urinary incontinence, and related responses. Anesthesiologists try to move patients quickly and smoothly through this stage.
- **Stage III. Surgical anesthesia** — The disappearance of the excitement and respiratory irregularities of Stage II mark the onset of this stage which extends until spontaneous respiration ceases. Gillespie[6] divided this stage into four planes numbered 1 to 4, in order of increasing depth of anesthesia. Eye movements, pupil size, certain reflexes, and respiration are monitored to determine the planes of Stage III. Most surgical procedures are carried out during planes 1 through 3 which are collectively named ''surgical anesthesia''.

- **Stage IV. Respiratory paralysis** — This stage begins with the failure of respiration and ends with circulatory failure and could be referred to as overdose.

These stages are clearly observed with diethyl ether which produces very slow deepening of anesthesia. Stages I and II combined are termed "induction time" which is often prolonged and difficult with ether. Newer, less blood-soluble agents and i.v. barbiturates produce such rapid induction that the characteristic stages described by Guedel are not clearly observed. Consequently, very precise methods must be used to administer anesthetics in order to achieve surgical anesthesia without overdosing the patient with its resultant respiratory and circulatory collapse.

Administration of Inhalational Anesthetics

Accurate control of the gases and vapors delivered to the patient is achieved through the use of anesthetic machines. These devices allow the anesthesiologist to deliver oxygen and anesthetic gas or vapor to the patient with control of carbon dioxide and water vapor concentrations in the anesthetic atmosphere. These systems may be of the open, nonrebreathing design in which the exhaled gases escape out into the room. Carbon dioxide absorption is unnecessary and new gases are supplied at a rate which meets the patient's needs. A second system design is the complete rebreathing system or closed system. Oxygen and anesthetic are delivered to the patient and exhaled gases are rebreathed after carbon dioxide is chemically removed, and oxygen and anesthetic are supplemented to replace that absorbed by the patient. These represent the two extremes, nonrebreathing and complete rebreathing systems. Other designs of breathing systems allow for partial rebreathing and may include carbon dioxide absorption equipment. Irrespective of the breathing system design, the anesthetic machine will likely include a vaporizer to vaporize liquid anesthetics, flowmeters, control valves, and perhaps gas warming and humidifying equipment. The reader is referred to other sources for a complete analysis of anesthetic equipment.[7]

PHARMACOKINETICS OF GENERAL ANESTHETICS

Volatile Anesthetics

Uptake, Distribution, and Elimination

Depth of anesthesia is determined by the amount of anesthetic in the brain. Factors determining brain concentration of anesthetic are cardiac output which controls the rate of blood flow through the brain and lungs, alveoli ventilation, and solubility of the anesthetic in tissues and blood. When administration of anesthetic is begun, its concentration in the blood increases toward that in the inhaled gas. If the agent has high blood solubility, several passages of blood through the lungs will be required to saturate the blood with anesthetic. Until this blood reservoir is nearly filled, appreciable amounts of anesthetic will not diffuse into other tissues including the brain. Consequently, anesthetics with relatively high blood solubility will demonstrate slow induction times. Conversely, agents with low blood solubility rapidly saturate the blood allowing for their quick transfer to brain tissue. These relatively blood-insoluble anesthetics have characteristically short induction times.

The most commonly used measure of blood solubility of anesthetics is the blood/gas partition coefficient. This property of anesthetics correlates very well with induction time as can be seen in Table 2. Relatively highly blood-soluble agents like diethyl ether, chloroform, and methoxyflurane have slow induction times, whereas blood-insoluble anesthetics like cyclopropane and nitrous oxide rapidly produce surgical anesthesia.

Once administration of anesthetic is stopped, elimination of the agent begins, mainly through the lungs and the factors that controlled uptake will now operate in reverse fashion during recovery. Blood-insoluble agents will rapidly leave the blood and be exhaled leading to rapid recovery. Highly soluble agents will be released from the blood slowly. For short

Table 2
PARTITION COEFFICIENTS AND INDUCTION-EMERGENCY CHARACTERISTICS OF VOLATILE ANESTHETICS[a]

Anesthetic	MAC[b] Man%	Partition coefficient at 37 ± 0.5°C Oil/Gas	Blood/gas	Induction-emergence
Cyclopropane	9.2	11.8	0.415	Fast
Ethylene	67.0	1.28	0.14	Fast
Diethyl ether	1.92	65.0	12.1	Slow
Divinyl ether	—	58.0	2.8	Fast
Chloroform	—	394.0	8.4	Slow
Ethyl chloride	—	4.05	3.0	Fast
Trichloroethylene	—	960.0	9.15, 9.85	Slow
Halothane	0.73—0.77	224.0	2.3	Fast
Fluroxene	3.4	47.7	1.37	Fast
Methoxyflurane	0.16	970.0	12, 13	Slow
Enflurane	1.68	98.0	1.9, 1.78	Fast
Isoflurane	1.15	98.0	1.41	Fast
Sevoflurane	2.6[c]	55.0	0.6	Fast
Aliflurane	1.22[c]	117.0	1.7	Fast
Nitrous oxide	105.0[d]	1.4	0.47	Fast

[a] Most partition coefficients taken from Reference 8 and 9. Properties for sevoflurane and aliflurane are from References 11, 12, and 13.
[b] Minimum alveolar concentration (MAC) are from Reference 10.
[c] MAC in man is predicted from oil/gas partition coefficient.
[d] A MAC value greater than 100% means hyperbaric conditions are required to reach 1 MAC.

periods of anesthesia, recovery time, like induction time, correlates very well with the blood/gas partition coefficient (Table 2). However, for long periods of anesthesia, tissue solubility also becomes important in determining recovery time. Adipose tissue can provide a large reservoir for highly fat-soluble anesthetics. Since adipose tissue is poorly perfused by blood, the rate of anesthetic elimination will be slow, giving an extended recovery period.

Elimination usually occurs through the lung but a small amount of anesthetic gases are also lost by diffusion through the skin and mucous membranes. Also, varying amounts of anesthetics are now known to be metabolized, leading to both volatile metabolites that are exhaled and water-soluble compounds that are excreted in the urine.

A more detailed analysis of anesthetic uptake and distribution can be found elsewhere.[9,14]

Metabolism

Volatile anesthetics, being hydrocarbons, ethers, and simple halogenated compounds, were formerly considered chemically stable substances. This chemical stability coupled with short exposure times of patients to anesthesia led to the idea that anesthetics were not metabolized; this false notion persisted until 1964 when Van Dyke and co-workers reported on the in vivo metabolism of several anesthetics to CO_2 and chloride ion.[15] Subsequent studies have shown that all volatile organic anesthetics (except, perhaps, cyclopropane) are metabolized to a variable extent with greatest lability to metabolism being demonstrated by methoxyflurane ($CHCl_2CF_2–O–CH_3$). Metabolic studies reveal that as much as 50% of methoxyflurane taken up by the body may be broken down before excretion. At the other extreme is the inorganic agent, nitrous oxide, which is thought not to be metabolized by human liver. Yet, a recent study has shown that nitrous oxide is metabolized by intestinal microflora to nitrogen gas, and further it has been postulated that some of the reactive, metabolic intermediates may be reabsorbed from the intestines leading to the toxic side effects of nitrous oxide.[16]

Metabolism does not account for termination of action of anesthetics since the bulk of it occurs after administration of the agent has been discontinued. However, metabolism may be responsible for some of the toxic effects of anesthetics through the formation of toxic metabolic intermediates or products (see Toxicity of Volatile Anesthetics section of this chapter).

Hydrocarbons are metabolized by expected oxidative pathways yielding ultimately CO_2. The one exception may be cyclopropane. Van Dyke and Chenoweth[17] suggested that cyclopropane was metabolized to CO_2 in rats, but a later investigation[23] was unable to detect any biotransformation of cyclopropane in miniature swine. Simple ethers are cleaved yielding the component alcohols and aldehydes which are then oxidized to the corresponding acids and eventually CO_2. Diethyl ether metabolism in rats yields 4 to 5% CO_2 and 2% urinary metabolites.[19]

The metabolism of halogenated anesthetics is much more complex because microsomal enzymes can cleave all types of carbon-halogen bonds including carbon-fluorine bonds. Table 3 shows the proposed metabolites for several halogenated anesthetics.

It had been assumed that these metabolic reactions were occurring in the liver but it has recently been discovered that pulmonary and renal microsomes can also readily perform many biotransformations of anesthetics.[18]

The degree to which an agent is metabolized is primarily dependent upon its chemical structure; however, Table 4 demonstrates that the length of time that an agent spends in the body is also important. The anesthetics in Table 4 are arranged in order of increasing amount metabolized. Notice that this is accompanied by increasing blood solubility and fat solubility as measured by their respective partition coefficients. Evidently, the ideal agent should have very low blood and fat solubilities which would minimize metabolism and at the same time give the agent very short induction and recovery times. In the following section it will be seen that the situation is not quite so simple since high lipid solubility is required for an anesthetic to possess high potency. Consequently, very low blood solubility is desirable but a balance with lipid solubility must be established to give good potency without undue adipose tissue accumulation of the agent.

I.V. Anesthetics

I.V. anesthetics produce loss of consciousness within 30 to 60 sec after i.v. injection. The duration of action of currently used agents is usually 10 to 30 min with their action being terminated due to redistribution of the anesthetic from brain into muscle tissue. I.V. anesthetics are extensively metabolized with their metabolites being excreted primarily in the urine. Specific information regarding individual i.v. anesthetics can be found in the monographs.

EVALUATION OF ANESTHETICS AND MAC

The procedure most often used to evaluate anesthetics and to determine their relative potencies involves the determination of an agent's MAC. MAC is the minimum alveolar concentration of anesthesia at 1 atm that produces immobility in 50% of those patients or animals exposed to a noxious stimulus.[24] MAC is expressed as a percent of 1 atm and therefore is a partial pressure. At equilibrium, the partial pressure of an anesthetic will be equal in all body tissues including the brain, the site of anesthetic action. Table 2 gives MAC values as determined in man for several anesthetics. The MAC listed for diethyl ether is 1.92%; that is, administration of 1.92% ether in oxygen prevents response (movement)

Table 3
PROPOSED METABOLITES OF FLUORINATED ANESTHETICS

Enflurane[18,20]

$$CHF_2OCF_2CHClF \longrightarrow \begin{matrix} HO{-}CF_2{-}CHClF \\ \text{and/or} \\ HOOC{-}CHClF \end{matrix} + CHF_2OCF_2COOH + F^- + Cl^- \longrightarrow HOOC{-}COOH + F^- + Cl^- + [CF_2O]$$

Fluroxene[20]

$$CF_3CH_2OCH{=}CH_2 \longrightarrow CO_2 + CF_3CH_2OH \longrightarrow CF_3COOH$$

Halothane[18,20]

$$CF_3CHClBr \longrightarrow CF_3CClBrOH \longrightarrow CF_3COCl + Br^- \longrightarrow CF_3COOH \text{ (major route)}$$

$$CF_3CHClBr \longrightarrow CF_3{-}CH_2Cl + CF_2{=}CHCl + Br^- + F^- \text{ (minor route, during hypoxia)}$$

Isoflurane[20,21]

$$CHF_2OCHClCF_3 \longrightarrow [CF_2HOH] + CF_3COOH + Cl^- \longrightarrow CF_3COOH + HCOOH + F^-$$

Methoxyflurane[18,20]

$$CH_3OCF_2CHCl_2 \longrightarrow \begin{matrix} CHCl_2{-}CF_2OH \\ \text{and/or} \\ CHCl_2{-}COOH \end{matrix} + CH_3OCF_2COOH + Cl^- + F^- + HCHO \longrightarrow HOOCCOOH + Cl^- + F^- + CO_2$$

Sevoflurane[11,20]

$$FCH_2OCH(CF_3)_2 \longrightarrow (CF_3)_2CH{-}OH + F^- + [HCHO]$$

Adapted from Jacoby, R. L. and Nieforth, K. A., in *Principles of Medicinal Chemistry*, 3rd ed., Foye, W. O., Ed., Lea & Febiger, Philadelphia, in press.

Table 4
BIOTRANSFORMATION AND SOLUBILITY OF COMMONLY USED VOLATILE ANESTHETICS[a]

Agent	Per cent of uptake recovered as metabolites	Partition coefficients at 37°C	
		Blood/gas	Fat/gas
Nitrous oxide	None	0.47	1.22
Isoflurane	0.17—0.2	1.4	94.5
Enflurane	2.4	1.9	105
Halothane	15—20	2.3	185
Methoxyflurane	50	13.0	890

[a] Data taken from References 21 and 22.

in 50% of the patients administered that mixture. Note that the MAC in Table 2 for nitrous oxide in 105% meaning that hyperbaric conditions are required to produce 1 MAC in man.

Generally, an animal is anesthetized with the anesthetic in oxygen and administration of the mixture is continued for 15 min to allow equilibrium to be reached between alveolar gas, arterial blood, and the brain. The animal is then stimulated (tail clamp or s.c. electrical current) to determine responsiveness. Concentrations are then increased or decreased to derive the anesthetic concentrations that permit and prevent movement. The concentration midway between these two is the MAC for the anesthetic.

MACs have been determined in several different animals (dog, horse, monkey, cat, rabbit, rat) with the tail clamp being used as the stimulus. With other species, different procedures and stimuli must be used. For example, the stimulus in mice or newts is a rotating chamber, and the endpoint is loss of righting reflex. A complete listing with references of MACs of various anesthetics in different species has recently been published.[10]

The determination in man is different since the stimulus is a surgical skin incision and only one observation can be obtained for each patient. Several patients are anesthetized with concentrations estimated from animal studies to be above and below that estimated to permit movement in response to surgical incision. The results are tabulated as percentage of patients responding and the MAC is the concentration which prevented 50% of the patients from responding.

The determination of MAC has been reviewed[9,10] and other procedures have recently been described in detail.[25] An example of modern testing procedures used for the experimental anesthetic, alifurane, has recently been published.[12,13]

Mice are frequently used for initial screening of agents for anesthetic properties and descriptions of the procedures have been reported.[26,27] Physical chemical properties are very important in the initial evaluation of potential anesthetics. Determinations are made of an agent's flammability, chemical stability, and reactivity to soda lime.[26] Measurement of an agent's oil/gas partition coefficient is also done early in the evaluation of a new agent, since this property is an excellent predictor of the relative anesthetic potency of an agent. Gas chromatography (GC) is commonly used to make this determination as has been recently described.[28]

MECHANISM OF ACTION OF GENERAL ANESTHETICS

Physical Theories of Anesthesia

Anesthetics have been used clinically for more than 130 years and the first theory of anesthesia was advanced by Meyer[29,30] and Overton[31] at the turn of the century. Yet, after accumulation of enormous amounts of data concerning a wide variety of substances capable of producing anesthesia, the mechanism of anesthesia is still unknown.

Anesthesia has been naively explained as being the temporary absence of "consciousness". Unfortunately, our meager understanding of the state of consciousness is of no help when attempting to formulate a theory of anesthesia. Of further complication is that anesthetics are nonspecific in their action, no receptor sites are involved, and consequently there are no structure activity relationships to help us envision the means by which anesthesia is produced. In spite of these difficulties, many theories of anesthesia have been advanced which have shaped our current concepts of anesthetic mechanisms at molecular and subcellular levels. Following is a review of currently recognized events that may be responsible for producing anesthesia.

It is generally accepted that anesthetics act by blocking synaptic transmission through their ability to decrease excitatory postsynaptic potential amplitude and to reduce the lifetime of open postsynaptic ionic channels.[32] Most agree that these effects of anesthetics are produced within the membrane of nerve cells and emphasis has been placed on hydrophobic interactions of anesthetics (highly lipid-soluble substances) with membrane constituents. However, there is no consensus as to what that specific site in the membrane might be. Many hypothesize that anesthetics interact with membrane lipid causing expansion, disordering (termed fluidization), or melting (that is, to trigger a change in lipid physical state) of the lipid bilayer.[33] These lipid perturbations supposedly affect the excitability of important proteins[34] or stabilize ion channels in the closed state.[35] Evidence has also accumulated indicating that anesthetics interact directly with excitable membrane proteins to stabilize them directly.[36] Finally, Pauling[37] and Miller[38] have postulated that anesthetics produce their effects in the aqueous phase of nerve cells through the formation of stable, crystal-like water structures called clathrates. Most evidence does not support this aqueous phase hypothesis, although the importance of anesthetic polar interactions is still being explored.[39]

The picture of anesthesia is then one of blocked nerve synapses through anesthetic molecules interacting with nerve membrane lipid and/or protein. The attachment of anesthetics to membrane constituents is most likely through hydrophobic bonding.

Lipid solubility is still viewed as the best predictor of potency for volatile anesthetics. In Table 2, nitrous oxide is the least potent agent having a MAC of 105% and it is also the least lipid soluble as measured by its oil/gas partition coefficient. Methoxyflurane, being the most lipid soluble (oil/gas partition coefficient) is the most potent anesthetic listed in the table. This excellent correlation between MAC (potency) and oil/gas partition coefficient is illustrated in Figure 1.[40] Note that the potency range of the agents plotted in Figure 1 is approaching five orders of magnitude. This is a remarkable linear relation!

It was the relationship between lipid solubility and potency that lead Meyer[29,30] and Overton[31] to formulate their theory of anesthesia over 80 years ago which stated that anesthesia resulted from an interaction of the anesthetic with lipids of the cell. Partition coefficients were determined in olive oil and were assumed to approximate anesthetic solubility in brain lipid although there was no direct evidence to support this view. To overcome this difficulty, Ferguson[41] in 1939 proposed to use of "thermodynamic activities". At equilibrium, as in a steady level of anesthesia, the thermodynamic activity of a substance is the same in extracellular fluids, brain, and even in the hydrophobic site of action in synapses. Thus, measurement of the partial pressure of an anesthetic in a gas mixture producing a steady level of anesthesia, and the saturated vapor pressure of the pure anesthetic, allows one to calculate the thermodynamic activity of that agent. Ferguson showed that the thermodynamic activity of a wide range of anesthetics was nearly constant although their equianesthetic concentrations varied widely.

In 1954 Mullins extended the hydrophobic theory of anesthesia with his critical volume hypothesis which states that potency of an anesthetic is related not only to its concentration

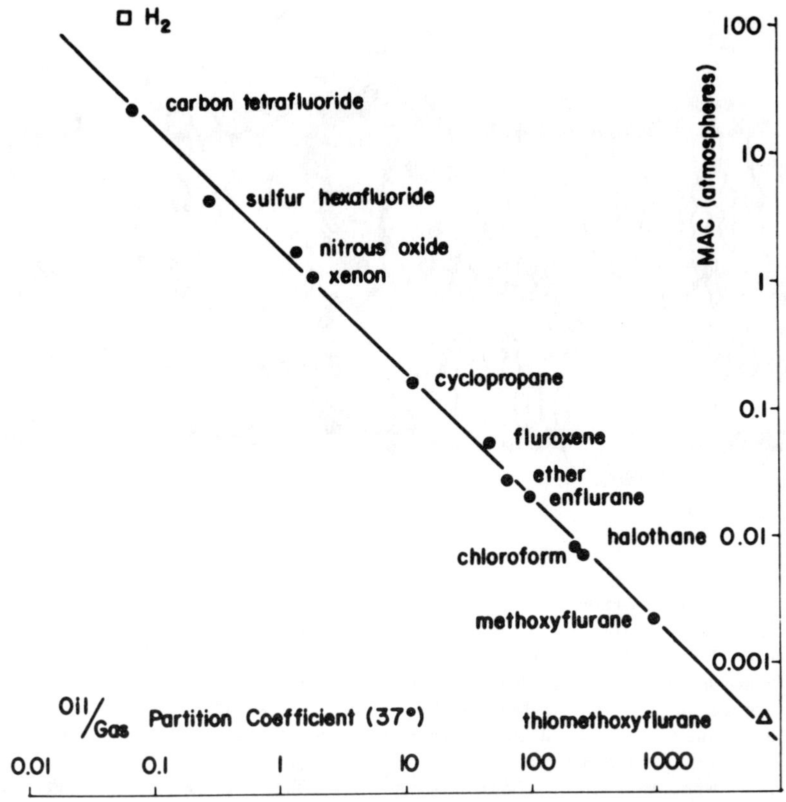

FIGURE 1. Relationship between MAC and oil/gas partition coefficient for several volatile anesthetics in the dog. MAC may be correlated with lipid solubility over a 74,300-fold difference in potency form 0.00035 atm(thiomethoxyflurane) to 26 atm(carbon tetrachloride). The line of perfect correlation is drawn through the points and corresponds to the equation: MAC X oil/gas partition coefficient equal 2.17 atm. Due to pressure reversal, the value for H_2 deviates from the correlation and was not used in estimating the 2.17-atm value. (From Tanifugi, Y., Eger, E. I., II, and Terrell, R. C., *Anesth. Analg.*, 56, 387, 1977. By courtesy of the International Anesthesia Research Society.)

in the membrane, but also to the volume occupied by the anesthetic.[42] This theory suggests that anesthesia is produced when a certain critical volume of anesthetic is reached within a membrane. Mullins proposed that anesthesia was dependent on the occupation of a critical volume of the nerve membrane by anesthetic molecules, resulting in a blockage of ion channels, thus resulting in a loss of excitability.

Out of these earlier theories of anesthesia grew our modern concepts of the interactions of an anesthetic with the cell membrane and its component parts. As was emphasized earlier, hydrophobic interactions are regarded to be of prime importance when considering the molecular mechanisms of anesthesia. How then are these physical theories of anesthesia related to inhibition of synaptic transmission?

The critical volume hypothesis of Mullins was extended to suggest that anesthetics ''expand'' membranes. The swelling caused by a volume increase in the membrane would result in compression of the membrane component parts. Excitable protein function would be inhibited or ion channels would remain closed.

The critical volume hypothesis is supported by pressure reversal of anesthesia. It has been well documented that application of pressure above 100 atm restores locomotion of orienting reflexes to animals anesthetized at 1 atm.[43] Also, additional studies have demonstrated this pressure reversal phenomenon in isolated neural preparations and in model membranes.[36,44]

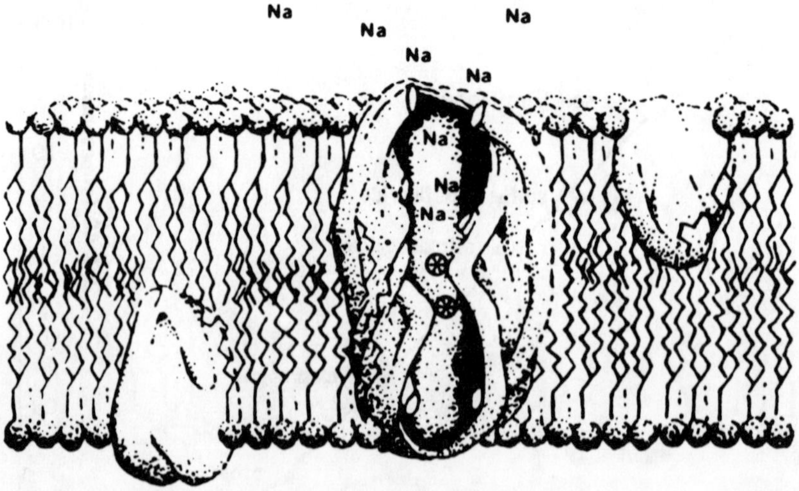

A

FIGURE 2. Anesthetic-membrane interactions: possible mechanisms of action. (A) Membrane at rest. The membrane is composed of a phospholipid bilayer embedded with function protein (ionophore channel). The channel is "gated" by charged (*) or bulky constituents. (B) A voltage-dependent conformational change of ionophore displaces the "gates," and creates an active (open) channel for ions (current) to flow. (C) Anesthetic molecules (large "A" enclosed in dark circle) bind to hydrophobic (and perhaps charged) sites within the membrane. Fluidization of bilayer and alteration of conformation of functional proteins result in an increase of membrane surface area (membrane expansion) with a decrease in membrane thickness. (D) A sufficient voltage-dependent conformation change of ionophore to form an open channel cannot occur because of anesthetic-induced perturbation of bilayer and functional proteins. Anesthetic may also effectively block the channel. (Figure and caption reproduced, with permission, from the Annual Review of Pharmacology and Toxicology, Volume 19 - © 1979 by Annual Reviews Inc.[44]

This suggests that pressure simply opposes the expansion or increased fluidity of the "anesthetized" membrane, thus restoring normal function.

In summary, the process of anesthesia is as follows: an anesthetic (molecular or high lipid solubility and appropriate size) accumulates in nerve membranes, interacting with lipid bilayer or protein to cause a swelling and/or fluidization of the membrane. This disordering in the membrane affects the excitability of proteins or directly stabilizes ion channels in the closed state. Application of pressure to the membrane can reorder its constituents, thus antagonizing anesthesia. The possible effects of anesthetics on membranes is shown in Figure 2. Questions which are yet unanswered are: do anesthetics interact with membrane lipid or proteins? How does this interaction inhibit membrane excitability? How is synaptic activity inhibited? And lastly, what is anesthesia; what is consciousness?

This is a very brief review of the theories of the mechanism of action of anesthesia. The reader is referred to other, more comprehensive reviews for in-depth discussions of membrane actions of anesthetics,[45] physical and biophysical mechanisms of anesthesia,[44,46] pressure reversal of anesthesia,[36] unitary theory of anesthesia,[34] and other more general reviews.[33,47]

Biochemical Theory of Anesthesia

Evidence is continuing to accumulate which suggests that anesthesia may be related to an inhibition of gamma-aminobutyric acid (GABA) catabolism in the brain. Cheng and Brunner[48] have shown in brain slices that many anesthetics inhibit the metabolic breakdown of the inhibitory neurotransmitter, GABA. This inhibition leads to increased brain concentrations of GABA and enhanced synaptic inhibition as a consequence of the elevated GABA

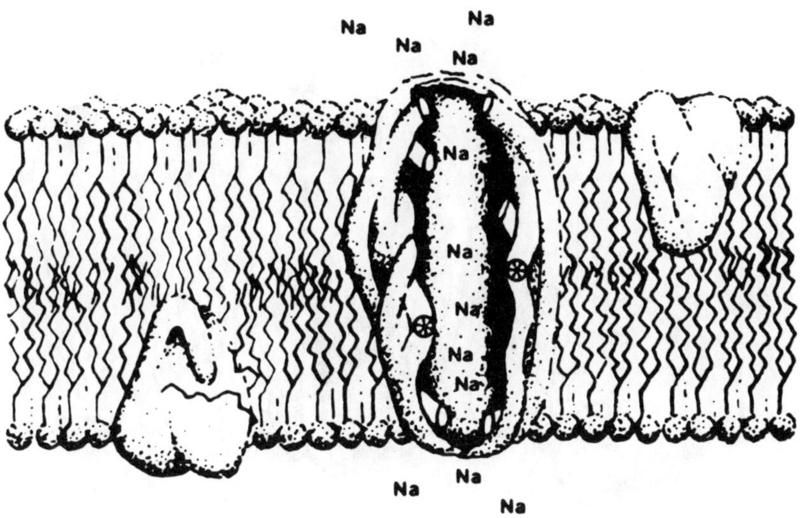

FIGURE 2B

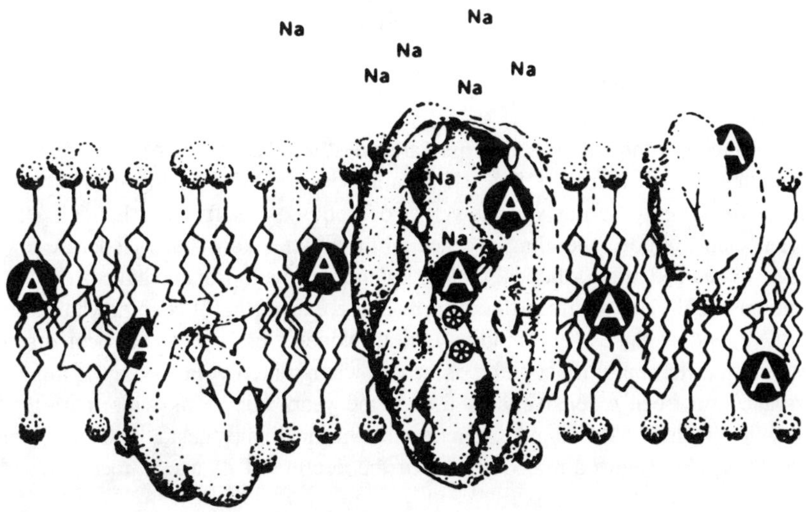

FIGURE 2C

levels. These findings have prompted Cheng and Brunner[48] to hypothesize that anesthesia may be related to inhibition of GABA disposal. Of the agents studied, they found that chloroform was the most potent inhibitor of GABA catabolism, followed by halothane, enflurane, ether, and thiopental. Other nonanesthetic agents, like ethanol, morphine, and phenytoin were found to be inactive or only weak inhibitors of GABA catabolism.

Although it has not been proven that anesthesia is a direct consequence of increased brain concentrations of GABA, interest in the area continues to grow as the significance of GABA as a neurotransmitter is recognized.[49]

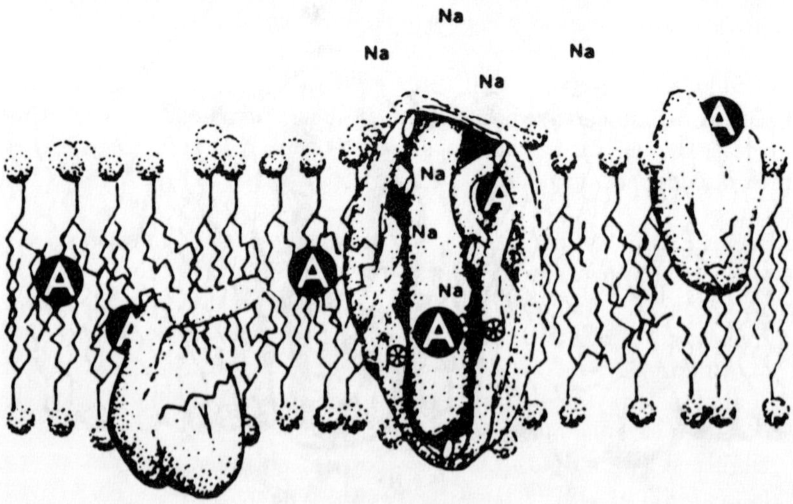

FIGURE 2D

STRUCTURE-ACTIVITY RELATIONSHIPS (SARs) OF VOLATILE ANESTHETICS

Nonhalogenated Anesthetics

Hydrocarbons

Alkanes, alkenes, and alkynes demonstrate anesthetic activity. Anesthetic potency increases as chain length increases in each homologous series. Toxicity also increases with increased chain length leaving very few hydrocarbons as useful anesthetics. Alkenes are more potent than the corresponding alkanes while alkynes above acetylene are chemically too unstable to be used as anesthetics. Only ethylene and cyclopropane have been extensively used in the past. Ethylene is much like nitrous oxide except that it is somewhat more potent (MAC is 67%) and produces very rapid induction. Nonetheless, its use has been discontinued due to its extreme flammability and explosiveness when mixed with oxygen or nitrous oxide. Cyclopropane, has been extensively used over the years due to its wide margin of safety and ability to produce very rapid induction of surgical anesthesia. Like ethylene, it is explosive when mixed with air accounting for the decline of its use in recent years.

Ethers

Simple ethers possess anesthetic activity. Potency and toxicity of the aliphatic ethers increase with increasing molecular weight. Introduction of unsaturation shortens induction and recovery time. Diethyl ether has a prolonged induction period while its unsaturated derivative, divinyl ether, presents such rapid induction as to be hazardous. Use of the simple, aliphatic ethers has diminished in recent years due to the accompanying risk of fire and explosion and the availability of better alternative agents.

Halogenated Anesthetics

Many halogenated hydrocarbons and ethers have been employed as anesthetics since the introduction of chloroform in 1847. Halogenation decreases or eliminates flammability of ethers and hydrocarbons and usually increases anesthetic potency. Unfortunately, these advantages of halogenation are often offset by increased renal and hepatic toxicity along with increased incidence of cardiac arrhythmias. These unfavorable cardiac effects are due to sensitization of the myocardium to epinephrine and other pressor amines.

Compounds containing only bromine or iodine are not useful as anesthetics due to their low volatility and low chemical stability. Some brominated compounds are acceptable as anesthetics if other halogens, chlorine or fluorine, are also contained in the same molecule as exemplified by halothane, CF_3–CHClBr.

Chlorinated hydrocarbons have been extensively studied as anesthetic agents. Chloroform, ethyl chloride, and trichloroethylene had been used clinically as anesthetics. However, their use has been discontinued due to their characteristic side effects: hepatotoxicity and cardiac arrhythmias.

In recent years, highly fluorinated agents have been introduced due to the favorable chemical and pharmacological properties they possess. These polyfluorinated hydrocarbons and ethers are generally nonflammable, easy to volatilize, and chemically stable. The incidence of renal and hepatic toxicity is low which is probably due to the relatively minor metabolism of fluorinated agents. As fluorine atoms replace other halogens, the frequency of catechol-induced arrhythmias decreases. Some agents containing fluorine as the sole halogen do not sensitize the myocardium to epinephrine.[50]

Some fluorinated hydrocarbons and ethers possess convulsant properties. Best known is the convulsant gas flurothyl($CF_3CH_2OCH_2CF_3$) which produces convulsions in low concentration and anesthesia at higher concentration. A structural isomer of flurothyl, iso-indoklon [$(CF_3)_2CHOCH_3$] is an anesthetic with no convulsant properties.[51] It has been suggested that "differences in solubility parameters between convulsants and anesthetics might influence the relative distribution of these compounds into the biophases associated with convulsive activity and anesthesia" (see Reference 51 for additional related references).

At the present time, clear SARs among anesthetics do not exist. Each potential anesthetic must be evaluated to determine its suitability for use in clinical anesthesia.

FLAMMABILITY OF VOLATILE ANESTHETICS

The use of flammable anesthetics has declined during the past 20 years to the point where their use is rare. A recent survey of anesthesiology teaching hospitals revealed that 99.7% of the time, nonflammable anesthetics were used and only 5 of 126 anesthesiology training programs used flammable agents in 1980. Of 171,000 anesthetic uses, only 408 involved flammable anesthetics with 271 uses of ether and 137 uses of cyclopropane.[51] Authors of a recent article reviewed the advantages and disadvantages for the continued use of flammable anesthetics and concluded that the use of these agents should be banned.[52] Based on the recommendations and standards developed by the National Fire Protection Association, Underwriter's Laboratories, and the Joint Commission on Accreditation of Hospitals, the use of flammable anesthetics probably adds approximately $9.4 million to the annual cost of surgical care.[52]

TOXICITY OF VOLATILE ANESTHETICS

Acute Toxicity

Toxic reactions are uncommon in patients undergoing surgical anesthesia due to short exposure time of the patient to anesthetic and the inertness of anesthetic agents. Yet, some toxic reactions are infrequently observed in patients; most notable is hepatic necrosis, polyuric renal failure, and rarely, malignant hyperthermia.

Chloroform, the first used halogenated anesthetic enjoyed considerable popularity in the latter part of the 19th century, but awareness of its toxic effects to the liver rapidly diminished its use. Today, chloroform is recognized as a true hepatotoxin and new halogenated agents are scrutinized for liver toxicity.

During the first years of clinical use of halothane, reports of postoperative hepatitis appeared and finally in 1961, the Committee on Anesthesia of the National Academy of Sciences-National Research Council initiated a retrospective study to evaluate reports of halothane-induced massive hepatic necrosis.[53] The report concluded that fatal hepatic necrosis following halothane anesthesia was rare (about 1 per 10,000) and that hepatic necrosis occurred with other anesthetics also. Cyclopropane showed the greatest incidence of necrosis, followed by halothane, nitrous oxide-barbituate, and ether.

The possible mechanisms leading to hepatotoxicity have been reviewed[54] and new evidence is appearing that supports the hypothesis that metabolism of halothane is responsible for its toxicity (see Reference 55 and references therein). This hypothesis suggests that halothane is metabolized by a reductive pathway (see Table 3) to produce reactive intermediates which bind to macromolecules producing direct toxicity or a delayed immunological response.

Renal toxicity of the fluorinated anesthetics is related to the metabolic release of inorganic fluoride. Levels of serum inorganic fluoride in excess of 40 $\mu M/\ell$ are associated with high-output renal failure due to interference in sodium transport at the proximal tubule.[54] The relative rates of metabolic defluorination of the commonly used fluorinated anesthetics are methoxyflurane > enflurane > isoflurane > halothane.[22] Only in the case of methoxyflurane are these nephrotoxic levels easily reached.

Malignant hyperthermia is a rare (1 per 14,000 anesthetic uses) but life-threatening complication of anesthesia. In genetically susceptible individuals, this complication can occur during anesthesia or up to several hours postoperatively with all potent inhalation anesthetics, muscle relaxants, or even stress alone triggering the complication. It involves the release of calcium from the sarcoplasmic reticulum of skeletal muscle, causing intense contraction and large amounts of heat. Untreated, the syndrome progresses rapidly to death.[56] Among the fluorinated anesthetics the following potential to precipitate this syndrome is observed: halothane > isoflurane > enflurane > methoxyflurane.[57] At this time the role of nitrous oxide in causing malignant hyperthermia is unknown. If susceptible patients must be anesthetized, i.v. anesthetics like thiopental, droperidol, or ketamine may be safe to use. Treatment for malignant hyperthermia includes intensive supportive therapy and i.v. administration of the muscle relaxant, dantrolene sodium.[56]

Chronic Toxicity

There is growing concern regarding health problems associated with work in the operating room and dental operatory due to continual exposure to low levels of waste anesthetic gas. Evidence continues to accumulate which links chronic anesthetic exposure to health risks. A recent survey of U.S. dental health professionals suggests that exposure to nitrous oxide may be associated with increased incidence of birth defects.[58] Some 20 other epidemiologic surveys also confirm the presence of an occupational hazard. Unfortunately, the animal data that has been gathered are inconsistent and species variations have prevented development of an experimental model. The literature has been critically reviewed regarding tract concentrations of anesthetic gases[59] and the involvement of regulatory agencies (FDA, OSHA, etc.) has recently been examined.[60] The entire controversial issue of toxicity associated with chronic exposure to anesthetics has been thoroughly summarized in a recent book.[54]

Drug Interactions

The most prominent drug interaction is between general anesthetics and other CNS depressant drugs like opiates, benzodiazepine central relaxants, barbiturates, and antihistamines. Administration of these CNS depressant drugs is commonly used as part of the preoperative preparation to reduce apprehension in the patient and to facilitate induction. The CNS depression caused by these preanesthetic medications is useful in limiting the dose of anesthetic required, thus minimizing the potential for toxic side effects associated with the general anesthetic (e.g., lower concentrations of methoxyflurane produce less renal toxicity). Although these are true drug interactions, most anesthesiologists view them as simply a predictable consequence of preanesthetic patient preparation.

Cyclopropane and halogenated anesthetics sensitize the heart to vasopressors, like epinephrine, resulting in arrhythmias. A dose of only 2.1 μg/kg of epinephrine produces extrasystoles in 50% of patients given 1.25 MAC halothane. The newer fluorinated agents, enflurane and isoflurane, are much less likely to demonstrate this adverse interaction with epinephrine,[61] although caution is recommended when using vasopressors with any halogenated anesthetic.

A frequently reported interaction is associated with the use of nondepolarizing muscle relaxants (e.g., *d*-tubocurarine) and volatile anesthetics. Modern anesthetics have sufficient neuromuscular blocking activity[21] to require that the dose of muscle relaxant be reduced to prevent paralysis of respiration during surgery or prolonged neuromuscular blockage during recovery.[62] Accidental muscle paralysis may be reversed by neostigmine.

A similar interaction occurs with neomycin and other aminoglycoside antibiotics that possess neuromuscular blocking activity. It has been suggested that these antibiotics be avoided in patients who are potential candidates for surgery in order to avoid the real possibility of respiratory paralysis due to exaggerated neuromuscular blockage.[63]

More detailed discussions of these, and other specific interactions can be found elsewhere.[64] Also, the effects of many different drugs on the determination of MAC has been extensively reviewed.[10]

THERAPEUTIC USE OF GENERAL ANESTHETICS

The typical anesthetic sequence of major surgery involves the administration of several types of drugs. Preanesthetic medications are prescribed to provide sedation and relieve apprehension before surgery. Commonly used agents for this purpose are barbiturates (pentobarbital and secobarbital), antihistamines (hydroxyzine), and antianxiety benzodiazepines (diazepam and lorazepam). The benzodiazepines also produce amnesia of the preoperative period in most patients and yet allows them to remain cooperative. The antihistamine, hydroxyzine, produces an antiemetic effect which is likewise beneficial.

A second group of preanesthetic medications are the opioids. Commonly used agents are morphine, meperidine, and fentanyl, which relieve preoperative pain. This helps relieve anxiety in the patient and promotes smooth induction of anesthesia.

These two groups of agents (sedatives and opiates) not only relieve anxiety and pain but they also reduce the amount of general anesthetic that will be required to produce surgical anesthesia.

Additional preanesthetic medication may also include the use of anticholinergic drugs (atropine or scopolamine) to prevent excessive respiratory tract secretions or more importantly, to counteract the vagal effects of hypotension and bradycardia. Patients may also be given neuromuscular blocking drugs (succinyl choline, tubocurarine plus others) to facilitate endotracheal intubation, relieve larynospasm, and provide muscle relaxation during surgical procedures.

Finally, an i.v. induction agent (e.g., thiopental or ketamine) would be administered to the patient resulting in loss of consciousness within 30 to 60 sec. For most procedures, surgical anesthesia is then maintained by administration of a potent, volatile anesthetic in a mixture of 70% nitrous oxide, 30% oxygen. Depth of anesthesia can then be easily maintained (by an experienced anesthesiologist) by varying the concentration of inhaled, volatile anesthetic.

Choice of volatile anesthetic is based upon the type of surgical procedure, health requirements of the patient, anesthetic equipment available and, importantly, the experience of the anesthesiologist. In this author's mind, only three agents are available that warrant consideration for routine use: enflurane, halothane, and isoflurane. Ether and cyclopropane can be excluded due to their flammability; nitrous oxide for its lack of potency and methoxyflurane for its attendant potential for nephrotoxicity. Wade and Stevens[22] have recently provided a

Table 5
COMPARATIVE ASSESSMENT OF ENFLURANE, HALOTHANE, AND ISOFLURANE

	Enflurane		Isoflurane		Halothane
Chemical stability	0	=[a]	+ + +		–
Low blood solubility	0		+ + +		–
Pleasant odor	–	=[a]	0		+ + +
Minimal respiratory depression	–		0	=[a]	+ + +
Minimal circulatory depression	–		+ + +		0
Low incidence of arrhythmias	0		+ + +		–
Produces muscle relaxation	0	=[a]	+ + +		–
Increases intracranial pressure/cerebral blood flow	0		+ + +		–
Precipitates seizure	–		0	=[a]	+ + +
Low amount metabolized	0		+ + +		–
Low toxicity	0		+ + +		–

Note: + + +, Superior; 0, intermediate; –, inferior.

[a] Equal sign indicates that the difference in the given property is minimal between the two agents.

Adapted from Wade, J. G. and Stevens, W. C., *Anesthesiology*, 60, 666, 1981. With permission.

comparative assessment of enflurane, halothane, and isoflurane, which is shown in Table 5.

DRUG MONOGRAPHS

Nonflammable Volatile Anesthetics

Enflurane

Pharmacokinetics

Enflurane is a pleasant smelling halogenated ether with low blood solubility (Table 6) which provides rapid induction. Alveolar concentration equal to 50% of the inspired concentration is reached after 5 min of enflurane administration[65] (see Figure 3). For induction, a 4% v/v concentration of enflurane is vaporized into oxygen or oxygen and nitrous oxide mixture. Surgical anesthesia is maintained with 1 to 3% v/v concentration. Enflurane is primarily excreted through the lungs and ony 2 to 3% of the enflurane taken up by the body is metabolized to difluoromethoxydifluoroacetic acid, fluoride ion, plus other metabolites[66] (see Table 3). Fluoride ion plasma levels do not reach those usually associated with nephrotoxicity.[67]

Toxicity

The cardiovascular system remains relatively stable during normal levels of enflurane anesthesia. Enflurane is a more potent respiratory depressant[68] than are halothane and isoflurane. Spontaneous respiration is usually sufficient at light levels of anesthesia but, at deeper levels, assisted or controlled ventilation is required. The agent can induce seizures in patients with abnormalities in their EEG or a history of seizure disorders. The agent is contraindicated in epileptic patients.[69] Studies indicate that enflurane does sensitize the heart to catecholamines in a variable fashion.[61] Sensitization is usually less pronounced than with halothane allowing for subcutaneous or submucosal injection of epinephrine during enflurane anesthesia.

Table 6
PROPERTIES OF NONEXPLOSIVE INHALED ANESTHETICS[a]

Generic Name (Trade Name)	Isoflurane (Forane)	Enflurane (Ethrane)	Halothane (Fluothane)	Methoxyflurane (Penthrane)	Nitrous Oxide
Structure	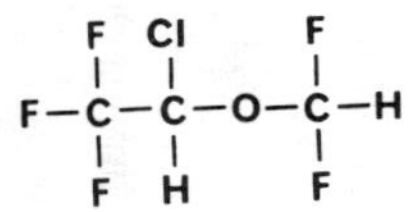	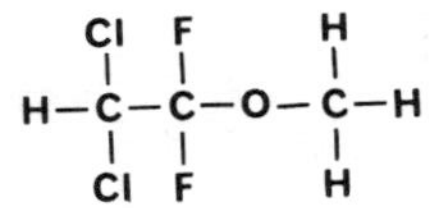	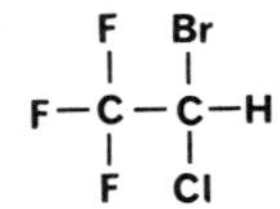	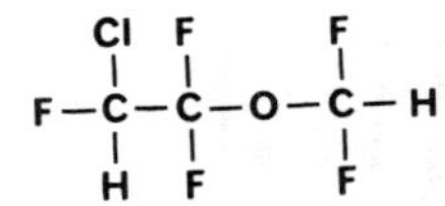	N_2O
Molecular weight (g)	184.5	184.5	197.4	165.0	44.0
Specific Gravity	1.50 (25°C)	1.52 (25°C)	1.86 (22°C)	1.41 (25°C)	–
Boiling Point (°C)	48.5	56.5	50.2	104.7	–
Vapor Pressure at 20°C	240.0	175.0	243.0	25.0	–
Preservative (if required, 0.01% W/W	None	None	Thymol	Butylated hydroxytolene	None
Stability to					
Metal	Nonreactive	Nonreactive	May react	May react	Nonreactive
Soda lime	Stable	Stable	Breaks down	Breaks down	Stable
Ultraviolet light	Stable	Stable	Decomposes	Decomposes	Stable
Partition coefficients at 37°C					
Blood/gas	1.4	1.9	2.3	13.0	0.47
Brain/gas	3.65	2.6	4.1	22.1	0.50
Fat/gas	94.5	105.0	185.0	890.0	1.22
Liver/gas	3.50	3.8	7.2	24.8	0.38
Muscle/gas	5.6	3.0	6.0	20.0	0.54
Oil/gas	97.8	98.5	224.0	930.0	1.4
Water/gas	0.61	0.8	0.7	4.5	0.47
Rubber/gas (at 23°C)	62	74	120	630	1.2
Minimum flammable concentrations (MFC) in 70% N_2O, 30% O_2 (%)	7.0	5.8	4.8	>vapor pressure	–
MAC in O_2[b]	1.15	1.58	0.75	0.16	110
MAC in 70% N_2O[b]	0.50	0.57	0.29	0.07	–
MFC/MAC in 70% N_2O	14.0	10	16.5	–	–
% of uptake recovered as metabolites (approx.)	0.2	2.4	15-20	50	>

[a] Data adapted from References 21 and 22.

[b] In 30- to 55-year old patients; percent of one atmosphere.[21]

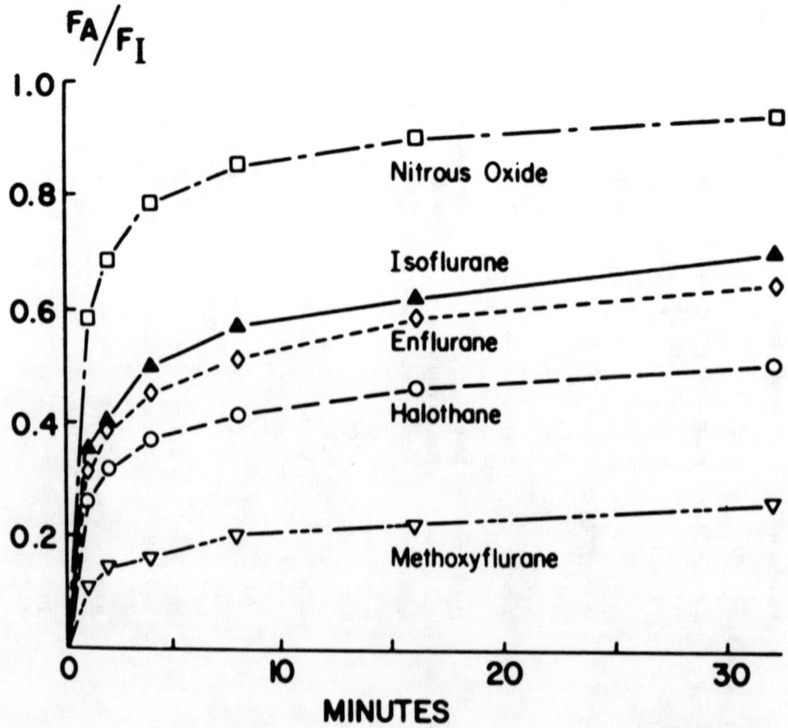

FIGURE 3. Graph of the ratio of F_A/F_I shows the rate at which the alveolar concentration (F_A) approaches the concentration of anesthetic being breathed (F_I). The ratio varies inversely with solubility. (From Eger, E., I, II, *Isoflurane (Forane): A Compendium and Reference*, Ohio Medical Products, Madison, Wis., 1981. With permission.)

Physical-Chemical and Other Data

- **Chemical name** — 2-Chloro-1-(difluoromethoxy)-1,1,2-trifluoroethane, or 2-chloro-1,1,2-trifluoroethyl difluoromethyl ether; $C_3H_2ClF_5O$ (184.50). See Table 6 for additional properties.

The liquid is available in 125- and 250-mℓ containers; trade name Ethrane (Ohio Medical).

Halothane

Pharmacokinetics

Halothane is the most commonly used halogenated hydrocarbon anesthetic which produces alveolar concentration equal to 50% of the inspired concentration in about 30 min[65] (see Figure 3). Induction can be achieved with 1.5 to 3% v/v halothane in oxygen or mixtures of nitrous oxide and oxygen in about 5 min, although the usual practice is to use an i.v. anesthetic (e.g., thiopental) to accomplish induction. Surgical anesthesia is then maintained with concentrations of 0.5 to 1.5% v/v halothane. Approximately 15 to 20% of the halothane taken up by the body is metabolized.[70,71] The principal metabolites, bromide ion and trifluoroacetic acid, are found in the urine (see Table 3). An unusual reductive route of metabolism of halothane (see Table 3), which may occur during hypoxia, has been implicated with the infrequent hepatic necrosis of halothane anesthesia.[55]

Toxicity

Halothane has been the most commonly used anesthetic for the past 20 years in all types of surgical procedures. Hepatitis that occurs postoperatively is infrequent (1 case in approximately 10,000 halothane administrations) and its frequency of occurrence is similar to that observed with other general anesthetics.[53] The mechanism of this toxic reaction may be related to the agent's metabolism as described above. Halothane does sensitize the heart to catecholamines to a greater extent than other potent anesthetics[61] making the concomitant use of epinephrine and halothane potentially dangerous. Halothane does not provide good analgesia nor muscle relaxation which then necessitates additional drug therapy during anesthesia.

Physical-Chemical and Other Data

- **Chemical name** — 2-Bromo-2-chloro-1,1,1-trifluoroethane; $C_2HBrClF_3$(197.38). See Table 6 for additional properties.
- **Stereochemistry** — Halothane contains one chiral center producing two optical isomers. The separate *d* and *l* optical isomers have been evaluated for anesthetic activity in two model systems. The activity of the two optical isomers did not significantly differ, thus supporting the theory that anesthesia involves physical phenomena rather than drug-receptor interactions.[72]
- **Dosage forms** — The light sensitive liquid is available in amber-colored bottles with 0.01% thymol added as a stabilizer. It is available in 250-mℓ containers generically and under the trade name, Fluothane (Ayerst) in 125- and 250-mℓ containers.

Isoflurane

Isoflurane, a structural isomer of enflurane, is the newest halogenated anesthetic. Its low blood/gas partition coefficient permits rapid induction and recovery rates (Table 6). An alveolar concentration equal to 50% of the inspired concentration is reached after 4 min with isoflurane[73] (see Figure 3). However, a very rapid induction cannot be achieved with isoflurane due to its mild pungency which can cause breathholding, coughing, or laryngospasm if too rapid an increase in inhaled concentration is attempted. For induction, low concentration (0.25 to 0.5%) of anesthetic is initiated and then every four to five breaths the concentration is increased to a peak of 3.5 to 4.0%. A concentration of 1 to 2% isoflurane in oxygen or a mixture of nitrous oxide and oxygen will maintain surgical anesthesia. Isoflurane is primarily excreted through the lungs and less than 0.2% of that taken up by the body is metabolized. Inorganic fluoride and trifluroacetic acid have been identified as end products of isoflurane metabolism[74] (see Table 3). This near absence of metabolism suggests that isoflurane may not be nephrotoxic or hepatotoxic. Serum inorganic fluoride levels are far below those associated with nephrotoxicity.[67]

Toxicity

Isoflurane was scheduled for release in 1975 but marketing was delayed because an animal study suggested that it was potentially carcinogenic.[75] Subsequent studies did not confirm that isoflurane is a hepatocarcinogen[76] and it is now accepted not to be carcinogenic nor mutagenic. The agent produces respiratory depression like all potent anesthetics and can trigger malignant hyperthermia. Isoflurane, like enflurane is much less likely than halothane to sensitize the heart to the arrhythmogenic effects of epinephrine.[61] Limited clinical experience with isoflurane indicates that it might be free of hepatic and renal toxicity. See two reviews for a complete discussion of this new anesthetic.[21,22]

Physical-Chemical and Other Data

- **Chemical name** — 1-Chloro-2,2,2-trifluoroethyl difluoromethyl ether; $C_2H_2ClF_2O$ (184.50). See Table 6 for additional properties.
- **Dosage forms** — The liquid is available in 100-mℓ amber-colored bottles under the trade name Forane (Ohio Medical).

Methoxyflurane

Pharmacokinetics

This is the most potent volatile anesthetic that is available. Once anesthesia has been induced, only about 0.5% methoxyflurane in oxygen is required to maintain surgical anesthesia. Induction with methoxyflurane is slow (20 to 30 min) and is achieved with a 2 to 3% concentration of anesthetic (Figure 3). Since an excitement stage is not uncommon with methoxyflurane induction, i.v. agents (e.g., thiopental) are usually used instead to induce anesthesia, followed by methoxyflurane in maintenance concentration. The very high lipid solubility of the agent causes considerable accumulation of methoxyflurane in adipose tissue. Slow diffusion from these sites accounts for prolonged recovery periods and extensive liver metabolism of methoxyflurane. As much as 50 to 70% of the absorbed dose is metabolized in the liver to fluoride ion, oxalic acid, and other organic fluorine-containing urinary metabolites[77] (see Table 3). Peak plasma concentration of fluoride ion is usually found 2 to 4 days after exposure to methoxyflurane.

Toxicity

If metabolism of methoxyflurane yields plasma concentration of fluoride ion in excess of 40 $\mu M/\ell$, direct damage to the renal tubules occurs.[54] As a result of this renal toxicity, the use of methoxyflurane as a general anesthetic is limited. The agent does produce profound analgesia and is still used in limited concentration during the first stage of labor for that purpose.

Physical-Chemical and Other Data

- **Chemical name** — 2,2-Dichloro-1,1-difluoro-1-methoxyethane or 2,2-dichloro-1,1-difluoroethyl methyl ether; $C_3H_4Cl_2F_2O$ (164.97). See Table 6 for additional properties.
- **Dosage forms** — The liquid is available in 15- and 125-mℓ bottles under the trade name, Penthrane (Abbott). It contained 0.01% w/w butylated hydroxytoluene as an antioxidant.

Nitrous Oxide

Pharmacokinetics

Nitrous oxide is rarely used as the sole agent to produce surgical anesthesia due to its low anesthetic potency. The MAC value for nitrous oxide is 105 to 110% and concentrations of 80% or more are required to produce anesthesia. At these concentrations the danger of causing hypoxic organ damage is great. The principal use of nitrous oxide is as a supplement to other more potent anesthetics, which allows the use of smaller doses of the potent agent and reduces the potential for toxic reactions from occurring. The concentration of potent anesthetic is reduced by one half to two thirds in the presence of 70% nitrous oxide in oxygen as is illustrated by the MACs in 70% nitrous oxide in Table 4. Nitrous oxide is also used in combination with i.v. anesthetics plus neuromuscular blocking agents and opiates to produce "balanced anesthesia".[78] Because of its low tissue and blood solubility, nitrous oxide is rapidly eliminated as expired gas through the lungs. Present methods have not detected any nitrous oxide liver metabolism.

Toxicity

A recent study has shown that nitrous oxide is metabolized by intestinal microflora to nitrogen gas.[16] The authors suggest that this metabolic route may include free radicals or other reactive intermediates which may be reabsorbed from the intestinal tract to produce carcinogenic or teratogenic effects. Nitrous oxide has also been shown to inhibit methionine synthetase in human liver. This depression of methionine synthetase activity is associated with an impairment of DNA synthesis (see Reference 79 and references therein).

Nitrous oxide is 34 times more soluble in blood than in nitrogen, and so during anesthesia, large volumes of nitrous oxide displace smaller volumes of nitrogen from enclosed body cavities. This can result in increased internal pressure or organ distension involving the middle ear, loops of intestine, lung, or skull.

Physical-Chemical and Other Data

The compressed gas is available generically in blue, steel cylinders. See Table 6 for the properties of nitrous oxide.

Flammable Volatile Anesthetics

Cyclopropane

Pharmacokinetics

Cyclopropane is readily absorbed on inhalation. Due to its low blood/gas partition coefficient (0.42) it produces rapid induction being given in 25 to 50% concentration in oxygen for that purpose. Surgical anesthesia is maintained with 10 to 25% cyclopropane in oxygen. Recovery is rapid due to relatively low tissue and blood solubility of cyclopropane (see Reference 9 for tissue partition coefficients). Studies are inconclusive regarding the metabolism of cyclopropane; one study in rats showed cyclopropane metabolized to CO_2[17] while a second study in miniature swine demonstrated no biotransformation of cyclopropane.[23]

Toxicity

Anesthesia with cyclopropane is frequently associated with cardiac arrhythmias as a consequence of myocardial sensitization to catecholamines and elevated plasma levels of endogenous catecholamines. Recovery from cyclopropane anesthesia is often accompanied by nausea and vomiting. Hepatic necrosis following cyclopropane anesthesia has been found to be more frequent than with halothane, ether, or barbiturate-nitrous oxide combination.[53]

Physical-Chemical and Other Data

- **Chemical name** — Cyclopropane; C_3H_6 (42.08).
- **Flammability** — Cyclopropane is highly flammable and forms explosive mixtures with air (2.4 to 10.4%) and oxygen (2.5 to 60%). Its explosiveness is responsible for its rare use. See earlier section of this chapter entitled Flammability of Anesthetics for a more complete discussion of this hazard associated with the use of cyclopropane.
- **Dosage forms** — The gas is available generically in orange, metal cylinders or orange labeled chrome cylinders.

Ether (Diethyl Ether)

Pharmacokinetics

Ether is readily absorbed on inhalation. The blood/gas partition coefficient is high (Table 2) resulting in slow induction and often, prolonged, difficult recovery. Ether concentrations of 10 to 30% are used for induction and 5 to 15% concentrations of ether are used to maintain surgical anesthesia. It appears that about 15% of the ether absorbed by the body is metabolized to CO_2 and urinary metabolites.

Toxicity

The major clinical disadvantages associated with ether anesthesia are slow induction and recovery, nausea and vomiting during recovery, increased salivation and bronchiole secretions, and rare hepatic necrosis. More importantly, its use is rare today due to its flammability

Physical-Chemical and Other Data

- **Chemical name** — Ethyl ether, 1,1′-oxybisethane; $C_4H_{10}O$ (74.12).

- **Chemical stability** — Ether tends to form explosive peroxides when exposed to air and light, especially when evaporation to dryness is attempted. Decomposition is retarded in the presence of copper or hydroquinone.
- **Flammability** — Ether is highly flammable and forms explosive mixtures with air (1.85 to 36.5%), oxygen (2 to 82%), and nitrous oxide (1.5 to 24%). See earlier section of this chapter entitled Flammability of Anesthetics for a discussion of this hazard.
- **Dosage forms** — The liquid is available generically in 0.25-, 0.5-, 1-, and 5-lb airtight sealed containers lined with copper to prevent peroxide formation.

I.V. Anesthetics

Thiopental Sodium

Mechanism of Action

General anesthesia is produced by barbiturates through the suppression of the reticular activating system in the brain stem and is discussed in detail in the chapter entitled "Barbiturate Sedatives and Hypnotics". It is unclear whether barbiturates produce anesthesia by the same molecular mechanism as that of volatile anesthetics (see References 46 and 79 for a discussion).

Pharmacokinetics

Thiopental is the most commonly used agent for induction of surgical anesthesia. Unconsciousness is produced in 30 to 60 sec following i.v. administration and consciousness returns in 20 to 30 min, if no additional anesthetic is given. The very rapid onset of action of thiopental is due to rapid transfer of the agent into the brain as a consequence of high cerebral blood flow and high lipid solubility of thiopental. Within 1 min following i.v. injection, brain and blood levels of thiopental begins to decrease as the drug "redistributes" into different tissues. The agent redistributes first into highly perfused tissues as kidney and liver and then into muscle and finally into fat.[80] It is this tissue redistribution of thiopental that terminates its anesthetic action. See Reference 81 for partition coefficients of i.v. barbiturates.

The total plasma concentration of thiopental necessary to produce anesthesia has been determined to be 39 μg/mℓ. This corresponds to a free fraction thiopental concentration of 6 μg/mℓ.[82] Thiopental is highly bound to plasma proteins (72 to 88%) and the binding appears to be concentration dependent, particularly at high total plasma concentrations. Plasma protein binding studies done with thiopental have yielded variable results which are discussed in a recent study by Morgan and colleagues.[83,84]

The agent is slowly, although completely metabolized by hepatic microsomal enzymes. The main metabolites are the result of ω and ω-1 oxidation of the 5-position 1-methylbutyl substituent and desulfuration,[85,86] as is shown in Figure 4.

Pharmacokinetic parameters for thiopental have been determined and representative mean values are presented in Table 7. The pharmacokinetics of thiopental are difficult to determine as demonstrated by the variability of reported parameters. Mean elimination half-life values ranging from 2.6 hr to 3.6 days have been reported. Morgan et al.[83] have critically reviewed these various studies on thiopental pharmacokinetics and contributed their own parameters. For reviews of the pharmacokinetics of thiopental, see References 87 and 88.

Use and Dosage

Thiopental sodium is by far the most frequently used i.v. anesthetic. For induction in adults, 50 to 75 mg (2 to 3 mℓ of 2.5% of freshly prepared solution) is initially administered followed by additional doses of 50 to 75 mg at 20- to 40-sec intervals until anesthesia is established. For maintenance, 25 to 50 mg is injected whenever the patient moves. Thiopental is only useful as the sole anesthetic for short surgical procedures. For those procedures

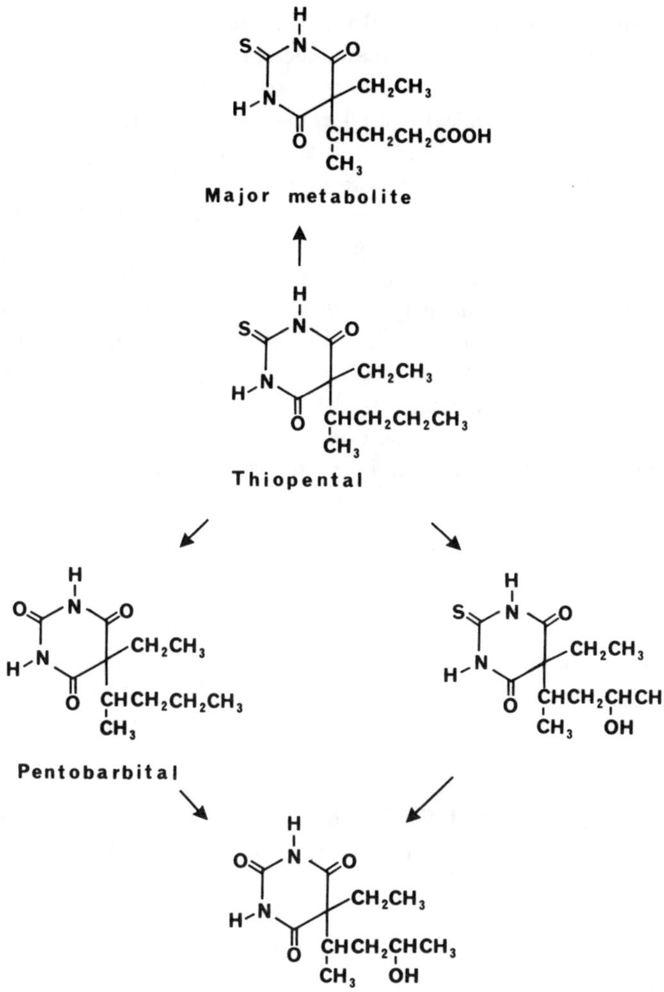

FIGURE 4. Biotransformation of thiopental in man.[85]

Table 7
MEAN VALUES OF ELIMINATION HALF-LIVES, APPARENT VOLUMES OF DISTRIBUTION (Vd) AND CLEARANCE (Cl) OF I.V. ANESTHETICS IN MAN

Compound	Half-life (hr)	Vd (ℓ/kg)	Cl (mℓ/min)	Ref.
Thiopental	6.2	1.6	144[a]	95
Methohexital	1.6	1.1	825	96
Ketamine	3.4	3.3[b]	1296	97

[a] Extrapolated from the mean body weight of patients studied.
[b] Calculated assuming that the mean body weight of patients studied was 65 kg.

Adapted from Duvaldestin, P., *Clin. Pharmacokinet.*, 6, 61, 1981. With permission.

lasting longer than 15 min, volatile anesthetics are used to maintain anesthesia. This agent is rarely given rectally to produce anesthesia because of its unpredictable absorption from the rectum.

In psychiatric practice, thiopental is used in narcoanalysis and narcosynthesis.

Toxicity

Thiopental produces rapid, pleasant induction of anesthesia and allows for fast recovery accompanied by little postanesthetic excitement or vomiting. Most of the complications associated with its use are minor and can be avoided by careful use of the agent. It is a potent respiratory depressant and may cause yawning, coughing, and laryngospasm immediately following injection. Patients are "titrated" with thiopental sodium to prevent overdosage which results in hypotension, respiratory depression, and prolonged recovery period. There is no specific antagonist available to counteract barbiturate overdose. Other general side effects, toxic reactions, and drug interactions of the barbiturates are discussed in the chapter entitled "Barbiturate Sedatives and Hypnotics". Comprehensive reviews have recently appeared concerning adverse effects of i.v. anesthetics[89] and drug interactions of anesthetics.[64,90]

Physical-Chemical and Other Data

Thiopental sodium is a white to off-white, crystalline powder or a yellowish to pale greenish yellow hygroscopic powder. It may have a disagreeable odor. It is soluble in water and alcohol. Its aqueous solution is alkaline to litmus and decomposes on standing. Precipitation occurs when its solutions are boiled. Thiopental sodium solutions are incompatible with acidic solutions. pKa = 7.4 (conjugate acid).

- **Chemical name** — 4,6(1H,5H)-Pyrimidinedione, 5-ethyldihydro-5-(1-methylbutyl)-2-thioxo-, monosodium salt; or Sodium 5-ethyl-5-(1-methylbutyl)-2-thiobarbiturate; $C_{11}H_{17}N_2NaO_2S$ (264.32).
- **Dosage forms** — Sterile thiopental sodium is available for injection in syringes (250, 400, and 500 mg), vials (0.5 and 1.0 g), and kits (1, 2.5, and 5 g).

Included with each gram of thiopental sodium is 60 mg of sodium carbonate which increases the alkalinity of i.v. solutions (pH = 10 to 11).

Thiopental rectal suspension contains 400 mg thiopental sodium per gram of suspension. The suspension vehicle contains mineral oil, dimethyldioctadecyl ammonium bentonite, and 24 mg of anhydrous sodium carbonate. Thiopental sodium is available under the trade name Pentothal (Abbott).

Methohexital Sodium

Mechanism of Action

See Thiopental Sodium.

Pharmacokinetics

Methohexital sodium is the second most commonly used i.v. barbiturate. As with thiopental, it produces anesthesia within 30 to 60 sec after i.v. administration but differs from thiopental in that it allows for a more rapid rate of recovery from anesthesia. Following i.v. injection, the patterns of distribution of the two barbiturates are similar and the anesthetic action of methohexital is terminated by redistribution of the drug from brain into muscle tissue, as with thiopental. Its plasma protein binding is similar to that of thiopental (72 to 86%) but its elimination half-life is much shorter, being only 1.6 hr. It is assumed that methohexital's high plasma clearance rate and short elimination half-life are the consequence

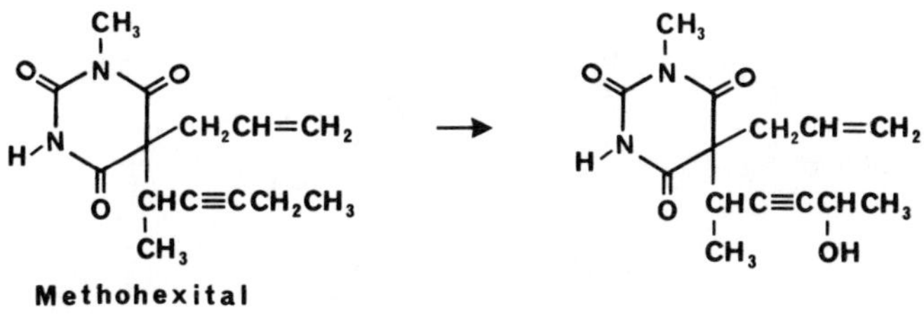

FIGURE 5. Biotransformation of methohexital in rat and dog.[101]

of rapid metabolism and low tissue accumulation.[91] Methohexital is rapidly and completely (> 99%) metabolized in the liver by ω-1 hydroxylation of the 5-position side chain to the major metabolite shown in Figure 5. The pharmacokinetics of methohexital have been reviewed[87,88,91] and mean pharmacokinetic parameters are given in Table 7.

Uses and Dosage

Methohexital is used for similar purposes as thiopental and is especially useful in procedures where brevity of action is desirable. In adults, induction of anesthesia is accomplished by i.v. administration of a 2-mℓ (1% solution) test dose, followed by 5 to 12 mℓ of 1% solution injected at the rate of 1 mℓ every 5 sec. Surgical anesthesia is then maintained by injection of 2 to 4 mℓ of 1% solution every 4 to 7 min as required.

Toxicity

The toxic reactions and side effects associated with methohexital are the same as those of thiopental.

Physical-Chemical and Other Data

Methohexital sodium is a white to off-white odorless powder. It is soluble in water and the aqueous solution is alkaline to litmus. Sterile water for injection solutions of methohexital sodium are stable at room temperature for at least 6 weeks. Its solutions are incompatible with acidic solutions; pKa = 8.2 (conjugate acid). Solutions of methohexital sodium are incompatible with silicone and should not be allowed to come in contact with rubber stoppers or parts of disposable syringes that have been treated with silicone. See References 81 for partition coefficients.

- **Chemical name** — 2,4,6(1H,3H,5H)-Pyrimidinetrione, 1-methyl-5-(1-methyl-2-pentynyl)-5-(2-propenyl)-, (±)-, monosodium salt; or Sodium 5-allyl-1-methyl-5-(1-methyl-2-pentynyl)-barbiturate; $C_{14}H_{17}N_2NaO_3$(284.29).
- **Synthesis** — The barbiturate is unusual in that it has two asymmetric centers giving four stereoisomers. These stereoisomers have been resolved into two diastereomeric pairs, α-± and the β-±, of which the α pair is used medicinally (the β-form causes undesirable side effects). The usual method of synthesis (malonic ester synthesis) is not used for methohexital because it yields mainly the unwanted β-form. An alternate synthesis yielding mainly the desired α-form is briefly described elsewhere.[92]
- **Dosage forms** — Sterile methohexital sodium is available under the trade name Brevital Sodium (Lilly) as a powder in 0.5-, 2.5-, and 5-g containing vials. Included with each gram of methohexital sodium is 60 mg of sodium carbonate making i.v. solutions alkaline (pH = 10 to 11).

Thiamylal Sodium

Mechanism of Action

See Thiopental Sodium.

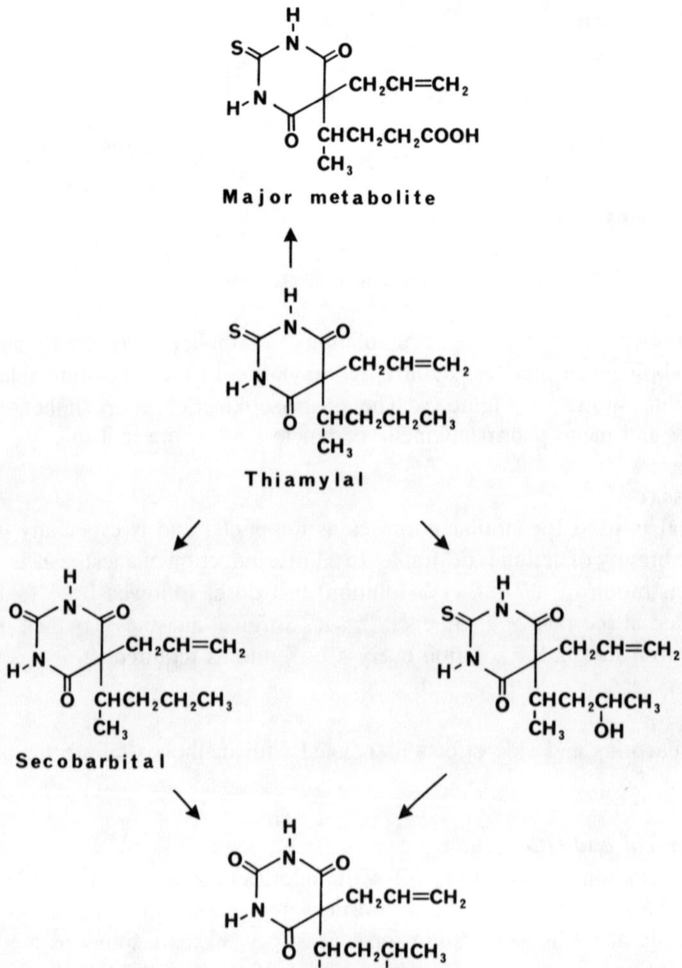

FIGURE 6. Biotransformation of thiamylal in man.[85]

Pharmacokinetics

The pharmacokinetics of thiamylal sodium have not been well studied, but clinically the drug cannot be distinguished from thiopental sodium. Studies show that thiamylal is extensively, but slowly metabolized as a thiopental.[85,86] The major metabolites of thiamylal are shown in Figure 6.

Uses and Dosage

Thiamylal sodium is used primarily to induce anesthesia and to maintain anesthesia for surgical procedures lasting less than 15 min. In adults for induction, a 2.5% solution of thiamylal is administered i.v. at a rate of 1 mℓ every 5 sec until the desired effect is obtained. A total dose of 3 to 6 mℓ is usually required for induction of anesthesia. A dose of 0.5 to 1 mℓ as required is used to maintain anesthesia.

Toxicity

The side effects and toxicity are the same as those of thiopental.

Physical-Chemical and Other Data

Thiamylal sodium is a pale yellow hygroscopic powder, having a disagreeable odor. It

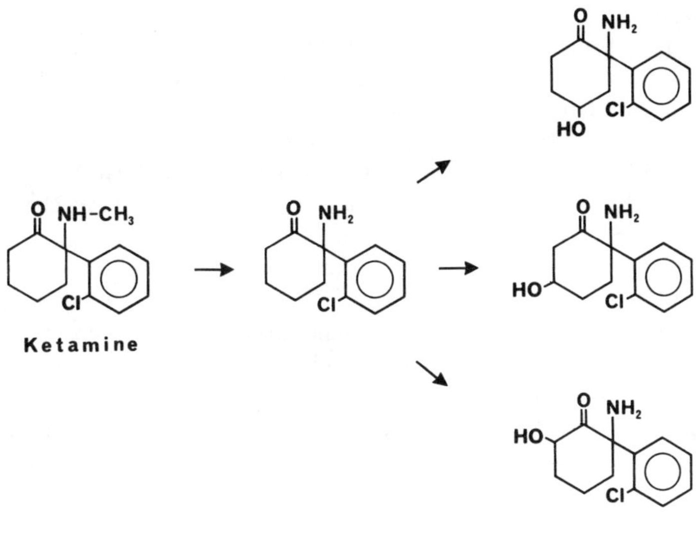

FIGURE 7. Biotransformation of ketamine in man.[93]

is soluble in water and the aqueous solution is alkaline to litmus and decomposes on standing. Thiamylal sodium solutions are incompatible with acidic solutions; its pKa is 7.3 (conjugate acid). See Reference 81 for partition coefficients.

- **Chemical name** — 4,6-(1H,5H)-Pyrimidinedione, dihydro-5-(1-methylbutyl)-5-(2-propenyl)-2-thioxo-, monosodium salt, or Sodium 5-allyl-5-(1-methylbutyl)-2-thiobarbiturate; $C_{12}H_{17}N_2NaO_2S$ (276.33).
- **Dosage forms** — The sterile powder for injection (0.5, 2.5, and 5 g) is available in vials under the trade name Surital (Parke Davis).

Ketamine Hydrochloride

Mechanism of Action

Ketamine is a nonbarbiturate anesthetic that produces a state known as "dissociative anesthesia" in which the patient may appear to be awake but is dissociated from the environment and does not respond to pain. The mechanism by which ketamine produces anesthesia and analgesia is unknown. The basic pharmacology and CNS effects of ketamine have recently been reviewed.[93]

Pharmacokinetics

The biodisposition of ketamine is similar to that of the thiobarbiturate, thiopental, in that it demonstrates a very rapid onset of action (less than 1 min after i.v. injection), and a short duration of anesthesia (5 to 10 min). Peak plasma levels are achieved within 1 min following i.v. administration and redistribution of ketamine from the brain to other tissues is primarily responsible for termination of its anesthetic effects. Ultimately, ketamine is eliminated by hepatic metabolism with up to 95% of a dose being recovered in the urine, mainly as water-soluble metabolites and their glucuronide conjugates. Ketamine is metabolized primarily by N-demethylation and hydroxylation of the cyclohexanone ring as shown in Figure 7. The hydroxylated metabolites are unstable at high temperature, and dehydrate to a cyclohexenone derivative. In the past, this unsaturated ketone was thought to be a metabolite of ketamine but is now suspected to be an artifact produced during gas chromatographic analysis of samples.[93]

The pharmacokinetic parameters of ketamine are reported in Table 7. The pharmacokinetics of ketamine have been reviewed in recent journals.[87,88,93]

Use and Dosage

Ketamine is used for induction of anesthesia by i.v. injection of 2 mg/kg of body weight over a period of 60 sec. It can also be administered i.m. for induction which requires a dose of 6.5 to 13 mg/kg of body weight. For maintenance of anesthesia, one half of the full induction dose is repeated as necessary. Ketamine can also be employed in combination with any of the volatile anesthetics. It is especially useful in children for minor surgical or diagnostic procedures or for repeated procedures that require anesthesia/analgesia, such as changing burn dressings.[92]

Toxicity

About 12% of the time, recovery after ketamine anesthesia is accompanied by unpleasant dreams and even hallucinations. These psychic disturbances occur more frequently in adults than in children 15 years of age or less. Benzodiazepines appear to be highly effective in preventing these unpleasant emergence reactions. The drug is contraindicated in patients with severe coronary artery disease, severe hypertension, or a history of stroke due to ketamine's cardiovascular stimulant properties. Other untoward effects are minimal.[89,93]

Physical-Chemical and Other Properties

Ketamine hydrochloride is a white, crystalline powder which melts with decomposition between 258 and 261°C. It is an acidic salt (pKa = 7.5) that is soluble in water, alcohol, and chloroform. Its water solution (10%) has a pH between 3.5 and 4.1.

- **Chemical name** — Cyclohexanone, 2-(2-chloro-phenyl)-2-(methylamino)-, hydrochloride or (±)-2-(*o*-chlorophenyl)-2-(methylamino)cyclo-hexanone hydrochloride; $C_{13}H_{16}ClNO \cdot HCl$ (274.19).
- **Stereochemistry** — Ketamine contains one chiral center producing two resolvable optical isomers. The current clinically used product is the racemic mixture although recent studies indicate that S(+)-ketamine produces more effective anesthesia with a lower incidence of psychic emergency reactions than does the R(−)-isomer.[94] The differences in pharmacological activity and toxicity of the optical isomers of ketamine have recently been reviewed.[93]
- **Dosage forms** — Ketamine hydrochloride is available in solution for i.v. or i.m. use in ampules containing 10 or 50 mg of ketamine base per milliliter under the trade names Ketaject (Bristol) and Ketalar (Parke Davis). The solution contains 0.1 mg/mℓ benzethonium chloride as a preservative.

OTHER DRUGS

Fentanyl Citrate and Droperidol, Figure 8 (Innovar)

Innovar is a fixed-dose combination which contains the narcotic analgesic, fentanyl (0.05 mg/mℓ) and the neuroleptic butyrophenone, droperidol (2.5 mg/mℓ). It is used to produce "neuroleptanalgesia". This condition is characterized as one of deep sedation, reduced motor activity, and profound analgesia. Patients are able to understand and answer simple questions and obey commands. Neuroleptanalgesia is particularly valuable for use during minor surgical or diagnostic procedures when the patient's cooperation is required. If 65% nitrous oxide in oxygen is also administered to the patient, unconsciousness results. This condition is referred to as "neuroleptanesthesia".

Pharmacokinetics of the two agents are markedly different.[87] Fentanyl, an exceptionally potent analgesic (50 to 100 times more potent than morphine) has a very rapid onset (1 to 2 min) and a short duration of action while droperidol has a slow onset (10 to 15 min) and a long duration of action. Consequently, the fixed-dose combination should not be used for

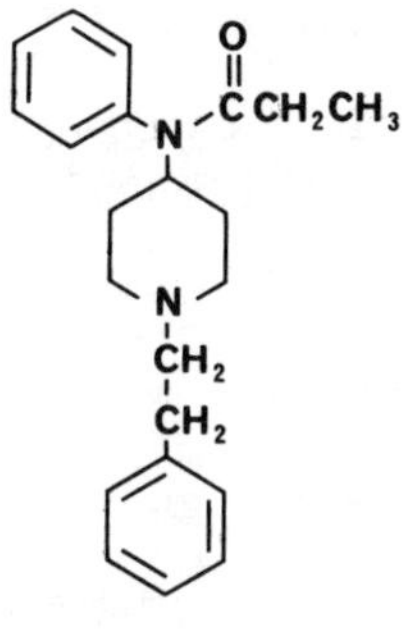

Fentanyl

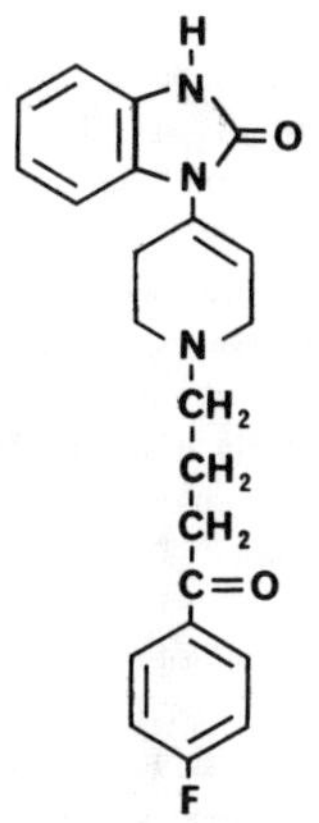

Droperidol

FIGURE 8. Structures of fentanyl and droperidol.

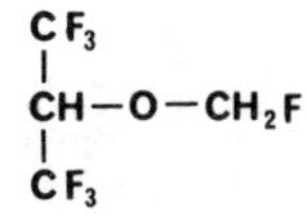

FIGURE 9. Structure of sevoflurane.

maintenance therapy. Some recommend that induction be started with a single dose of droperidol (0.15 mg/kg), and 6 to 8 min later, fentanyl (0.002 to 0.003 mg/kg) be given in increments over a period of 6 to 8 min. If neuroleptanesthesia is intended, nitrous oxide is started when drowsiness develops.

The side effects and toxicity of fentanyl are similar to those of the narcotic analgesics (chapter entitled "Narcotic Analgesics, Antagonists, and Antitussive Agents") and those of droperidol are like those of antipsychotic agents (chapter entitled "Antipsychotic Agents").

Sevoflurane

Sevoflurane, fluoromethyl 2,2,2-trifluoro-1-(trifluoromethyl)ethyl ether (Figure 9) is a promising new, nonflammable volatile anesthetic. It has a blood/gas partition coefficient of 0.60[95] and an oil-gas partition coefficient of 55.[96] In the rat, it has a MAC of 2.5%,[96] and a predicted MAC in man of 2.6% based on its oil/gas partition coefficient.[11] Initial studies in rats,[97] dogs,[97] and man[11] indicate that 1.5 to 2.5% of the absorbed sevoflurane is metabolized in hexafluoroisopropanol and fluoride ion (Table 3). Preliminary human studies suggest that sevoflurane produces excellent anesthesia with no systemic toxicity.[11]

REFERENCES

1. **Seeman, P.,** The membrane actions of anesthetics and tranquilizers, *Pharmacol. Rev.*, 24, 583, 1972.
2. **Davison, M. H. A.,** *The Evolution of Anesthesia,* Williams & Wilkins, Baltimore, 1965.
3. **Keys, T. E.,** *The History of Surgical Anesthesia,* Dover, New York, 1963.
4. **Sykes, W. S.,** *Essays on the First Hundred Years of Anesthesia,* E. & S. Livingstone, Edinburgh, 1960.
5. **Jacoby, R. L. and Nieforth, K. A.,** Volatile anesthetics, in *Principles of Medicinal Chemistry,* 2nd ed., Foye, W. O., Ed., Lea & Febiger, Philadelphia, 1981, 146.
6. **Gillespie, N. A.,** The signs of anesthesia, *Curr. Res. Anesth. Analg.*, 22, 275, 1943.
7. **Schreiber, P.,** *Anesthetic Equipment,* Springer-Verlag, Berlin, 1972.
8. **Patel, A. R.,** General anesthetics, in *Medicinal Chemistry,* 3rd ed., Burger, A., Ed., Interscience, New York, 1970, 1319.
9. **Eger, E. I.,** *Anesthetic Uptake and Action,* Williams & Wilkins, Baltimore, 1974, 82.
10. **Quasha, A. L., Eger, E. I., II, and Tinker, J. H.,** Determination and applications of MAC, *Anesthesiology,* 53, 315, 1980.
11. **Holaday, D. A. and Smith, F. R.,** Clinical characteristics and biotransformation of sevoflurane in healthy human volunteers, *Anesthesiology,* 54, 100, 1981.

12. **Munson, E. S., Schick, L. M., Chapin, J. C., Kuskins, L. G., and Navarro, A. A.,** Determination of the minimum alveolar concentration (MAC) of aliflurane in dogs, *Anesthesiology*, 51, 545, 1979.
13. **Holaday, D. A., Jardines, M. C., and Greenwood, W. H.,** Uptake and biotransformation of aliflurane (1-chloro-2-methoxy-1,2,3,3-tetrafluorocyclopropane, compound 26-P) in man, *Anesthesiology*, 51, 548, 1979.
14. **Smith, T. C., Cooperman, L. H., and Wollman, H.,** History and principles of anesthesiology, in *The Pharmacological Basis of Therapeutics*, 6th ed., Gilman, A. G., Goodman, L. S., and Gilman, A., Eds., Macmillan, New York, 1980, chap. 13.
15. **Van Dyke, R. A., Chenoweth, M. B., and Van Poznak, A.,** Metabolism of volatile anesthetics. I. Conversion *in vivo* of several anesthetics to $^{14}CO_2$ and chloride, *Biochem. Pharmacol.*, 13, 1239, 1964.
16. **Hong, K., Trudell, J. R., O'Neil, J. R., and Cohn, E. N.,** Metabolism of nitrous oxide by human and rat intestinal contents, *Anesthesiology*, 52, 16, 1980.
17. **Van Dyke, R. A. and Chenoweth, M. B.,** Metabolism of volatile anesthetics, *Anesthesiology*, 26, 348, 1965.
18. **Blitt, C. D., Gandolfi, A. J., Soltis, J. J., and Brown, B. R.,** Extrahepatic biotransformation of halothane and enflurane, *Anesthesiology*, 60, 129, 1981 and references therein.
19. **Van Dyke, R. A.,** Introduction: metabolic pathways and anesthetic detoxification, in *Toxicity of Anesthetics*, Fink, B. R., Ed., Williams & Wilkins, Baltimore, 1968, 62.
20. **Van Stee, E. W.,** Toxicity of inhalation anesthetics and metabolites, *Ann. Rev. Pharmacol.*, 16, 67, 1976.
21. **Eger, E. I., II,** Isoflurane: a review, *Anesthesiology*, 55, 559, 1981.
22. **Wade, J. G. and Stevens, W. C.,** Isoflurane: an anesthetic for the eighties?, *Anesth. Analg.*, 60, 666, 1981.
23. **Halsey, M. J., Sawyer, D. C., Eger, E. I., II, Bahlman, S. H., and Impelman, D. J. K.,** Hepatic metabolism of halothane, methoxyflurane, cyclopropane, Ēthrane and Forane in miniature swine, *Anesthesiology*, 35, 43, 1971.
24. **Eger, E. I., II, Saidman, L. J., and Bandstater, B.,** Minimum alveolar anesthetic concentration: a standard of anesthetic potency, *Anesthesiology*, 26, 756, 1965.
25. **Dobkin, A. B.,Ed.,** *Developments of New Volatile Inhalation Anesthetics*, Elsevier/North-Holland, Amsterdam, 1979.
26. **Wallin, R. F., Napoli, M. D., and Regan, B. M.,** Laboratory investigation of a new series of inhalation anesthetic agents. The halomethyl polyfluoroisopropyl ethers, in *Cellular Biology and Toxicity of Anesthetics*, Fink, B. R., Ed., Williams & Wilkins, Baltimore, 1972, 289.
27. **Koblin, D. D., Eger, E. I., II, Johnson, B. H., Collins, P., Terrell, R. C., and Speers, L.,** Are convulsant gases also anesthetics?, *Anesth. Analg.*, 60, 464, 1981.
28. **Koblin, D. D., Eger, E. I., II, Johnson, B. H., Collins, P., Harper, M. H., Terrell, R. C., and Speers, L.,** *Anesthesiology*, 54, 314, 1981.
29. **Meyer, H. H.,** Zur theorie de alkoholnarkose. I. Mitt, welche Eigenschaft der Änasthetika bedingt ihre Narkotische wirkung?, *Arch. Exp. Pathol. Pharmakol.*, 42, 109, 1899.
30. **Meyer, H. H.,** Zur theorie due alkoholnarkose. III. Mitt der einfluss wechselnder Temperatur auf Wirkungsstärke und Teilungskoefficient der Narkotika, *Arch. Exp. Pathol. Pharmacol.*, 46, 338, 1901.
31. **Overton, E.,** Studien Über die Narkose Zugleich ein Beitrag zur Allgmeinen Pharmacologie, G. Fischer, Jena, 1901.
32. **Torda, T. A. and Gage, P. W.,** Postsynaptic effect of i.v. anesthetic agents at the neuromuscular junction, *Br. J. Anaesth.*, 49, 771, 1977.
33. **Miller, K. W.,** Towards the molecular bases of anesthetic action, *Anesthesiology*, 46, 2, 1977 and references therein.
34. **Trudell, J. R.,** A unitary theory of anesthesia based on lateral phase separations in nerve membranes, *Anesthesiology*, 46, 5, 1977.
35. **Lee, A. G.,** Model for action of local anesthetics, *Nature (London)*, 262, 545, 1976.
36. **Seeman, P.,** Anesthetics and pressure reversal of anesthesia, *Anesthesiology*, 47, 1, 1977 and references therein.
37. **Pauling, L.,** A molecular theory of anesthesia, *Science*, 134, 15, 1961.
38. **Miller, S. L.,** A theory of gaseous anesthetics, *Proc. Natl. Acad. Sci. U.S.A.*, 47, 1515, 1961.
39. **Sandorfy, C.,** Intermolecular interactions of anesthesia, *Anesthesiology*, 48, 357, 1978.
40. **Tanifugi, Y., Eger, E. I., II, and Terrell, R. C.,** Some characteristics of an exceptionally potent inhaled anesthetic: thiomethoxyflurane, *Anesth. Analg.*, 56, 387, 1977.
41. **Ferguson, J.,** The use of chemical potentials as indices of toxicity, *Proc. R. Soc. Lond.*, B127, 387, 1939.
42. **Mullins, L. J.,** Some physical mechanisms in narcosis, *Chem. Rev.*, 54, 289, 1954.
43. **Halsey, M. J. and Wardley-Smith, B.,** Pressure reversal of narcosis produced by anaesthetics, narcotics, and tranquilizers, *Nature (London)*, 257, 811, 1975.
44. **Roth, S. H.,** Physical mechanisms of anesthesia, *Ann. Rev. Pharmacol. Toxicol.*, 19, 168, 1979 and cited references therein.

45. **Seeman, P.,** The membrane actions of anesthetics and tranquilizers, *Pharmacol. Rev.*, 24, 583, 1972.
46. **Kaufman, R. D.,** Biophysical mechanisms of anesthetic action, *Anesthesiology,* 46, 49, 1977.
47. **Miller, K. W.,** General anesthesia, in *Burger's Medicinal Chemistry* Part III, 4th ed., Wolff, M. E., Ed., John Wiley & Sons, New York, 1981, chap. 50.
48. **Cheng, S-C. and Brunner, E. A.,** Effects of anesthetic agents on synaptosomal GABA disposal, *Anesthesiology,* 55, 34, 1981 and references therein.
49. **Krogsgaard-Larsen, P. and Christensen, A. V.,** GABA agonists and antagonists, *Ann. Rep. Med. Chem.*, 15, 41, 1980 and references therein.
50. **Clayton, J. W.,** Fluorocarbon toxicity and biological action, in *Toxicity of Anesthetics,* Fink, B. R., Ed., Williams & Wilkins, Baltimore, 1968, chap. 8.
51. **Duncalf, D.,** Flammable anesthetics are nearing extinction, *Anesthesiology,* 56, 217, 1982.
52. **Fineberg, H. V., Pearlman, L. A., and Gabel, R. A.,** The case for abandonment of explosive anesthetic agents, *N. Engl. J. Med.*, 303, 613, 1980.
53. Summary of the National Halothane Study, *JAMA,* 197, 775, 1966.
54. **Cohen, E. N.,** Mechanisms of toxicity, in *Anesthetic Exposure in the Workplace,* P. S. G. Publishing Co., Littleton, Mass., 1980, chap. 5.
55. **Gourlay, G. K., Adams, J. F., Cousins, M. J., and Hall, P.,** Genetic differences in reductive metabolism and hepatotoxicity of halothane in three rat strains, *Anesthesiology,* 55, 96, 1981.
56. Medical Letter, Dantrolene for malignant hypothermia during anesthesia, July 25, 1980.
57. **Britt, B. A., Endrenyi, L., Frodis, W., Scott, E., and Kalow, W.,** Comparison of effects of several inhalation anesthetics on caffeine-induced contractures of normal and malignant hyperthermic skeletal muscle, *Can. Anaesth. Soc. J.*, 24, 12, 1980.
58. **Cohen, E. N., Brown, B. W., Wu, M. L., Whitcher, C. E., Brodsky, J. B., Gift, H. C., Greenfield, W., Jones, T. W., and Driscoll, E. J.,** Occupational disease in dentristry and chronic exposure to trace anesthetic gases, *J. Am. Dent. Assoc.*, 101, 21, 1980.
59. **Ferstandig, L. L.,** Trace concentrations on anesthetic gases: A critical review of their disease potential, *Anesth. Analg.*, 57, 328, 1978.
60. **Mazze, R. I.,** Waste anesthetic gases and the regulatory agencies, *Anesthesiology,* 52, 248, 1980.
61. **Johnston, R. R. and Eger, E. I., II, and Wilson, C.,** A comparative interaction of epinephrine with enflurane, isoflurane and halothane in man, *Anesth. Analg.*, 55, 709, 1976.
62. **Wand, B. E.,** Decrease in dose requirement of d-trubocurarine by volatile anesthetics, *Anesthesiology,* 51, 298, 1979.
63. **Pettinger, C. B., Eryasa, Y., and Adamson, R.,** Antibiotic-induced paralysis, *Anesth. Analg.*, 49, 487, 1970.
64. **Grogono, A. W. and Seltzer, J. L.,** A guide to drug interactions in anesthetic practice, *Drugs,* 19, 279, 1980.
65. **Munson, E. S., Eger, E. I. II, Tham, M. K., Embro, W. J.,** Increase in anesthetic uptake, excretion, and blood solubility in man after eating, *Anesth. Analg.*, 57, 224, 1978.
66. **Chase, R. D., Holaday, D. A., Fiserova-Bergerova, V., Saidman, L. J., and Mack, F. E.,** The biotransformation of Ethrane in man, *Anesthesiology,* 35, 262, 1971.
67. **Cousins, M. J., Greenstein, L. R., Hitt, B. A., and Mazze, R. I.,** Metabolism and renal effects of enflurane in man, *Anesthesiology,* 44, 44, 1976.
68. **Calverley, R. K., Smith, N. T., Jones, C. W., Prys-Roberts, C., Eger, E. I., II,** Ventilatory and cardiovascular effects of enflurane anesthesia during spontaneous ventilation in man, *Anesth. Analg.*, 57, 610, 1978.
69. **Opitz, A., Brecht, S., and Stenyel, E.,** Enflurane anesthesia in epileptus, *Anesthesiology,* 26, 329, 1977.
70. **Rehder, K., Forber, J., Alter, H., Hessler, O., and Stier, A.,** Halothane biotransformation in man: a quantitative study, *Anesthesiology,* 28, 711, 1967.
71. **Cascorbi, H. F., Blake, C. A., Helrich, M.,** Differences in the biotransformation of halothane in man, *Anesthesiology,* 35, 262, 1970.
72. **Kendig, J. J., Trudell, J. R., and Cohen, E. N.,** Halothane, stereoisomers, *Anesthesiology,* 39, 518, 1973.
73. **Cromwell, T. H., Eger, E. I., II, Stevens, W. C., and Dolan, W. M.,** Forane uptake, excretion, and blood solubility in man, *Anesthesiology,* 35, 401, 1971.
74. **Hitt, B. A., Mazze, R. I., Cousins, M. J., Edmunds, H. N., Barr, G. A., and Trudell, J. R.,** Metabolism of isoflurane in Fischer 344 rats and man, *Anesthesiology,* 40, 62, 1974.
75. **Corbert, T. H.,** Cancer and congenital anomalies associated with anesthetics, *Ann. N.Y. Acad. Sci.*, 271, 58, 1976.
76. **Eger, E. I., II, White, A. E., Brown, C. L., Biava, C. G., Corbett, T. H., and Stevens, W. C.,** A test of the carcinogenicity of enflurane, isoflurane, halothane, methoxyflurane, and nitrous oxide in mice, *Anesth. Analg.*, 57, 678, 1978.

77. **Holaday, D. A., Rudofsky, S., and Treuhaft, P. S.,** Metabolic degradation of methoxyflurane in man, *Anesthesiology,* 33, 579, 1970.
78. General anesthetics, in *AMA Drug Evaluations,* 4th ed., John Wiley & Sons, New York, 1980, 323.
79. **Koblin, D. D., Waskell, L., Watson, J. E., Stokstad, E. L. R., and Eger, E. I., II,** Nitrous oxide inactivates methionine synthetase in human liver, *Anesth. Analg.,* 61, 75, 1982.
80. **Saidman, L. J.,** Mechanisms of barbiturate action, in *Anesthetic Uptake and Action,* Eger, E. I., II, Ed., Williams & Wilkins, Baltimore, 1974, 17.
81. **Bush, M. T., Berry, G., and Hume, A.,** Ultra-short-acting barbiturates as oral hypnotic agents in man, *Clin. Pharmacol. Ther.,* 7, 373, 1966.
82. **Becker, K. E.,** Plasma levels of thiopental necessary for anesthesia, *Anesthesiology,* 49, 192, 1978.
83. **Morgan, D. J., Blackman, G. L., Paull, J. D., and Wolff, L. J.,** Pharmacokinetics and plasma binding of thiopental. I. Studies in surgical patients, *Anesthesiology,* 54, 468, 1981.
84. **Morgan, D. J., Blackman, G. L., Paull, J. D., and Wolf, L. J.,** Pharmacokinetics and plasma binding of thiopental. II. Studies at cesarean section, *Anesthesiology,* 54, 474, 1981.
85. **Carroll, F. I., Smith, D., Mark, L. C., Brand, L., and Perel, J. M.,** Determination of optically active thiopental, thiamylal, and their metabolites in human urine, *Drug Metab. Disp.,* 5, 343, 1977.
86. **Hirtz, J. L.,** *Analytical Metabolic Chemistry in Drugs,* Garrett, E. R., transl. ed., Marcel Dekker, New York, 1971, chap. 8.
87. **Ghoneim, M. M. and Korttila, K.,** Pharmacokinetics of intravenous anesthetics: implications for clinical use, *Clin. Pharmacokinet.,* 2, 344, 1977.
88. **Duvaldestin, P.,** Pharmacokinetics in intravenous anaesthetic practice, *Clin. Pharmacokinet.,* 6, 61, 1981.
89. **Clarke, R. S. J.,** Adverse effects of intravenously administered drugs used in anaesthetic practice, *Drugs,* 22, 26, 1981.
90. **Smith, N. T., Miller, R. D., and Carbascio, A.,** *Drug Interactions in Anesthesia,* Lea & Febiger, Philadelphia, 1981.
91. **Breimer, D. D.,** Clinical pharmacokinetics of hypnotics, *Clin. Pharmacokinet.,* 2, 93, 1977.
92. **Swinyard, E. A.,** General anesthetics, in *Remington's Pharmaceutical Sciences,* 16th ed., Osol, A., Ed., Mack Publishing, Easton, Pa., 1980, chap. 54.
93. **White, P. F., Way, W. L., and Trevor, A. J.,** Ketamine — its pharmacology and therapeutic uses, *Anesthesiology,* 56, 119, 1982.
94. **White, P. F., Ham, J., Way, W. L., and Trevor, A. J.,** Pharmacology of ketamine isomers in surgical patients, *Anesthesiology,* 52, 231, 1980.
95. **Wallin, R. F., Regan, B. M., Napoli, M. D., and Stern, I. J.,** Sevoflurane, a new inhalational anesthetic agent, *Anesth. Analg.,* 54, 758, 1975.
96. **Cook T. L., Beppu, W. J., Hitt, B. A., Kosek, J. C., and Mazze, R. I.,** Renal effects and metabolism of sevoflurane in Fisher 344 rats: an *in vivo* and *in vitro* comparison with methoxyflurane, *Anesthesiology,* 43, 70, 1975.
97. **Martis, L., Lynch, S., Napoli, M. D., and Woods, E. F.,** Biotransformation of sevoflurane in dogs and rats, *Anesth. Analg.,* 60, 186, 1981.
98. **Ghoneim, M. M. and VanHamme, M. J.,** Pharmacokinetics of thiopentone: effect of enflurane and nitrous oxide anesthesia and surgery, *Br. J. Anaesth.,* 50, 1237, 1978.
99. **Breimer, D. D.,** Pharmacokinetics of methohexitone following intravenous infusion in humans, *Br. J. Anaesth.,* 48, 643, 1976.
100. **Wieber, J., Gugler, R., Hengstmann, J. H., and Deugler, H. J.,** Pharmacokinetics of ketamine in man, *Anaesthesist,* 24, 260, 1975.
101. **Welles, J. S., McMahon, R. E., and Doran, W. J.,** *J. Pharm. Exp. Ther.,* 139, 166, 1963.

BARBITURATE SEDATIVES AND HYPNOTICS

Ronald J. Kotun and John D. Alvin

BARBITURATES: AN OVERVIEW

The barbiturates are general CNS depressants used by a substantial proportion of the population primarily for the treatment of insomnia. They have also been quite extensively used as sedatives, but since the evolution of the benzodiazepines, a safer class of compounds, the use of barbiturates as a sedative has begun to decline. However, their use as anticonvulsants and induction agents to anesthesia continues.

Chemistry

Barbituric acid is a 2,4,6-trioxohexahydropyrimidine. The barbituric acid derivatives are colorless crystalline solids which are relatively insoluble in water but soluble in ethanol and nonpolar solvents. Sodium salts of the barbiturates are readily soluble in water and generate unstable, alkaline solutions. The carbonyl group at position 2 is acidic and is susceptible to "keto-enol" tautomerism because it lies between two highly electronegative nitrogens. The equilibrium tends to favor the "enol" tautomer in alkaline solutions resulting in the salt (Figure 1).

Barbituric acid lacks central depressant activity, but alkyl groups at position 5 give the compound its sedative-hypnotic properties and other pharmacological properties.[1] The substituents at position 5 definitely exhibit a structural-activity relationship. It has been shown that increasing the chain length of the alkyl substituents at position 5 enhances the potency of the barbiturates (up to five or six carbons).[2] An alkyl group of greater than seven carbons results in convulsant activity.[1] A branched substituent has briefer actions than straight chained alkyl groups. An unsaturated group has briefer effects than a saturated group. Alkyl substituents bound to nitrogens at positions 1 and 3 have stimulant properties and shorten the duration of action. Thiobarbiturates (barbiturates with a sulfur at position 2 instead of an oxygen) are more stable, more lipophilic, and have a diminished duration of action relative to the oxybarbiturates.[2] If the lipid solubility of the barbiturate is enhanced, the duration of action is shorter, the onset of action is earlier, the metabolism is faster and the potency is greater. The excitatory effect of certain 5,5-disubstituted barbiturates having an asymmetric center may be due to one of the isomers (see the monograph on pentobarbital). Table 1 lists chemical structures and dosages of barbiturates, while Table 2 lists their solubility data and pKa values.

Pharmacology

Barbiturates primarily affect the CNS. They can reversibly depress the activity of all excitable tissue; however not all are equally affected. They selectively depress the reticular activating system and leave the major sensory and motor pathways relatively uninfluenced at therapeutic concentrations. The barbiturates cover a wide spectrum of activity ranging from sedation and sleep to anesthesia, coma, and ultimately death.

After a hypnotic dose of a barbiturate, drowsiness may occur for only a few hours. However, depression, referred to as a "hangover", may occur the following day. Because of the depressant effects, one does not have full control of his faculties; therefore one should not drive or perform any intellectual function which may impair the safety of himself or others. Vertigo, diarrhea, nausea, and vomiting are also characteristic untoward effects of the barbiturates.

The barbiturates exhibit a low degree of selectivity and a low therapeutic index except for phenobarbital. Regarding sleep, the barbiturates tend to decrease sleep latency, the

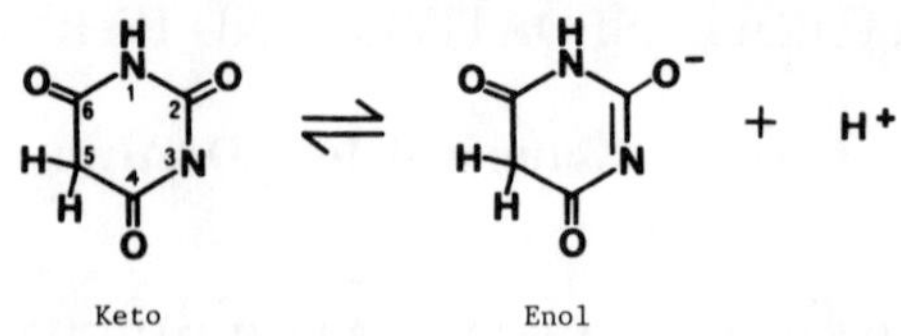

FIGURE 1. Keto-enol tautomerism of barbituric acid.

frequency of awakening and body movement. However, they prolong the latent period before REM sleep but decrease the true number and activity of REM occurrences.[3]

Barbiturates also depress the respiratory drive. Hypnotic doses diminish the neurogenic drive similar to the depression caused by normal sleep. Higher doses depress the chemical drive located in the medulla and render the chemoreceptors insensitive to low pH. Toxic doses will render the chemoreceptors at the carotid and aortic arches insensitive to pO_2 and consequently obliterate the hypoxic drive resulting in respiratory paralysis.

Oral sedative-hypnotics exert very little or no effect on the cardiovascular system. However, the ultrashort acting i.v. anesthetics (i.e., thiopental) tend to decrease cardiac output and renal plasma flow. To compensate for the decreased cardiac output, total peripheral resistance increases. Heart rate only slightly increases. However, there is an increase in blood flow and blood volume to the extremities, but cerebral blood flow decreases.

Barbiturates decrease the tone of the intestinal smooth muscle and inhibit the gastric secretory processes. Barbiturates also enhance the activity of the liver by stimulating the production of enzymes, proteins, and lipids in the smooth endoplasmic reticulum to increase the rate of metabolism of several drugs and endogenous substrates. The barbiturates may also stimulate the rate of their own metabolism to establish a drug-dispositional tolerance and a cross-tolerance to other drugs.[1] Induction can also potentiate the activity of other enzymes such as glucuronyl transferase and the bilirubin-binding Y proteins. Due to the induction of these proteins, barbiturates are used in the treatment of hyperbilirubinemia and kernicterus in the neonate.[22] They have also been beneficial in the treatment of cholestasis by enhancing the metabolism and excretion of bile salts.[23]

Table 3 summarizes the pharmacokinetic parameters of the barbiturates. Not only does it list half-lives of the agents, it also provides information regarding onset and duration of action, volume of distribution, clearance, extent of protein binding, and how much of the drug is excreted unchanged.

Mechanism of Action

Although the barbiturates are one of the most widely used class of drugs, little is known regarding their mechanisms of action. Several theories have been postulated, but it is very difficult to identify and confirm the actual mechanism. Lipophilicity and neurotransmitter release have been the primary targets involved in the actions of the barbiturates.

It has been proposed that the barbiturates are capable of penetrating the lipid membrane and altering its structure. The structural changes may interfere with the ion channels and the membrane-bound enzymes.[4] The barbiturates were also found to be able to increase the fluidity of the hydrophobic portion of the synaptic membrane in a manner similar to that of ethanol.[5,6] These small changes in the fluidity of the membrane can result in large changes in the permeabilities of ions. The large influence that fluidity has on ion permeability implies that there exists a mechanism for potentiating weak membrane perturbations produced by low concentrations of barbiturates.[7] Fluidity changes may also affect the surface charge of brain membranes independent of calcium availability and cause an increase in neurotransmitter release.[8]

Sedative barbiturates inhibit excitatory transmission and enhance inhibitory transmission.[9]

Table 1
CHEMICAL STRUCTURE AND RECOMMENDED DOSES FOR DESIRED PHARMACOLOGICAL RESPONSE

		Structure		Recommended doses (mg)		
Generic	Trade	R_5	R_5'	Sedative	Hypnotic	Anticonvulsant
Amobarbital	Amytal	CH_3CH_2–	$(CH_3)_2CHCH_2CH_2$–	20—40	100—200	200—400
Aprobarbital	Allonal	CH_2=$CHCH_2$	$(CH_3)_2CH$	40—80	40—160	—
Barbital	Veronal	CH_3CH_2–	CH_3CH_2–	—	300	—
Butabarbital	Butisol	CH_3CH_2–	$CH_3CH_2(CH_3)CH$–	15—30	100—200	—
Hexobarbital[a]	Evipal	CH_3–	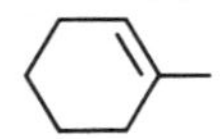	200—400[c]	—	—
Mephobarbital[a]	Mebaral	CH_3CH_2–	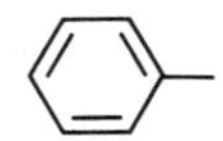	30—100	100	30—100
Methohexital[a]	Brevital	CH_2=$CHCH_2$–	$CH_3CH_2C{\equiv}C(CH_3)CH$–	70—100[c]	—	—
Pentobarbital	Nembutal	CH_3CH_2–	$CH_3(CH_2)_2(CH_3)CH$–	30	100	>100
Phenobarbital	Luminal	CH_3CH_2	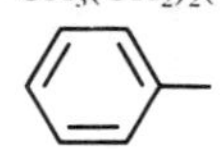	15—30	100—200	15—30
Probarbital	Ipral	CH_3CH_2	$(CH_3)_2CH$	250	—	—
Secobarbital	Seconal	CH_2=$CHCH_2$–	$CH_3CH_2(CH_3)CH$–	30—60	100—200	—
Talbutal	Lotusate	CH_2=$CHCH_2$	$CH_3CH_2(CH_3)CH$	30	120	—
Thiopental[b]	Pentothal	CH_3CH_2–	$CH_3(CH_2)_2(CH_3)CH$–	50—75[c]	—	75—125

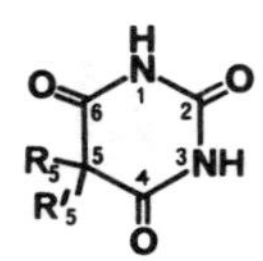

[a] CH_3 bound to N at position 1 instead of H.
[b] Thiobarbiturate; S replaces 0 at position 2.
[c] Anesthetic doses.

Table 2
SOLUBILITIES OF BARBITURATES IN SOME COMMON SOLVENTS AND THEIR pKa VALUES

	H_2O	EtOH	Ether	$CHCl_3$	pKa
Amobarbital	1/1500	1/5	1/6	1/20	7.9
Aprobarbital	1/350	1/2.5	1/5	1/40	7.5
Barbital	1/160 1/12[a]	1/15	1/40	1/70	7.9
Butabarbital	Slight	Readily soluble	Readily soluble	Soluble	7.9
Hexobarbital	1/3000 1/250[a]	1/45	1/80	1/4	8.3
Mephobarbital	Slight	1/240	1/200	1/40	7.7
Methohexital	Very slight	Slight	Slight	Slight	8.3
Pentobarbital	Slight	Readily soluble	Readily soluble	—	8.1
Phenobarbital	1/1000	1/15	1/40	1/50	7.4
Probarbital	Slight	Readily soluble	Readily soluble	Soluble	8.0
Secobarbital	Slight	Very soluble	Very soluble	—	7.9
Talbutal	Insoluble	Soluble	Soluble	Soluble	7.8

Note: Values given as part of barbiturate per given parts of solvent.

[a] In boiling H_2O.

Table 3
PHARMACOKINETIC PROPERTIES

Generic	Onset (min)	Duration	Half-life (hr)	Volume of distribution (ℓ/kg)	% Excreted unchanged	Clearance mℓ/kg/min	Protein binding (%)
Amobarbital	20—30	Interm	8—42[1]	1.00[28]	<1ℓ	1.49[28]	40—60
Aprobarbital	—	Interm	—	—	9—18[37]	—	55—70
Barbital	30—60	Long	—	—	100	—	—
Butabarbital	20—30	Interm	34—42[1]	0.78[29]	5—9[29]	0.25[29]	—
Hexobarbital	<1%	Ultrashort	2.7—7[1]	1.10[30]	<1	3.60[30]	—
Mephobarbital	30—60	Long	—	—	<1	—	—
Methohexital	0.5	Ultrashort	3—8[31]	1.13[31]	<1	12.14[31]	73
Pentobarbital	20—30	Short	15—48[1]	0.81[32]	<1	1.08[32]	50
Phenobarbital	20—40	Long	79—93[33]	0.60[34]	24[33]	0.05[34]	20—40
Probarbital	—	Interm	—	—	—	—	—
Secobarbital	20—30	Short	19—34[1]	0.49[35]	5[35]	0.20[35]	37
Talbutal	—	Interm	—	—	—	—	—
Thiopental	0.5	Ultrashort	3—8	1.90[36]	<1	3.20[36]	70

Inhibition of excitatory transmission is due to either a decrease in the postsynaptic effects of excitatory neurotransmitters, presynaptic inhibition, or both. Barbiturates have also been found to enhance the effects of GABA, reverse the effects of GABA antagonists, and mimic the effects of GABA.[10-12] The GABA synapse is highly sensitive and selective to barbiturates; hence, it may be important with regard to the anesthetic and anticonvulsant effect of barbiturates.[9]

Barbiturates inhibit the accumulation of calcium by nerve terminals,[9] therefore there should be an inhibition of neurotransmitter release. It has been demonstrated that barbiturates inhibit the release of acetylcholine,[13] norepinephrine,[14] GABA,[15] and glutamate.[15] Certain areas of the brain appear to be more sensitive than others regarding the release of specific neurotransmitters and regarding inhibition by specific barbiturates as well as by their specific enantiomers. However, there is no concrete evidence that these effects influence sedation or anesthesia.[9]

Barbiturates also enhance the tonic release of acetylcholine at the neuromuscular junction.[16] This may explain why skeletal muscle twitches during induction of anesthesia. The enhanced release may be due to a decreased surface charge on the synaptic vesicle.

The barbiturates may produce CNS effects by facilitating GABA-mediated inhibitory neurotransmission[17] or enhancing the binding of GABA.[18,19] The relative activity of the barbiturates to enhance GABA binding is in direct relation with its anesthetic capabilities and ability to augment the postsynaptic inhibitory response to GABA. Barbiturates may also increase the number of GABA receptors.[20]

Clinical Indications

Since the evolution of the benzodiazepines, the use of the barbiturates as sedative-hypnotics has significantly declined. They lack specificity in the CNS, have a lower therapeutic index, and contain a greater potential for abuse and the development of tolerance. They also interact with several more drugs.

Although use as a daytime sedative has considerably faded, some barbiturates are still employed for sedation with "inefficacious combinations" for GI disorders, hypertension, asthma, and coronary artery disease.[1] They have also been used to moderate the irritability and uneasiness in children that suffer with colic, whooping cough, and nausea and vomiting of functional origin. The sedative action has also been employed to make children less apprehensive prior to any medical or dental procedures.[1]

Barbiturates will antagonize the effects of CNS stimulants such as *d*-amphetamine and theophylline. They also work well to deter convulsions including those due to status epilepticus, drug toxicity of the CNS, eclampsia, tetanus, and cerebral hemorrhage. The ultra-short-acting barbiturates are useful as i.v. anesthetics and the short-acting barbiturates are useful as adjuncts to other agents in obstetrical anesthesia. However, the use of barbiturates in obstetrics should be avoided because of the possibility that the neonate may experience respiratory depression. Anesthetic doses of barbiturates will attenuate cerebral edema caused by cerebral ischemia, head injury, or surgery.[21]

Untoward Effects

Occasionally, a barbiturate user may experience overt excitement. He may awaken feeling euphoric and highly energetic, but when he challenges himself to normal daytime functions, he finds his performance to be impaired and consequently becomes irritable and angry. Barbiturates suppress REM and stage 4 sleep; however, withdrawal of the drug will cause a rebound in REM sleep and result in an increased frequency of nightmares and contribute to the patient's irritability.

Barbiturates should never be administered to patients experiencing pain. Because the barbiturates depress inhibitory centers, there is a removal of inhibition at pain centers and a concomitant exacerbation of one's reaction to pain. The depression of inhibitory centers may cause overt excitement instead of sedation or hypnosis. The *N*-methyl barbiturates and phenobarbital share this property.

Barbiturates are capable of interacting with ethanol, antihistamines, and other depressants as well as with MAO inhibitors to potentiate the CNS depression. Microsomal induction will increase the rate of metabolism of various exogenous agents (i.e., oral anticoagulants, digitoxin) as well as endogenous substances (i.e., corticosteroids). They facilitate the removal of vitamin K and clotting factors II and VIII to cause hemorrhaging. They also facilitate the removal of vitamin D to cause hypocalcemia. Barbiturates, especially in the elderly, may increase the probability of obtaining fractures.[24] The increased rate of disappearance of endogenous steroids results in endocrine disturbances. Barbiturates are also capable of competitively inhibiting the metabolism of certain drugs (i.e., tricyclics). They also compete with other weak acids for binding sites on albumin. The barbiturates are contraindicated in those suffering from porphyria because barbiturates enhance the synthesis of porphyrins. They are also contraindicated in those suffering from pulmonary insufficiency because they cause apnea, cough, laryngospasm, and respiratory depression. A rapid i.v. administration may cause cardiovascular collapse before the onset of anesthesia.

Pharmacodynamic and drug-dispositional tolerance develop from the use of barbiturates. These tolerances decrease the duration of action and the intensity of action of the barbiturates. Consequently, higher doses are needed to elicit the same response prior to the development of the tolerances. As blood levels of barbiturates decline, symptoms of CNS depression disappear at levels that are associated with the presentation of extreme CNS depression.[25] Hence, the therapeutic index increases as tolerance increases.[1] Tolerance to mood and sedation occurs more readily than to the anticonvulsant properties of barbiturates.[1] However, the tolerance has an upper limit. When the abuser surpasses the limit, he presents the obvious intoxication symptoms.[26] Chronic use and abuse further leads to the development of a cross-tolerance to other CNS depressants.

Those who maintain the barbiturate intoxication are characteristically sluggish, have slurred speech, have poor memory, are inattentive, are agitated, are paranoid, and are suicidal.[25] Withdrawal from short-acting barbiturates initially shows improvement by receding from intoxication. Eventually, the patient becomes restless, anxious, and weak. He experiences nausea, vomiting, and orthostatic hypotension. During the first day, he has hand tremors, an increased blink reflex, and begins pleading for the drug. The symptoms peak at the second or third day of withdrawal. The likelihood of convulsions exists.[1]

Symptoms resulting from the withdrawal of long-acting barbiturates tend to peak on the seventh or eighth day. Seizures may occur, but postictally several of the patients begin to improve. Others become delirious, agitated, and hypothermic. This leads to cardiovascular collapse and eventually death.

Babies that are born to mothers who are addicted to barbiturates also experience withdrawal.[27] The neonates are irritable, have tremors, have a high-pitched cry, breathe rapidly, have hyperactive reflexes, and experience increased yawning, vomiting, sneezing, and fever. These symptoms are very similar to babies born to opiate-addicted mothers.

MONOGRAPHS

Following is a set of monographs summarizing the pharmacology and metabolism of the barbiturates frequently used in clinical situations. Monographs for thiopental and methohexital may be found in the chapter titled "General Anesthetics" and the monograph on mephobarbital may be found in the chapter titled "Anticonvulsant Agents."

Amobarbital (5-Ethyl-5-Isopentylbarbituric Acid)

Other names: Amylobarbitone
Pentymalum

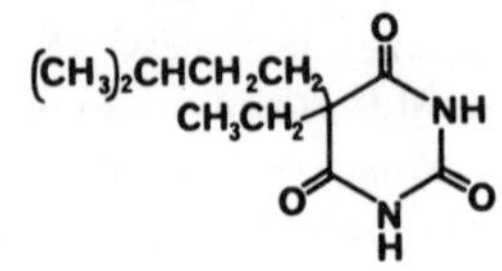

Amobarbital is an intermediate-acting barbiturate used in sedative doses to alleviate tension and anxiety and in hypnotic doses to induce for anesthesia. The sodium salt has been

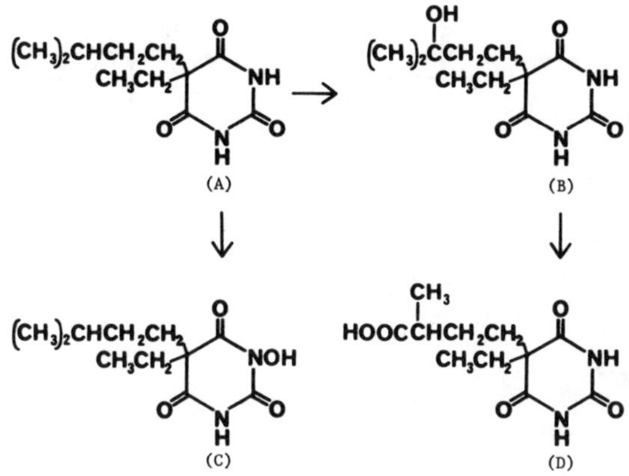

FIGURE 2. Metabolic scheme of amobarbital.

administered i.v. and i.m. to control anxiety in hypnotic doses and to control seizures caused by chorea, meningitis, tetanus, eclampsia, adverse reactions to procaine or cocaine, and strychnine or picrotoxin poisoning. The salt has also been indicated for managing epileptic seizures and manic and catatonic reactions. Amobarbital has also been useful in narcotherapy and as a diagnostic aid in schizophrenia. If chronic administration of amobarbital proves to be ineffective after 2 weeks of treatment, prolonged administration should be discontinued and another drug for the treatment of insomnia should be considered.

Amobarbital (A) undergoes the reactions that are typical of barbiturate metabolism (Figure 2). The predominant metabolic pathway involves the hydroxylation of the isopentyl moiety to yield 5-(3′-hydroxy-3′-methylbutyl)-5-ethylbarbituric acid (B).[38] This 3′-hydroxy metabolite accumulates in the body in cases when patients are suffering from renal impairment. The generation of this metabolite is less evident in neonates and the elderly. Recently, it has been shown that the isopentyl side chain can undergo further oxidation to generate 5-(3′-carboxybutyl)-5-ethyl barbituric acid (D).[39] To a lesser extent, amobarbital undergoes N-hydroxylation (C).[40] Because amobarbital is moderately lipophilic, the drug can readily penetrate the placental barrier. Amobarbital has also been detected in breast milk and saliva at concentration which paralleled serum levels.[41-43]

Aprobarbital (5-Allyl-5-Isopropylbarbituric Acid)

Other names: Aprobarbitone
Allylisopropylmalonylurea
Allylpropymal

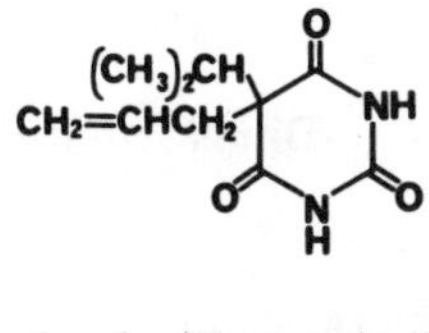

Aprobarbital is an intermediate-acting barbiturate indicated for sedation and the induction of sleep when conditions mandate the use of a sedative-hypnotic agent. It has been found especially suitable for daytime sedation and the treatment of evening insomnia.

The metabolism of this agent (Figure 3) is not as clearly established as other barbiturates. Approximately 9 to 18% of the drug is excreted unchanged.[37] The allyl side chain undergoes hydroxylation to yield the diol (B) to constitute 12% of the urinary metabolites.[44] N-hy-

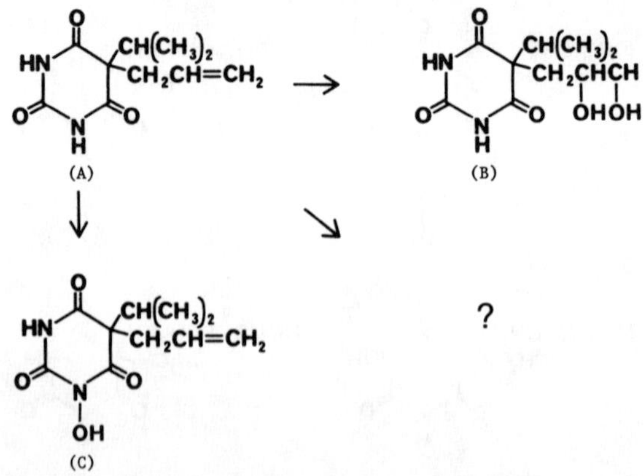

FIGURE 3. Metabolic scheme of aprobarbital.

droxylation (C), which many believed to be the principal metabolic pathway, only accounts for 4% of the urinary metabolites.[44] The remainder of the metabolites remain to be elucidated.

Butabarbital (5-Ethyl-5-Sec-Butylbarbituric Acid)

Other names: Secbutabarbitone
Secbutabarbital
Secumalum

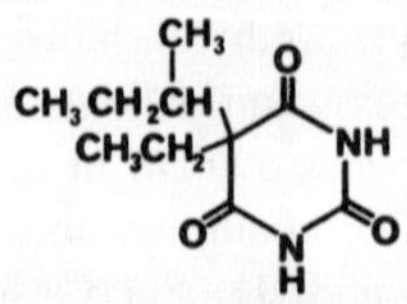

Butabarbital is an intermediate-acting barbiturate predominantly used as a sedative-hypnotic. However, prolonged administration of butabarbital is not recommended since it has been demonstrated to be ineffective after 14 days of ingestion. If insomnia persists, one should abstain from the drug for one or more weeks prior to the commencement of retreatment. The best remedy is the utilization of an alternative drug.

Metabolism of butabarbital (Figure 4) follows regular barbiturate oxidations of the side chains. The major metabolite of butabarbital (A) is the carboxylic acid, 5-(2-carboxy-1-methylethyl)-5-ethylbarbituric acid (D). Some minor metabolites include the 2′-hydroxy derivative (B) as well as the 2′-oxo derivative (C). Approximately 5 to 10% of the drug is excreted unchanged.[45]

Hexobarbital (5-Cyclohex-1′-Enyl-1,5-Dimethylbarbituric Acid)

Other names: Hexobarbitone
Enhexymalum
Enimal
Methexenyl
Methylhexabarbital
Narcosarum
Methylcyclohexenylmethylbarbituric acid

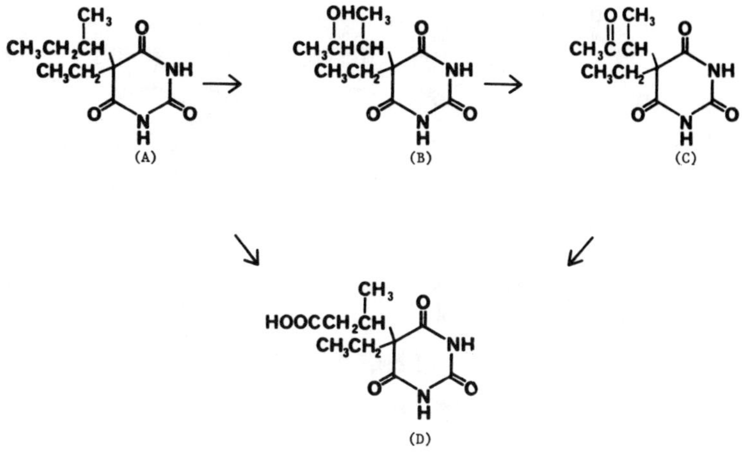

FIGURE 4. Metabolic scheme of butabarbital.

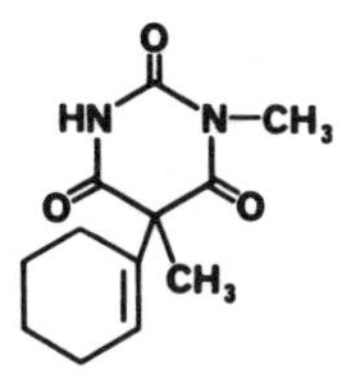

Hexobarbital is an ultrashort duration barbiturate that was once employed like thiopental to induce anesthesia. However, the evolution of newer and better agents has rendered the clinical use of hexobarbital essentially obsolete. Although it is seldomly used clinically, hexobarbital utilization is quite extensive in laboratory research. The lipophilicity of the agent allows it to be absorbed and distributed rapidly. Hypnotic activity is readily attainable at convenient doses when administered intraperitoneally to produce a sleep of moderate duration. Hexobarbital is also a model substrate to study the microsomal enzymes of the liver.[46]

Hexobarbital (Figure 5) (A) undergoes simple oxidation to generate a 3′-hydroxy metabolite (B) which further oxidizes to the 3′-keto derivative (C).[47] Oxidation may also occur at the 6′-position to yield the corresponding 6′-hydroxy (D) and 6′-keto (E) derivatives.[48] A small fraction of the hexobarbital undergoes N-demethylation to yield a more polar norhexobarbital (F). Likewise, the norhexobarbital generates hydroxy (G) and keto (H) metabolites at the 3′-position.[48]

Hexobarbital also undergoes a more obscure epoxide-diol pathway.[49] A 1′,2′-epoxide (I) was isolated in rat urine, but not in human urine.[50] However, researchers isolated and identified 1,5-dimethylbarbituric acid (K) in human urine.[50] This end product is evidence of the existence of the epoxide-diol pathway. The 1′,2′-diol[50] would be an unstable intermediate that could not be isolated. Researchers were unable to isolate the epoxide in human urine probably because the epoxide is an excellent substrate for epoxide hydratase and is rapidly converted to K via the diol. The compound, 1,5-dimethyl-5-hydroxybarbituric acid (L) has also been isolated in human urine. It has been proposed that its presence depends on the pH of the urine.

The principal metabolic route for hexobarbital in humans is the oxidation to the hydroxy and keto forms. Approximately 55% of the dose exists in the form of A, B, and C. The lesser important route is the epoxide-diol pathway where approximately 23% of the dose exists in the form of compounds I, J, K, and L.

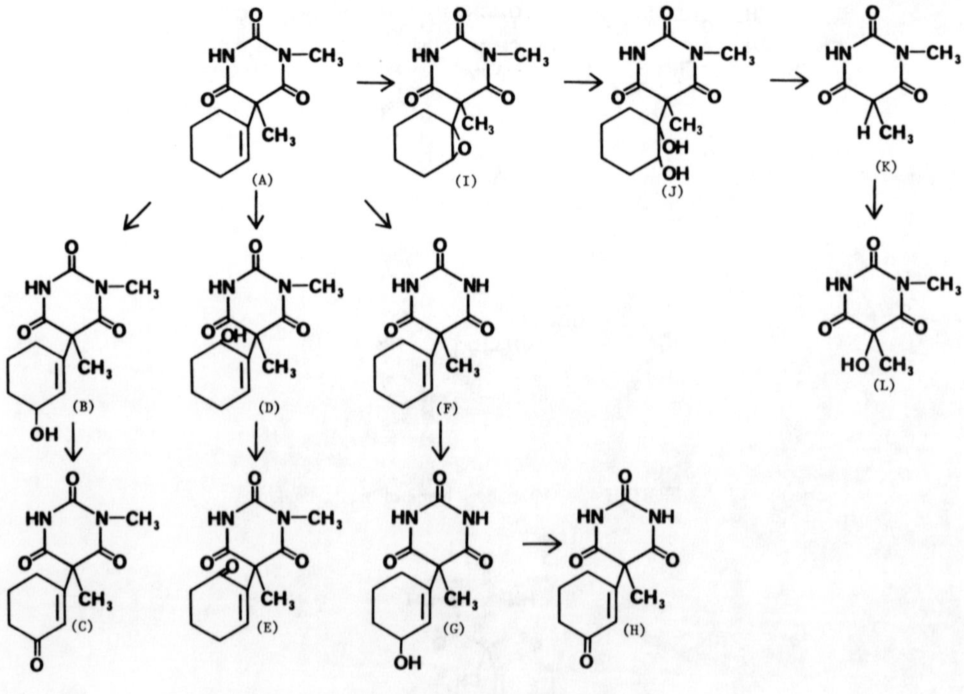

FIGURE 5. Metabolic scheme of hexobarbital.

Pentobarbital [5-Ethyl-5-(1-Methylbutyl)Barbituric Acid]

Other names: Pentobarbitone
Ethaminal

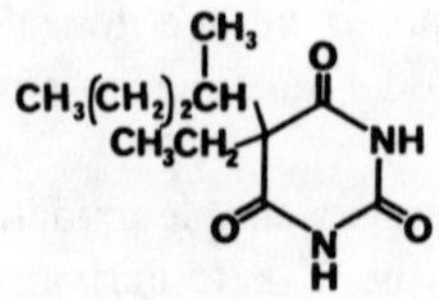

Pentobarbital is a short-acting barbiturate used in the treatment of insomnia and also in alleviating mild to moderate degrees of anxiety. It has also been used to produce amnesia in obstetrical cases so the mother is less likely to remember the unpleasant aspects of childbirth. However, this latter use is not highly recommended because of the susceptibility of the child to respiratory depression. Paradoxically, pentobarbital has been used to sedate the patient who has difficulty with a respirator. Although the uptake and accumulation of pentobarbital is not comparable to its sulfur analog, thiopental, it has been suggested for utilization as a basal anesthetic for surgery. Finally, it has also been found to be quite effective as an anticonvulsant for tetanus and strychnine poisoning.

The hepatic microsomes are responsible for the metabolism of pentobarbital. Oxidation of the penultimate carbon of the methylbutyl side chain yields a 3′-hydroxymetabolite.[51] When a racemic mixture was administered to humans, approximately 5.4 times more of the levorotatory alcohol than the dextrorotatory alcohol was excreted in the urine.[38] In Figure 6, it can be seen that the enantiomers (A and B) give rise to two pairs of diastereomers (C and F, D and E). Studies in dogs have shown that the diastereomeric pair, C and F, has a net (+)-rotation and conversely, the other pair of diastereomers has a net (−)-rotation.[52,53] In naive and tolerant mice, it has been demonstrated that the hydroxy metabolites of pen-

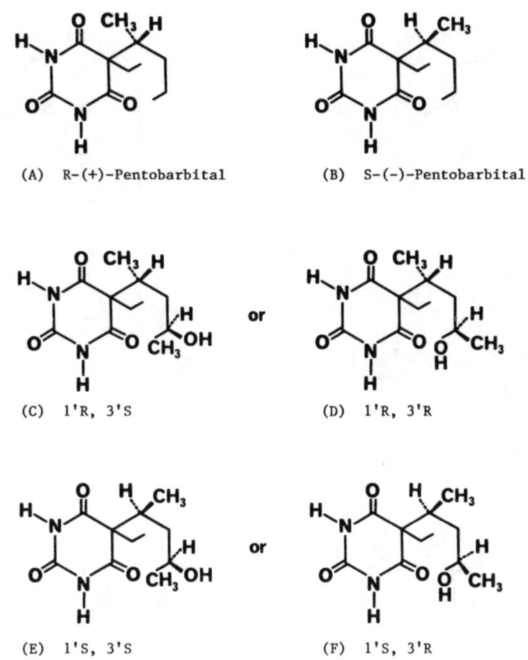

FIGURE 6. Enantiomers and diastereoisomeric oxidation products of pentobarbital.

tobarbital are antagonistic to narcosis.[54] Whether this antagonistic effect is stereochemically specific has not been established. Transient periods of excitation following the oral administration of subhypnotic doses of pentobarbital as the commercial racemate are caused by the R (+) isomer.[55] Following hydroxylation, it may be possible for pentobarbital to be further metabolized to yield a carboxy metabolite. Pentobarbital also undergoes N-hydroxylation[56] and may undergo ring scission.

Phenobarbital (5-Ethyl-5-Phenylbarbituric Acid)

Other names: Phenobarbitone
Fenobarbital
Phenemalum
Phenylethylmalonylurea

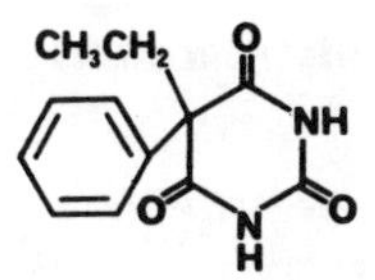

Phenobarbital is a long-acting barbiturate predominantly used prophylactically to treat grand mal, cortical-focal, and psychomotor seizures. It is administered alone or in combination with phenytoin. The sodium salt is used in the control of status epilepticus convulsions resulting from eclampsia, or prophylactically for febrile convulsions.

Phenobarbital is also useful in producing sedation; its use has significantly diminished since the evolution of the benzodiazepines. The smallest effective dose decreases the irritability and nervousness without impairing motor function. However, it does cause drowsiness

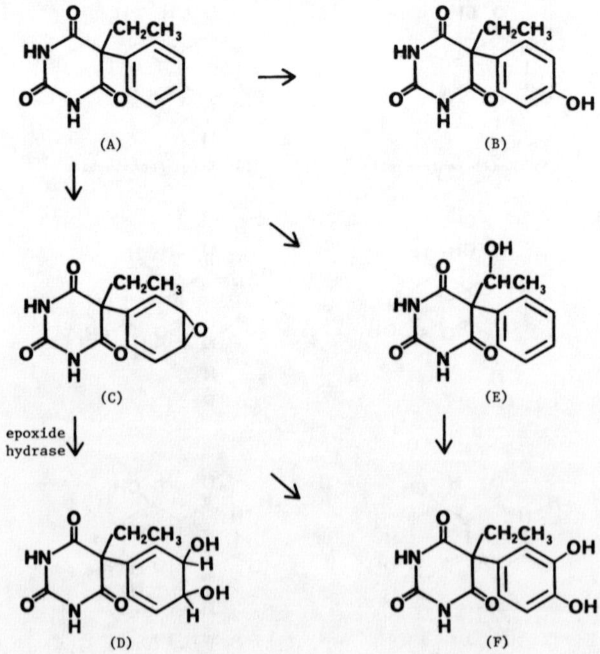

FIGURE 7. Metabolic scheme of phenobarbital.

and lethargy. Its use as an anti-anxiety agent has also yielded to the safer and more expensive benzodiazepines, although some practitioners believe they are equally effective.

Phenobarbital is safer than most drugs for the prevention of seizures. Skin rashes may occur in those who are hypersensitive to barbiturates. Those who have renal or hepatic insufficiencies should not receive phenobarbital because it will accumulate and potentiate the CNS depression.

The liver readily metabolizes phenobarbital (A) (Figure 7) to its *p*-hydroxy derivative (B).[57] Some investigators have deemed the metabolite to be inactive,[1] but others have found it to have some anticonvulsant activity.[58] The phenyl substituent at position 5 loses its aromaticity to become a cyclohexadienyl diol.[59] Phenobarbital (A) undergoes epoxidation to yield an arene oxide (C), whereby it can be oxidized by the microsomal enzyme, epoxide hydrase, to generate a dihydrodiol (D). The ethyl group also undergoes oxidation to give a hydroxyethyl derivative (E). The arene oxide (C) may regain its aromaticity to become an aromatic diol (F).[60]

Secobarbital [5-Allyl-5-(1′-Methylbutyl)Barbituric Acid]

Other names: Quinalbarbitone
Meballymal

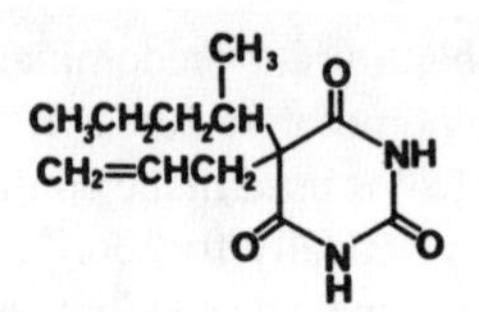

Secobarbital is a short-acting barbiturate employed for its sedative-hypnotic properties. It produces rapid relaxation and sleep for short duration. Secobarbital is effective in preparing

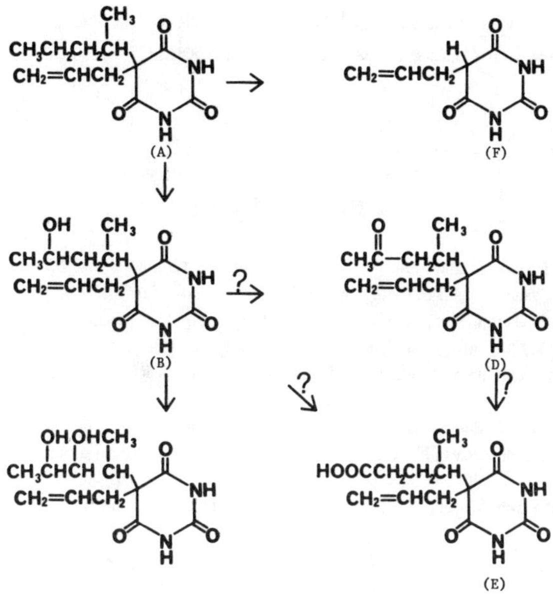

FIGURE 8. Metabolic scheme of secobarbital.

patients for general spinal and regional anesthesia as well as for intubation. It also reduces the tension and apprehension one has prior to dental procedures. Furthermore, its use as an anticonvulsant has been established in cases of convulsions attributed to tetanus, status epilepticus, strychnine poisoning, or toxic reactions to local anesthetics.

The metabolism of secobarbital (A) (Figure 8) has been well established. A racemic modification of secobarbital will generate several oxidized products.[61] The major metabolite is a diol, 5-allyl-5-(1′-methyl-2′,3′-dihydroxybutyl)barbituric acid (C). The second most abundant biotransformation product is 5-allyl-5-(1′-methyl-3′-carboxypropyl)barbituric acid (E).[2] The one of least importance appears to be the diastereomeric hydroxy metabolite (B), 5-allyl-5-(3′-hydroxy-1-methylbutyl)barbituric acid. Based upon chromatographic methods, Waddell et al.[62] designated one spot "hydroxysecobarbital S" to represent the slow migrator and the other spot "hydroxysecobarbital F" to represent the fast migrator. The S racemate has a melting point of 164 to 167°C and comprises 80% of the mixture. The F racemate has a melting point of 176 to 177°C and consequently constitutes 20% of the mixture. Also present is an oxo metabolite, 5-allyl-5-(3′-oxo-1-methylbutyl) barbituric acid (D), which may be an extension of the hydroxy metabolite. One other metabolite that was isolated and certified arose via the loss of the sec-amyl substituent to simply give 5-allylbarbituric acid (F).[62,63]

Talbutal (5-Allyl-5-Secbutylbarbituric Acid)

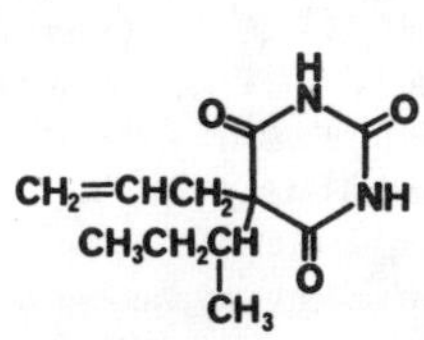

Talbutal is an intermediate-acting barbiturate generally used in combination with other agents such as aspirin, acetaminophen, phenacetin, caffeine, and codeine. Naturally, as a barbiturate, it is responsible for the sedative effect of these commercial preparations. It alleviates tension and acts as an anxiolytic and muscle relaxant. Hence, these commercial preparations are quite effective against tension headaches.

REFERENCES

1. **Harvey, S. C.,** Hypnotics and sedatives, in *Goodman and Gilman's The Pharmacological Basis of Therapeutics,* Gilman, A. G., ed., 6th Ed., Macmillan, New York, 1980, 349.
2. **Daniels, T. C. and Jorgerson, E. C.,** in *Textbook of Organic Medicinal and Pharmaceutical Chemistry,* Wilson, C. O., Ed., 5th ed., J. B. Lippincott, Philadelphia, 1966, 372.
3. **Feinberg, I., Hibi, S., Cavness, C., and March, J.,** Absence of REM rebound after barbiturate withdrawal, *Science,* 185, 534, 1974.
4. **Seeman, P.,** The membrane actions of anesthetics and tranquilizers, *Pharmacol. Rev.,* 24, 583, 1972.
5. **Harris, R. A. and Schroeder, F.,** Interaction of barbiturates with brain membranes, *Soc. Neurosci. Abstr.,* 6, 772, 1980.
6. **Harris, R. A. and Scheveder, F.,** Perturbation of synaptic membranes by ethanol, *Fed. Proc.,* 39, 745, 1980.
7. **Pang, K. Y., Chang, T. L., and Miller, K. W.,** On the coupling between anesthetic induced membrane fluidization and cation permeability in lipid vesicles, *Mol. Pharmacol.,* 15, 729, 1979.
8. **Matthews, K. K. and Nordmann, J. J.,** The synaptic vesicle. Calcium ion binding to the vesicle membrane and its modification by drug action, *Mol. Pharmacol.,* 12, 778, 1976.
9. **Ho, I. K. and Harris, R. A.,** Mechanism of action of barbiturates, *Ann. Rev. Pharmacol. Toxicol.,* 21, 83, 1981.
10. **Nicoll, R. A.,** Pentobarbital: differential postsynaptic actions on sympathetic ganglion cells, *Science,* 182, 720, 1978.
11. **Schlosser, W. and Franco, S.,** Modification of GABA-mediated depolarization of the cat ganglion by pentobarbital and two benzodiazepines, *Neuropharmacology,* 18, 377, 1979.
12. **Bowery, N. G. and Droy, A.,** Reversal of the action of amino acid antagonists by barbiturates and other hypnotic drugs, *Br. J. Pharmacol.,* 63, 197, 1978.
13. **Richter, J. A. and Waller, M. B.,** Effects of pentobarbital on the regulation of acetylcholine content and release in different regions of rat brain, *Biochem. Pharmacol.,* 26, 609, 1977.
14. **Haycock, J. W., Levy, W. B., and Cotman, C. W.,** Pentobarbital depression of stimulation-secretion coupling in brain. Selective inhibition of depolarization-induced calcium-dependent release, *Biochem. Pharmacol.,* 26, 159, 1977.
15. **Cutler, R. W. P. and Dudzinski, D. S.,** Effect of pentobarbital on uptake and release of [^{3}H]GABA and [^{14}C] glutamate by rat brain slices, *Brain Res.,* 88, 415, 1974.
16. **Thomson, T. D. and Turkanis, S. A.,** Barbiturate-induced transmitter release of frog neuromuscular junction, *Br. J. Pharmacol.,* 48, 48, 1973.
17. **Evans, R. H.,** Potentiation of the effects of GABA by pentobarbital, *Brain Res.,* 171, 113, 1979.
18. **Willow, M. and Johnston, G. A. R.,** Enhancement of GABA binding by pentobarbitone, *Neurosci. Lett.,* 18, 323, 1980.
19. **Asano, T. and Ogasawara, N.,** Chlorie-dependent stimulation of GABA and benzodiazepine-receptor binding by pentobarbital, *Brain Res.,* 25(1), 212, 1981.
20. **Asano, T. and Ogasawara, N.,** Stimulation of GABA-receptor binding by barbiturates, *Brain Res.,* 77(4), 355, 1982.
21. **Smith, A. R.,** Barbiturate protection in cerebral hypoxia, *Anesthesiology,* 47, 285, 1977.
22. **Steir, L., Khanna, N. N., Levy, G., and Yaffee, S. J.,** Effects of phenobarbital on hyperbilirubinemia and glucuronide formation in newborns, *Am. J. Dis. Child.,* 120, 26, 1970.
23. **Steine, A., Thaler, M., and Amuara, W. H.,** Effects of phenobarbital on bile salt metabolism in cholestosis due to interhepatic bile duct hyperplasia, *Pediatrics,* 51, 992, 1973.
24. **Young, R. E., Ramsay, L. E., and Murray, T. S.,** Barbiturates aid serum calcium in the elderly, *Postgrad. Med.,* 53, 212, 1977.
25. **Smith, C. M.,** The pharmacology of sedative/hypnotics, alcohol, and anesthetics: sites and mechanisms of action, in *Drug Addiction I: Morphine, Sedative/Hypnotics and Alcohol Dependence,* Martin, W. R., Ed., Vol. 45 (Part 1), Springer-Verlag, Berlin, 1977, 413.
26. **Isbell, H. and White, W. M.,** Chemical characteristics of addiction, *Am. J. Med.,* 14, 558, 1953.
27. **Desmond, M. M., Schwareke, R. P., Wilson, G. S., Yosanuga, S., and Burgdorff, I.,** Material barbiturate utilization and neonatal withdrawal symptomatology, *J. Pediatr.,* 80, 190, 1972.

28. **Balasubramanian, K., Lucas, S. B., Marver, G. E., and Simons, P. J.,** The kinetics of amylobarbitone metabolism in healthy men and women, *Br. J. Pharmac.*, 39, 564, 1970.
29. **Breimer, D. D.,** Pharmacokinetics of butabarbital after single and multiple oral doses in man, *Eur. J. Clin. Pharmacol.*, 10, 263, 1976.
30. **Breiner, D. D., Hohoff, C., Zilly, W., Richter, E., and vanRossum, J. M.,** Pharmacokinetics of hexobarbital in man after intravenous infusion, *J. Pharmacokinet. Biopharm.*, 3, 1, 1975.
31. **Breimer, D. D.,** Pharmacokinetics of methohexitone following intravenous infusion in humans, *Br. J. Anesth.*, 48, 643, 1976.
32. **Reiderberg, M. M., Lowenthal, D. T., Briggs, W., and Gasparo, M.,** Pentobarbital elimination in patients with poor renal function, *Clin. Pharmacol. Ther.*, 20, 67, 1976.
33. **Alvin, J., McHorse, T., Hoyumpa, A., Bush, M. T., and Schenker, S.,** The effect of liver disease in man on the disposition of phenobarbital, *J. Pharmacol. Exp. Ther.*, 192, 224, 1975.
34. **Butler, T. C., Makaffee, C., and Waddell, W. J.,** Phenobarbital: studies of elimination, accumulation, tolerance and dosage schedules, *J. Pharmacol. Exp. Ther.*, 111, 425, 1954.
35. **Clifford, J. M., Cookson, J. H., and Wickhum, P. E.,** Absorption and clearance of secobarbital, heptabarbital, methaqualone and ethinamate, *Clin. Pharmacol. Ther.*, 14, 376, 1974.
36. **Burch, P. G. and Starski, D. R.,** Decreased protein binding and thiopental kinetics, *Clin. Pharmacol. Ther.*, 32, 212, 1982.
37. **Lous, P.,** Plasma levels and urinary excretion of 3 barbituric acids after oral administration to man, *Acta Pharmacol. Toxicol.*, 10, 147, 1954.
38. **Maynert, E. W.,** The alcoholic metabolites of pentobarbital and amobarbital in man, *J. Pharmacol. Exp. Ther.*, 150, 118, 1965.
39. **Baldeo, W., Gilbert, J. N. T., Powell, J. W., Jones, L. V., and Whitehouse, M. J.,** A new metabolite of amylobarbitone: 5-(3′-carboxylbutyl)-5-ethylbarbituric acid, *J. Pharm. Pharmac.*, 29, 254, 1977.
40. **Tang, B. K., Inaba, T., and Kalow, W.,** N-hydroxyamobarbital: the second major metabolite of amobarbital in man, *Drug Metab. Disp.*, 3, 479, 1975.
41. **Krauer, B., Draffan, G. H., Williams, F. M., Clare, R. A., Dollery, C. T., and Hawkins, D. F.,** Elimination kinetics of amobarbital in mothers and their newborn infants, *Clin. Pharmac. Ther.*, 14, 442, 1973.
42. **Draffan, G. H., Dollery, C. T., Davies, D. S., Krauer, B., Williams, F. M., Clare, R. A., Trudinger, B. J., Dailing, M., Sertel, H., and Hawkins, D. F.,** Maternal and neonatal elimination of amobarbital after treatment of the mother with barbiturates during late pregnancy, *Clin. Pharmac. Ther.*, 19, 271, 1979.
43. **Inaba, T. and Kalow, W.,** Salivary excretion of amobarbital in man, *Clin. Pharmac. Ther.*, 18, 558, 1978.
44. **Gilbert, J. N. T., Nelmes, P. T. J., and Powell, J. W.,** The synthesis and urinary estimation of N-hydroxyaprobarbitone, *J. Pharm. Pharmac.*, 30, 173, 1978.
45. **Gilbert, J. N. T., Powell, J. W., and Templeton, J.,** A study of the human metabolism of secbutabarbitone, *J. Pharm. Pharmac.*, 27, 923, 1975.
46. **Bush, M. T. and Weller, W. L.,** Metabolic fate of hexobarbital (HB), *Drug Metab. Rev.*, 1, 249, 1972.
47. **Takenoshita, R. and Toki, S.,** Rabbit liver 3-hydroxyhexobarbital dehydrogenase, *J. Biol. Chem.*, 249, 5428, 1974.
48. **Frey, V. H., Sudendey, F., and Krause, D.,** Vergliechende Untersuchungen uber Stoffwechsel, ausscheidung und nachweis von Schlafmittein avs der Barbitursaure-reihe, *Arzneim. Forsch.*, 9, 294, 1959.
49. **Vermuelen, N. P. E., van Bladeren, P. J., Breimer, D. D., and van der Ger, A.,** Hexobarbital-1′,2′-epoxide: a possible intermediate in the metabolism of hexobarbital, *Biochem. Pharmacol.*, 27, 135, 1978.
50. **Vermuelen, N. P. E., Bakker, B. H., Schvitink, J., van der Ger, A., and Breimer, D. D.,** The epoxide-diol pathway in the metabolism of hexobarbital in rat and man, *Xenobiotica*, 9, 289, 1979.
51. **Brodie, B. B., Burns, J. J., Mark, L. C., Lief, P. A., Bernstein, E., and Papper, E. M.,** The fate of pentobarbital in man and dogs and a method for its estimation in biological materials, *J. Pharmacol. Exp. Ther.*, 109, 26, 1953.
52. **Palmer, K. H., Fowler, M. S., Wall, M. E., Rhodes, S. L., Waddell, W. J., and Baggett, J.,** The metabolism of R(+)- and RS-pentobarbital, *J. Pharmacol. Exp. Ther.*, 170, 355, 1969.
53. **Palmer, K. H., Fowler, M. S., and Wall, M. E.,** Metabolism of optically active barbiturates. II. S(−)-pentobarbital, *J. Pharmacol. Exp. Ther.*, 175, 38, 1970.
54. **Yamamoto, I., Ho, I. K., and Loh, H. H.,** The antagonistic effects of 5-ethyl-5-(3-hydroxy-1-methylbutyl) barbituric acid on pentobarbital narcosis in both naive and tolerant mice, *Life Sci.*, 22, 1103, 1978.
55. **Huang, L. Y. M. and Barker, J. C.,** Pentobarbital: stereospecific actions of (+) and (−) isomers revealed on cultured mammalian neurons, *Science*, 207, 195, 1980.
56. **Tang, B. K., Inaba, T., and Kalow, W.,** N-hydroxylation of pentobarbital in man, *Drug Metab. Disp.*, 5, 71, 1977.
57. **Craig, C. R. and Shideman, F. E.,** Metabolism and anticonvulsant properties of mephobarbital and phenobarbital in rats, *J. Pharmacol. Exp. Ther.*, 176, 35, 1971.

58. **Craig, C. R., Hirano, K., and Shideman, K. R.,** Anticonvulsant activity of a metabolite of phenobarbital, *Fed. Proc.*, 19, 280, 1960.
59. **Grover, P. L., Hewer, A., and Sires, P.,** Formation of K-region epoxides as microsomal metabolites of pyrene and benzo(a)pyrene, *Biochem. Pharmacol.*, 21, 2713, 1972.
60. **Freuderthal, R. I. and Carroll, F. I.,** Metabolism of certain commonly used barbiturates, *Drug Metab. Rev.*, 2, 265, 1976.
61. **Gilbert, J. N. T., Hetherington, W. L., Powell, J. W., and Whalley, W. B.,** Metabolism of quinalbarbitone, *J. Pharm. Pharmac.*, 27, 343, 1975.
62. **Waddell, W. J.,** The metabolic fate of 5-allyl-5-(1-methylbutyl)-barbituric acid (secobarbital), *J. Pharmacol. Exp. Ther.*, 149, 23, 1965.
63. **Cochin, J. and Daly, J. W.,** The use of thin-layer chromatography for the analysis of drugs. Isolation and identification of barbiturates and nonbarbiturate hypnotics for urine, blood and tissues, *J. Pharmacol. Exp. Ther.*, 139, 154, 1963.

NONBARBITURATE SEDATIVES AND HYPNOTICS

Norman R. Boisse and John L. Neumeyer

INTRODUCTION

The sedative-hypnotic class of drugs used in medical practice is comprised of a large collection of compounds of diverse chemical structure. These drugs include the barbiturates, benzodiazepines, alcohols, aldehydes and derivatives, cyclic amides, and imides. Despite the structural heterogeneity of this class of drugs, the sedative-hypnotics are remarkably similar in their physiological effects at sedative-hypnotic doses. They produce daytime sedation in anxious or emotionally agitated patients and/or promote sleep in insomniacs.

The history of the sedative-hypnotic drug class has its origin with Neolithic man who discovered alcohol. Even though it is no longer used as a prescription sedative, alcohol continues to be used in society as a self-medicament to relieve anxiety, promote sleep, and as a socially acceptable recreational drug. However, public health problems posed by acute and chronic alcohol intoxication make alcohol abuse the number one drug problem in America. Consequently, physicians are reluctant to recommend patient use of alcohol as a sedative-hypnotic.

The next important era in the historical development of this drug class occurred at the turn of the century with the introduction of the barbiturates. The short-acting barbiturates, particularly pentobarbital, secobarbital, and amobarbital have enjoyed extensive use for over 50 years. The chemistry and pharmacology of these drugs is discussed in the Antianxiety Agents chapter. The barbiturates are no longer as widely used because of the development of safer and equally or more efficacious alternatives. Only phenobarbital, a long-acting barbiturate, continues to be widely used though mostly as an anticonvulsant drug. The impetus for the search for safer sedative-hypnotics derived from the demonstration by Isbell[1] in 1950 of barbiturate dependence in man under controlled conditions. Consequently, throughout the 1950s the pharmaceutical industry synthesized and marketed numerous sedative-hypnotics devoid of the barbituric acid ring system in an attempt to produce similar drugs devoid of dependence liability. These drugs include diazepam, ethinamate, ethchlorvynol, glutethimide, meprobamate, and methyprylon. However, the realization that all sedative-hypnotic drugs regardless of structure carry some risk for tolerance and physical dependence came about following numerous clinical reports in the medical literature. Only the benzodiazepine class of compounds such as diazepam offers substantial advantage over other chemical classes of sedative-hypnotics from the standpoint of safety from acute overdose or abuse, tolerance, and dependence. Accordingly, they are currently the most widely used drugs in clinical medicine. Dosages of the more commonly used nonbarbiturate sedative hypnotics are listed in Table 1.

PHYSIOLOGY OF SLEEP

Definition and Characteristics

Sleep is a state of diminished organismal activity and reactivity with the environment. It is characterized by periodic, recurrent interruptions of mental and sensory-motor functions of the brain. However, not all brain regions and functions are equally affected by the sleep state; circulation, respiration, and body temperature regulation are not so affected. Polygraphic recordings of the electroencephalogram (EEG), electromyogram (EMG), and electroculogram (EOG) in man have demonstrated that sleep is an active process. Sleep and wakefulness states are organized in a circadian fashion, normally with a 24-hr cycle which

Table 1
DOSAGES OF MORE COMMONLY USED NON-BARBITURATE SEDATIVE-HYPNOTICS

Chemical class	Generic name[a]	Hypnotic doses (usual oral)	Sedative dose (daytime)	Active metabolites
Benzodiazepines	Clorazepate dipotassium (5)	—	15—60 mg total daily	Nor-diazepam
	Diazepam (2)	10 mg	2—10 mg, BID-QID	Nor-diazepam, temazepam
	Flurazepam (3)	15, 30 mg	—	*N*-desalkylflurazepam
	Lorazepam (6)	2, 4 mg	0.51—1 mg, BID-TID	—
	Temazepam (4)	15, 30 mg	—	—
Alcohols, Aldehydes, and derivatives				
	Chloral hydrate (15)	500, 1000 mg	250 mg TID	Trichloroethanol
	Triclofos sodium (19)	1500 mg	—	Trichloroethanol
	Paraldehyde (22)	4—8 mℓ	—	—
	Meprobamate (23)	800 mg	400 mg TID-QID	—
Cyclic amides and imides				
	Glutethimide (26)	250—500 mg	125—250 mg 2—3x daily	—
	Methyprylon (27)	200—400 mg	—	—

[a] The number in the parentheses refers to the structure, which can be found in the text.

is only in part determined by environmental sensory cues (Zeitgeber). Moreover, when all environmental cues are held constant, the brain cycles between sleep and wakefulness with a somewhat longer period of 25 hr which is inferred to reflect the pace of the organism's biological clock.

Furthermore, the sleep state is characterized by a temporal sequence of well-defined phases which are defined in terms of quantitative and qualitative activity states monitored by EEG, EMG, and EOG. On the basis of these polygraphic recordings, two distinct phases of sleep have been distinguished. One is slow wave (SW) sleep characterized by EEG synchronization (increased amplitude and reduced frequency); it is the first phase in the transition from wakefulness to sleep and four stages can be distinguished based on the degree of EEG synchronization. The other is called paradoxical or rapid eye movement (REM) sleep; it is paradoxical in that the EEG shows marked desynchronization (high frequency and low amplitude) which is most characteristic of the aroused state. In addition, characteristic REMs occur which are believed to be driven by phasic electrical bursts called P-G-O waves because they can be measured in the pons, lateral geniculate, and occipital cortex. During REM sleep, EMGs become silent; sympathetic activity is elevated as manifested by increased heart rate and blood pressure and cessation of GI movements; and dreams most often occur. During nonREM sleep, the muscles are relaxed and parasympathetic tone dominates Stage 4 of SW sleep which is most synchronized in the EEG. NonREM sleep is also the most difficult to interpret.

Typically, the architecture of a normal night's sleep consists of alternating periods of nonREM and REM sleep. REM sleep occupies only 20 to 50% of sleep time. The greatest part of the night is spent in stage 2 of SW sleep. REM sleep appears five to seven times in a cycling fashion. Both the frequency and duration of REM episodes increases with time during the night so that most REM occurs in the early morning hours just prior to awakening. In contrast, stages 3 and 4 of SW sleep occur mostly in the first few hours of the night.

The characteristics of a night sleep change with age. There is a sharp decline in sleep time during adolescence, it is fairly stable during middle age and further declines with old age.[2]

Electrophysiology of Sleep-Wakefulness Cycling

The Belgian neurophysiologist Frederick Bremer first demonstrated in the 1930s that the forebrain isolated by a midcollicular lesion in the midbrain (cerveau isolē) failed to cycle and continuously displayed high voltage low frequency activity characteristic of SW sleep. If the brain lesions were more caudal between medulla and spinal cord (encēphale isolē), the forebrain displayed normal sleep-wakefulness cycling. In 1949, Moruzzi and Magoun compared the effects of lateral tegmental lesion to destroy classical ascending sensory pathways to midline lesions at the same level of the midbrain to cut rostral reticular formation projections to nonspecific thalamic nuclei. They concluded that the reticular formation activated the cortex to keep it awake; a reduction in the activity of this so-called ascending reticular activating system (ARAS) accompanied the onset of sleep. Later, others found that when midbrain lesions were made more caudal than those for the cerveau isolē experimental subjects (cats) remained permanently awake. This finding suggests an important role for the rostral reticular formation in wakefulness and for the caudal reticular formation to induce sleep. Pharmacological studies to paralyze function in the caudal brainstem by selective regional administration of barbiturates have resulted in arousal from sleep and further support the concept of a hypnogenic center in this region. This center is believed to be the nuclei of the raphe. An additional hypnogenic center has been found in the medulla; electrical stimulation of the nucleus solitarius promotes sleep.[2,3]

Neurochemistry of Sleep-Wakefulness Cycling

Endogenous Hypnogogues

Even before the birth of the concept of neurochemical synaptic transmission, by 1900 there was evidence for the existence of humoral substances in the biological fluids of sleep-deprived animals that could put a normal animal to sleep. The search for possible molecular candidates for the sleep-promoting substance was dormant until the 1960s when two major laboratories — Monnier at the University of Basel and Pappenheimer at Harvard University — isolated and purified two different materials, DSIP[4,5] and factor S,[6] respectively.

DSIP, or delta sleep-inducing peptide is a nonapeptide, molecular weight 849 with a known amino acid sequence which was dialyzed from sleeping donor rabbits. Low doses (30 nmol i.v.) of DSIP in cats increased both SW and REM sleep; however a high dose (300 nmol i.v.) had no effect. The absence of dose-effect for DSIP on sleep has been interpreted by Monnier to suggest that DSIP acts as a neuromodulator of sleep rather than as a neurotransmitter.

Pappenheimer's factor S was isolated from human urine and cerebrospinal fluid and brain tissue of laboratory animals. Factor S consists of glutamic acid, alanine, diaminopimelic acid, and muramic acid in a 2:2:1:1 molar ratio with a molecular weight of 922. This peptide promotes SW sleep.

It is not likely that circulating levels of these sleep-promoting substances play a role in regulating sleep-wakefulness (S/W) cycling because there is independent S/W cycling by Siamese twins with separate nervous systems but shared circulation. However, the continued characterization of endogenous sleep-inducing molecules and structure-activity investigations may hold the key for a significant breakthrough in the pharmacological management of insomnia.[7]

Serotonergic Neurons in the Raphe Nuclei

Jouvet has proposed that serotonin in the raphe initiates and maintains nonREM sleep and may play a role in "priming" REM sleep. In support of this concept, destruction of the raphe or inhibition of the biosynthesis of serotonin by parachlorophenylalanine (PCPA) greatly decreases total sleep time. The PCPA effect can be reversed by small doses of 5-hydroxytryptophan (5-HTP), but not the surgical lesion. However, not all have confirmed these observations and there is evidence for partial restoration of sleep homeostasis with time after surgical or chronic pharmacological lesioning.[2,8,9] There are several excellent reviews on the monaminergic theory of S/W regulation.[8-10]

Noradrenergic Neurons in the Brain Stem

Jouvet found that lesions of the locus ceruleus of the brain stem, a region rich in adrenaline, abolished the muscle atonia that normally accompanies REM sleep. Most drugs that affect noradrenergic transmission also affect EEG and behavioral signs of REM. However, this region does not appear to direct REM sleep because α-methyltyrosine, which suppressed the biosynthesis of norepinephrine and locus ceruleus activity, failed to abolish REM in animal experiments.[9,10]

Cholinergic Neurons in the Brain Stem

There is also evidence for a cholinergic link in the expression of REM sleep. Direct pontine injections of carbachol, a cholinergic agonist, into nucleus reticularis oralis and nucleus reticularis caudalis produced a desynchronized EEG, REM, and muscle atonia in the cat. Similarly, systemic injection of physostigmine, a cholinesterase inhibitor, induces REM during nonREM sleep in man.[9,10]

INSOMNIA AND RELATED SLEEP DISORDERS

Insomnia is characterized by a distressing feeling of inadequate quantity or quality of sleep. It is a subjective complaint, not an illness. Insomnia afflicts 8 to 15% of the adult population. Because insomnia is associated with many different disorders, the Association of Sleep Disorders Centers has developed a nosology of sleep disorders in 1979 called DIMS or disorders of initiating and maintaining sleep. Transient situational insomnia may occur with any of a number of sudden life changes like hospitalization, bereavement, or personal stress. Many sleep specialists believe insomnia is most commonly associated with psychological problems including anxiety, depression, hypochondriasis, mania, schizophrenia, or personality disorder. Sedative-hypnotics are not effective in these patients. Suicidal risk is also of concern with psychiatric patients. Insomnia may be associated with prolonged use of hypnotics and/or alcohol. Sleep-induced cessation of breathing may be central or obstructive in origin. It is important to recognize that the use of sedative-hypnotics in these patients is not only inappropriate, it is outright dangerous. There is no effective drug therapy for central sleep apnea, but obstructive sleep apnea can be treated with a tracheostomy which is left open at night. Insomnia may also occur with nocturnal myoclonus which is characterized by frequent disturbances of sleep due to bilateral muscular jerks occurring at 20- to 40-sec intervals. A related disorder, the restless leg syndrome, is associated with discomfort in the thighs while resting which can be relieved by movement. No accepted treatment has yet been developed for these two neurological disorders. Insomnia can also result from disruption of the 24-hr sleep-wake cycle, called the phase-lag syndrome. Jet lag is another variant of this problem. Insomnia can also accompany any medical or surgical condition. Finally, some patients called ''primary insomniacs'' have persistent insomnia of unknown origin; many of these patients appear to have learned poor sleeping habits.

Although most adults sleep 7 to 9 hr per night, there are many who sleep much less yet never complain of insomnia. On the other hand, insomniacs tend to exaggerate their sleep problems compared to what is objectively measured in the sleep lab.

In one study the sleep times of 122 insomniacs were found to be slightly less but substantially overlapping the control values. It appears that insomniacs do not have substantially and reliably less sleep than healthy control subjects. Accordingly, in evaluating the course of the insomnia and its response to medical treatment, the physician must rely primarily on the subjective complaint rather than on objective measures obtained in the sleep lab.[11-15]

EVALUATION OF SEDATIVE-HYPNOTICS IN MAN

Traditional clinical trials have permitted the evaluation of sedative-hypnotic safety and efficacy in large and diverse patient populations. Measurements of drug response are usually based on the patient's response in a questionnaire designed to measure sleep complaint; some also utilize nurse observers. The large number of patients also facilitates detection of side effects. Traditional clinical trials continue to have an important place in the evaluation of newer sedative-hypnotics under development.

Sleep laboratory studies provide objective measurements of several physiological correlates of sleep in a controlled sleep lab environment. Most sleep labs can only evaluate four to eight subjects at one time; consequently, sleep studies are conducted with much smaller patient samples than are clinical trials. The sleep lab provides valuable data on hypnotic effects on the staging of sleep. Unfortunately, the clinical significance of the effects measured has yet to be established. There is a ''first-night'' effect in sleep staging that accompanies adaptation to the artificial environment of the lab that must be controlled in drug studies. Consequently, sleep lab evaluations of hypnotic drugs are a valuable complement rather than a replacement for clinical trials.[16,17]

Time-lapse photography has been proposed as a sensitive, clinically meaningful measurement of sleep. Moreover, movements during both REM and nonREM sleep are minimal, but they occur most often in the transition between these two stages. Video cameras could conceivably be used both in a hospital as well as in-home. At this time, this approach appears to have the greatest value combined with traditional sleep lab measures in researching sleep disorders and sleep disturbances associated with a host of diseases and pharmacological and environmental disturbances.[18]

ANIMAL TESTING OF HYPNOTICS

Historically, duration of loss of righting reflex and physiologic correlates of somnolence and sleep have been used to evaluate hypnotic agents in lab animals. The degree of sedative activity may be estimated by scoring reductions in the spontaneous motor activity, changes in posture, and extent of eye closure in the rat.[19] Comprehensive gross neurological testing has been employed to characterize hypnotics in both the rat and mouse.[20] EEG and polygraphic recordings in unrestrained rats may find greater utility in the future, particularly with the recent advances in telemetry. This approach has not been extensively exploited in small laboratory animals. A host of sedative-hypnotic drugs have been evaluated for their EEG effects in monkeys. Quantitative techniques used to analyze the EEG like spectral analysis have revealed subgroupings of sedative-hypnotics into heterogeneous classes.[21]

SITE AND MECHANISM OF ACTION

The sedative-hypnotic drugs may have a common site of action to depress neural activity in the reticular formation and thus reduce the output of the ascending reticular activating system (ARAS). Such an action has been well documented for the barbiturates and benzodiazepines, but not for most other hypnotics. However, because of the well-established role of the ARAS in S/W cycling and the similar pharmacological profile of nonbarbiturate hypnotics, all hypnotics may share a common depressant action at the reticular formation.

While hypnotic agents are believed to exert their actions at synapses rather than on axons, it is not clear which neurotransmitter system is most vulnerable. Moreover, despite the explosive growth in our understanding of neurotransmitters and neurotransmission in the brain, little can be said with confidence because the hypnotics as a class have never been systematically evaluated to delineate similarities and differences in mechanisms of action. The barbiturates and benzodiazepines have been shown to enhance GABA-mediated synaptic inhibition in a variety of test systems. For the benzodiazepines, these effects are mediated by specific receptors.

SEDATIVE-HYPNOTIC TOLERANCE AND PHYSICAL DEPENDENCE

Tolerance

It is generally accepted that tolerance accompanies the chronic use of hypnotic drugs. Moreover, most sedative-hypnotics lose their efficacy in treating sleep disorders by 2 weeks of use; none has been shown to be effective for more than 28 days.[15,17,22,23]

There are two major organs potentially involved in the tolerance (chronic ineffectiveness) to hypnotic drugs — the liver and the brain. The induction of microsomal enzymes by the liver results in dispositional tolerance which can be identified by an increased rate of drug metabolism or decreased biologic half-life of the drug in the bloodstream. All of the short-acting barbiturates display this property. Glutethimide is also known to induce hepatic mixed function oxidases. The benzodiazepines as a class and chloral hydrate appear to be devoid of enzyme induction effects.

The other source of tolerance, the brain sensitivity to the drug, is called CNS or functional tolerance. This source of tolerance is more difficult to evaluate in man for ethical concerns. Ideally, it requires the establishment of elevated brain levels for the same hypnotic response or less hypnotic response for the same or greater brain level compared to the first administration (acute). There is evidence for functional tolerance, based on blood levels in man for barbiturates, benzodiazepines, as well as other hypnotics.[24-26]

Unfortunately, neither dispositional nor functional tolerance have been systematically evaluated across the major chemical subgroups of sedative-hypnotics. The most pertinent data would be obtained if blood levels of drug were monitored for concentration and analyzed kinetically in subgroups of patients participating in clinical trials or sleep lab studies.

Cross-tolerance across chemical classes of sedative-hypnotics is generally assumed. However, it has been only incompletely described. It may be difficult to predict cross-tolerance because its manifestation depends on so many factors such as previous older and recent drug history, magnitude and relative ratio of blood levels of the drugs combined, diet, and response criterion. Regardless, the potential for tolerance exists in those patients who regularly use alcohol, barbiturates, or any of the nonbarbiturate sedative-hypnotics for either diminished hypnotic response (cross-tolerance) or an exaggerated hypnotic response due to the additive or possibly synergistic effects of residual cumulative drug.

Physical Dependence

Physical dependence, defined by the occurrence of spontaneous manifestations of nervous hyperexcitation of the organism upon abrupt withdrawal, can occur with all conventional sedative-hypnotics regardless of chemical class. Moreover, the induction of physical dependence requires dosages of hypnotic drugs that exceed the usual boundaries of recommended dosages and produce overt depression of the nervous system. The degree of dependence is dose and time dependent. From the standpoint of relative dose, the benzodiazepines appear to have some advantage over the barbiturate and nonbarbiturate nonbenzodiazepine sedative-hypnotics by virtue of a more shallow dose effect relationship for anesthetic and lethal doses. However, most of the manifestations of withdrawal from all sedative-hypnotics are remarkably similar. They are collectively called the CNS depressant withdrawal syndrome. The syndrome may include muscle twitches and tremors, myoclonic jerks, convulsions, hallucinations, insomnia, anxiety, loss of appetite, excessive perspiration, weight loss, and even death following convulsions or abrupt cardiovascular collapse. Physical dependence can only be disclosed by withholding the drug and observing for withdrawal; accordingly, the most severe of the withdrawal syndromes have been seen with the short-acting sedative-hypnotics like amobarbital, pentobarbitol, secobarbital, meprobamate, glutethimide, and lorazepam. On the other hand, withdrawal syndromes seem to occur less frequently and are usually mild for sedative-hypnotics with long half-lives or active metabolites with long half-lives like the original benzodiazepines or phenobarbital. Accordingly, the medical management of the physically dependent subject is best accomplished by either maintaining the drug or substituting one with a longer half-life and then gradually reducing the dosage typically over a 4-week period.[27,28]

Because many of the manifestations of the withdrawal syndrome are the opposite of the direct depressant effects, Jaffe[28] has suggested that withdrawal signs might be the result of a rebound phenomenon emanating from the same neuronal systems depressed by the drug. A growing body of neurophysiological data support this hypothesis for chronic barbiturate, alcohol, and benzodiazepine treatments in laboratory animals. The significance of these observations is that the locus of the physical dependency process can be identified by the sites of direct action. For the benzodiazepines, this site may be the recently discovered specific CNS receptor. The enhancement of GABA-mediated synaptic inhibition in the CNS may be a common pharmacologic property of the CNS depressants.

Table 2
SEDATIVE-HYPNOTIC BENZODIAZEPINES

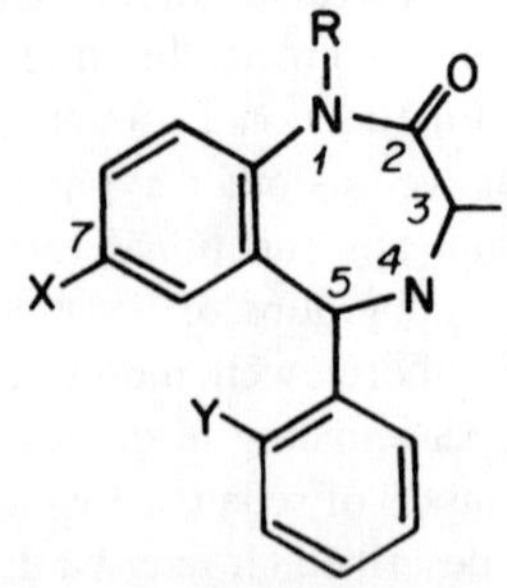

Structure 1

	R	R′	Y	X
Diazepam (2)[a]	CH_3	H	H	Cl
Flurazepam (3)[a]	$-(CH_2)_2N(C_2H_5)_2$	H	F	Cl
Temazepam (4)[a]	CH_3	OH	H	Cl
Clorazepate dipotassium (5)[a]	H	COO^-K^+	H	Cl
Lorazepam (6)[a]	H	OH	Cl	Cl
Flunitrazepam	CH_3	H	F	NO_2
Nitrazepam	H	H	H	NO_2

[a] The number in parentheses refers to the structure, which can be found in the text.

BENZODIAZEPINES

Mechanism of Action

The benzodiazepines (Structure 1 in Table 2) that have been employed as hypnotics in medicine in the U.S. are diazepam (Structure 2), flurazepam (Structure 3), lorazepam (Structure 6), clorazepate dipotassium (Structure 5), and temazepam (Structure 4). Benzodiazepines available in other countries are flunitrazepam and nitrazepam (Table 2). All active benzodiazepines are believed to have a common and unique component to their mechanism of action. The site at which they initiate the enhancement of GABA-mediated synaptic inhibition is a stereospecific and saturable binding site of high affinity (nanomolar) in CNS neuron membranes. This receptor was discovered independently by two groups in 1977.[29-32] For several drug response criteria (anxiolytic in man, skeletal muscle relaxation or anticonvulsant activity in animals), an excellent correlation has been disclosed between binding affinity (or dissociation constant, K_D) and ED_{50} dose.[33] There is no published data known which attempts to correlate K_Ds (or K_Is) to hypnotic efficacy. However, in 1981 Roche investigators reported the discovery of a specific receptor antagonist, RO 15-1788 (Structure 7), a compound which antagonized all of the CNS effects of benzodiazepines in all animal species tested, including anesthesia in monkeys.

This compound is apparently devoid of substantial intrinsic pharmacologic efficacy in classical test systems for anxiolytic — sedative-hypnotic — anticonvulsant activity.[34-38] Therefore, the recently discovered specific benzodiazepine receptors may very well be involved in the expression of benzodiazepine hypnotic action. It has been suggested that there may exist in the brain endogenous ligands for the benzodiazepine receptor. Numerous compounds and uncharacterized fractions from the brain have the capability of displacing tritium-labeled diazepam from its receptor.[39,40] These include the purines hypoxanthine (Structure 8) and inosine (Structure 9), nicotinamide (Structure 10) and many β-carbolines,

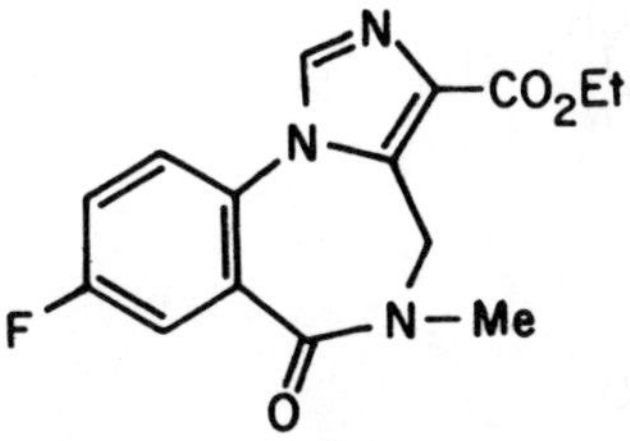

Structure 7

RO15-1788

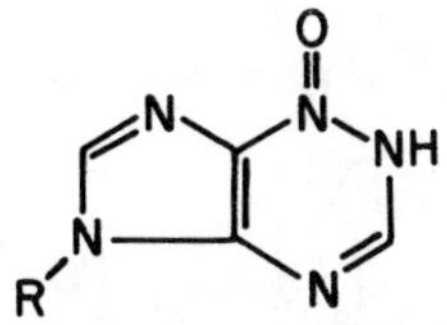

Structure 8

R= H-hypoxanthine

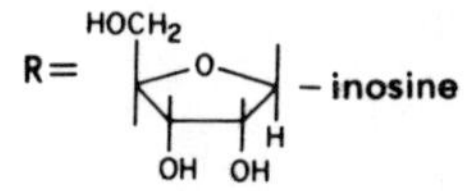

Structure 9

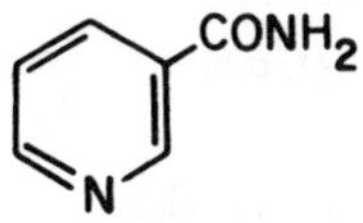

Structure 10

nicotinamide

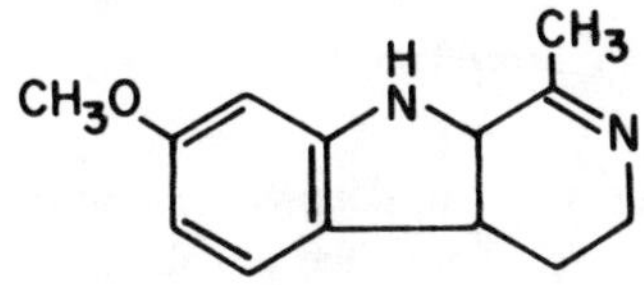

Structure 11

harmaline

a family of compounds derived from indole, including harmaline (Structure 11). Depending on both the compound and the physiological test system, these candidates for the putative endogenous ligand either mimic, antagonize, or produce effects opposite to those of diazepam. Furthermore, none of these materials have yet been demonstrated to be present in the brain in sufficient concentration to serve a physiological role at the receptor. Nevertheless, this receptor system and/or its endogenous ligand might provide a neurobiological explanation for some subtypes of insomnia.[41] Interestingly, one group of investigators has found evidence for the presence of a benzodiazepine-binding inhibitor in the brain of awake but not sleeping hamsters.[42]

A more detailed discussion of the molecular mechanism of action of benzodiazepines, particularly in relation to anxiolytic and anticonvulsant action, is found elsewhere in this volume.

Structure-Activity Relationships (SARs)

The more active members of this series have a 1,3-dihydro-5-aryl-2H-1,4-benzodiazepine-2-one framework (Structure 1) with a Cl or NO_2 at position 7, a CH_3 or $CH_2CH_2N(C_2H_5)_2$ group at position 1 and a phenyl or *o*-fluorophenyl as the aryl group at position 5 (Table 2). Recently an additional heterocyclic ring has been fused to position 1-2 as in estazolam (Structure 12) and triazolam (Structure 13) or at positions 4-5 as in ketazolam (Structure 14) to give more potent compounds. The CNS phamacological activities of the benzodiazepines correlate with the net charge on the carbonyl oxygen atom and the dipole moment of the molecule.[43]

Uses and Doses

Benzodiazepines are used at nighttime as hypnotics and during the day as anxiolytics. It would be misleading to describe benzodiazepines as producing daytime sedation. Moreover, doses of nonbenzodiazepine drugs needed to relieve anxiety usually produce sedation. However, for the benzodiazepines, sedation is usually experienced only for the first few doses. Tolerance develops to the sedation but apparently not to the anxiolytic effect.

As hypnotics, benzodiazepines are said to impart a sense of deep or refreshing sleep. They decrease latency to sleep onset, reduce the number of awakenings, and increase total sleep time. In contrast to barbiturates and most other sedative hypnotics, low doses of benzodiazepines may not reduce total REM time. Benzodiazepines substantially and con-

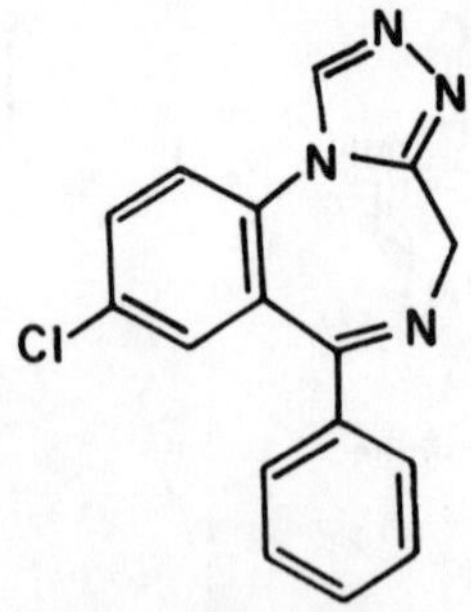

Structure 12
estazolam

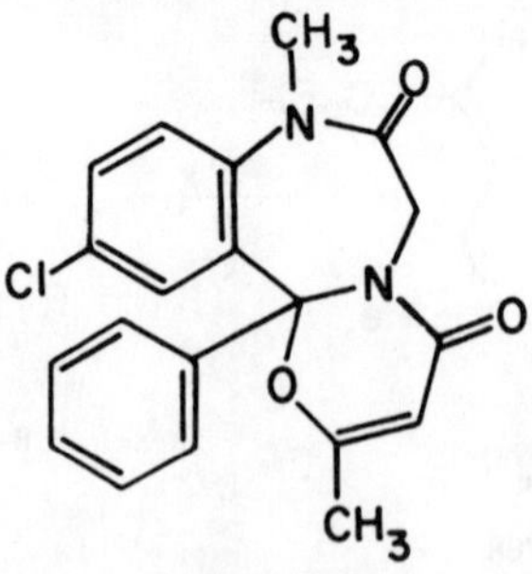

Structure 14
ketazolam

sistently reduce time spent in deep SW sleep (stages 3 and 4) while increasing time spent in stage 2 of SW sleep. Tolerance develops to EEG effects of benzodiazepines on sleep, particularly for suppression of REM sleep. However, while most sedative-hypnotics lose their efficacy to promote sleep in 2 weeks, flurazepam has been shown to be chronically effective for at least 28 days.[46,18]

By virtue of both efficacy and safety, the benzodiazepines are currently the drugs of choice in treating both insomnia as well as anxiety in lower doses. The daytime anxiolytic and nighttime hypnotic doses of the benzodiazepines used as hypnotics in medical practice are listed in Table 1. Flurazepam and temazepam have been marketed as hypnotic drugs but not as anxiolytic-sedative drugs during the day.

Pharmacokinetics

When prescribed for the treatment of insomnia or daytime excitement, the benzodiazepines are usually administered orally. All of the benzodiazepines utilized as sedative-hypnotics are effectively absorbed by the GI tract. Diazepam, the most lipid soluble of the benzodiazepines, enters the blood and the brain most rapidly with nearly 100% bioavailability to produce a noticeable subjective effect sometimes called a ''high'', ''buzz'', or ''spaced-out'' feeling.[47,48] However, all of the other hypnotic benzodiazepines have been reported to produce a sedative-like effect especially when beginning chronic dosing.

The hypnotic benzodiazepines are usually grouped according to their duration of action. The long-acting benzodiazepines that have been used as hypnotics are diazepam, chlordiazepoxide, flurazepam, and clorazepate; these drugs are long-acting by virtue of accumulation of slowly eliminated active metabolites. The short-acting benzodiazepines employed as hypnotics are lorazepam and temazepam. The half-lives of the benzodiazepines and their active metabolites are discussed in the chapter on anxiolytics.

Diazepam, chlordiazepoxide, and clorazepate are generally prescribed as anxiolytics; however, they are effective as hypnotics and are occasionally used to treat patients suffering from both anxiety in the day and insomnia at night. The major metabolic pathway for diazepam involves N-demethylation to nordiazepam followed by oxidation to oxazepam which is excreted as a glucuronide conjugate. A minor metabolic pathway is oxidation of diazepam to temazepam followed by glucuronidation and excretion. Nor-diazepam accumulates during chronic diazepam dosing and accounts for most of the drug activity.[49]

Chlorodiazepoxide undergoes N-demethylation by the liver oxidative enzymes to give nor-chlordiazepoxide which is further oxidized to demoxepam, then nor-diazepam, and finally oxazepam. Nor-chlordiazepoxide, demoxepam, and nor-diazepam accumulate to varying degrees to contribute to the pharmacologic activity with chronic treatments.[50-52]

Clorazepate is decarboxylated by gastric acid to yield nor-diazepam which is then absorbed from the small intestine. The absorption of nor-diazepam is impaired by concurrently ad-

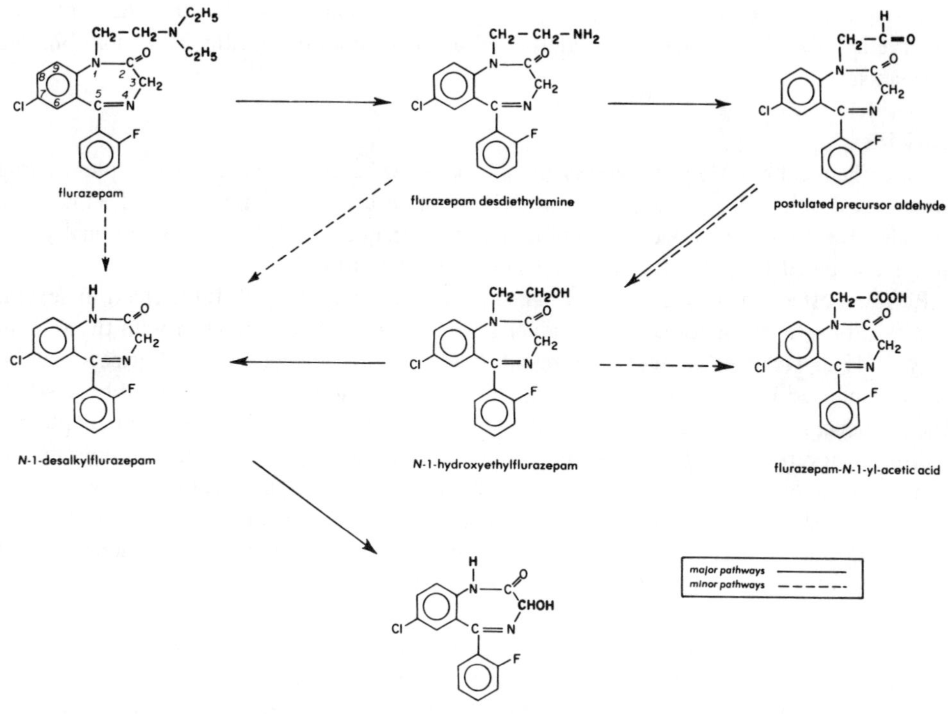

FIGURE 1. Postulated metabolic pathway of flurazepam (3) in man.

ministered antacids.[53,54] Clorazepate levels in the blood are transient and variable. Nor-diazepam is the main active metabolite which accumulates during chronic clorazepate administration. Nor-diazepam is further metabolized by hydroxylation at the C-3 position to form oxazepam and at the para-position of the C-5 phenyl to yield para-hydroxylated metabolites which are then excreted in the urine in both free and conjugated glucuronide form.[55] An age-related decline in the capacity for hepatic hydroxylation of nor-diazepam has been found in men but not in women.[56,57]

Flurazepam HCL, the most widely prescribed benzodiazepine hypnotic in the U.S., is rapidly and extensively metabolized prior to elimination. The postulated biotransformation pathways in man are shown in Figure 1. The *N*-1-desalkyl metabolite of flurazepam is active and accumulates to levels five- to sixfold greater than following acute administration by virtue of its long elimination half-life (47 to 100 hr). This metabolite probably accounts for most of the pharmacologic activity of orally administered flurazepam.[58,59] Indeed, the hypnotic efficacy of flurazepam increases on the second and third night of consecutive administration. Unlike most other hypnotics, flurazepam does not lose its efficacy when administered nightly for 4 weeks.[60] The principal urinary metabolite is the conjugated hydroxyethyl analogue of flurazepam.[58]

Lorazepam and temazepam are short-acting hypnotics by virtue of a 3-hydroxyl group in the benzodiazepine nucleus which facilitates inactivation by conjugation and rapid elimination into the urine. Most of the drug is excreted in the urine as glucuronide of the parent drug. Other minor metabolites of lorazepam are hydroxylorazepam, a quinazolinone derivative, and a quinazoline carboxylic acid.[61,62] For temazepam, a small but substantial quantity of *N*-desmethyl-temazepam conjugate is also excreted in the urine; however, no active metabolites have been found in the blood.[63] By virtue of their short half-lives (10 to 12 hr), neither lorazepam nor temazepam exhibit drug accumulation with continuous treatment.

All benzodiazepines are extensively (80 to 99%) bound to plasma protein (serum albumin).

The displacement of benzodiazepines from plasma proteins by other drugs has not been observed.[45] This finding has been interpreted to indicate a nonspecific, reversible binding mechanism.

Toxicities

As a chemical class, the benzodiazepines appear to be the safest sedative-hypnotic drug group yet developed by virtue of the wide separation between therapeutic and lethal dose in man. Overdosage with benzodiazepines has been frequent, but death is exceptionally rare unless combined with alcohol or other CNS depressant drugs.[64]

Residual effects on daytime performance have been reported with flurazepam in several studies. Some have suggested that residual effects are more likely to occur with flurazepam ($t_{1/2}$ = 47 to 100 hr) than with temazepam ($t_{1/2}$ = 10 hr). Indeed, residual effects may be less pronounced with temazepam, but they do not appear to be absent as would be desired. The incidences of adverse reactions like drowsiness, dizziness, lethargy, confusion, euphoria, and diarrhea reported in broad patient surveys are similar for flurazepam and temazepam. Differences in patient/subject population, dose equivalence, and the specific test for residual effect (i.e., EEG changes, psychomotor tasks, and subjective evaluation) impede effective drug comparisons. Accordingly, there is a need for a more critical, comprehensive, and systematic comparative investigations of all benzodiazepines for residual effects over the clinically relevant dose range.

A number of short-acting benzodiazepines have been shown to cause a worsening of sleep disturbance beyond the original baseline upon abrupt withdrawal. This condition has been labeled "rebound" insomnia. It is distinguished from nonbenzodiazepine hypnotic drug withdrawal insomnia in that it occurs with only a short-term (1 to 2 week) course of single nightly doses to develop, is not accompanied by REM rebound, and is not associated with the sedative addiction complex.[65,66] Nevertheless, the occurrence of "rebound" insomnia upon abrupt withdrawal at therapeutic dosages demonstrates the occurrence of physical dependence in a physiological sense to chronic benzodiazepine administration. This factor may contribute to reinforcement of nightly reliance upon benzodiazepines in addition to the loss of effectiveness or hypnotic tolerance with time that has been demonstrated for several short-acting benzodiazepines. Flurazepam is the only benzodiazepine for which chronic (28-day) effectiveness has been demonstrated for objective EEG criteria.[67,68]

An anterograde amnesia has been produced by certain benzodiazepines. Memory impairment has been most often reported with lorazepam. However, the relative incidence of memory deficit by long- and short-acting hypnotic benzodiazepines has not been compared. Dose, route of administration, and rapidity of onset of drug action may be important determinants of anterograde amnesia as the incidence is i.v.>i.m.>oral for diazepam. Reduced memory performance is the result of impaired acquisition rather than reduced retention.[60,69,70] Some tolerance to memory impairment was reported to 3 weeks once daily morning administration of diazepam (0.2 mg/kg); no residual memory effects were found after withdrawal.[71]

Physicochemical Data

Diazepam (Structure 2), **Valium®**, $C_{16}H_{13}ClN_2O$ (7-chloro-1,3-dihydro-1-methyl-5-phenyl-2H-1,4-benzodiazepin-2-one) appears as off-white crystals (m.p. 125 to 126°C) and is insoluble in water. pKa = 3.4 (base).

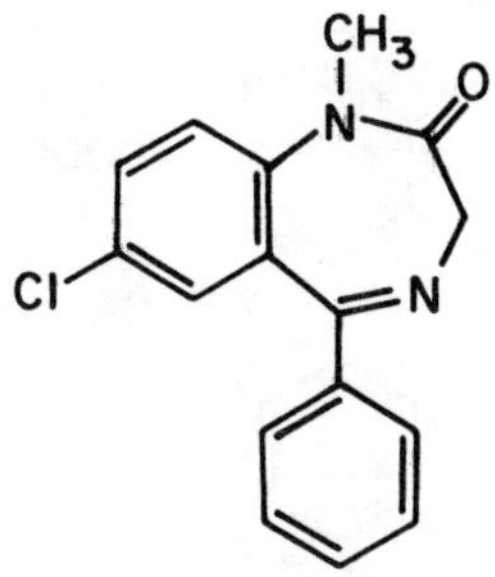

Structure 2

Flurazepam hydrochloride (Structure 3), **Dalmane®**, $C_{21}H_{23}ClFN_3O{\cdot}HCl$ (7-chloro-1-[2-(dimethylamino)ethyl]-5-(*o*-fluorophenyl)-1,3-dihydro-2H-1,4-benzodiazepin-2-one hydrochloride) appears as pale yellow crystals (m.p. 215.5 to 217.5°C) and is freely soluble in ethanol and water. pKa = 8.16 (acid), 1.90 (base).

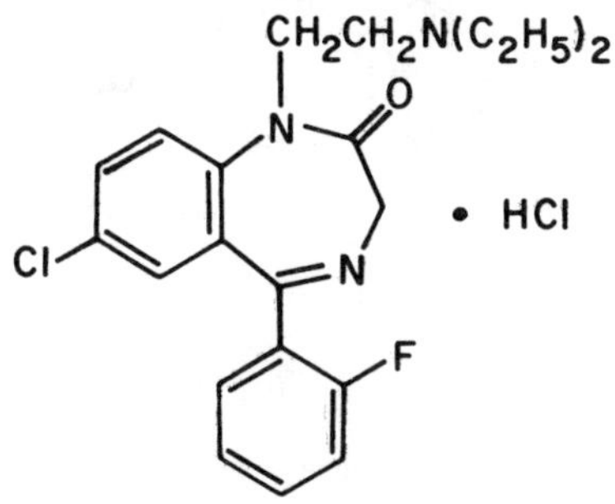

Structure 3

Temazepam (Structure 4), **Restoril®**, $C_{16}H_{13}ClN_2O_2$(7-chloro-1,3-dihydro-3-hydroxy-1-methyl-5-phenyl-2H-1,4-benzodiazepin-2-one) appears as an odorless white crystalline powder (m.p. 156 to 162°C) which is insoluble in water.

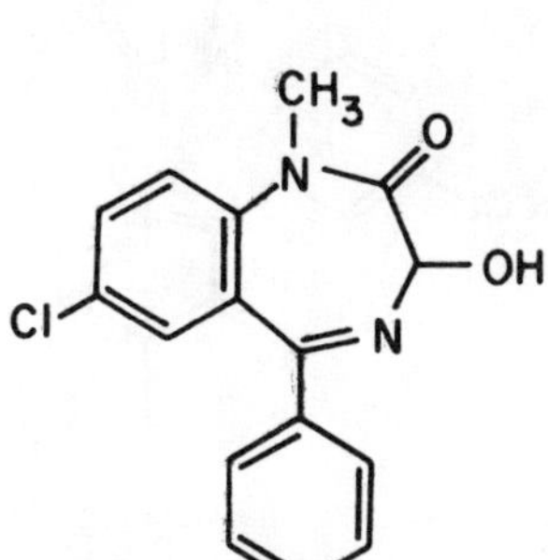

Structure 4

Chlorazepate dipotassium (Structure 5), **Tranxene®**, $C_{16}H_{11}ClN_2O_4K_2$(7-chloro-2,3-dihydro-2,2-dihydroxy-5-phenyl-lH-1,4-benzodiazepine-3-carboscylic acid, potassium salt) appears as a pale yellow powder and is very soluble in water.

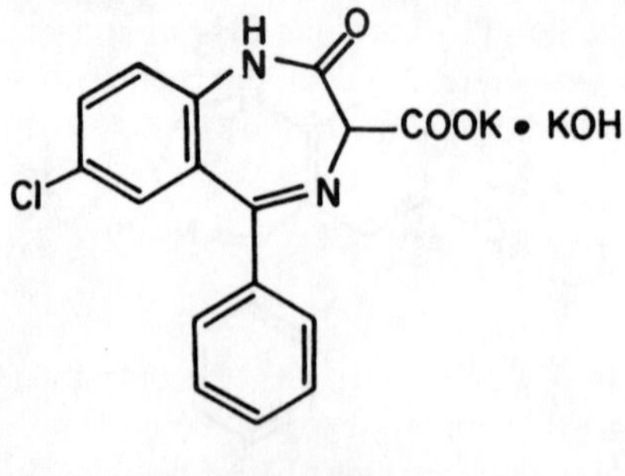

Structure 5

Lorazepam (Structure 6), **Ativan®**, $C_{15}H_{10}Cl_2N_2O_2$(7-chloro-5-(*O*-chlorophenyl)-1,3-dihydro-3-hydroxy-2H-1,4-benzodiazepin-2-one) appears as a white powder (m.p. 166 to 168°C) and is insoluble in water. pKa = 11.5 (acid), 1.3 (base).

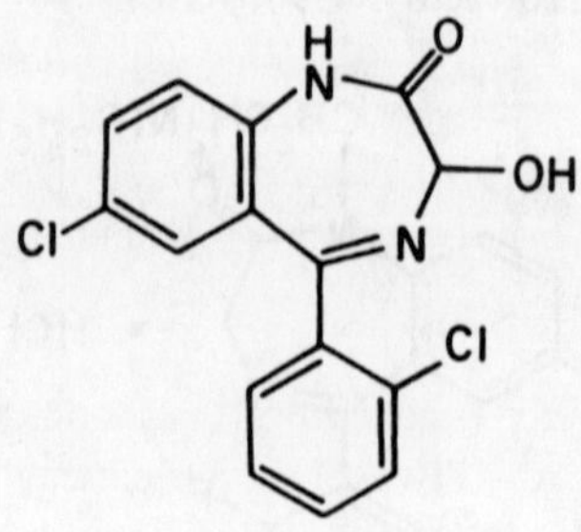

Structure 6

Triazolam (Structure 13), **Halcion®**, $C_{17}H_{12}Cl_2N_4$(8-chloro-6-(o-chlorophenyl)-1-methyl-4H-s-triazolo[4,3-a][1,4]benzodiozepine) appears as a white crystalline powder (m.p. about 240°C) and is poorly soluble in water.

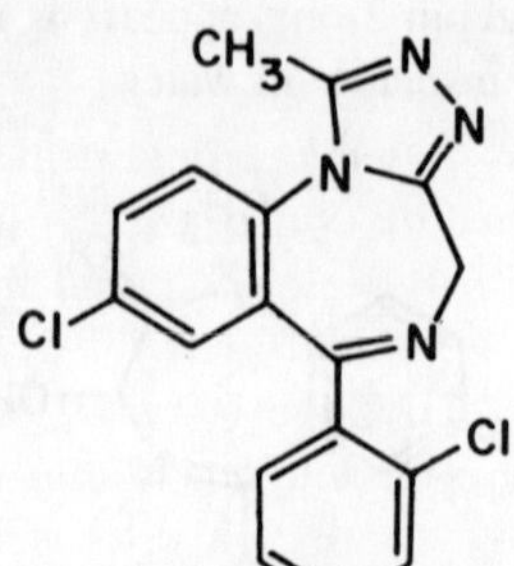

Structure 13

ALCOHOLS, ALDEHYDES, AND DERIVATIVES

Ethyl alcohol (CH_3CH_2OH) in the form of wine or beer has continued to be used since the earliest days of history to induce sleep. However, its danger for abuse and chronic toxicity restricts its value as a therapeutic agent. The more selectively toxic hypnotics were introduced into medical practice in the mid-19th century when chloral hydrate (Structure 15) was first prepared by Liebig in 1832 and employed as a hypnotic by Liebreich in 1869. Subsequently other alcohols and aldehydes and their derivatives were introduced into medical practice. Currently, besides chloral hydrate, the following agents are in use: ethchlorvynol (Structure 20), ethinamate (Structure 21), chloral betaine (Structure 16), pentaerythritol chloral (Structure 17), chlorhexadrol (Structure 18), triclofos sodium (Structure 19), and paraldehyde (Structure 22).

Chloral Hydrate

Mechanisms of Action

Chloral hydrate, ($CCl_3CH(OH)_2$ Structure 15), the oldest synthetic hypnotic in medical use, is the prototype and most commonly used of the chloral derivatives. Other hypnotics chemically related to chloral hydrate include chloral betaine (Structure 16), pentaerythritol chloral (Structure 17), chlorhexadrol (Structure 18), and triclofos sodium (Structure 19), which are prescription drugs and the veterinary hypnotic and anesthetic α-chloralose. What these agents have in common is that they are ultimately biotransformed to trichloroethanol which is responsible for their pharmacologic activity. Very little is known regarding the specific sites and mechanisms underlying the hypnotic actions of trichloroethanol. During the first week of nightly administration, chloral hydrate reduces sleep latency and number of nocturnal awakenings. Most EEG studies find chloral hydrate spares REM sleep like benzodiazepines.[17,64] However, part of the hypnotic effect in sleep lab studies is usually gone by 2 weeks of nightly use, in that the number of awakenings is increased.[72,73] Thus, the hypnotic effect of chloral hydrate exhibits a tolerance that may limit medical use.

Uses and Doses

The chloral derivatives are only used as oral hypnotics in the U.S. The manufacturer recommended dose is 0.5 to 1.0 g. However, some patients may require as much as 2.0 g to promote sleep. Chloral hydrate is administered in liquid capsule form or aqueous solution. Chloral betaine and triclofos lack the disagreeable taste of chloral hydrate and are less irritating to the GI tract.

Pharmacokinetics

Chloral hydrate is readily absorbed from the GI tract and rapidly reduced to trichloroethanol by enzymes in the liver.[74] The half-life of chloral hydrate is on the order of minutes and chloral hydrate is not usually detectable in the blood. Accordingly, the chloral derivatives are prodrugs for trichloroethanol. The half-life of trichloroethanol (TCE) is about 8 hr in man. TCE undergoes conjugation with glucuronic acid to form the inactive metabolite urochloralic acid which is eliminated by urinary excretion with a half-life of 6 to 8 hr.[75]

A small and variable amount (5 to 47%) of TCE and chloral hydrate is oxidized to form trichloroacetic acid (TCA). TCA has a half-life of 4 days. Consequently, it accumulates with nightly administration to a level of up to 100 mg/ℓ of blood.[76] The effects of TCA are unknown, except for displacement of protein-bound drug from their binding sites.[77]

Historically, it has been assumed that chloral hydrate is reduced by alcohol dehydrogenase in the liver to form TCE.[78] However, recent biochemical studies of the enzymology of this reduction indicate that aldehyde reductase catalyzes this reaction in vivo.[79] Moreover, alcohol dehydrogenase activity is very sensitive to the NAD/NADH ratio of the incubation and this enzyme is inactive when incubated at the in vivo coenzyme ratio. Pyrazole and isobutyramide, inhibitors of alcohol dehydrogenase, failed to inhibit the formation of TCE when chloral hydrate was added to rat liver slices. In the presence of ethanol, however, both enzymes contribute to the reduction of chloral hydrate. With ethanol, alcohol dehydrogenase is converted from an oxidizing form (enzyme-NAD complex) into a reducing enzyme form (enzyme-NADH complex), which prior to dissociation of NADH can reduce aldehydes to alcohols.

Toxicities

Chloral hydrate is irritating to mucous membranes. Epigastric distress, nausea, occasional vomiting, and flatulence are commonly reported side effects in treatment. This problem can be attenuated by diluting the drug with water or milk. Intoxicating doses of chloral hydrate have produced gastric necrosis. Less gastric irritation accompanies use of chloral betaine and triclofos.

Chloral hydrate is the only hypnotic capable of displacing other drugs from protein binding sites. The effect is mediated by the metabolite TCA, which is tightly bound to proteins. For patients receiving the anticoagulants warfarin or bishydroxycoumarin, chloral hydrate can precipitate a transient excessive hypoprothombinemia by increasing the free concentration of active anticoagulant in the circulation.[77]

Complex interactions can occur when chloral hydrate is combined with alcohol because of combined depressent actions and enchanced reduction of chloral hydrate to TCE by liver alcohol dehydrogenase.[80] Detailed studies of the interaction of hypnotic and of toxic effects of chloral hydrate and ethanol in mice disclosed simple addition, potentiation, or antagonism, depending on dose ratios administered. The myth of synergy for this so-called "Mickey Finn" combination is not supported by low dose combination studies in man.[81]

In acute overdose, chloral hydrate enhances the automaticity of the cardiac pacemaker cells culminating in supraventricular and ventricular tachyarrhythmias.[81,82]

Physicochemical Data

Chloral hydrate (Structure 15), **Noctec®**, $C_2H_3Cl_3O_2$, (2,2,2-trichloro-1,1-ethanediol) occurs as colorless or white crystals (m.p. 55°C) having an aromatic, penetrating, and slightly acrid odor and a bitter caustic taste. It is very soluble in water and in ethanol. pKa = 10.04.

$$Cl_3CCH(OH)_2$$

Structure 15

Chloral betaine (Structure 16), **Beta-Chlor®**, is a chemical complex of chloral hydrate and betaine. The complex does not have the disagreeable taste of chloral hydrate.

$$Cl_3CCH(OH)_2 \cdot (CH_3)_3\overset{+}{N}CH_2COO^-$$

Structure 16

Pentaerythritol chloral (Structure 17), **Petrichloral®**, $C_{13}H_{16}Cl_{12}O_8$, (1,1′-[2,2-bis-[(2,2,2-trichloro-1-hydroxyethoxy)methyl]-1,3-propanediylbis(oxy)]bis[2,2,2-trichloro-ethanol]) occurs as a yellow glass (m.p. 52 to 54°C) and is soluble in water. It lacks the acrid odor and the bitter taste of chloral.

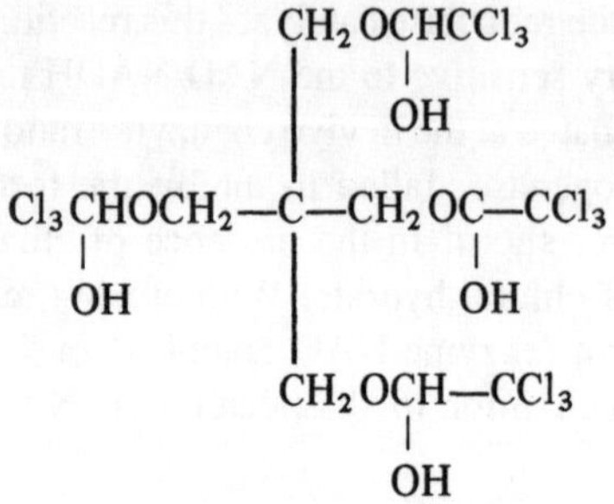

Structure 17

Chlorhexadrol (Structure 18), **Lora®; Mecoral®; Medodorm®**, $C_8H_{15}Cl_3O_3$, (2-methyl-4(2,2,2-trichloro-1-hydroxyethoxy)-2-pentanol) occurs as crystals (m.p. 102 to 104°C) and is readily soluble in alcohol and chloroform. It has a slightly bitter taste.

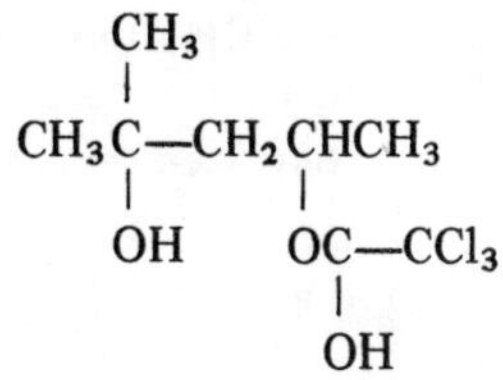

Structure 18

Triclofos sodium (Structure 19), **Triclos®**, $C_2H_3Cl_3NaO_4P$ (2,2,2-trichloroethanol dihydrogen phosphate monosodium salt) is soluble in water, and is free of the unpleasant odor and taste of chloral hydrate.

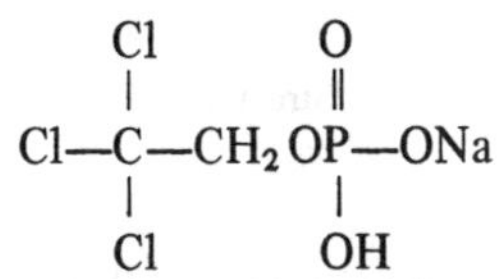

Structure 19

Ethchlorvynol (Structure 20) **Placidyl®**, C_7H_9Cl0, (1-chloro-3-ethyl-1-penten-4-yl-3-ol) occurs as a water immiscible liquid witha pungent aromatic odor (d^{25}_4 1.065 to 1.070; b.p.$_{760}$ 173 to 174°C).

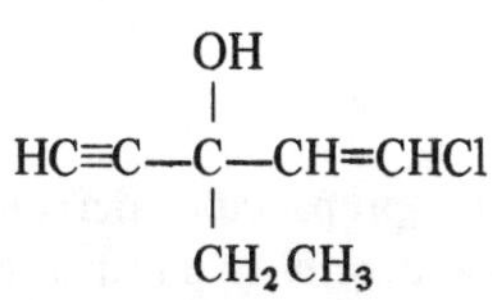

Structure 20

Ethinamate (Structure 21), **Valmid®**, $C_9H_{13}NO_2$, (1-ethynylcyclohexanol carbamate) is a low melting crystalline solid (m.p. 96 to 98°C; b.p.$_3$ 118 to 122°C) and is soluble in water at 25°C to the extent of 0.25%.

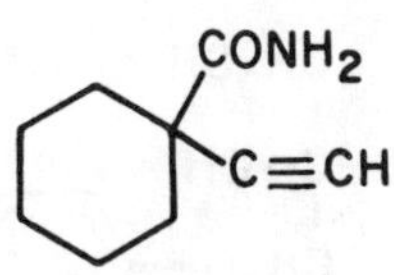

Structure 21

Paraldehyde (Structure 22), $C_6H_{12}O_3$, (2,4,6-trimethyl-*s*-trioxane) exists as a colorless liquid (b.p. 124°C, solidifies at 12°C) and has a characteristic aromatic odor and disagreeable taste. It is soluble in 8 parts of water at 25°C and in 17 parts of water at 100°C.

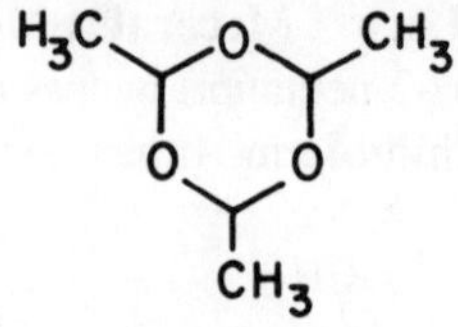

Structure 22

Meprobamate (Structure 23)

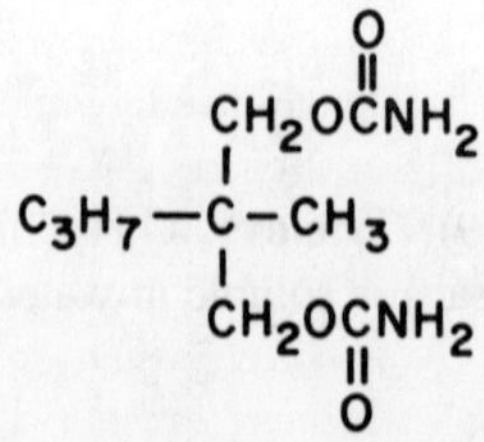

Structure 23

Mechanism of Action

Very little information is known regarding the cellular mechanism of meprobamate action. It was originally developed by Berger in 1954 as a mephenesin-like skeletal muscle relaxant with a longer duration of action. However, some congeners of meprobamate produce muscle relaxation with less sedation (see chapter entitled "Drugs Used to Treat Neuromuscular Disorders: Antiparkinsonism Agents and Skeletal Muscle Relaxants").

In clinical medicine, the pharmacological profile of meprobamate is indistinguishable from the barbiturates.[84] Meprobamate has no advantage over other commonly used agents as a sedative-hypnotic.

SARs

Mephenesin (Structure 24), a 1,2-propanediol derivative, was known to act within the CNS by depressing spinal interneural activity and had been tried in various types of psychiatric disorders. It was of limited value, however, because of its short duration of action and unreliable absorption. Following the observation that the short duration of action was a result of rapid oxidation of the primary hydroxyl group, many analogues in which this functional group was modified were prepared. Over 1200 compounds were investigated before the 1,3-propanediol derivative of meprobamate (Structure 25) was selected for clinical application as an anxiolytic and hypnotic.

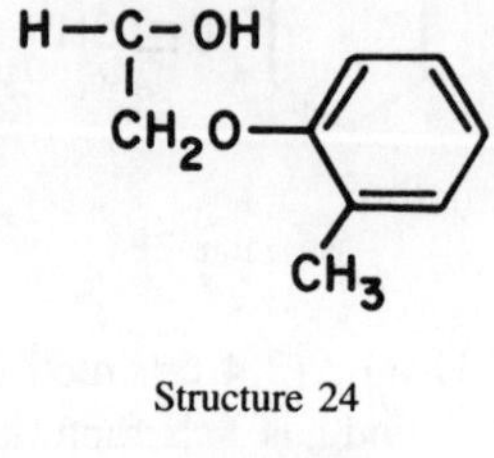

Structure 24

mephenesin

Uses and Doses

Meprobamate is more extensively used as an anti-anxiety agent or daytime sedative than as a hypnotic. The usual hypnotic and sedative doses are listed in Table 1. The physician acceptance and prescription use were phenomenal when meprobamate was first introduced. With the availability of a greater number of benzodiazepines, the use of meprobamate has declined in medical practice.

Pharmacokinetics

Meprobamate is well absorbed by the oral route. Peak blood levels are achieved in 2 to 3 hr and it is eliminated with a half-life of about 10 hr.

As much as 80 to 90% of a dose is metabolized by the liver; the remainder is excreted unchanged in the urine. The metabolites of meprobamate are inactive because they do not cross the blood-brain barrier by virtue of their polarity. The metabolites in man are products of oxidation, hydroxylation, and conjugation reactions which include meprobamate-*N*-mono-β-D-glucopyranosiduric acid, 2-methyl-2-(2′-hydroxypropyl)-1,3-propanediol dicarbamate and its *O*-glucuronide conjugate.[85-87]

Based on the apparent increase in urinary excretion of the hydroxy metabolite during chronic treatment for over 1 month, it has been suggested that meprobamate may increase its own rate of metabolism.[88] However, others have reported extended half-lives (24 to 48 hr) during chronic treatment.[89] Consequently, it is not often clear precisely how the metabolism and kinetics of meprobamate are altered with prolonged treatment.

Meprobamate is not significantly protein bound.[86] It is widely distributed throughout the body, traverses the placenta, and has been found in the milk of nursing mothers.

Toxicities

The most common patient complaint in the use of meprobamate as a sedative is drowsiness and ataxia. Meprobamate may impair learning and motor coordination and prolong reaction times especially in high doses. There is a low incidence of allergic reactions; allergic reaction most commonly occur in patients with a history of allergic and skin disorders. There have been a small number of reports of various blood dyscrasias.[64]

Meprobamate has had a history of abuse and both psychological and physical dependence, which has probably limited the continued popularity of this drug in medical practice.[90,91]

Meprobamate and ethanol may have additive CNS depressant effects.[86] Meprobamate can increase the rate of metabolism of other drugs by microsomal enzyme induction in animals.[92]

Physicochemical Data

Meprobamate (Structure 23), **Equanil®**, **Miltown®**, $C_9H_{18}N_2O_4$ (2-methyl-2-propyl-1,3-propanediol dicarbamate) is an odorless, colorless, crystalline powder (m.p. 103 to 107°C) with a characteristic bitter taste. The solubility in water is 1 part in 240.

CYCLIC AMIDES AND IMIDES

Several cyclic amides and imides have proved to be effective as sedative hypnotic drugs. Of these, glutethimide (Structure 26) and methyprylon (Structure 27) bear a close structural relationship to the barbiturates (Structure 25).

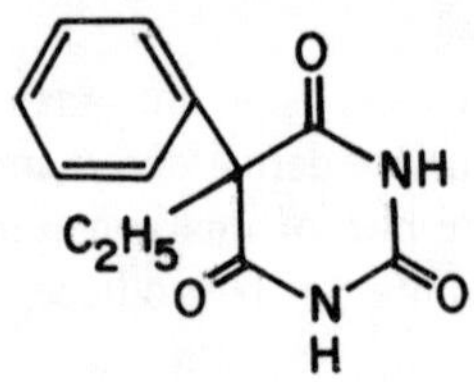

Structure 25

phenobarbital

Glutethimide

Mechanism of Action

The pharmacology of glutethimide resembles that of the short-acting barbiturates. Its liability for abuse and physical dependence is also barbiturate-like. In addition, glutethimide has significant anticholinergic activity as manifested by mydriasis, inhibition of salivary secretion and intestinal motility, and prevention of motion sickness.[64]

Sleep lab studies indicate that glutethimide also suppresses REM, but may do so in doses that do not reduce the frequency of nocturnal awakenings or increase total sleep time.[17] Glutethimide has no distinct advantage over other sedative-hypnotics.

Uses and Doses

Glutethimide is primarily used as a hypnotic at 250 to 500 mg. The sedative dose range is 125 to 250 mg one to three times daily.

Pharmacokinetics

Glutethimide is erratically absorbed by the oral route. It is about 50% bound to plasma proteins. By virtue of its high oil/water partition coefficient, it readily penetrates the blood-brain and placental barriers. Glutethimide, a racemate, is hydroxylated by the liver to form 2- and 4-hydroxy metabolites. The 4-hydroxy metabolite is pharmacologically active and may contribute to intoxication with overdose. The half-life of glutethimide is estimated to be 5 to 22 hr.[93] Glutethimide is an effective inducer of the liver microsomal enzyme system.[17]

Toxicities

Undesirable effects are rare; they include "hangover", excitement, blurred vision, gastric distress, headache, skin rash, and occasional blood dyscrasias.

Acute intoxication with glutethimide is complicated by the antimuscarinic actions and by a significant prolongation of the elimination half-life (~100 hr). Chronic use has been associated with tolerance as well as psychological and physical dependence.[17,64]

Physicochemical Data

Glutethimide (Structure 26), **Doriden®**, $C_{13}H_{15}NO_2$(3-ethyl-3-phenyl-2,6-piperidinedione) is a white powder (m.p. 84°C) and is insoluble in water. The dl form is used as a hypnotic. pKa = 4.52.

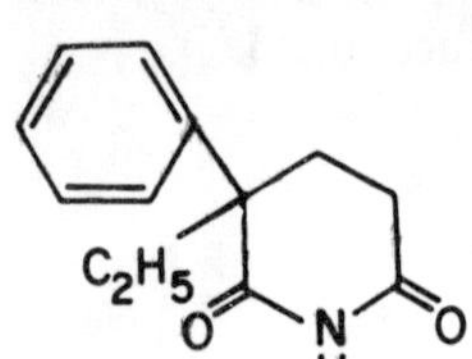

Structure 26

glutethimide

Methyprylon

Mechanism of Action

Methyprylon has been described to produce a hypnotic effect that is indistinguishable from a barbiturate. Like a barbiturate, it can be very toxic in an overdose.[94] Methyprylon has been shown to suppress REM sleep; however, there is a paucity of information on the EEG effects of this drug. Nevertheless, it has compared favorably to placebo in non-EEG sleep studies in terms of induction and maintenance of sleep.[13]

Uses and Doses

Methyprylon is promoted as a prescription hypnotic drug. The usual hypnotic dose is 200 or 400 mg before retiring.

Pharmacokinetics

Methyprylon is extensively metabolized; only 3% is excreted unchanged. Both glucuronides and unconjugated metabolites have been found in the urine in animals.[95] The plasma half-life is 4 hr.[96] Methyprylon has been reported to increase hepatic microsomal enzyme activity.

Toxicities

While low in incidence, untoward effects include nausea, vomiting, diarrhea, headache, and ''hangover''. Manifestations of overdose are like that of barbiturates and they include somnolence, confusion, coma, miosis, respiratory depression, and hypotension. One patient recovered from a 30-g overdose by hemodialysis.[97]

Physicochemical Data

Methyprylon (Structure 27), **Noludar®**, $C_{10}H_{17}NO_2$(3,3-diethyl-5-methyl-2,4-piperidinedione) is a white powder (m.p. 74 to 77°C) with a bitter taste; moderately soluble in water and alcohol. pKa = 12.0.

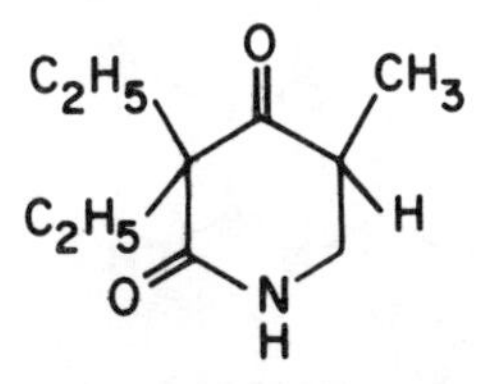

Structure 27

methyprylon

Methaqualone (Structure 28)

Mechanism of Action

As with most sedative-hypnotics, the molecular and cellular bases for methaqualone's therapeutic effects remain unknown. Methaqualone also possesses anticonvulsant, antispasmodic, antitussive, weak antihistaminic, and local anesthetic activity. It is not analgesic. EEG effects of methaqualone are variable and reports are contradictory.[17,64]

Uses and Doses

Methaqualone has been withdrawn from the market in the U.S. and is very popular in drug cultures. It is known as the ''love drug'' because it is claimed to remove inhibitions of sexual expression.

Pharmacokinetics

Methaqualone is completely absorbed orally within 2 hr. A dose of 300 mg produces serum levels of 3 μg/mℓ.[98] With chronic nightly administration of 300 mg for 28 days, there is no evidence of accumulation or change in absorption or elimination. The half-life of methaqualone is 19 to 40 hr.[99,75] It is highly protein bound in plasma.[100]

Methaqualone is primarily hydroxylated by hepatic microsomal enzymes followed by O-methylation or glucuronide conjugation. However, methaqualone is only a weak inducer of mixed function oxidases in man. It has no effect on the anticoagulant activity of warfarin.[101]

Toxicities

Fatigue, dizziness, and torpor have been reported following daytime sedation. Transient paresthesias[102] during the onset and "hangover" during morning recovery from nighttime hypnotic doses have been reported. GI discomfort, nausea, vomiting, diarrhea, sweating, and skin rash have been described.[64]

With overdosage, methaqualone shows a profile of generalized CNS depression (amnesia, coma, cardiovascular and respiratory depression) as well as evidence of CNS stimulation (hypertonus, hyperreflexia, myoclonus, convulsion, restlessness, and excitement). Most fatalities have involved concurrent consumption of alcohol. Methaqualone produces a bleeding tendency in acute overdosage by inhibiting platelet aggregation.[103] This effect, or thrombocytopenia, may be related to the observation of retinal hemorrhage during a suicide attempt.[104]

Physicochemical Data

Methaqualone hydrochloride (Structure 28), **Quaalude®, Parest®, Sopor®**, $C_{16}H_{14}N_2O$, (2-methyl-3-*o*-tolyl-4(3H)-quinazolinone) is a white crystalline powder (m.p. 120°C) practically insoluble in water and soluble in ethanol.

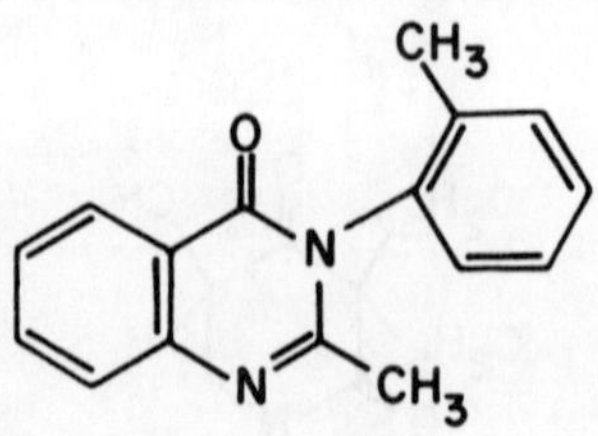

Structure 28

methaqualone

ACKNOWLEDGMENT

The authors wish to thank Carl Regillo for help in preparing the manuscript, Noralie Barnett for the art work, and Kimberly Rapoza for typing this chapter. The partial support of BRSG Grant SO7 RR 05830-03 awarded by the National Institutes of Health is gratefully acknowledged.

REFERENCES

1. **Isbell, H., Attschal, S., Kornetsky, C. H., Eiseman, A. J., Flanary, H. G., and Fraser, H. F.,** Chronic barbiturate intoxication, *Arch. Neurol. Psychiatr.*, 64, 1, 1950.
2. **Kelley, D. D.,** Physiology of sleep and dreaming, in *Principles of Neural Science*, Kandel, E. R. and Schwartz, J. H., Eds., Elsevier/North-Holland, New York, 1981, chap. 40.

3. **Bremer, F.,** Cerebral hypnogenic centers, *Ann. Neurol.,* 2, 1, 1977.
4. **Monnier, M., Koller, T., and Graber, S.,** Humoral influences of induced sleep and arousal upon electrical brain activity of animals with crossed circulation, *Exp. Neurol.,* 8, 264, 1963.
5. **Schoenenberger, G. A. and Monnier, M.,** Characterization of a delta-encephalogram (-sleep)-inducing peptide, *Proc. Natl. Acad. Sci. U.S.A.,* 74, 1282, 1977.
6. **Pappenheimer, J. R., Miller, T. B., and Goodrich, C. A.,** Sleep-promoting effects of cerebrospinal fluid from sleep-deprived goats, *Proc. Natl. Acad. Sci. U.S.A.,* 58, 513, 1967.
7. **Drucker-Colin, R.,** Endogenous sleep peptides, in *Psychopharmacology of Sleep,* Wheatley, D., Ed., Raven Press, New York, 1981, 53.
8. **Puizillout, J. J., Gaudin-Chazal, G., Sayadi, A., and Vigier, D.,** Serotonergic mechanisms and sleep, *J. Physiol. (Paris),* 77, 415, 1981.
9. **Gillin, J. C., Mendelson, W. B., Sitaram, N., and Wyatt, R. J.,** The neuropharmacology of sleep and wakefulness, *Ann. Rev. Pharmacol. Toxicol.,* 18, 563, 1978.
10. **Koella, W. P.,** Neurotransmitters and sleep, in *Psychopharmacology of Sleep,* Wheatley, D., Ed., Raven Press, New York, 1981, 19.
11. **Soldatos, C. R., Kales, A., and Kales, J. D.,** Management of insomnia, *Ann. Rev. Med.,* 30, 301, 1979.
12. **Kales, A. and Kales, J. D.,** Evaluation, diagnosis and treatment of clinical conditions related to sleep, *JAMA,* 213, 2229, 1970.
13. **Mendelson, W. B.,** *The Use and Misuse of Sleeping Pills,* Plenum, New York, 1980, chap. 1.
14. **Weitzman, E. D.,** Sleep and its disorders, *Ann. Rev. Neurosci.,* 4, 381, 1981.
15. Institute of Medicine, Report of a Study, Sleeping Pills, Insomnia, and Medical Practice, National Academy of Sciences, Washington, D.C., 1979, chap. 1 and 2.
16. **Kales, A., Scharf, M. B., Soldatos, R., and Bixler, E. O.,** Clinical evaluation of hypnotic drugs: contributions from sleep laboratory studies, *J. Clin. Pharmacol.,* 19, 329, 1979.
17. **Mendelson, W. B.,** *The Use and Misuse of Sleeping Pills,* Plenum, New York, 1980, chap. 4.
18. **Hobson, J. A., Sapgna, T., and Earls, P.,** Dreamstage: An Experimental Portrait of the Sleeping Brain, Hoffmann-La Roche, Inc., Nutley, N.J., 1977.
19. **Ling, R. K. S.,** Animal techniques for evaluating hypnotics, in *Animal and Clinical Pharmacologic Techniques in Drug Evaluation,* Nodine, J. and Siegler, P., Eds., Yearbook Medical Publishers, Chicago, 1964, 291.
20. **Irwin, S.,** Anti-neurotics: practical pharmacology of the sedative-hypnotics and minor tranquilizers, in Psychopharmacology — A Review of Progress 1957—1967, Efron, D. H., Ed., U.S. Government Printing Office, Washington, D.C., 1968, 185.
21. **Gehrmann, J. E. and Killam, K. F.,** EEG changes following the administration of sedative-hypnotic drugs, in *Hypnotics — Methods of Development and Evaluation,* Kagan, F., Harwood, K., Rickels, K., Rudzik, A. D., and Sorer, H., Eds., Spectrum Publications, New York, 1975, 241.
22. **Kales, A., Bixler, E. O., Tan, T., Scharf, M. B., and Kales, J. D.,** Chronic hypnotic-drug use: ineffectiveness, drug-withdrawal insomnia, and dependence, *JAMA,* 227, 513, 1974.
23. **Cooper, J. R.,** Sedative-Hypnotic Drugs: Risks and Benefits, National Institute on Drug Abuse, U.S. Department of Health, Education and Welfare, Washington, D.C., 1977.
24. **Boisse, N. R. and Okamoto, M.,** Ethanol as a sedative-hypnotic; comparison with barbiturate and non-barbiturate sedative-hypnotics, in *Alcohol Tolerance and Dependence,* Rigter, H. and Crabbe, J. C., Eds., Elsevier/North-Holland, Amsterdam, 1980.
25. **Kalant, H., LeBlanc, A. E., and Gibbons, R. J.,** Tolerance to, and dependence on, some non-opiate psychotropic drugs, *Pharmacol. Rev.,* 23, 135, 1971.
26. **Smith, C. M.,** The pharmacology of sedative/hypnotics, alcohol and anesthetics. Sites and mechanisms of action, in *Drug Addiction I,* Martin, W. R., Ed., Springer-Verlag, New York, 1977, 413.
27. **Essig, C. F.,** Addiction to nonbarbiturate sedative and tranquilizing drugs, *Clin. Pharmacol. Ther.,* 5, 334, 1964.
28. **Jaffe, J. F.,** Drug addiction and drug abuse, in *The Pharmacological Basis of Therapeutics,* 6th ed., Gilman, A. G., Goodman, L. S., and Gilman, A., Eds., Macmillan, New York, 1980, chap. 23.
29. **Squires, R. F. and Braestrup, C.,** Benzodiazepine receptors in rat brain, *Nature (London),* 266, 732, 1977.
30. **Mohler, H. and Okada, T.,** Benzodiazepine receptor: demonstration in the central nervous system, *Science,* 198, 849, 1977.
31. **Tallman, J. F., Paul, S. M., Skolnick, P., and Gallagher, D. W.,** Receptors for the age of anxiety: pharmacology of the benzodiazepines, *Science,* 207, 274, 1980.
32. **Müller, W. E.,** The benzodiazepine receptor — an update, *Pharmacology,* 22, 153, 1981.
33. **Braestrup, C. and Squires, R. F.,** Pharmacological characterization of benzodiazepine receptors in the brain, *Eur. J. Pharmacol.,* 78, 263, 1978.
34. **Hunkeler, W., Mohler, H., Pieri, L., Polc, P., Bonetti, E. P., Cumin, R., Schaffner, R., and Haefely, W.,** Selective antagonists of benzodiazepines, *Nature (London),* 290, 514, 1981.

35. **Darragh, A., Scully, M., Lambe, R., and Brick, I.,** Investigation in man of the efficacy of a benzodiazepine antagonist, Ro 15-1788, *Lancet,* 2, 8, 1981.
36. **Polc, P., Laurent, J.-P., Scherschlict, R., and Haefely, W.,** Electrophysiological studies on the specific benzodiazepine antagonist Ro 15-1788, *Naunyn-Schmiedeberg's Arch. Pharmacol.,* 316, 317, 1981.
37. **Mohler, H., Burkard, W. P., Keller, H. H., Richards, J. G., and Haefely, W.,** Benzodiazepine antagonist Ro-15 1788; binding characteristics and interaction with drug-induced changes in dopamine turnover and cerebellar c-GMP levels, *J. Neurochem.,* 37, 714, 1981.
38. **Mohler, H. and Richards, J. G.,** Agonist and antagonist benzodiazepine receptor interaction *in vitro, Nature (London),* 294, 763, 1981.
39. **Skolnick, P. and Paul, S. M.,** Benzodiazepine receptors, in *Annual Reports in Medicinal Chemistry,* Vol. 16, Hess, H. J., Ed., Academic Press, New York, 1981, chap. 3.
40. **Marangos, P. J., Paul, S. M., Goodwin, F. K., and Skolnick, P.,** Minireview: putative endogenous ligands for the benzodiazepine receptor, *Life Sci.,* 25, 1093, 1979.
41. **Skolnick, P., Mendelson, W. B., and Paul, S. M.,** Benzodiazepine receptors in the central nervous system, in *Psychopharmacology of Sleep,* Wheatley, D., Ed., Raven Press, New York, 1981, 117.
42. **Poddar, M. K., Urquhart, D., and Sinba, A. K.,** Diazepam binding in brain after sleep and wakefulness, *Brain Res.,* 193, 519, 1980.
43. **Blair, T. and Webb, G. A.,** Electronic factors in the structure activity relationship of some 1,4-benzodiazepin-2-ones, *J. Med. Chem.,* 20, 1206, 1977.
44. **Bellatuono, C., Reggi, V., Tognoni, G., and Garattini, S.,** Benzodiazepines: clinical pharmacology and therapeutic use of benzodiazepines, *Drugs,* 19, 195, 1980.
45. **Sellers, E. M.,** Clinical pharmacology and therapeutics of benzodiazepines, *Can. Med. Assoc. J.,* 118, 1533, 1978.
46. **Kales, A., Bixler, E. O., Kales, J. D., and Scharf, M. B.,** Comparative effectiveness of nine hypnotic drugs: sleep laboratory studies, *J. Clin. Pharmacol.,* 16, 207, 1977.
47. **Bliding, A.,** Effects of different rates of absorption of two benzodiazepines on subjective and objective parameters — significance for clinical use and risk of abuse, *Eur. J. Clin. Pharmacol.,* 7, 201, 1974.
48. **Greenblatt, D. J. and Shader, R. I.,** Pharmacokinetic understanding of antianxiety drug therapy, *South. Med. J.,* 71, 2, 1978.
49. Valiumb Diazepam/Roche CIV, a Monograph, Hoffmann-LaRoche, Inc., Nutley, N.J., 1980.
50. **Boxenbaum, H. G., Geitner, K. A., Jack, M. L., Dixon, W. R., Spiegel, H. E., Symington, J., Christian, R., Moore, J. D., Weissman, L., and Kaplan, S. A.,** Pharmacokinetic and biopharmaceutic profile of chlordiazepoxide HCl in healthy subjects: single-dose studies by the intravenous, intramuscular and oral routes, *J. Pharmacokinet. Biopharm.,* 5, 3, 1977.
51. **Boxenbaum, H. G., Geitner, K. A., Jack, M. L., Dixon, W. R., and Kaplan, S. A.,** Pharmacokinetic and biopharmaceutic profile of chlordiazepoxide HCl in healthy subjects. Multiple-dose oral administration, *J. Pharmacokinet. Biopharm.,* 5, 25, 1977.
52. **Dixon, R., Brooks, M. A., Postma, E., Hackman, M. R., Spector, S., Moore, J. D., and Schwartz, M. Z.,** N-desmethyldiazepam: a new metabolite of chlordiazepoxide in man, *Clin. Pharmacol. Ther.,* 20, 450, 1976.
53. **Abruzzo, C. W., Macasilb, T., Weinfeld, R., Rider, J. A., and Kaplan, S. A.,** Changes in the oral absorption characteristics in man of dipotassium chlorazepate at normal and elevated gastric pH, *J. Pharmacokinet. Biopharm.,* 5, 377, 1977.
54. **Shader, R. I., Georgotas, A., Greenblatt, D. J., Hermatz, J. L., and Divoll, M. D.,** Impaired absorption of desmethyldiazepam from clorazepate by magnesium aluminum hydroxide, *Clin. Pharmacol. Ther.,* 24, 308, 1978.
55. Tranxeneb (clorazepate dipotassium), Drug Monograph, Abbott Labs, North Chicago, Ill., 1981.
56. **Shader, R. I., Greenblatt, D. J., Ciraulo, D. R., Divoll, M., Hermatz, S., and Georgotas, A.,** Effect of age and sex on disposition of desmethyldiazepam formed from its precursor clorazepate, *Psychopharmacology,* 75, 193, 1981.
57. **Ochs, H. R., Greenblatt, D. J., Allen, M. D., Hermatz, J. S., Shader, R. I., and Bodem, G.,** Effect of age and Billroth gastrectomy on absorption of desmethyldiazepam from clorazepate, *Clin. Pharmacol. Ther.,* 26, 449, 1979.
58. **Schwartz, M. A. and Postma, E.,** Metabolism of flurazepam, a benzodiazepine in man and dog, *J. Pharm. Sci.,* 59, 1800, 1970.
59. **Kaplan, S. A., de Silva, J. A. F., Jack, M. L., Alexander, K., Strojny, N., Weinfeld, R. E., Puglisi, C. V., and Weissman, L.,** Blood level profile in man following chronic oral administration of flurazepam hydrochloride, *J. Pharm. Sci.,* 62, 1932, 1973.
60. **Kales, A.,** Benzodiazepine hypnotics: carryover effectiveness, rebound insomnia, and performance effects, in Benzodiazepines: A Review of Research Results, 1980, NIDA Research Monograph 33, a RAUS Review Report, Szara, S. I. and Ludford, J. P., Eds., Superintendent of Documents, U.S. Government Printing Office, Washington, D.C., 1981, 61.

61. **Greenblatt, D. J., Schillings, R. T., Kyriakopoulos, A. A., Shader, R. I., Sisenwine, S. F., Knowles, J. A., and Ruelius, H. W.,** Clinical pharmacokinetics of lorazepam. I. Absorption and disposition of oral ^{14}C-lorazepam, *Clin. Pharmacol. Ther.*, 20, 329, 1976.
62. **Greenblatt, D. J.,** Clinical pharmacokinetics of oxazepam and lorazepam, *Clin. Pharmacokinet.*, 6, 89, 1981.
63. **Schwarg, H. J.,** Pharmacokinetics and metabolism of temazepam in man and several animal species, *Br. J. Clin. Pharmacol.*, 8, 23S, 1979.
64. **Harvey, S. C.,** Hypnotics and sedatives, in *The Pharmacological Basis of Therapeutics*, 6th ed., Gilman, A. G., Goodman, L. S., and Gilman, A., Eds., Macmillan, New York, 1980, chap. 17.
65. **Kales, A., Scharf, M. B., and Kales, J. D.,** Rebound insomnia: a new clinical syndrome, *Science*, 201, 1639, 1978.
66. **Kales, A., Scharf, M. B., Kales, J. D., and Soldatos, C. R.,** Rebound insomnia: a potential hazard following withdrawal of certain benzodiazepine drugs, *JAMA*, 241, 1692, 1979.
67. **Kales, A., Bixler, E. O., Scharf, M., and Kales, J. D.,** Sleep laboratory studies of flurazepam: a model for evaluating hypnotic drugs, *Clin. Pharmacol. Ther.*, 19, 576, 1976.
68. **Dement, W. C., Carskadon, M. A., Mitler, M. D., Phillips, R. L., and Zarcone, V. P.,** Prolonged use of flurazepam: a sleep laboratory study, *Behav. Med.*, 5, 25, 1978.
69. **Greenblatt, D. J. and Shader, R. I.,** *Benzodiazepines in Clinical Practice*, Raven Press, New York, 1974.
70. **Bixler, E. O., Sharf, M. B., Soldatos, C. R., Mitsky, D. J., and Kales, A.,** Effects of hypnotic drugs on memory, *Life Sci.*, 25, 1379, 1979.
71. **Ghoneim, M. M., Mewalt, S. P., Berie, J. L., and Hinrichs, J. V.,** Memory and performance effects of single and 3-week administration of diazepam, *Psychopharmacology*, 73, 147, 1981.
72. **Kales, A., Kales, J. D., Scharf, M. B., and Tan, T. L.,** Hypnotics and altered sleep patterns. II. All-night EEG studies of chloral hydrate, flurazepam, and Methaqualone, *Arch. Gen. Psychiatr.*, 23, 219, 1970.
73. **Kales, A., Allen, C., Scharf, M. B., and Kales, J. D.,** Hypnotic drugs and their effectiveness. All-night studies of insomniac subjects, *Arch. Gen. Psychiatr.*, 23, 226, 1970.
74. **Marshall, E. K. and Owens, A. H.,** Absorption, excretion and metabolic fate of chloral hydrate and trichloroethanol, *Bull. Johns Hopkins Hosp.*, 95, 1018, 1954.
75. **Breimer, D. D.,** Clinical pharmacokinetics of hypnotics, *Clin. Pharmacokinet.*, 2, 93, 1977.
76. **Owens, A. H. and Marshall, E. K.,** Further studies on the metabolic fate of chloral hydrate and trichloroethanol, *Bull. Johns Hopkins Hosp.*, 97, 320, 1955.
77. **Greenblatt, D. J. and Miller, R. R.,** Hypnotics, in *Drug Effects in Hospitalized Patients*, Miller, R. R. and Greenblatt, D. J., Eds., John Wiley & Sons, New York, 1976, 171.
78. **Greenblatt, D. J. and Shader, R. I.,** The clinical choice of sedative-hypnotics, *Ann. Int. Med.*, 77, 91, 1970.
79. **Schultz, J. and Weiner, H.,** Alteration of the enzymology of chloral hydrate reduction in the presence of ethanol, *Biochem. Pharmacol.*, 28, 3379, 1979.
80. **Gessner, P. K. and Cabona, B. E.,** A study of the interaction of the hypnotic effects and of the toxic effects of chloral hydrate and ethanol, *J. Pharmacol. Exp. Ther.*, 174, 247, 1970.
81. **Sellers, E. M., Carr, G., Bernstein, J. G., Sellers, S., and Koch-Weser, J.,** Interaction of chloral hydrate and ethanol in man. II. Hemodynamics and performance, *Clin. Pharmacol. Ther.*, 13, 50, 1972.
82. **Gustafson, A., Svensson, S. E., and Ugander, L.,** Cardiac arrythmias in chloral hydrate poisoning, *Acta Med. Scand.*, 201, 227, 1977.
83. **Marshall, A. J.,** Cardiac arrhythmias caused by chloral hydrate, *Br. Med. J.*, 2, 994, 1977.
84. **Greenblatt, D. J. and Shader, R. I.,** Meprobamate: a study of irrational drug use, *Am. J. Psychiatr.*, 127, 1297, 1971.
85. **Meyer, M. C. and Straughn, A. B.,** Meprobamate bioavailability monography, *J. Am. Pharm. Assoc.*, 17, 173, 1977.
86. **Carpenter, J. A., Marshman, J. A., and Ginnins, R. J.,** The effects of alcohol and meprobamate applied singly and jointly in human subjects. I. Theoretical considerations and literature review, *J. Studies Alcohol*, 123, 1, 1975.
87. **Kaplan, S. A. and Jack, M. L.,** Pharmacokinetics and metabolism of anxiolytic, in *Handbook of Experimental Pharmacology*, Vol. 55 (Part 2), Hoffmeister, F. and Stille, G., Eds., Springer-Verlag, Berlin, 1981, 321.
88. **Douglas, J. F., Ludwig, B. J., and Smith, M.,** Studies on the metabolism of meprobamate, *Proc. Soc. Exp. Biol. Med.*, 112, 436, 1963.
89. **Hollister, L. E. and Levy, G.,** Kinetics of meprobamate elimination in humans, *Chemotherapia*, 9, 20, 1964.
90. **McClane, T. K. and Martin, W. R.,** Subjective and physiological effects of morphine, pentobarbital, and meprobamate, *Clin. Pharmacol.*, 20, 192. 1976.

91. **Allen, M. D., Greenblatt, D. J., and Noel, B. J.,** Meprobamate overdosage: a continuing medical problem, *Clin. Toxicol.*, 11, 501, 1977.
92. **Greenblatt, D. J. and Shader, R. I.,** The clinical choice of sedative-hypnotics, *Ann. Int. Med.*, 77, 91, 1972.
93. **Curry, S. H., Riddell, D., Gordon, J. S., Simpson, P., Binns, T. B., Rondel, R. K., and McMartin, D.,** Disposition of glutethimide in man, *Clin. Pharmacol. Ther.*, 12, 849, 1971.
94. **Pelligrino, E. D. and Henderson, R. R.,** Clinical toxicity of methyprylon, *J. Med. Soc. N.J.*, 54, 515, 1957.
95. **Randall, L. O., Iliev, V., and Brandman, O.,** Metabolism of methylprylon, *Arch. Int. Pharmacodyn.*, 106, 388, 1956.
96. **Pancorbo, A. S., Palagi, P. A., Piecoro, J. J., and Wilson, H. D.,** Hemodialysis in methyprylon overdose. Some pharmacokinetic consideration, *JAMA*, 237, 470, 1977.
97. **Xanthaky, G., Freireich, A. W., Matusiak, W., and Lukash, L.,** Hemodialysis in methyprylon poisoning, *JAMA*, 198, 1212, 1966.
98. **Smyth, R. D., Lee, J. K., Polk, A., Chemburkar, P. B., and Savacool, A. M.,** Bioavailability of methaqualone, *J. Clin. Pharmacol.*, 13, 391, 1973.
99. **Nayak, R. K., Smyth, R. D., Chamberlain, J. H., Polk, A., DeLong, A. F., Herzog, T., Chemburkar, P. B., Joslin, R. S., and Reavy-Cantwell, N. H.,** Methaqualone pharmacokinetics after single- and multiple-dose administration in man, *J. Pharmacokinet. Biopharm.*, 2, 107, 1974.
100. **Christensen, J. M. and Holfort, S.,** Methaqualone in human serum and cerebrospinal fluid after oral intake, *J. Pharm. Pharmacol.*, 27, 538, 1975.
101. Research Monograph: Quaalude® (Methaqualone) Lemmon Co., Sellersville, Pa., 1981.
102. **Marks, P. and Sloggen, J.,** Peripheral neuropathy caused by methaqualone, *Am. J. Med. Sci.*, 272, 323, 1976.
103. **Mills, D. G.,** Effects of methaqualone on blood platelet function, *Clin. Pharmacol.*, 23, 685, 1978.
104. **Teese, M.,** Retinal hemorrhage caused by overdose of methaqualone (Quaalude), *Am. J. Ophthalmol.*, 91, 20, 1981.

ANTIPSYCHOTIC AGENTS

J. A. Vida and P. A. Tenthorey

Antipsychotic drugs are used for the treatment of psychosis and a variety of organic brain disorders as follows:

1. Acute treatment of psychosis, such as agitation, delusions, confusion, hallucinations, inappropriate behavior and irrational thinking, schizophrenia, mania, and psychotic depression.
2. Long-term maintenance of individuals with schizophrenia and prevention of exacerbation of the condition of chronic schizophrenia.

PHARMACOLOGIC EVALUATION OF ANTIPSYCHOTIC AGENTS

A battery of tests is used to evaluate antipsychotic properties in animals. In this treatment we will describe the following, most often used tests: (1) preliminary screening methods and (2) tests for antipsychotic potency.

Preliminary Screening Methods

Although it is difficult to derive reproducible results that would reveal a dose-response relationship, simple observations in rats can be used as an efficient preliminary screening method for antipsychotic activity.

- **Catalepsy** — Most antipsychotic agents produce a characteristic long lasting state of cataleptic immobility in rats.[1]
- **Ptosis** — The palpebral ptosis-producing ability of compounds in the rat is a reliable measure of the antipsychotic activity.[2]
- **Taming** — Although antianxiety agents also produce these effects, the ability of compounds to tame aggressive animals such as monkeys[3] and minks[4] reveals potential antipsychotic activity.
- **Spontaneous motor activity** — Although other CNS depressants also reduce spontaneous motor activity in mice and rats, this reduction may be due to antipsychotic activity. The observation can be quantitated by devices, such as photocell activity cages,[5] jiggle cages,[6] and rotating drums.[7]
- **Observation of exploratory behavior** — Antipsychotics produce characteristic behavior, i.e., ambulation, rearing, preening, and emotional defecation that is different from the effects of other CNS agents.[8]
- **Blockade of fighting behavior** — Rats/mice can become hostile as a result of septal lesions, isolation, or foot shock (in pairs of animals).[9,10] Antipsychotics reduce fighting behavior commensurate with their potencies. Antianxiety agents are also active in this unconditioned, unlearned behavioral test.

Tests for Antipsychotic Potency

The following tests are based on motivated behavior: the inhibitory effects of the drug on the learned, conditioned, and predictable behavior of rats and dogs are measured. In tests using a positive reinforcing stimulus the animals are rewarded (receive food) while in tests using a negative reinforcing stimulus, the animals are punished (electric shock). The response of the animal precedes the stimulus in order to obtain it (food) or avoid it (shock).

Conditioned Avoidance Response (Pole-Climbing) Test[11,12]

Rats are placed in a cage with a grid floor through which electric shocks (unconditioned stimulus) may be administered. Animals are trained by sounding a buzzer (conditioned stimulus) to escape the shock by climbing a pole. Well-trained animals will climb the pole in response to the buzzer only. It should be of interest that Courvoisier used this test when the unique properties of chlorpromazine were discovered. Antipsychotics selectively block the conditioned response, although much higher doses of the antipsychotics will also block the unconditioned response. Neither barbiturates nor antianxiety agents block the conditioned response selectively without also blocking the unconditioned response.

Conditioned Avoidance Response (Jumping Box or Shuttle Box) Test[13]

Rats, dogs, or monkeys are placed into a two-compartmental box with a low barrier in between. The animal avoids the shock when a buzzer sounds (conditioned response) by climbing or jumping over the barrier. Antipsychotics block the conditioned avoidance responses in a rank order that corresponds to their clinical potencies.

Reinforced Behavioral Response Test (Skinner Box)[14]

A simple action of an animal is brought under control by positive reinforcement (reward) or negative reinforcement (punishment). A positive reinforcement is a stimulus (food or water) for which a hungry animal will produce a response. A negative reinforcement is an aversive stimulus (electric shock, noise, strong light) that will provoke a response by the animal in order to avoid or escape the punishment. Trained animals will perform responses with characteristic frequency that corresponds to the reinforcement schedule. The reinforcement schedules can be variable-interval, fixed-interval, or a combination reinforcement schedule requiring a certain number of responses within an interval. In the Skinner box, the animal stands on a grid-floor, through which electric shock can be delivered. The animal presses a lever on the wall of the box to avoid a shock. Alternatively, pressing the lever provides the reward (food or water). Antipsychotics will decrease the overall response rates of the animals.

Continuous Avoidance Test (Sidman Test)[15]

The rats are trained to postpone a shock for a specified interval (20 sec) by pressing a lever (avoidance). If the lever is not pressed within a specified interval, the animal receives a shock. In addition, the animal is also trained to discriminate between two levers. Responses on one lever may postpone shocks (avoidance), while responses on a second lever terminate the shock (escape) when it occurs. Antipsychotics decrease average response rates. Some anxiolytics also decrease the response rates but in much higher doses than antipsychotics.

Antagonism of Apomorphine-Induced Emesis[16]

This is a very sensitive test in dogs for determining antipsychotic activity. By varying the time interval between drug administration and apomorphine injection, relative ED_{50} values and times of peak effects can be obtained. This model, along with the amphetamine-induced stereotypies in rats model have been instrumental for demonstrating the activities of long-acting injectable neuroleptics. Even in these two sensitive models, much higher doses of the injectable antipsychotics had to be used than the classical nonesterified antipsychotics to elicit a response.

Antagonism of Amphetamine-Induced Hyperactivity[17,12]

Antipsychotics antagonize the agitation induced in rats by injection of amphetamine. It is known that the lethality of amphetamine is much more pronounced when given to groups of animals, as opposed to that in an individual animal. Antipsychotics protect against the enhanced lethality of amphetamine in grouped animals. This model proved to be useful for determining the potencies of long-acting injectable antipsychotic esters.

MECHANISM OF ACTION OF ANTIPSYCHOTIC DRUGS

Depletion of Monoamine Stores

The first antipsychotic drug, reserpine, was introduced in 1931 for the treatment of psychoses. However, it was much later that the discovery was made that reserpine causes depletion of monoamine stores in brain and other tissues. First serotonin (5-HT), then norepinephrine were implicated. However, after the discovery of dopamine in the brain, depletion of dopamine by reserpine in the brain has been firmly established.[18] Later it was possible to protect the stores of individual monoamines selectively against the action of reserpine.[19] For example administration of DOPA reversed the dopamine depletion caused by reserpine. These investigations led to the conclusion[20] that although serotonin contributes to the reserpine syndrome, dopamine is responsible for the most important features of the reserpine syndrome[21] (including severe depression).

Antipsychotics That Do Not Deplete Monoamine Stores

Chlorpromazine was synthesized by Charpentier in 1950 at Rhone-Poulenc Laboratories in France. Shortly thereafter, the usefulness of chlorpromazine in the treatment of psychotic patients was discovered by Delay and Deniker in France.

When the pharmacologic properties of chlorpromazine were extensively investigated, it turned out that chlorpromazine does not cause any depletion of monoamines from stores in the brain and other tissues. Other phenothiazines shared this property of chlorpromazine. When the butyrophenones (haloperidol) were discovered, it became evident that the pharmacological profile of butyrophenones was similar to that of phenothiazines and dissimilar to that of reserpine. With the exception of reserpine, none of the other antipsychotic agents caused depletion of monoamines from stores in the brain and other tissues.

On the other hand, unlike reserpine, other antipsychotic agents shared a common feature, the blockade of dopamine receptors. In 1963, Carlsson and Lindquist found that chlorpromazine and haloperidol have a specific action on catecholamine metabolism. These scientists reported that low doses of chlorpromazine and haloperidol increased the accumulation of the O-methylated basic metabolites of norepinephrine and dopamine following monoamine oxidase inhibition.

Antipsychotic agents had apparently influenced the metabolism of catecholamines (Scheme 1). It was known that increased catecholamine levels (both norepinephrine and dopamine) obtained after the administration of L-DOPA increased the level of wakefulness. On the other hand, it was also observed that administration of chlorpromazine and haloperidol reduced the awakening action of norepinephrine and dopamine. Carlsson and Lindquist postulated that chlorpromazine and haloperidol act by blocking dopamine and norepinephrine receptors. The increase in metabolites (normetanephrine and 3-*O*-methyldopamine) was the result of increased release and metabolism of the catecholamine in the brain as a compensatory mechanism to overcome the receptor blockade. Since the increased release and metabolism of catecholamines was not coupled with decreased catecholamine brain levels, it was also suggested by Carlsson and Lindquist that chlorpromazine and haloperidol stimulated catecholamine synthesis.

Based on the results obtained with chlorpromazine and haloperidol, the generalization was made that antipsychotic agents that do not deplete monamine stores (other than reserpine) block dopamine and norepinephrine receptors and due to feedback activation of the catecholamine containing neurons, such antipsychotics accelerate the synthesis and metabolism of catecholamines.

Carlsson and Lindquist's hypothesis has proven to be correct.

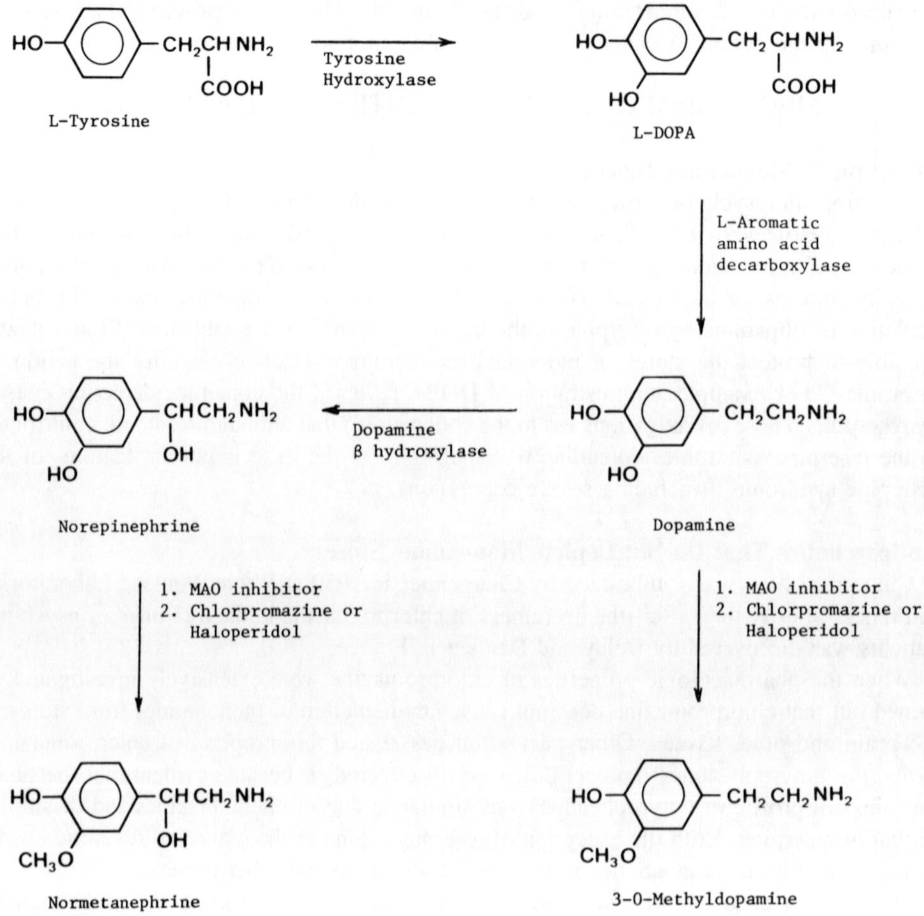

Scheme 1

Dopamine Receptor Blockade

Antipsychotic agents antagonize dopamine and dopamine agonists, e.g., apomorphine in a dose-dependent manner.[22-24] There seems to be a correlation between the antipsychotic potency and the dopamine antagonistic effect. These data were obtained in rats by administering antipsychotic agents after excessive stimulation with dopamine or dopamine agonists. Excessive stimulation of the dopamine receptors in the rats' precipitates inadequate behavior in a discrimination test. Antipsychotic agents normalize the rats behavior. When the antipsychotic agents were given in the same doses as before without first giving dopamine stimulants, the behavior of rats was also impaired. This shows that either dopamine agonists or dopamine antagonists may compete for the dopamine receptor site and produce excessive stimulation of the dopamine receptors.[25,26] The fact that psychotic patients tolerate higher doses of antipsychotic agents can be taken as an indication that psychotic patients' dopaminergic receptor activity is elevated.

Antipsychotic agents stimulate the synthesis and turnover of dopamine[27-31] and stimulate the release of dopamine.[32]

These effects can be rationalized on the basis of dopamine receptor blocking activities of antipsychotic agents. As a result of the dopamine receptor site blockade, the dopamine neurons receive a message through neuronal feedback that the receptor sites are depleted of dopamine. Accordingly, the dopamine neurons proceed to fire more rapidly, which in turn leads to further synthesis of dopamine. A correlation exists between the dopamine synthesis

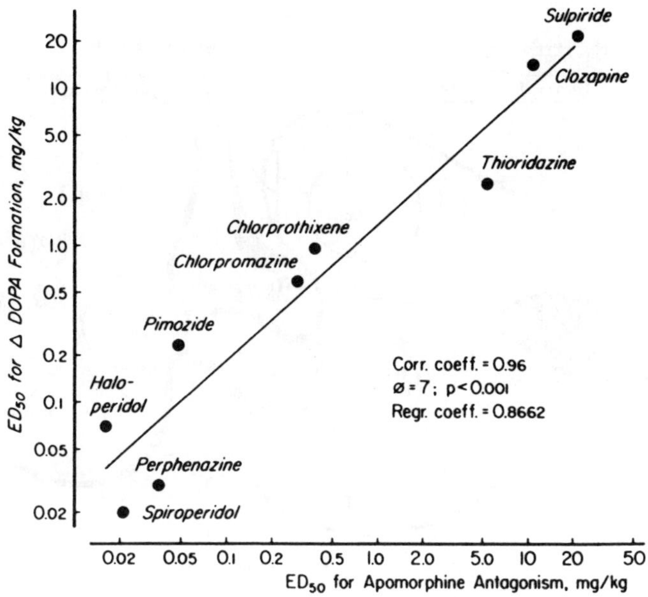

FIGURE 1. Ed_{50} for neuroleptic-induced increase in DOPA formation (after inhibition of the aromatic L-aminoacid decarboxylase) vs. ED_{50} for antagonism of apomorphine-induced sterotypies. ED_{50S} for apomorphine antagonism are the most recent data from C. Niemegeers of the Janssen Laboratories (personal communication). Apomorphine was injected, i.v., in a dose of 1.25 mg/kg, and the animals were scored, i.a., for sterotypies for the period 5 to 60 min after injection. The neuroleptics were given s.c. 30 min before apomorphine, except for pimozide, which was injected 4 hr beforehand. (From Carlsson, A., in *Psychopharmacology: A Generation of Progress,* Lipton, M. A., DiMascio, A., and Killam, K. F., Eds., Raven Press, New York, 1978, 1059. With permission.)

stimulating effect of antipsychotics in the striatum and their apomorphine antagonistic activity in the rat, which is taken as a measure of antipsychotic potency (see Figure 1). Interestingly, the ED_{50} values for the two types of activities were very similar suggesting that the receptor affinities of neuroleptics to the receptors involved are similar.

α-Adrenoceptor Blockade

Chlorpromazine, thioridazine, perphenazine, and fluphenazine are quite potent α-adrenoceptor blocking agents. They stimulate norepinephine synthesis and turnover in the brain. Several other antipsychotic drugs are, however, devoid of α-adrenoceptor activity. The correlation between antipsychotic activity and α-adrenoceptor activity is poor.

Anticholinergic Activity

Antipsychotic agents competitively antagonize muscarinic acetylcholine receptors.[34,35] Direct measurement of muscarinic acetylcholine receptors has been accomplished by measuring the binding of radioactive [^{3}H]-3-quinuclidinyl benzilate (QNB) to membrane preparations from the CNS. The binding of radioactive QNB, a potent antagonist of muscarinic cholinergic receptors, represents a specific binding to muscarinic acetylcholine receptors.[36]

It was postulated that the affinity for the muscarinic acetylcholine receptor of antipsychotic drugs in the brain correlates inversely with their tendency to elicit extrapyramidal side effects.[36]

The most powerful muscarinic antagonist drug among the antipsychotics, i.e., the drug

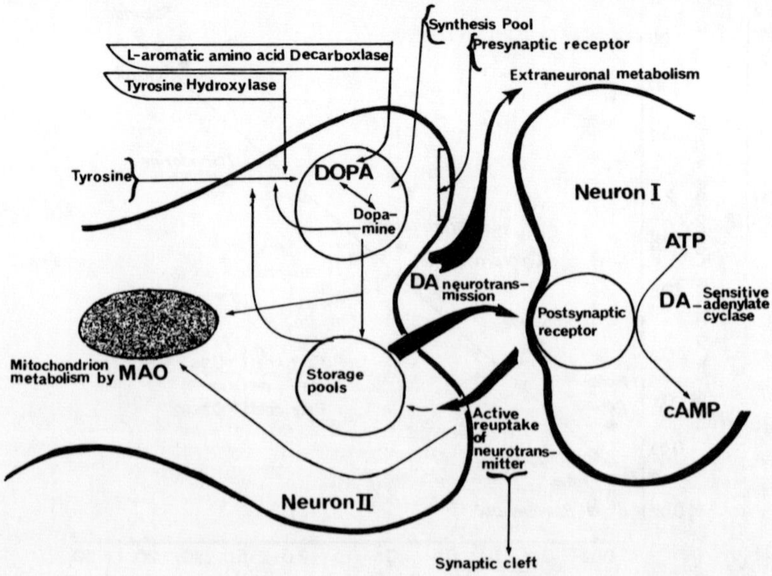

FIGURE 2. The dopaminergic synapse. Dopamine is synthesized from tyrosine in two steps and then stored in synaptic vesicles (storage pools). When an action potential depolarizes the presynaptic membrane, dopamine is released into the synaptic cleft. Dopamine reaches the postsynaptic receptor by diffusion across the synaptic cleft. Interaction of dopamine with the postsynaptic receptor produces a series of events, including changes in membrane permeability to ions and formation of cyclic AMP which, in turn activates other intracellular reactions. The neurotransmitting signal is terminated as dopamine is removed from the synaptic cleft partly by extreneuronal metabolism, partly by postsynaptic receptor uptake and partly by active reuptake into the presynaptic terminal where it is metabolized by monoamine oxidase (MAO) or restored for future use. Rapid regulation of dopamine synthesis may occur through a presynaptic feedback regulation. A buildup of dopamine (and norepinephrine) inhibits the activity of tyrosine hydroxylase, which catalyses the first step in the synthesis. On the other hand, increased dopamine neurotransmission reduces the amount of dopamine in the storage pools, removing the inhibition: tyrosine hydroxylase activity increases and more dopamine is synthesized.

with the greatest affinity for the acetylcholine receptor, clozapine, is also characterized by the least extrapyramidal side effects.

It is interesting to note that extrapyramidal side effects of antipsychotic drugs, in contrast to their antipsychotic activity, are efficiently counteracted by anticholinergic agents.

The Dopaminergic Synapse

Since the functional effects of dopamine are antagonized by antipsychotic drugs, the significance of dopaminergic transmission through the dopaminergic synapse is well understood (Figure 2).

Dopamine exerts its neurotransmitter role primarily in the caudal reticular activating and limbic systems, which are rich in dopaminergic neurons. The biosynthesis of dopamine from the amino acid tyrosine takes place in presynaptic nerve terminals in two steps.

Presynaptic structures (vesicles or granules) serve for the uptake and storage of dopamine. During nerve impulse, dopamine is released from the vesicles into the synaptic cleft and interacts with postsynaptic receptors to either facilitate or inhibit the action of the second neuron. Dopamine stimulation induces conformational changes in the dopamine receptor which produce changes in membrane permeability to ions and induce a cascade of intracellular reactions, including accumulation of cyclic AMP. The increased levels of cyclic AMP may

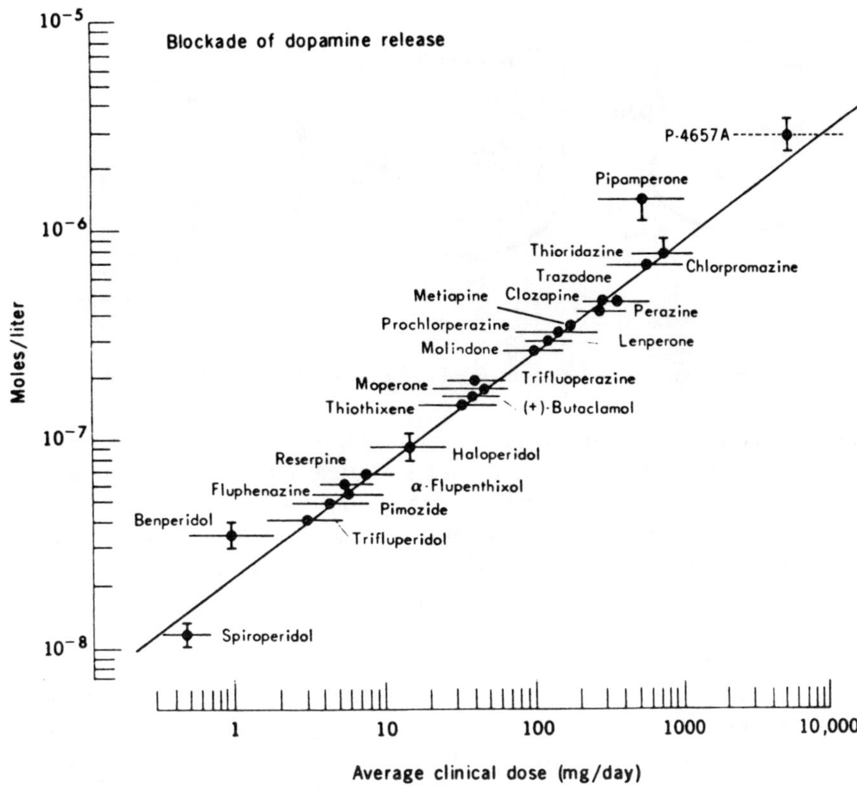

FIGURE 3. The neuroleptic IC_{50} values (the neuroleptic concentrations which inhibited the stimulated release of dopamine by 50% correlate with the average clinical doses (in milligrams per day) for controlling schizophrenia. The horizontal bars indicate the range of clinical doses. The vertical bars show the 20% variation in the IC_{50} values. The correlation includes such diverse compounds as phenothiazines, butyrophenones, reserpine, pimozide, trazodone, clozapine, and (+) - butaclamol. *Trans*-thiothixene (P-4567A) is of the order of onehundredth the potency of its cis-isomer. (From Seeman, P. and Lee, T., *Science,* 188, 1217, 1975. By courtesy of A.A.A.S., copyright 1975).

produce a series of physiological events in nonneuronal tissues. Dopamine turnover, synthesis, and metabolism in dopaminergic neurons is governed by an intraneuronal compensatory feedback mechanism that is regulated by dopamine levels at both presynaptic and postsynaptic receptors. For dopamine and several other neurotransmitters, there is a specific reuptake mechanism which transfers dopamine from the synaptic cleft back into the presynaptic cell. The transmitter can either undergo metabolic degradation or be stored in vesicles for reuse.

Presynaptic Dopamine Receptors

Antipsychotic drugs, especially butyrophenones, display very potent blocking effects on stimulation-evoked release of dopamine from presynaptic terminals.[37] A correlation between in vitro and in vivo activity of several antipsychotic drugs was established using neostriatal preparations as an in vitro presynaptic model.[38,39] It was also found that preincubation of rat neostriatal slices with tritiated dopamine produces an overflow of tritium and the stimulated-evoked overflow is reduced by subsequent administration of apomorphine.[40,41]

A direct quantitative correlation between the clinical antipsychotic activity of various antipsychotic drugs and their ability to block the presynaptic impulse-coupled release of tritiated dopamine in rat neostriatal slices was established.[42] Antipsychotic drugs block the presynaptic coupling between impulse and neurosecretion. The correlation is shown in Figure 3.[42] This observation is in direct contradiction with previously obtained results with

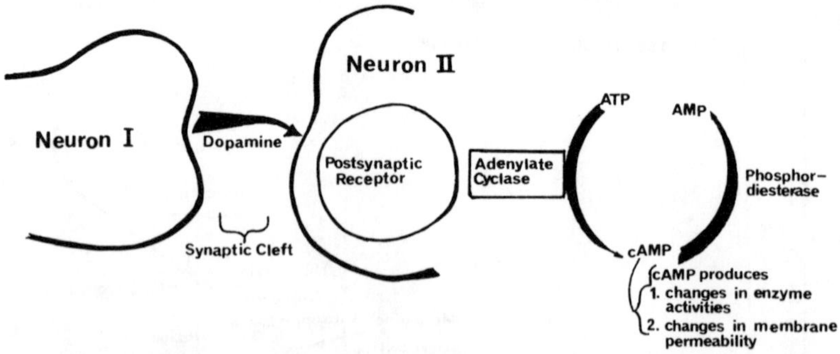

FIGURE 4. Postsynaptic events triggered by dopamine-sensitive adenylate cyclase.

pimozide and chlorpromazine, since in that experiment, both produced enhancement rather than reduction of the stimulation-induced release of tritiated-dopamine.[40] The latter result would indicate a postsynaptic localization of the ^{3}H-neuroleptic binding sites and not a presynaptic localization. Subsequently it has been suggested that ^{3}H-dopamine-agonists bind to presynaptic receptors and ^{3}H-neuroleptics bind to postsynaptic dopamine receptors that are not linked to adenylate cyclases (D2 receptors).

Postsynaptic Dopamine Receptors

A dopamine-sensitive adenylate cyclase has been found[43] in the striatum of the rat brain, an anatomic area known to be rich in dopamine terminals. The dopamine-sensitive adenylate cyclase, which appears to be postsynaptically located, displays agonist specificity that is characteristic for the dopamine receptor. The dopamine-sensitive adenylate cyclase converts ATP to cyclic AMP, which in turn initiates a chain of events that is characteristic for the stimulation of dopamine receptor. Cyclic AMP is inactivated by phosphodiesterases; these enzymes in turn can be inhibited by the methylxanthines (caffeine, theophylline, and aminophylline) (Figure 4).

There is a good correlation between the pharmacologic potencies of phenothiazines and their inhibitory effects on the dopamine-sensitive adenylate cyclase. There are, however, major discrepancies in the butyrophenone series.[44] Spiroperidol and pimozide, two butyrophenones that are weaker than chlorpromazine in inhibiting the dopamine-sensitive adenylate cyclase[44] are considerably more potent in vivo than chlorpromazine based on behavioral and clinical data.[45] This discrepancy has been used to question the validity of the hypothesis that antipsychotic drugs produce their therapeutic effects by a blockade of the postsynaptic dopamine receptor.[46] As previously described, it has been proposed that the antipsychotic drugs might act instead presynaptically by inhibiting dopamine release.[42]

Dopamine receptor binding has been demonstrated in brain tissues by labeling the dopamine receptor with both the agonists ^{3}H-dopamine and ^{3}H-apomorphine[47] and the antagonist ^{3}H-haloperidol.[48] The regional distribution of receptor binding is associated with postsynaptic receptors.

The best evidence for a common mechanism of action for antipsychotic drugs is provided by the outstanding correlations between inhibition of ^{3}H-haloperidol binding that is known to occur by blockade of postsynaptic dopamine receptors and in vivo antipsychotic assays. These in vivo antipsychotic assays include: (1) apomorphine induced stereotyped behavior, (2) amphetamine induced stereotyped behavior, and (3) apomorphine induced emesis.

The blockade of apomorphine-induced stereotyped behavior in rats is a measure of the antipsychotic potency. It is believed to originate from the blockade of dopamine receptors by the antipsychotic drugs and inhibition of the dopamine receptor stimulating effect of apomorphine.[45] It was found that the affinities of antipsychotic drugs for ^{3}H-haloperidol

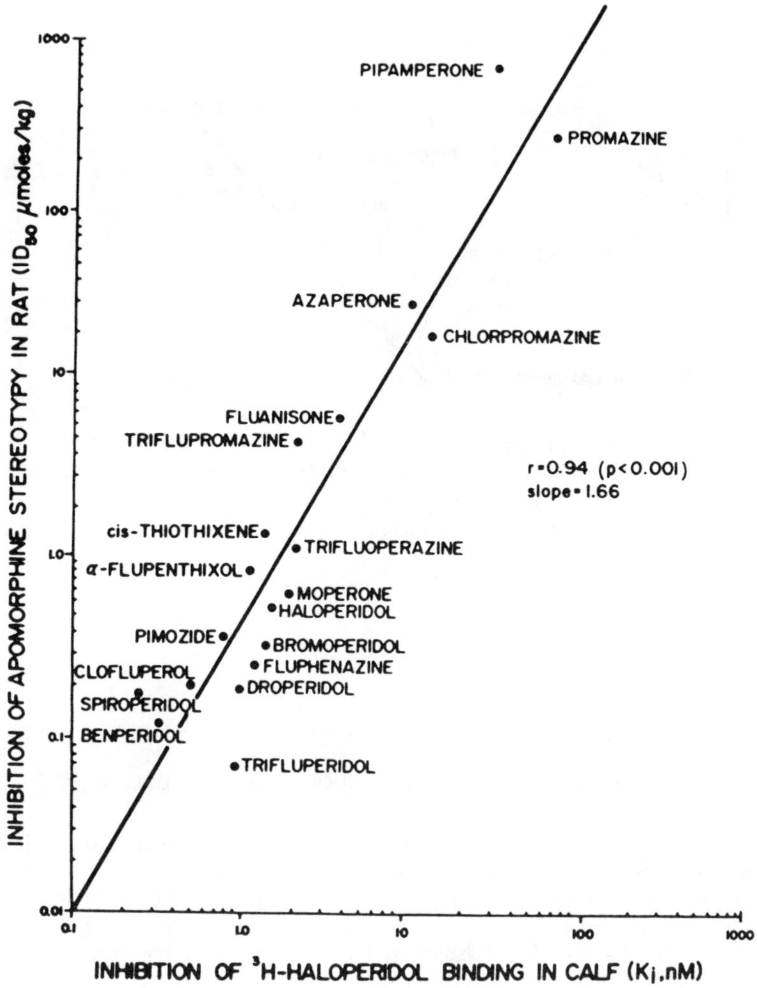

FIGURE 5. Antischizophrenic drugs: correlation between affinities for ^{3}H-haloperidol binding sites and antagonism of apomorphine stereotypy in the rat. The correlation coefficient $r = 0.94$, is significant at the $p < 0.001$ level. (From Creese, I. and Snyder, S. H., in *Psychopharmacology: A Generation of Progress,* Lipton, M. A., DiMascio, A., and Killam, K. F., Eds., Raven Press, New York, 1978, 377. With permission.

binding sites correlate very well with their potency in antagonizing the apomorphine induced stereotypy[46] (Figure 5). The animal data were taken from Reference 45.

It has been found that clinical potency in man of antipsychotic drugs as determined by their average oral effective dose correlates markedly with competition for ^{3}H-haloperidol binding[49,50] (Figure 6).

These data support the theory that antipsychotic agents stimulate the firing and the transmitter release, metabolism, and synthesis of dopamine in dopaminergic neurons.

Multiple Dopamine Receptors

The first subclassification of dopamine receptors into D-1 and D-2 types was provided on the basis of their effects on cyclic AMP synthesis.[51] It was postulated that D-1 dopamine receptors are involved in the dopamine-stimulated adenylate cyclase enzyme activity while D-2 dopamine receptors are not.[52] The classification was later revised for the D-2 receptors.[53] Accordingly the D-2 receptors display reduced response in synthesizing cyclic AMP as a

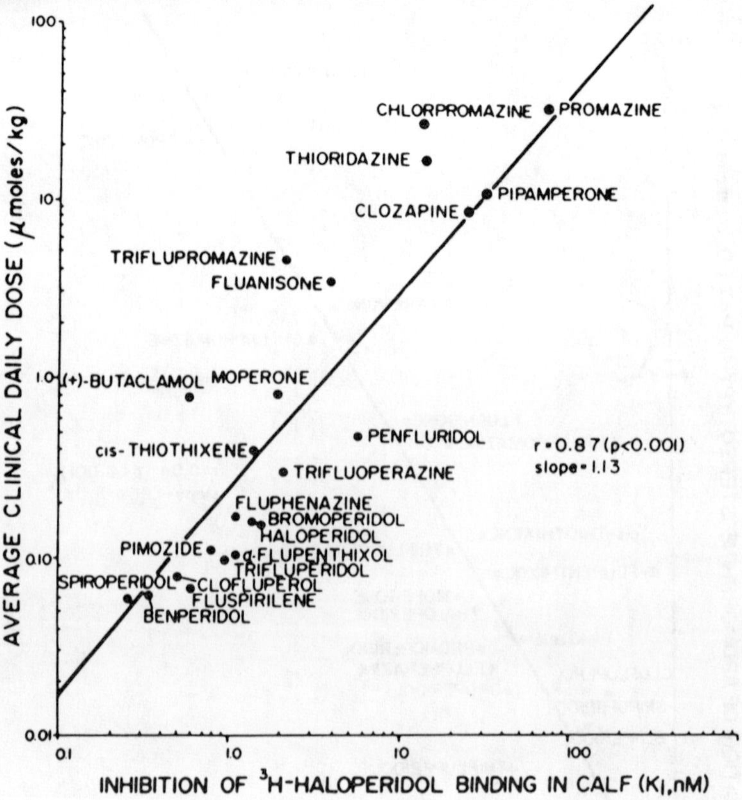

FIGURE 6. Antischizophrenic drugs: correlation between affinities for ^{3}H-haloperidol binding and clinical potencies. Clinical data were derived from published results (reviewed in Reference 49). The correlation coefficient, $r = 0.87$, is significant at the $p < 0.001$ level. (From Creese, I. and Snyder, S. H., in *Psychopharmacology: A Generation of Progress*, Lipton M. A., DiMascio, A., and Killam, K. F., Eds., Raven Press, New York, 1978, 377. With permission.)

result of β-adrenergic agonist stimulation. It was found that dopaminergic ergots (bromocryptine and lisuride) and apomorphine are powerful D-2 receptor agonists, while sulpiride, molindone, and the antiemetic drug metoclopromide are selective D-2 receptor antagonists. Selective antagonists for D-1 receptors have not been found.

Although antipsychotic drugs are potent antagonists of dopamine receptors in many tissues, these drugs are of only limited use in discriminating between the different types of dopamine receptors.[52] Butyrophenones, for example, are weak D-1 antagonists but are also D-2 antagonists, although less potent than would be expected from their potency as antipsychotic drugs[54] or antagonists of the dopamine receptor in the anterior pituitary.[55] Other antipsychotic compounds do not show differential dopamine receptor antagonistic properties between the anterior pituitary and striatal adenyl cyclase (D-1) receptor sites.

The criteria for the classification of dopamine receptors are shown in Table 1.

Clinical Relevance of Dopamine Receptor Classification

Schizophrenia

Drugs that stimulate D-2 receptors can produce delusions and hallucinations, while D-2 receptor blockers help in reducing schizophrenia. It was suggested that in schizophrenia, D-2 activation predominates. The dopamine brain levels in schizophrenics are normal and the site of the dopamine receptors has not been identified.

Table 1
CRITERIA FOR THE CLASSIFICATION OF DOPAMINE RECEPTORS

Name	D-1	D-2
Cyclase linkage	Yes	No
Location of prototype receptor	Bovine parathyroid	Mammotroph of anterior pituitary
Dopamine	Agonist (μmolar potency)	Agonist (nmolar potency)
Apomorphine	Partial agonist or antagonist	Agonist (nmolar potency)
Dopaminergic ergots	Potent antagonist (nmolar potency) Weak agnoist (μmolar potency)	Agonist (nmolar potency)
Selective antagonist	None known as yet	Metoclopramide sulpiride
Radiolabeled ligand	*cis*-flupenthixol[a]	Dihydroergocryptine

[a] Radiolabeled *cis*-flupenthixol can be used as a ligand specific for the dopamine receptor linked to adenylyl cyclase in the rat striatum.[56] Its affinity for the dopamine receptor in the anterior pituitary has not been measured.

Kebabian, J. W. and Calne, D. B., *Nature (London)* 277, 93, 1979. With permission.)

Dyskinesia (Choreoathetosis)

The dopamine brain concentrations are normal in choreoathetoid dyskinesia. Nonselective D-1 and D-2 agonists induce dyskinesia in Parkinsonism patients. Conversely, nonselective dopamine blocking drugs (phenothiazines, butyrophenones, etc.) alleviate dyskinesia. It is not known whether D-1 or D-2 agonists produce less dyskinesia than levodopa.[57] Prolonged therapy with antipsychotic drugs, which are dopaminergic blocking agents, often produces tardive dyskinesia. Paradoxically, the diskinesia is exacerbated when the dose of drug is reduced. There is no reasonable explanation for this phenomenon.

Hyperprolactinemia

Dopamine blocking drugs (butyrophenones, phenothiazines, etc.) induce hyperprolactinemia. Prolactin is believed to be secreted by the anterior pituitary. As a result of D-2 receptor activation the release of prolactin is inhibited. D-2 receptor blockade by antipsychotic drugs, on the other hand, stimulates prolactin release. The hyperprolactinemic syndrome can be treated with selective D-2 agonists, especially bromocriptine.

STRUCTURAL FEATURES OF ANTIPSYCHOTIC AGENTS

General Structural Features

In spite of the considerable diversity of the chemical structures of potent antipsychotic drugs, Janssen postulated[58] that all antipsychotic drugs contain a fragment, shown:

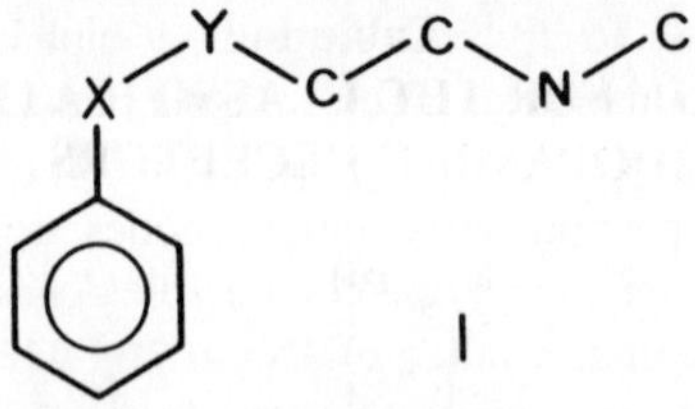

I

where X = C, N, O, or S and Y = C, N, or O.

In fragment (I) a benzene or an isosteric aromatic ring is connected through a four-membered chain to a basic amino group. X and Y may be but are not necessarily carbon atoms. Also, X and Y cannot be heteroatoms at the same time and X − Y may become a double bond only when X and Y are identical.

The tricyclic antipsychotics were represented by Janssen with the general formula II, fragment I being part of the general formula II.

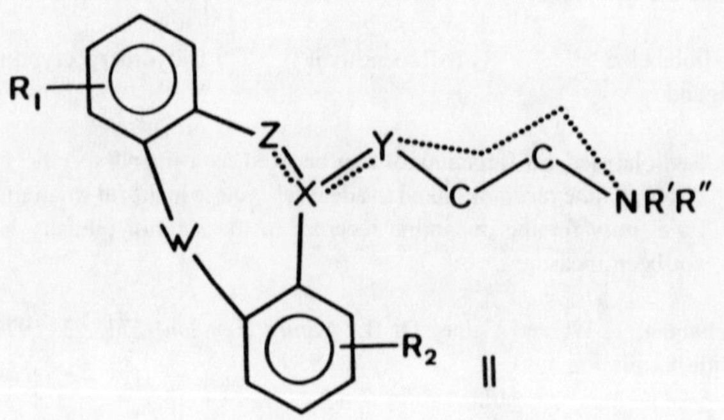

II

where W = C, N, O, or S; Z = C, N, O, or S; R′ and R″ = alkyl or part of a ring; R_1 and R_2 are usually electron-withdrawing groups; X and Y = as defined above.

Formula II represents either a six-membered (Z is missing) or seven-membered central ring to which two aromatic external rings are attached (mostly benzene, thiophene, or pyridine). In order to elicit antipsychotic activity W must always be present. Similarly X must always be present. Consequently the central ring must be at least six membered.

If the central ring is seven membered, a double bond may be present between X and Z. The central ring usually consists of the following chemical classes: benzene, pyran, thiopyran, pyridine, 1,4-oxazine, 1,4-thiazine, cycloheptene, oxepine, thiepine, azepine, 1,4-oxazepine, 1,4-thiazepine, and 1,4-diazepine.

The tricyclic ring structure constitutes a lipophylic part while the hydrophylic character in the antipsychotic molecule is conferred by the presence of the basic aminoalkyl side chain (see fragment I). The hydrophylic residue is always attached to the central ring.

PHARMACOKINETICS AND METABOLISM

Most antipsychotics are quite lipophilic compounds which are rapidly absorbed and distributed into body tissues. Particularly high levels of antipsychotics are found in the brain. Highest plasma levels of tricyclic antipsychotics are obtained by i.v. administration, closely followed by i.m. administration, while oral administration gives considerably lower blood levels. The difference is due to differences in the rate of metabolic conversions of tricyclic drugs to inactive metabolites.

After the initial distribution of drug into body tissues, plasma levels of phenothiazines fall rapidly. This rapid fall with a plasma half-life of 2 hr is followed by a slow disappearance rate with a plasma half-life of 16 hr.[59] As a result, a biphasic curve is obtained for the disappearance rates of phenothiazines. In the plasma, most of phenothiazine is protein bound.[60]

The exact plasma levels of phenothiazine antipsychotics are variable and fall into a wide range (5 to 500 μg/mℓ) in patients receiving comparable doses. In order to achieve clinical improvement with chlorpromazine, a range of 150 to 300 μg/mℓ plasma level is required. Variations in the plasma levels are caused by many factors, including rate of metabolism. Approximately 60 to 70% of the chlorpromazine dose is rapidly removed from portal circulation by the liver. Chlorpromazine is a very susceptible substrate for liver microsomal enzymes. As a result, a large number of metabolites are formed.

The general metabolic pathways for tricyclic antipsychotic drugs are shown in Table 2.[61] The metabolic pathways are heavily influenced by such factors as species, age, sex, and interaction with other drugs.

LONG-ACTING INJECTABLE ANTIPSYCHOTIC DRUGS

Long-acting injectable antipsychotic drugs or depot antipsychotics have been introduced into clinical therapy to enhance patient compliance and to avoid possible problems associated with absorption and metabolism. The currently available long-acting antipsychotics are shown in Table 3.[62-67] The structures are shown below:

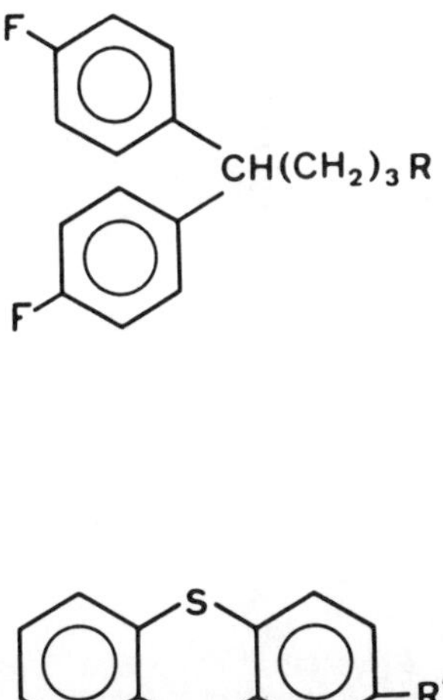

Penfluridol R =

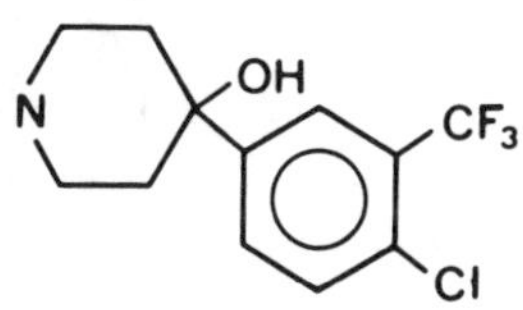

Fluspirilene R =

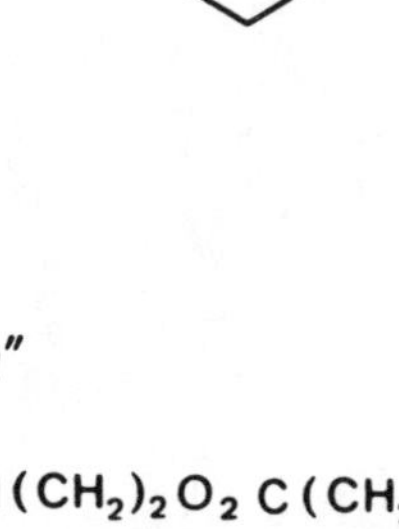

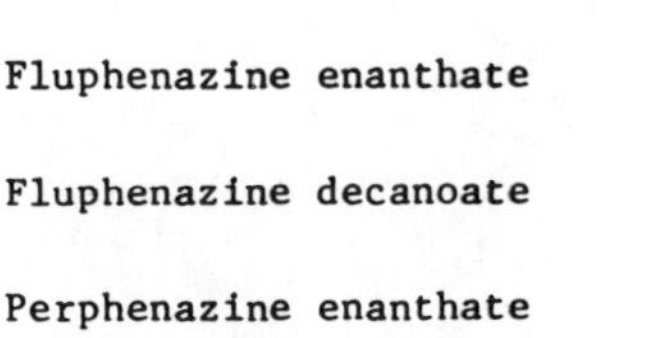

	R′	R″
Fluphenazine enanthate	CF_3	$N(CH_2)_2O_2C(CH_2)_5CH_3$
Fluphenazine decanoate	CF_3	$N(CH_2)_2O_2C(CH_2)_8CH_3$
Perphenazine enanthate	Cl	$N(CH_2)_2O_2C(CH_2)_5CH_3$
Pipotiazine undecylenate	$SO_2N(CH_3)_2$	$CH(CH_2)_2O_2C(CH_2)_8CH{=}CH_2$
Pipotiazine palmitate	$SO_2N(CH_3)_2$	$CH(CH_2)_2O_2C(CH_2)_{14}CH_3$

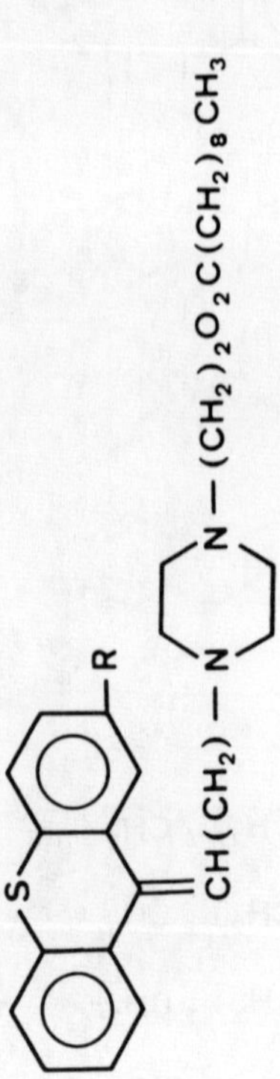

Flupentixol decanoate $R = CF_3$

Clopenthixol decanoate $R = Cl$

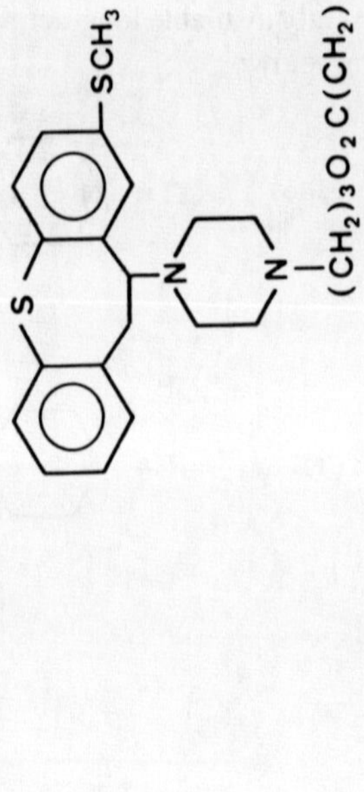

Oxyprothepin decanoate

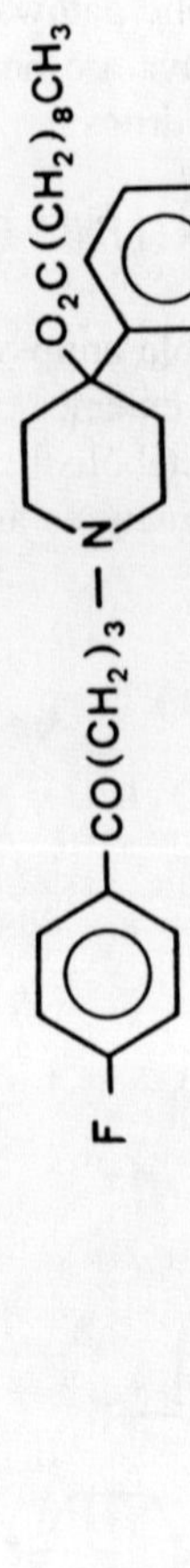

Haloperidol decanoate

Table 2
MAJOR METABOLIC PATHWAYS FOR ANTIPSYCHOTIC DRUGS

Pathway	Example
Oxidation	
Dealkylaton	
N-dealkylation	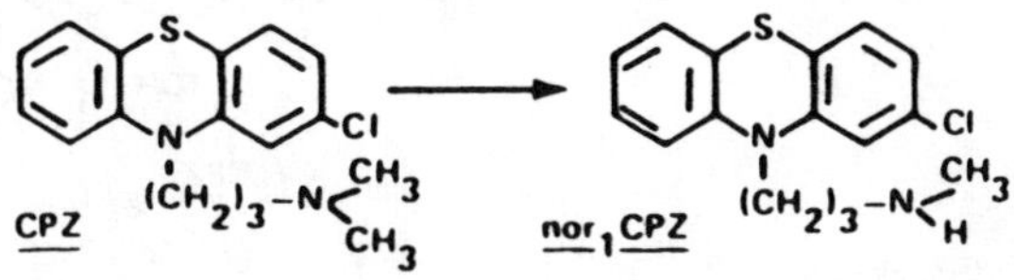
O-dealkylation	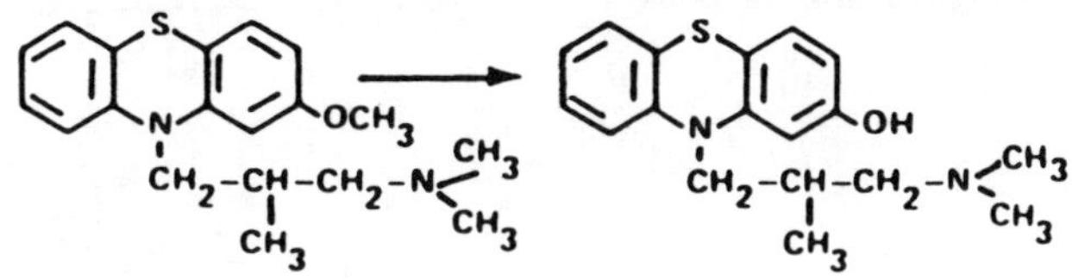 Methotrimeprazine
S-dealkylation	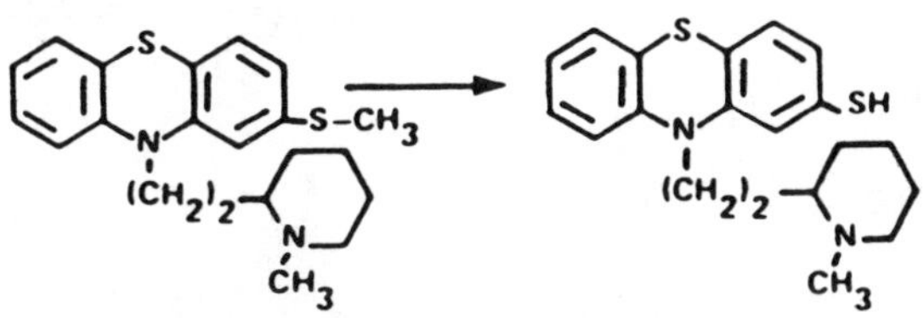 Thioridazine
Hydroxylation	
Aromatic hydroxylation	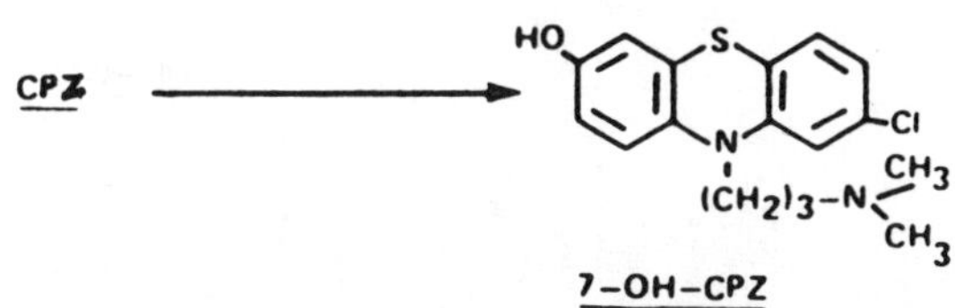
N-oxidation and N-hydroxylation	
N-oxidation	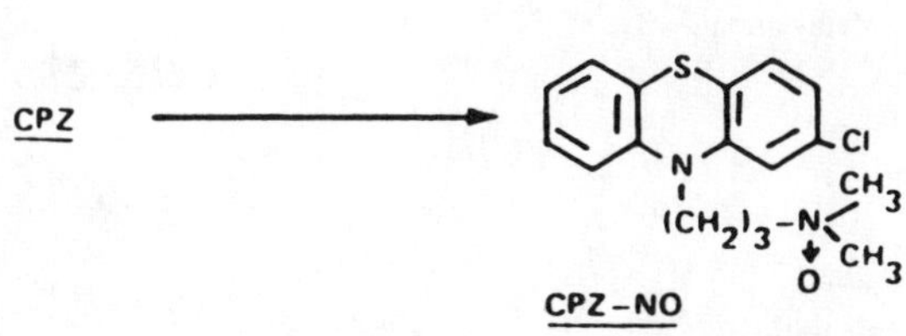
N-hydroxylation	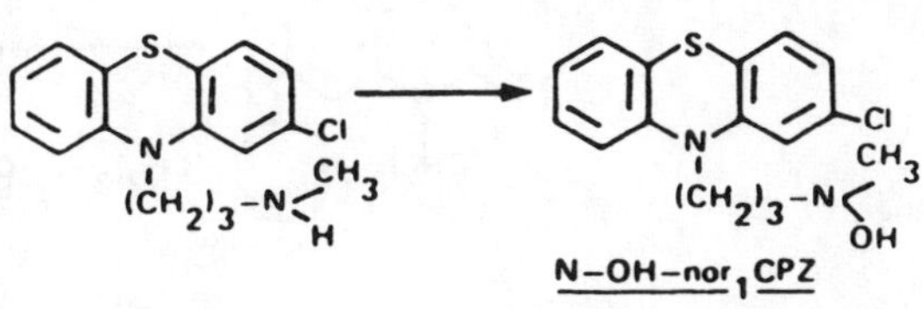
Sulfoxidation	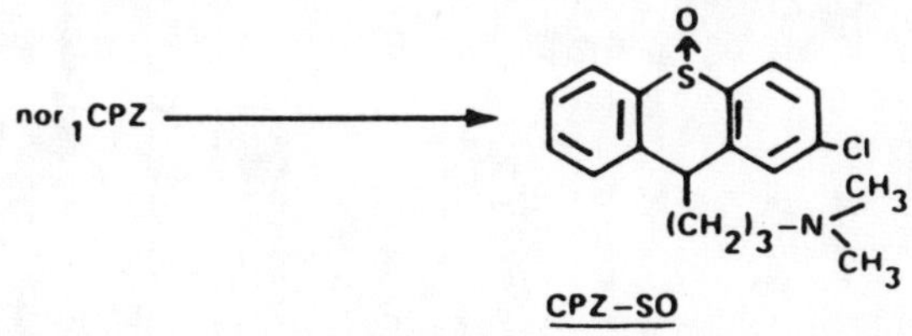

Table 2 (continued)
MAJOR METABOLIC PATHWAYS FOR ANTIPSYCHOTIC DRUGS

Pathway	Example
Amine deamination	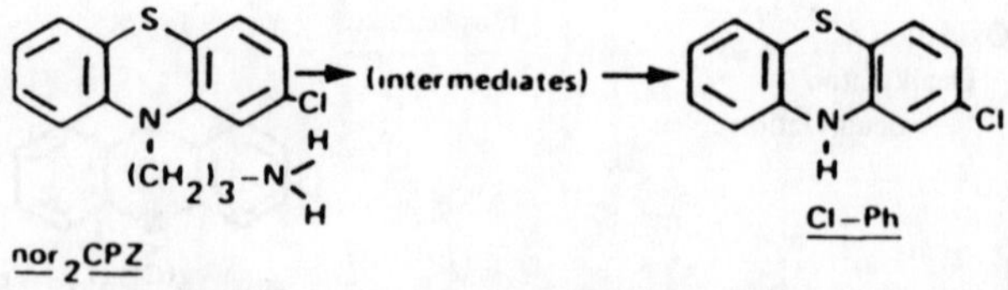
Conjugation *O*-glucuronide formation	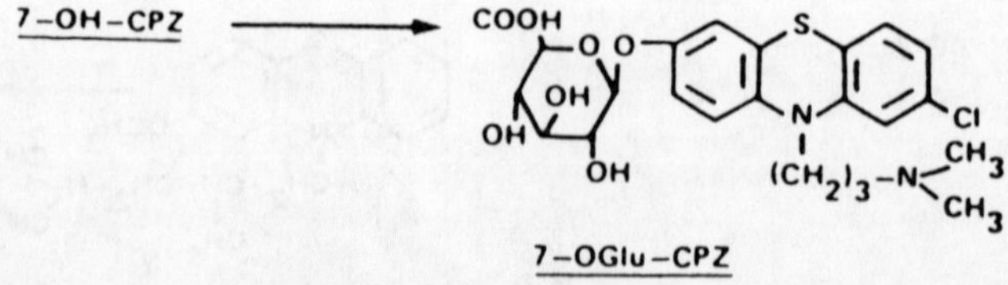
N-glucuronide formation	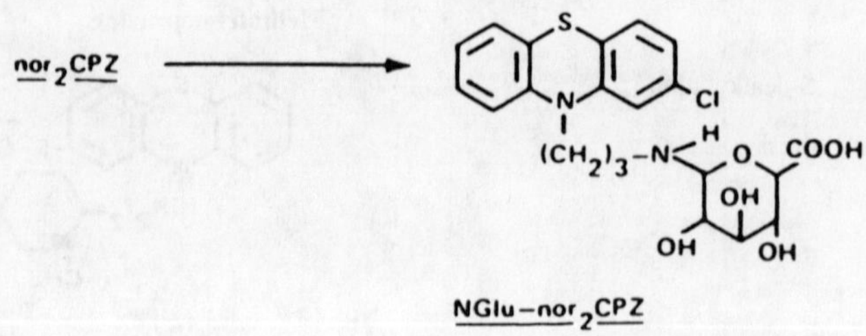
Sulfate conjugation	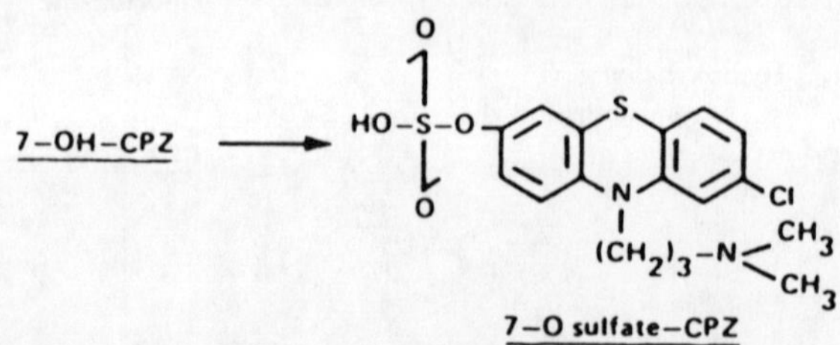
Methylation	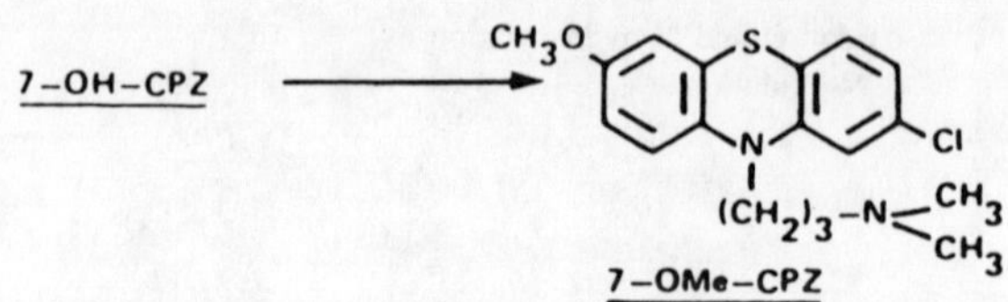
Hydrolysis	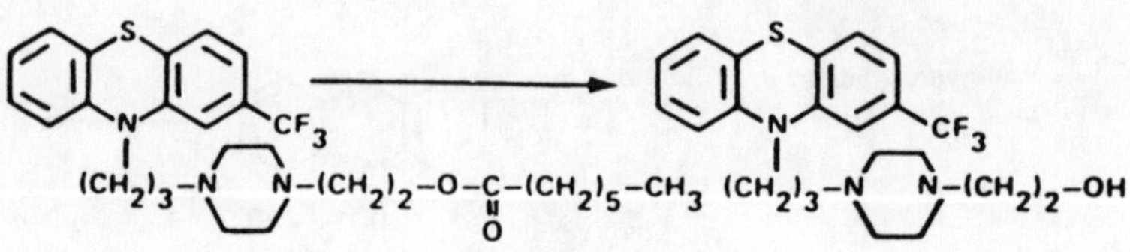

Table 2 (continued)
MAJOR METABOLIC PATHWAYS FOR ANTIPSYCHOTIC DRUGS

Pathway	Example	
	Fluphenazine Enanthate	Fluphenazine
Acetylation		

Adapted from Usdin, E., in *Psychopharmacology: A Generation of Progress,* Lipton, M. A., Dimascio, A., and Killam, K. F., Eds., Raven Press, New York, 1978, 896. With permission.

Table 3
LONG-ACTING NEUROLEPTICS

Generic name	Route of administration	Mean duration of action (week)	Dosage range (mg)
Penfluridol	p.o.	1	20—120
Fluspirilene	i.m.	1	2—12
Fluphenazine enanthate	i.m.	1—2	25—100
Perphenazine enanthate	i.m.	1—2	50—200
Flupentixol decanoate	i.m.	1—2	20—60
Pipotiazine undecylenate	i.m.	1—2	100—450
Fluphenazine decanoate	i.m.	2—3	25—100
Pipotiazine palmitate	i.m.	4	50—600
Clopenthixol decanoate	i.m.	2—3	200—400
Oxyprothepin decanoate	i.m.	NA	NA
Haloperidol decanoate	i.m.	4	50—150

NA, not available.

It has been claimed that due to the high oil-water partition ratios, the ester prodrugs injected i.m. or s.c. in oily solution are slowly released from the site of injection into the systemic circulation. The esters are then rapidly cleaved by esterases to generate the parent classical antipsychotic drug.

The main advantages of long-acting injectable antipsychotics are improved patient compliance, reduced period of hospitalization, and lower relapse rate (due to steady plasma levels). The difficulties are in determining the interval between injections and the choice of the therapeutic dose.

COMMON ADVERSE EFFECTS OF ANTIPSYCHOTIC DRUGS

Sedative Effects

Chlorpromazine, chlorprothixene, and thioridazine are the drugs most often producing sedation. Tolerance gradually develops to the use of these drugs and drowsiness is not usually a long-lasting problem if moderate doses of the drug are used. The piperazinyl phenothiazines, such as trifluoperazine, fluphenazine, and the piperazinyl thioxanthene, thiothixene, produce very little sedation. Haloperidol in usual doses produces mild sedation, but in higher doses the sedative effects are strong. Loxapine has a moderate sedative effect while molindone is practically devoid of sedative effects.

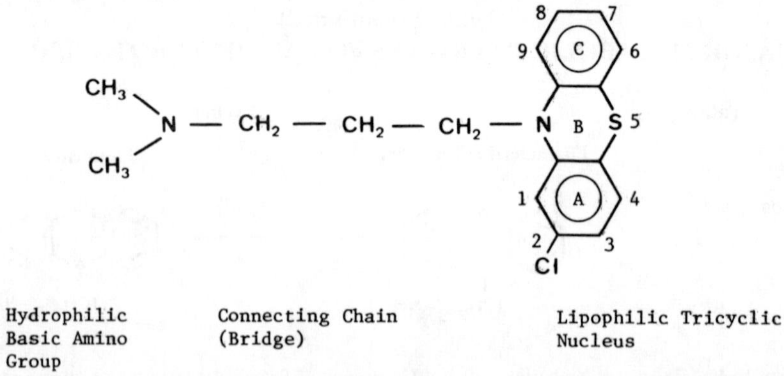

FIGURE 7. Structural features of chlorpromazine.

Hypotensive Effects

The hypotensive effects of antipsychotic drugs are due to their α-adrenergic blocking effects, that produce peripheral vasodilation, decrease in peripheral resistance, and ensuing fall of blood pressure. Patients may develop postural hypotension and may faint. Chlorpromazine, chlorprothixene, and thioridazine are potent α-adrenergic blocking agents and are most likely to produce hypotension of clinical significance. The piperazine phenothiazines such as trifluoperazine and fluphenazine and the piperazine thioxanthene, thiothixene, produce much less hypotension. Haloperidol is essentially devoid of cardiovascular effects. Loxapine and molindone have a moderate hypotensive potential.

Anticholinergic Effects

Of the antipsychotic drugs approved for sale in the U.S., thioridazine is the most potent anticholinergic. Other phenothiazines, except the ones containing a piperazine structure, are also potent anticholinergics. The common effects are blurred vision, dry mouth, constipation, urinary retention, decreased perspiration, and increased heart rate. Loxapine and molindone have moderate anticholinergic effects, while haloperidol rarely causes anticholinergic effects.

Extrapyramidal Effects

The antipsychotic drugs with the greater anticholinergic action tend to produce fewer extrapyramidal effects. Clozapine is practically devoid of causing extrapyramidal syndrome (EPS). Thioridazine produces very little EPS, chlorpromazine and chlorprothixene have moderate EPS effects, and the more potent antipsychotic drugs, trifluoperazine, fluphenazine, thiothixene, and haloperidol are more likely to produce EPS. Loxapine and molindone have a moderate EPS producing capacity. The EPS produced by antipsychotics are sometimes controlled by administration of antiparkinsonism agents or anticholinergics.

DRUG MONOGRAPHS

Phenothiazines

SARs

SAR studies reveal that biologic activity is most heavily influenced by structural modifications in three regions (Figure 7):

1. The lipophilic tricyclic nucleus
2. The connecting chain (bridge) separating the tricyclic nucleus from the basic amino group
3. The hydrophilic basic amino group

The Lipophilic Tricyclic Nucleus

The antipsychotic activity of phenothiazines is influenced by the nature and position of the ring substituents. The antipsychotic activity is increased by the introduction of electronegative substituents in the 2-position in approximately the following order:

$$OH < H \approx CONHNH_2 \approx OCH_3 < C(CH_3)_3 < CH_3 \approx CH(CH_3)_2 < COCH_3 \approx$$
$$COOCH_3 \approx SO_2N(CH_3)_2 \approx SCH_3 \approx Br \approx Cl \approx SCF_3 < SO_2CF_3 \approx CF_3$$

Substitution at position 1, 3, or 4 or simultaneous substitution in both aromatic rings usually produces a marked loss of antipsychotic activity.

The Connecting Chain (Bridge)

A three-carbon chain is optimal for connecting the two nitrogen functions. If the connecting chain consists of two carbons, antipsychotic activity is decreased.

Introduction of methyl groups into position 1 of the propylene bridge decreases antipsychotic activity and may confer antidepressant activity. If an oxygen is introduced into the 1-position of the propylene bridge (conversion of an aminoalkyl group to an aminoacyl group), the resulting 3-amino propionyl-phenothiazines have mostly antidepressant activities.

Introduction of a methyl group into the 2-position of the propylene bridge has an inconsistent influence on antipsychotic potency. When groups larger than methyl are introduced into the 2-position of the propylene bridge, antipsychotic activity is markedly decreased.

Basic Amino Group

A tertiary amino group is necessary for eliciting maximal antipsychotic potency. The primary amine didemethyl chlorpromazine and the secondary amine demethylchlorpromazine are much less potent than the tertiary amine chlorpromazine.

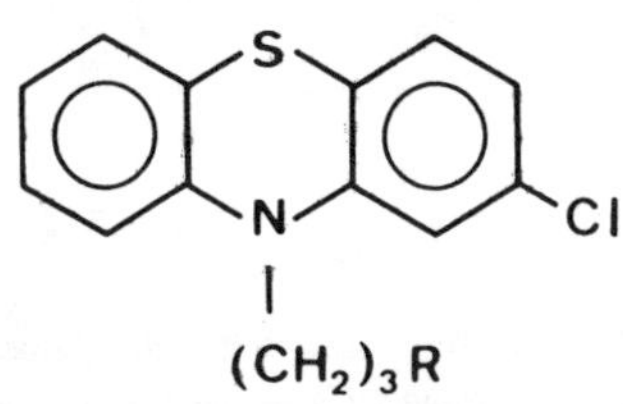

Didemethyl chlorpromazine — $R = NH_2$

Demethyl chlorpromazine — $R = NHCH_3$

Chlorpromazine — $R = N(CH_3)_2$

Antipsychotic potency is usually decreased as the size of the *N*-alkyl substituent is increased. The size of the terminal piperazine nitrogen substituent or the substituent in position 4 of the piperidine ring can be varied considerably without a loss in antipsychotic activity in the piperazine series.

Stereochemical and Conformational Features of Phenothiazines

Conformational analysis revealed that, in their respective preferred conformations, chlorpromazine and dopamine are superimposable.[68] The side chain on the nitrogen is tilted towards the ring that is substituted in the 2-position of ring A. This is a general finding for phenothiazines. If the phenothiazines lacks a substituent in the 2-position of one of the rings, the side chain extends fully away from the tricyclic structure and mimics the conformation

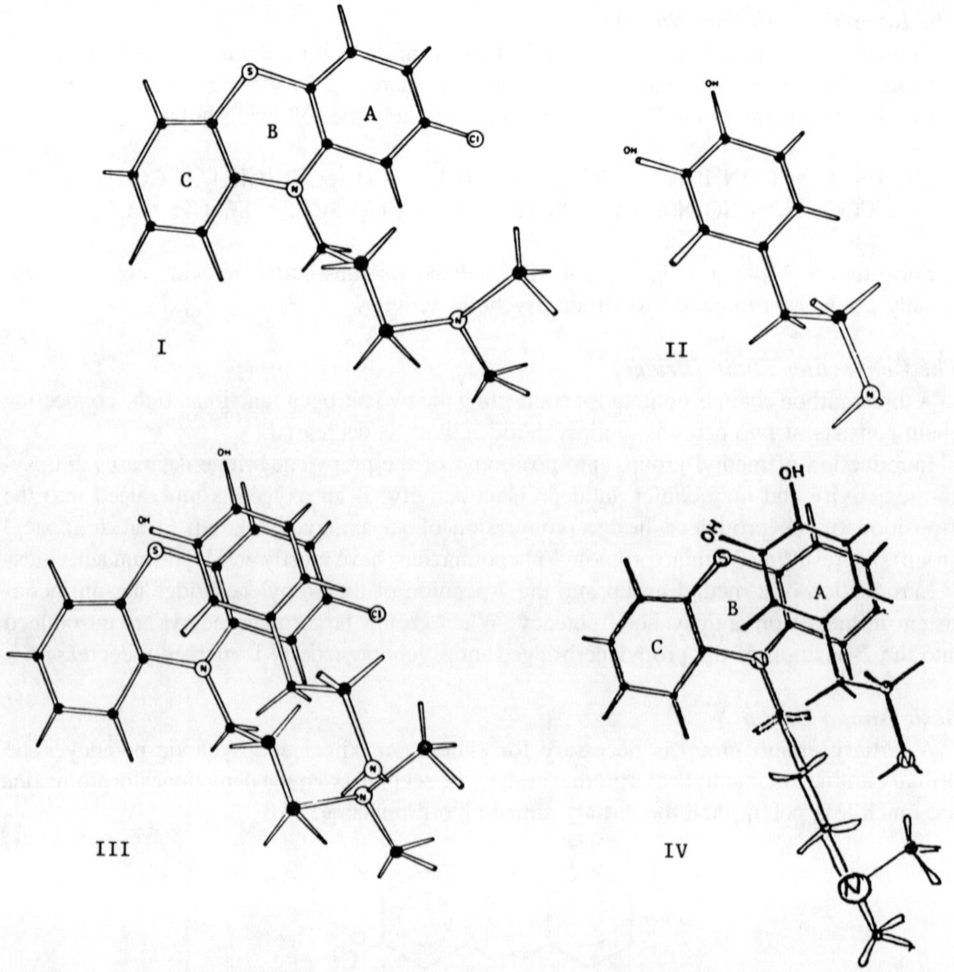

FIGURE 8. Conformation of chlorpromazine (I), dopamine (II), their superimposition (III) determined by X-ray crystallographic analysis. The A, B, and C in I and IV designate rings. Another conformation is pictured by IV in which the dimethyl aminoalkyl side chain is fully extended owing to the lack of substituent in ring A. The degree of overlap with dopamine in IV is much less than in III. (Adapted from Horn, A. S. and Snyder, S. H., *Proc. Natl. Acad. Sci. U.S.A.*, 68, 2835, 1971. With permission.)

of dopamine to a lesser degree (Figure 8). Mepazine and promazine, two phenothiazines lacking a ring substituent display less antipsychotic activity than those having a substituent in the 2-position of ring A.

Nonplanarity of the tricyclic nucleus is essential for antipsychotic activity. The two flat benzene rings are at a slight right angle along the imaginary N-S axis (see Figure 9). Planar compounds such as acridines and anthracenes are inactive.

Classification

Phenothiazines are conveniently classified into three groups based on the nature of the basic amino group.

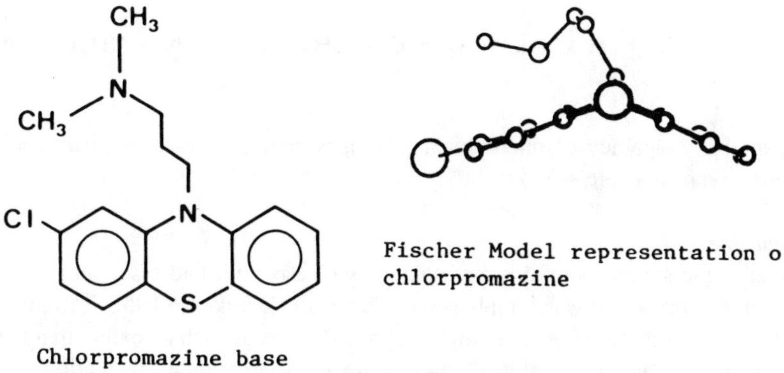

FIGURE 9. (Adapted from Pauling, P., Ramsay, W., and Ingold, C., in *Psychotherapeutic Drugs*, Part 1, Usdin, E. and Forrest, I.S., Eds., Marcel Dekker, New York, 1976, 59. With permission.)

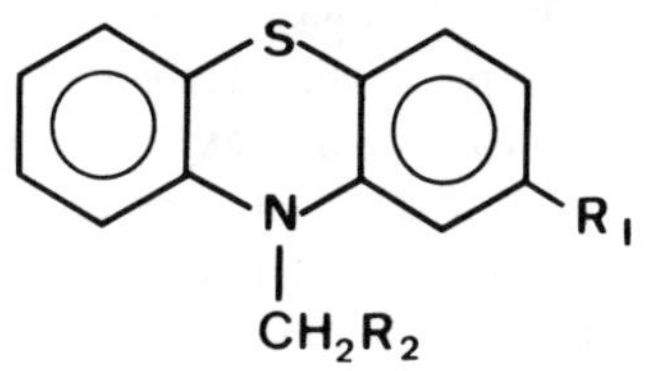

Aminoalkyl derivatives

Promazine	R_1 = H	R_2 = $CH_2CH_2N(CH_3)_2$
Chlorpromazine	R_1 = Cl	R_2 = $CH_2CH_2N(CH_3)_2$
Methopromazine	R_1 = OCH_3	R_2 = $CH_2CH_2N(CH_3)_2$
Acetylpromazine	R_1 = $COCH_3$	R_2 = $CH_2CH_2N(CH_3)_2$
Triflupromazine	R_1 = CF_3	R_2 = $CH_2CH_2N(CH_3)_2$
Trimeprazine	R_1 = H	R_2 = $CH(CH_3)CH_2N(CH_3)_2$
Levomepromazine	R_1 = OCH_3	R_2 = $CH(CH_3)CH_2N(CH_3)_2$

Piperidylalkyl derivatives

Pipamazine R_1 = Cl

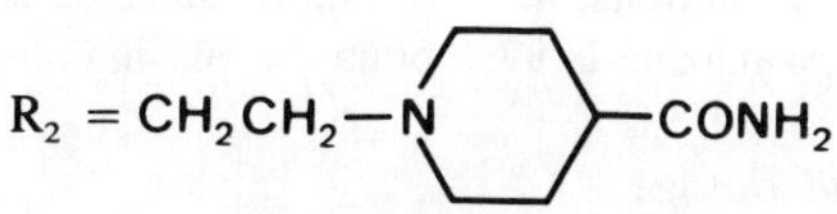

Thioridazine R_1 = SCH_3

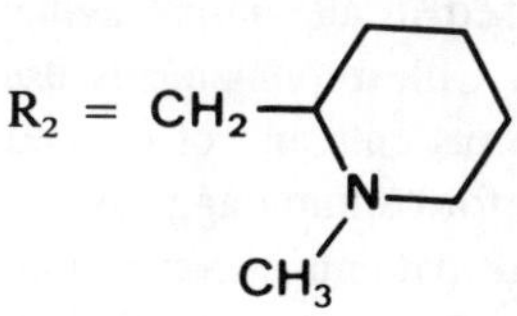

Piperazinylalkyl derivatives

Prochlorperazine	R_1 = Cl	R_2 =
Trifluoperazine	R_1 = CF_3	R_2 =
Thioproperazine	R_1 = $SO_2N(CH_3)_2$	R_2 =
Butaperazine	R_1 = $CO(CH_2)_2CH_3$	R_2 =

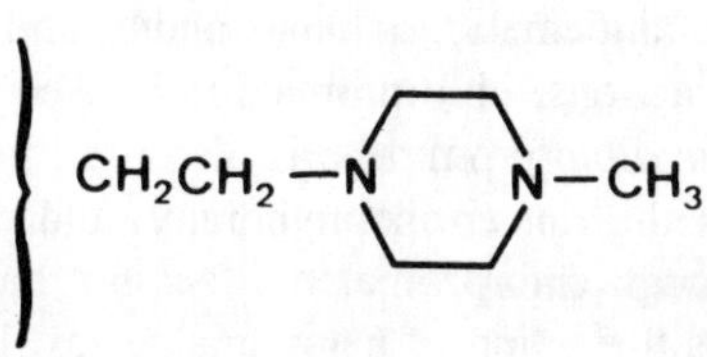

Perphenazine	R_1 = Cl	R_2 =
Fluphenazine	R_1 = CF_3	R_2 =
Acetophenazine	R_1 = $COCH_3$	R_2 =

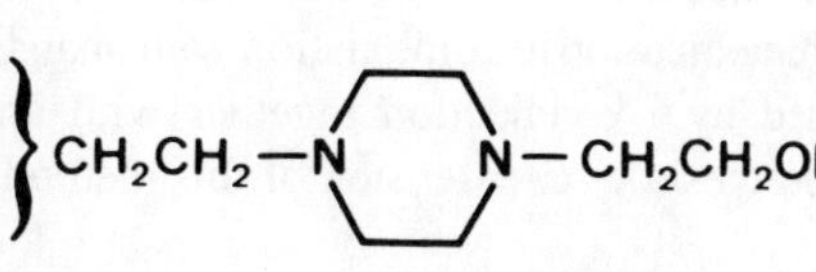

Dixyrazine R_1 = H R_2 =

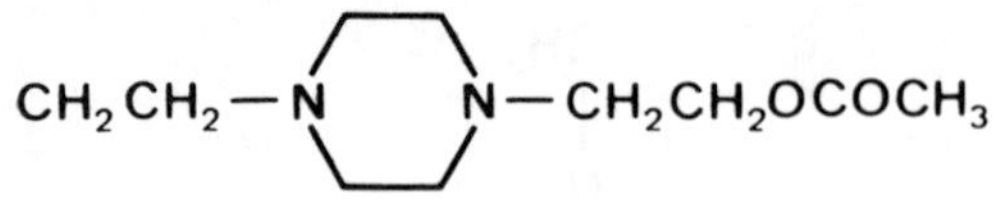

The comparative potencies of phenothiazine antipsychotics in various pharmacologic test methods are given in Table 4.

Chlorpromazine

Shortly after the synthesis of chlorpromazine by Charpentier and associates,[72] the clinical usefulness of the compound was established by Delay and Deniker in 1952.[73] Chlorpromazine was able to calm psychotic patients in highly agitated states to such an extent that the patients became cooperative and were able to participate in occupational, recreational, and psychotherapies.

Sections on the mechanism of action and SARs are dealt with in the text.

Pharmacokinetics

Chlorpromazine is readily absorbed from the GI tract and widely distributed in all tissues of the body. Chlorpromazine and its metabolites are excreted in the urine and the bile. The most common paths of the metabolism are hydroxylation of the nucleus, oxidation of the sulfur atom, and dealkylation of the dialkylamino group, followed by conjugation of the polar metabolites with glucuronic acid or sulfuric acid.

Chlorpromazine has 168 potential metabolites, 70 of which have been identified. Most polar metabolites (conjugated species and sulfoxides) are excreted in the urine while nonpolar metabolites (for example demethylated metabolites, hydroxylated species, and parent compound) are excreted in the bile.

Some typical metabolic pathways are shown in Table 2.

Chlorpromazine has been detected in the urine several months (3 to 6) after discontinuing long-term treatment. The therapeutic ranges of chlorpromazine are 50 to 300 ng/mℓ (adults) and 40 to 80 ng/mℓ (children).

The plasma half-life on the average is 9 hr, ranging from 2 to 24 hr. The plasma steady state is achieved in 5 to 7 days; therefore it is not advisable to change doses more often than once a week. Because of the slow disappearance of chlorpromazine from the plasma, single dose administration of chlorpromazine is practiced quite often.

Chlorpromazine is 99% bound to plasma albumin.

Uses and Dosage

Chlorpromazine is used in alleviating agitation, anxiety, and tension of psychoneurotic and psychotic patients. Chlorpromazine is useful in the treatment of the manic phase of acute manic-depressive patients and of excited schizophrenic patients. In mental hospitals chlorpromazine is used for calming aggressive and hostile patients. As a result of treatment with chlorpromazine, the patients become more cooperative and suitable for psychotherapy. Chlorpromazine has sedative and tranquilizing effects. To a lesser degree, chlorpromazine also has antiemetic, antispasmodic, and hypotensive actions. Due to its sedative and antiemetic actions, chlorpromazine is used in the treatment of acute alcoholism and delirium tremens. Chlorpromazine is also used in controlling vomiting associated with uremia, gastroenteritis, cancer chemotherapy, and radiation sickness.

Although chlorpromazine does not have a potent-enough analgesic action per se, it potentiates the action of many analgesics. Due to this effect, chlorpromazine is often used for premedication in combination with analgesics intramuscularly. Chlorpromazine has also been used by i.v. injection together with analgesics to produce a basal narcosis (potentiated anesthesia). An extension of this method was described as induced hypothermia or artificial

Table 4
CHLORPROMAZINE INDEXES (CIs)[a] OF PHENOTHIAZINE NEUROLEPTICS IN VARIOUS PHARMACOLOGICAL TEST SYSTEMS[b]

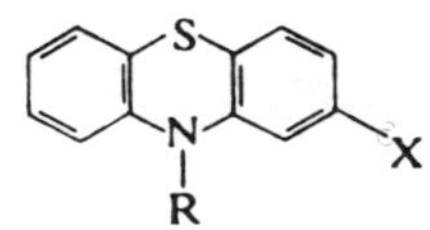

Name	X	R	Amphetamine antagonism	Ptosis production	Catalepsy production	Jumping-box test	Apomorphine antagonism	Open field test — ambulation	Tryptamine antagonism	Epinephrine antagonism
Promazine	H	c	0.034	0.33	0.19	0.09	<0.08	0.09	0.065	0.95
Chlorpromazine	Cl	c	1.0	1.0	1.0	1.0	1.0	1.0	1.0	1.0
			(1.1)[d]	(10.0)[e]	(7.5)[e]	(0.93)[d]	(6.5)[d]	(4.5)[d]	(1.3)[d]	(1.6)[d]
Methopromazine	OCH_3	c	0.18	2.5	1.2	1.0	0.14	0.85	0.15	2.5
Acetylpromazine	$COCH_3$	c	5.0	8.3	0.94	1.3	0.46	1.3	0.22	16.0
Triflupromazine	CF_3	c	3.8	3.3	4.2	3.6	3.6	3.5	1.5	2.1
Trimeprazine	H	f	0.038	0.5	0.3	0.23	<0.08	0.25	0.29	0.062
Levomepromazine	OCH_3	f	0.5	2.0	1.5	1.16	0.27	0.56	0.26	1.23
Prochlorperazine	Cl	g	2.3	1.0	1.9	1.16	3.2	2.8	0.57	0.094
Trifluoperazine	CF_3	g	4.4	2.0	18.8	7.8	11.9	11.5	0.16	0.08
Thioproperazine	$SO_2N(CH_3)_2$	g	8.5	1.25	5.8	5.5	13.8	12.1	0.065	0.32
Butaperazine	$CO(CH_2)_2CH_3$	g	2.82	3.1	3.8	1.37	26.0	3.0	0.045	0.73
Perphenazine	Cl	h	6.9	5.55	24.2	9.3	20.3	20.4	1.2	1.0
Fluphenazine	CF_3	h	11.0	8.3	46.8	37.2	50.0	45.0	0.52	0.70
Acetophenazine	$COCH_3$	h	0.52	7.2	5.0	3.0	29.5	3.0	0.05	2.26
Dixyrazine	H	i	1.33	0.67	1.07	0.47	1.55	0.64	0.18	0.061
Thiopropazate	Cl	j	12.5	5.88	20.8	8.5	43.3	20.4	1.2	0.36
Pipamazine	Cl	k	1.0	8.32	4.7	1.43	0.65	2.82	1.47	1.78
Thioridazine	SCH_3	l	0.15	2.0	0.58	0.047	<0.04	0.37	0.081	1.0

[a] CI's were obtained by dividing ED (mg/kg, s.c., in rats) value for chlorpromazine by the comparable value for the test compound.
[b] Adapted from Reference 70 by courtesy of John Wiley & Sons, Inc., copyright 1981.
[c] $R = (CH_2)_3N(CH_3)_2$.

Table 4 (continued)
CHLORPROMAZINE INDEXES (CIs)[a] OF PHENOTHIAZINE NEUROLEPTICS IN VARIOUS PHARMACOLOGICAL TEST SYSTEMS[b]

[d] ED_{50}(mg/kg, s.c., in rats) for chlorpromazine in the indicated test.
[e] Ptosis score (ED_4) for chlorpromazine,[71] catalepsy score (ED_3) for chlorpromazine.[71]
[f] R = $CH_2CH(CH_3)CH_2N(CH_3)_2$.
[g] 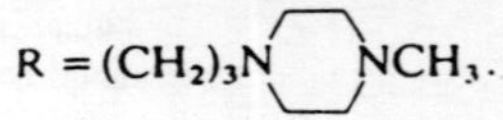

[h] 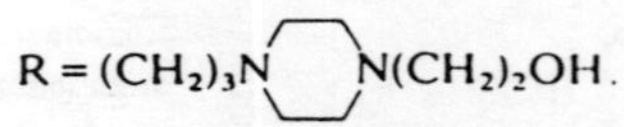

[i] 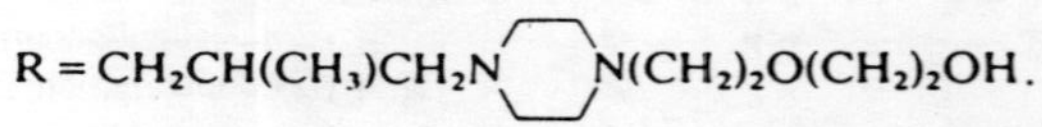

[j] 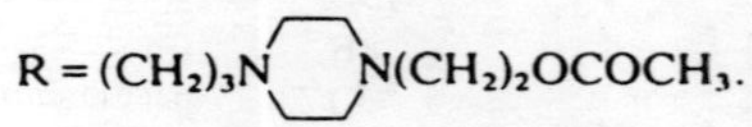

[k] 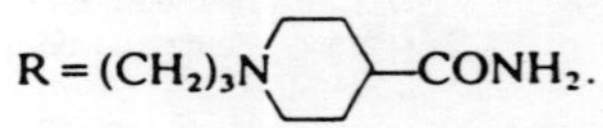

[l] 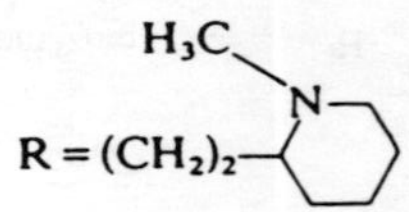

hibernation. This state can be induced by an i.v. injection of a mixture of chlorpromazine, promazine, and pethidine in conjunction with the use of surface cooling agents such as water jackets and ice bags.

Chlorpromazine hydrochloride is administered orally, rectally, by deep i.m. injection (not s.c.) as a 1% solution or by i.v. infusion. In chronic conditions the oral starting dose is 25 to 50 mg three times a day and increasing as necessary. The maintenance dose is 25 to 100 mg three times daily. In severe psychotic conditions much higher doses may be necessary. The usual dose by injection is 25 to 50 mg three times a day. For rectal administration suppositories containing 100 mg chlorpromazine base are used, which has the same effect as that obtained with 40 to 50 mg orally administered chlorpromazine hydrochloride or 25 mg intramuscularly administered drug.

For preoperative medication 50 mg chlorpromazine hydrochloride mixed with 100 mg pethidine hydrochloride and 50 mg promethazine hydrochloride is used. For children, one third to one half of the adult dose is given.

Toxicity

Chlorpromazine produces a wide variety of side effects usually in the early stages of treatment when high doses of the drug are employed. The side effects are seldom of sufficient severity to warrant the discontinuance of the treatment. The common side effects include drowsiness (characteristic of sedatives), dryness of the mouth, constipation (occasionally diarrhea), pallor of the skin (all associated with anticholinergic effects), dizziness, headaches, heartburn, polyuria, weakness, tiredness, lowering of body temperature (occasionally pyrexia), increase in heart rate, restlessness, photosensitivity, skin rashes, postural hypotension with tachycardia. Injections may cause irritation at the site of injection.

More severe side effects may occur occasionally and may cause discontinuance of medication with chlorpromazine. These include obstructive-type jaundice, severe and sometimes fatal blood dyscrasias including agranulocytosis, hypoplastic anemia, and leucopenia.

High doses of chlorpromazine produce extrapyramidal effects (Parkinsonian symptoms) which are spontaneously reversible on withdrawal of the drug or reducing the dose. These effects may also be controlled by concomitant administration of an anti-Parkinsonian drug.

Drug Interactions

Chlorpromazine enhances the effects of CNS depressants such as alcohol and barbiturates, antihypertensive agents, and anticholinergic agents such as atropine.

Physical Chemical and Other Data

Chlorpromazine — 2-Chloro-10-(3-dimethylaminopropyl) phenothiazine. $C_{17}H_{19}ClN_2S$. Molecular weight = 318.88.

Oily liquid. Amine odor. Alkaline reaction (pKa 9.2); b.p.$_{0.8}$ 200 to 205°C. Almost insoluble in water. Very soluble in alcohol, benzene, chloroform, ether, and dilute mineral acids. Hydrochloride $C_{17}H_{19}ClN_2S{\cdot}HCl$. Molecular weight = 355.3. A white or off-white powder with a slight odor and very bitter taste; m.p. 194 to 197°C. Soluble in water, alcohol, and chloroform. Insoluble in benzene and ether. It darkens on exposure to light.

Chlorpromazine is available in 100-mg suppositories. Chlorpromazine hydrochloride tablets are available in 10, 25, 50, 100, and 200 mg strengths, suppositories in 25 and 100 mg, injection 25 mg in 1 mℓ, 50 mg in 2 mℓ, and 250 mg in 10 mℓ.

Trade name: Thorazine (Smith Kline & French).

Thioridazine

Thioridazine, a representative of the piperidylalkyl phenothiazine class is very often the drug of choice in patients particularly susceptible to extrapyramidal effects if the psychotic symptoms can be controlled within the dosage considered safe, the upper limit being 800 mg/day. Higher doses are inadvisable due to possible danger of retinal damage and visual impairment.[74]

Sections on the mechanism of action and SARs are included in the text.

Pharmacokinetics

Thioridazine is well absorbed from the GI tract and is widely distributed throughout the body. A large number of metabolites have been identified, some of them biologically active, therefore it is uncertain that simply measuring levels of the parent drug is adequate. Thioridazine produces plasma concentrations that seem inordinately high as compared with those produced by comparable doses of chlorpromazine. There are no data on steady-state levels of thioridazine. It is known, however, that mesoridazine, a major active thioridazine metabolite is present in higher percentage in the unbound form than that of the parent drug. Therefore the range of therapeutic concentrations has not been defined for thioridazine.

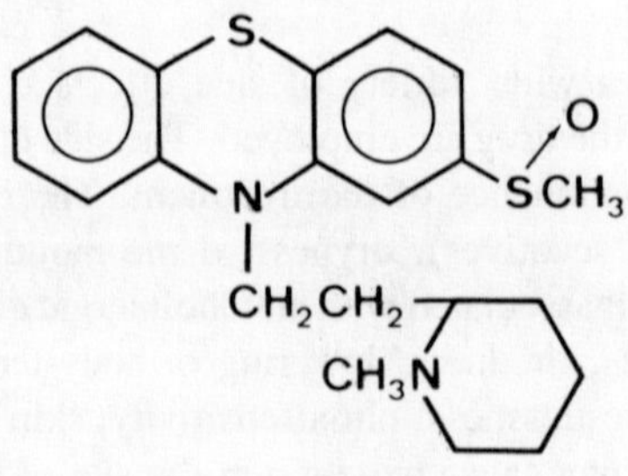

Mesoridazine

The comparative milligram doses of thioridazine and mesoridazine, equivalent to 100 mg of chlorpromazine, as defined by double blind studies are 97 ± 7 and 56 ± 6 mg, respectively.[75]

The main metabolic pathways are sulfoxidation of thioridazine to side-chain sulfoxide (mesoridazine) and sulfone as well as to ring sulfoxide. Demethylation is of minor importance.

Attempts have been made to correlate the ratios of mesoridazine to thioridazine-5-sulfoxide (side-chain vs. ring sulfoxide) levels in the plasma with the patient's improvements as measured by clinical global ratings. Based on the correlation, responder profiles (side-chain sulfoxide higher than ring sulfoxide) and nonresponder profiles (ring sulfoxide higher than side-chain sulfoxide) have been postulated.[76]

Uses and Dosage

Thioridazine has antipsychotic activities similar to those of chlorpromazine, but it has little antiemetic or hypothermic activity. It is used mainly for the control of mania, agitation, and for the treatment of schizophrenia. The clinical observation was made that thioridazine is superior to chlorpromazine in severely ill patients.[77] Thioridazine produces less extrapyramidal effects than other phenothiazines and is preferred if extrapyramidal effects are troublesome. Thioridazine may be used in the management of anxiety and in children with behavior disorders.

Thioridazine is used for the treatment of severe mental illness. In adults the usual initial dose is 50 to 100 mg three times a day with a gradual increment to a maximum of 800 mg daily if necessary. This is later reduced to a maintenance dose of 100 to 200 mg daily. In children the usual dose is 1 mg/kg body weight daily in divided doses.

Toxicity

Toxic effects are similar to those produced by chlorpromazine, but are usually milder and occur less frequently than with chlorpromazine. The side effects include drowsiness, dizziness, dryness of the mouth, skin rashes, leucopenia, amenorrhea, galactorrhea, nasal stiffness, gastric irritation, nocturnal confusion, postural hypotension, tachycardia, and transient edema. None of the side effects are serious enough to warrant discontinuance of therapy with thioridazine.

Drug Interactions

Drug interactions are similar to those produced by chlorpromazine.

Physical Chemical and Other Data

Thioridazine—2-methylmercapto-10-[2-(*N*-methyl-2-piperidyl)ethyl]phenothiazine. $C_{21}H_{26}N_2S_2$. Molecular weight 370.56. Crystals from acetone m.p.72 to 74°C. pKa 9.5. Thioridazine hydrochloride $C_{21}H_{26}N_2S_2 \cdot HCl$. Molecular weight 407. Crystals from acetone, m.p. 158 to 160°C. Soluble in water, alcohol, and chloroform; insoluble in ether. It must be protected from light.

Thioridazine hydrochloride is available in solution 3 g in 100 mℓ and in tablets of 10, 15, 25, 50, 100, 150, and 200 mg strengths.

Trade name: Mellaril (Sandoz).

Trifluoperazine

Along with other representatives of the piperazinylalkyl phenothiazines, fluphenazine and perphenazine, trifluoperazine produces relatively minor sedative effects, which may be beneficial to withdrawn or depressed patients. In addition the hypotensive actions of these three drugs are considerably less than those of thioridazine, chlorpromazine, or chlorprothixene. Furthermore, perphenazine, trifluoperazine, and fluphenazine have less pronounced anticholinergic effects than chlorpromazine and are less likely to induce cholinergic blockade in therapeutic doses. As a consequence, however, these drugs are more likely to induce extrapyramidal effects. Sections on the mechanism of action and SARs are included in the text.

Pharmacokinetics

Trifluoperazine is rapidly absorbed from the GI tract and is widely distributed in the body.

Trifluoperazine undergoes metabolic processes similar to those of chlorpromazine that are characteristic for the phenothiazine ring, e.g., sulfoxidation and hydroxylation of the phenothiazine ring. In addition metabolism involving the piperazine ring also takes place. This includes removal of the alkyl group of the piperazine ring, formation of N-oxides and cleavage of the piperazine ring that yields ethylenediamine derivatives.

The comparative milligram dose of trifluoperazine equivalent to 100 mg of chlorpromazine, as defined by double blind studies is 2.8 ± 0.4 mg.[75]

Uses and Dosage

The uses of trifluoperazine are similar to those of chlorpromazine. Trifluoperazine has a rapid onset of action and long-lasting effects. Trifluoperazine has potent antiemetic properties, weak α-adrenergic blocking, antihistaminic, and antispasmodic activities. It is devoid of sedative effects and may produce stimulating activity. When used for the treatment of acute mania, delirium tremens, or schizophrenia it is often necessary to start treatment with intramuscularly administered injections of 1 to 2 mg strength repeated every 4 to 6 hr. Maintenance therapy usually consists of 15 to 20 mg orally administered trifluoperazine daily. When used for the treatment of psychoneurosis, nausea or vomiting, 2 to 4 mg trifluoperazine daily is usually sufficient. It is not advisable to exceed 6 mg daily orally or 3 mg daily intramuscularly for the latter indications.

Toxicity

Toxic effects are similar to those produced by chlorpromazine. Minor side effects include drowsiness, agitation, dizziness, insomnia, nasal congestion, and sweating. Higher than 30 mg daily doses often produce extrapyramidal effects as manifested by akathisia or dystonia (muscle rigidity of the face, back, shoulder). These symptoms are usually reversible on reducing the dose or by coadministration of an antiparkinsonian drug.

Agranulocytosis, thrombocytopenia, and leucopenia have been reported occasionally.

Drug Interactions

Drug interactions are similar to those produced by chlorpromazine.

Physical Chemical and Other Data

Trifluoperazine — 2-trifluoromethyl-10-[3′-(1-methyl-4-piperazinyl) propyl]phenothiazine. $C_{21}H_{24}F_3N_3S$. Molecular weight 407.49. Oily substance b.p.$_{0.6}$ 202 to 210°C. pKa 8.3.

Dihydrochloride $C_{21}H_{24}F_3N_3S{\cdot}2HCl$. Cream colored powder from ethanol, m.p. 242 to 243°C. Soluble in water, slightly soluble in alcohol, chloroform. It must be protected from light.

Trifluoperazine hydrochloride is available in injections 20 mg in 10 mℓ and tablets of 1, 2, 5, and 10 mg.

Trade name: Stelazine (Smith, Kline & French).

Thioxanthenes

SARs

SARs for thioxanthenes are similar to those of phenothiazines. Substitution in 2-position is preferable; 3-dimethylaminopropylidene derivatives with chloro or trifluoromethyl groups in 1-, 3-, and 4-positions are less potent neuroleptics than the corresponding 2-substituted analogues. Highly active compounds carry a trifluromethyl or a chloro group in 2-position. Lower alkyl substituents in 2-position are compatible with neuroleptic activity, but larger substituents like n-C_5H_{11} or BzO are not. A double bond in 9-position, and a Z configuration with regard to the substituted ring, are particularly important. Hydrogenation and concomitant loss of isomerism, or the choice of the wrong configuration (E) lead to inactive or very weak compounds. The basic amino group at the end of the propylidene chain may be a part of a piperazine or piperidine ring, or simply dimethylamine. Examples of all three classes have found clinical use as antipsychotic drugs.

Stereochemical and Conformational Features

Due to the double bond present in the thioxanthenes, geometrical isomerism exists in this series. The 2-substituted active antipsychotic agents are *cis*.[78] It has been found that *cis*-flupentixol is a potent antagonist of dopamine sensitive adenylate cyclase (D-1 receptor marker) while the *trans* isomer is very weak in blocking dopamine effects.[24,79]

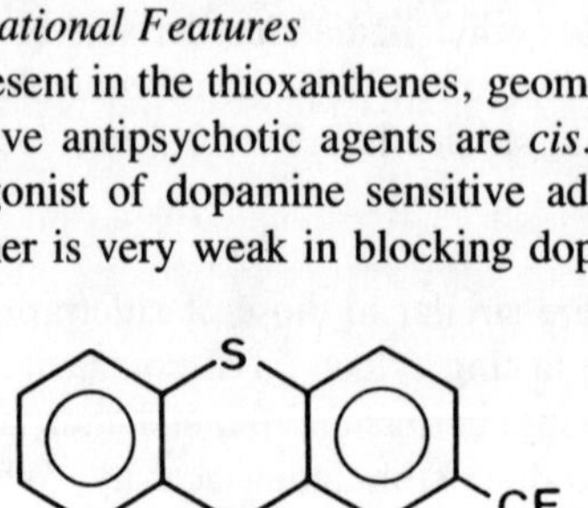

Cis-flupentixol 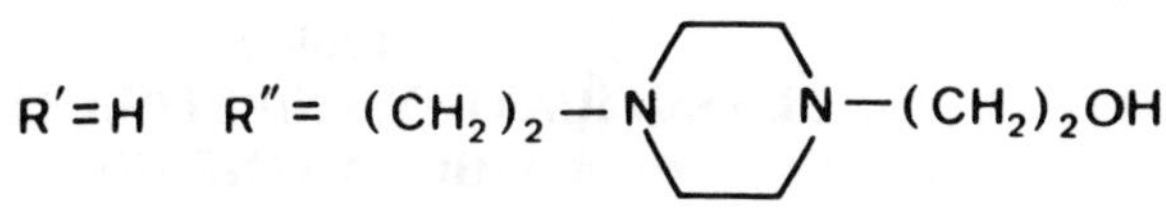

Trans-flupentixol 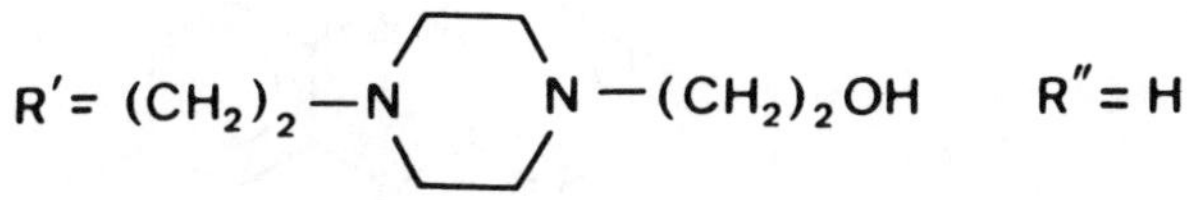

Similarly, *cis*-chlorprothixene is a potent dopamine antagonist, while the *trans* isomer is ineffective.

Classification

The classification of thioxanthenes is shown in Table 5.

Thiothixene

Pharmacokinetics

Thiothixene is rapidly absorbed following oral administration. Early plasma half-life was found to be 210 min, late half-life about 34 hr. One of the metabolites was identified as the *N*-demethyl analogue. Patients with adequately controlled schizophrenia had plasma levels of 10 to 22 ng/mℓ, determined 2 to 2.5 hr after a repeated oral dosage of 15 to 60 mg daily (2 to 4 divided doses per day).[80] Some enzyme induction was indicated by lower plasma levels after 8 weeks of treatment.[81]

Uses and Dosage

Thiothixene is used in the treatment of schizophrenia and paranoid states. In socially withdrawn patients, thiothixene was reported to be more effective than chlorprothixene or clopenthixol.[82] In boys with behavior disorders, a greater improvement was observed with thiothixene treatment than with chlorprothixene.[83] Thiothixene and perphenazine were effective in the treatment of schizophrenic patients who had relapses after attempted placebo therapy.[84] In chronic schizophrenic patients, improvement was reported in 21 out of 25 cases after a 3-day treatment with thiothixene i.m.[85]

The dosage of thiothixene should be individually adjusted, starting with low doses, until a satisfactory response is observed. An initial daily dosage of 6 mg, in three doses, may be appropriate for mild conditions. Usual dosages have been reported as 5 to 10 mg[86] or 20 to 30 mg/day.[87] Maximal oral dosage is 60 to 80 mg/day.

Toxicity

Side effects of thiothixene are similar to those of phenothiazine and involve CNS, cardiovascular, and endocrine systems.

Extrapyramidal symptoms, such as pseudoparkinsonism, akathisia, and dystonia have been reported. Tardive dyskinesia may appear in patients on long-term therapy, especially in female, elderly patients on high doses. Thiothixene may cause insomnia, restlessness, and agitation.

Physical-Chemical Data

Thiothixene base is a white to tan-colored almost odorless crystalline powder, m.p. 147 to 152°C. It is practically insoluble in water, soluble in alcohol, and very soluble in chloroform. The dihydrochloride exists as dihydrate (molecular weight 552.6) or in anhydrous form, $C_{23}H_{29}N_3O_2S_2 \cdot 2HCl$ (molecular weight 516.5).

Chemical name — Z-*N*,*N*-Dimethyl-9-[3-(4-methyl-1-piperazinyl)propylidene] thioxanthene-2-sulfonamide.

Table 5
CLASSIFICATION AND STRUCTURES OF THIOXANTHENES

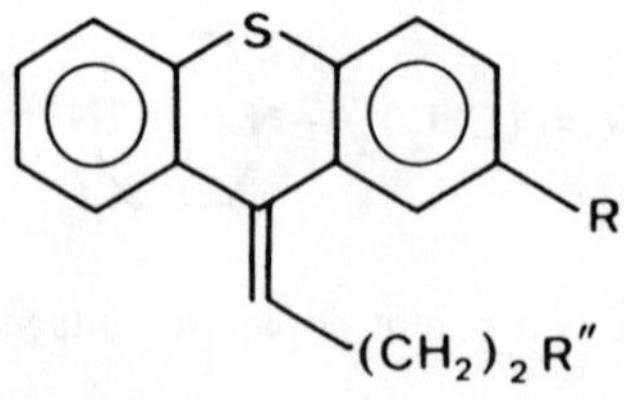

	R′	R″
Aminoalkyl derivatives		
Prothixene	H	$N(CH_3)_2$
Chlorprothixene	Cl	$N(CH_3)_2$
Piperazinylalkyl derivatives		
Thiothixene	$SO_2N(CH_3)_2$	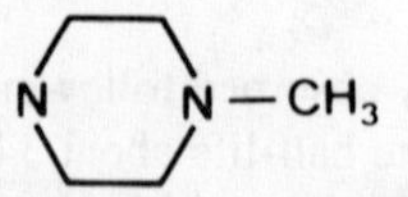
Clopenthixol	Cl	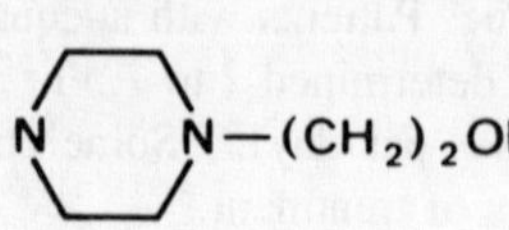
Flupentixol	CF_3	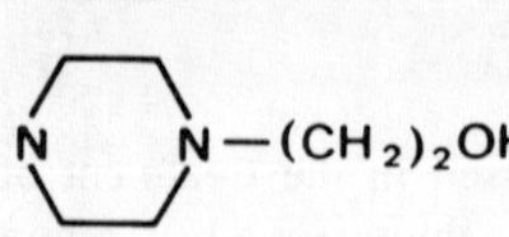
Flupentixol decanoate	CF_3	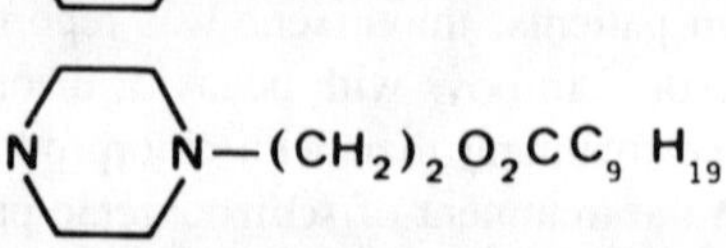

Dosage forms (U.S.): Capsules (1, 2, 5, 10, 20 mg), concentrate (hydrochloride, equivalent to 5 mg base/mℓ), injection (hydrochloride, equivalent to 2 mg or 5 mg base/mℓ).
Trade name — Navane (Roerig, Pfizer).

Butyrophenones and Diphenylbutylpiperidines

SARs of Butyrophenones

The relationships between chemical structure and antipsychotic activity in the class of butyrophenones have been extensively reviewed.[88,89] Janssen and co-workers[90] published the synthesis of haloperidol (1,L=F, R_1=OH, R_2=p−Cl−C_6H_4) and seven related compounds(1) in 1959 as a part of their investigations of CNS depressant agents related to meperidine(2).

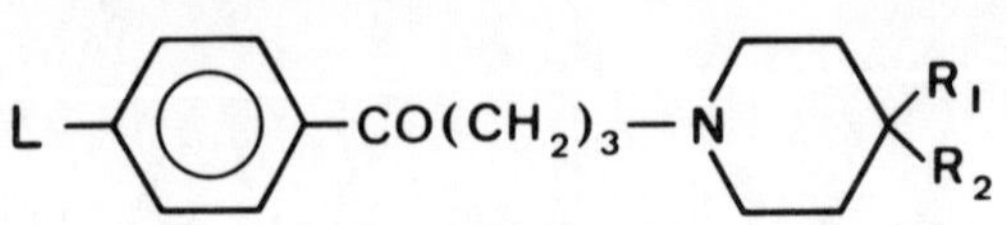

1

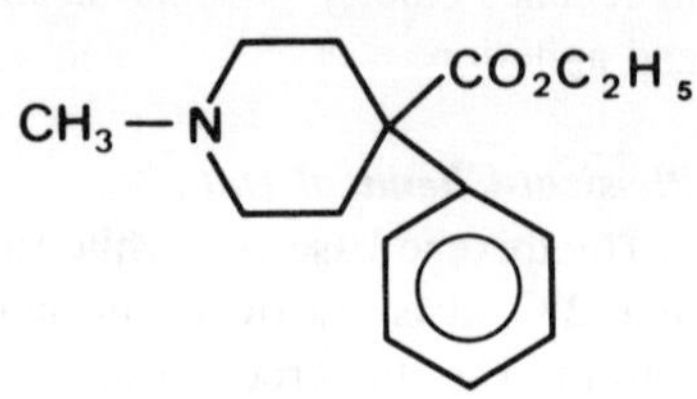

2

Extensive pharmacological testing of *1* and related butyrophenones in a large battery of animal models established the antipsychotic spectra of these compounds in comparison with many phenothiazine reference compounds.[71,90,91] Structures and pharmacological test results of a series of butyrophenones of clinical interest are shown in Table 6. The following SARs resulted from these investigations.

For high neuroleptic potency, a fluorine substituent is required in the para position of the benzene ring. Unsubstituted analogues are several times less active, and substitutions in other positions on the ring further reduce activity. Anisoperidone, a 4-methoxysubstituted butyrophenone of clinical relevance (see Table 6), has only 1/16 of the potency of the corresponding unsubstituted compound in the rat learned avoidance-escape test.[91]

Replacement of the keto group by a thioketone,[92] by olefinic, phenoxy, or hydroxyl groups[93] reduces neuroleptic potency. Replacement of the keto group by a *p*-flurobenzene group, however, leads to the clinically useful drugs pimozide, penfluridol, fluspirilene and related 1,1-diarylbutanamines with antipsychotic activity (see below).

The side chain connecting the aromatic ring with the basic amino group consists ideally of a 4-carbon chain. Shortening, lengthening, or branching decreases activity.

The basic amino group in position 4 of the intermediate chain is ideally a part of a piperidine, piperazine, or tetrahydropyridine function; substituted pyrrolidine or hexamethylenimines are considerably less potent. Properly designed substitution in 4-position of the piperidine ring is highly important for antipsychotic activity.

R_1 is preferably H, OH, CON(Alk)$_2$, CO_2Alk; but alkyl, ether, hydroxyalkyl, and other groups are also compatible with neuroleptic activity. R_2 is ideally a meta or para substituted phenyl or an aniline group and may be joined to the other substituent to form a spiro derivative. These spiro derivates include spiroperidol, the most potent neuroleptic agent known, and other potent agents such as benperidol and droperidol. Spiroperidol (spiperone) is often used in receptor-binding studies.

SARs of Diphenylbutylpiperidines

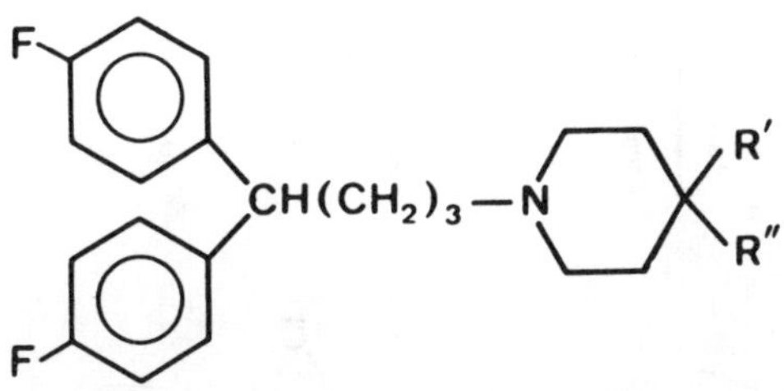

SARs of diphenylbutylpiperidines are similar to those of butyrophenones: parafluoro substituents on both phenyl rings, an unbranched propylene bridge connecting the basic nitrogen to the benzhydryl function, and a piperidine ring with appropriate substituents in 4-position, are required for high antipsychotic activity. The structures of penfluridol, fluspirilene and pimozide are shown in Figure 12. Penfluridol and fluspirilene are both long-acting antipsychotic drugs, but neither has achieved a major role commercially.

Stereochemical and Conformational Features

Butyrophenones

The general features of butyrophenone-antipsychotics are represented by the following structure:

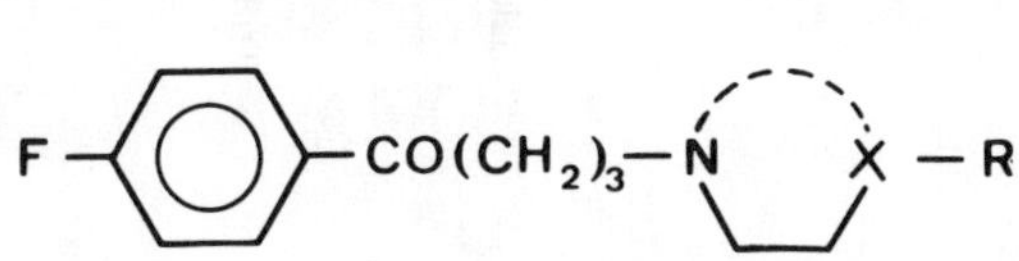

where X = C, O, N.

Table 6
CHLORPROMAZINE INDEXES OF BUTYROPHENONE NEUROLEPTICS IN VARIOUS PHARMACOLOGICAL TEST SYSTEMS

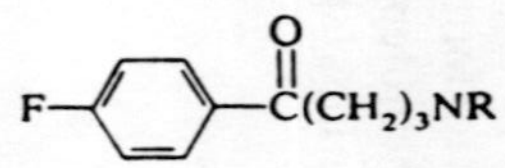

		Dog					Rat		
		Jumping-box test[a,b]		Apomorphine test[b,c] peak effect (hr)					
Compound	NR	Oral	Subcutaneous	Potency	Onset	Duration	Jumping-box test[b,d]	Anti-amphetamine test[b,e]	Antinor-epinephrine test[b,f]
4-Phenylpiperidine Derivatives									
Haloperidol	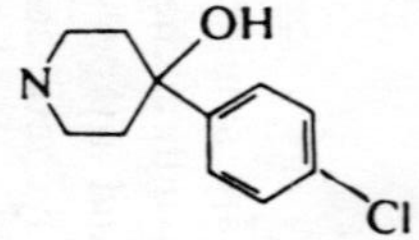	50	50	47	<1/2	6	17	35	1/4
Trifluperidol	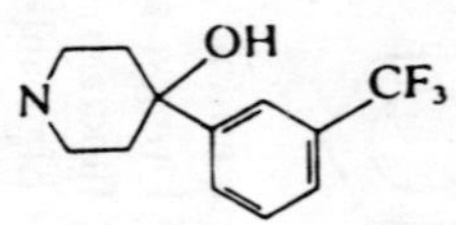	50	50	140	3/4	3	33	33	1.6
Methylperidol	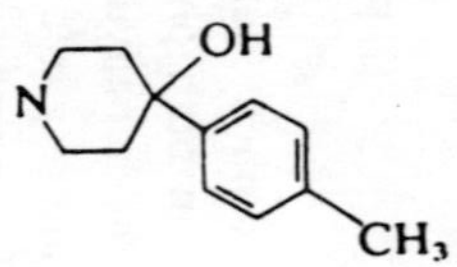	17	25	47	3/4	4	10	33	1/2

Paraperidide	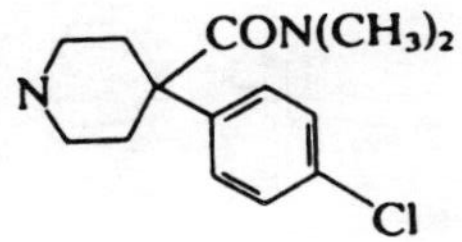	50	25	70	<$^1/_2$	6	5	10	$^1/_{14}$
Haloperidide	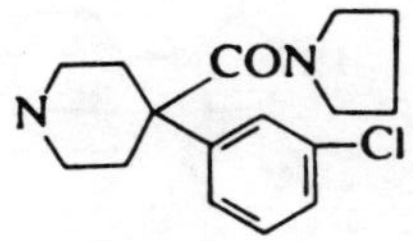	8	42	350	<$^1/_2$	12	5	10	$^1/_{20}$
Methylperidide	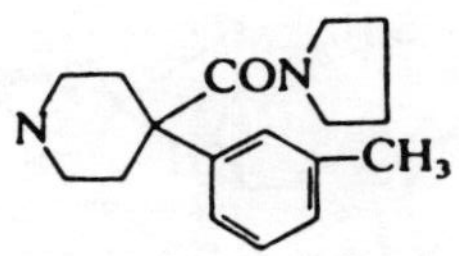	5	125	350	<$^1/_2$	6	10	10	$^1/_{10}$
Floropipamide	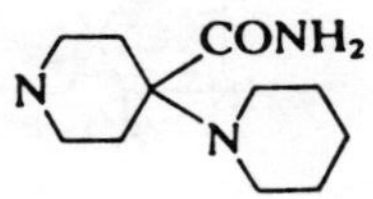	2.5	1.2	$^1/_2$	<$^1/_2$	4	$^1/_{10}$	$^1/_3$	$^1/_3$
Anisoperidone[g]	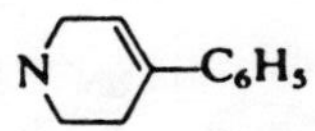	$^1/_8$	<$^1/_{30}$	$^1/_2$	<$^1/_2$	$^3/_4$	$^1/_2$	1	1.5
Aceperone	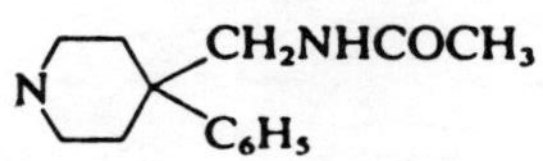	$^1/_2$	$^1/_8$	1	<$^1/_2$	3	$^1/_{10}$	$^1/_{160}$	$^1/_2$

Table 6 (continued)
CHLORPROMAZINE INDEXES OF BUTYROPHENONE NEUROLEPTICS IN VARIOUS PHARMACOLOGICAL TEST SYSTEMS

		Dog					Rat		
		Jumping-box test[a,b]		Apomorphine test[b,c] peak effect (hr)					
Compound	NR	Oral	Subcutaneous	Potency	Onset	Duration	Jumping-box test[b,d]	Anti-amphetamine test[b,e]	Antinor-epinephrine test[b,f]
		Piperidine Derivatives with Imidazoline Functions							
Spiroperidol	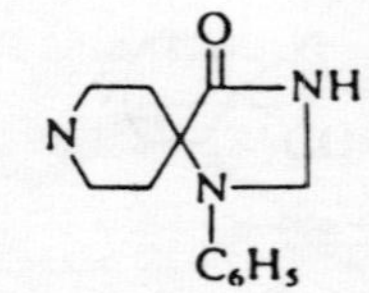	500	500	2800	4	8	100	33	1/2
Benperidol	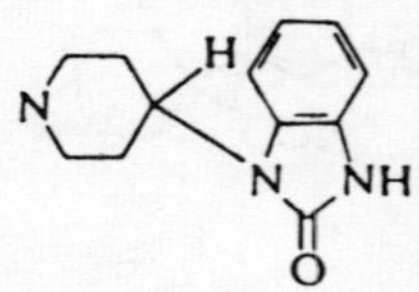	500	500	1400	3/4	2	33	33	1.6
Droperidol	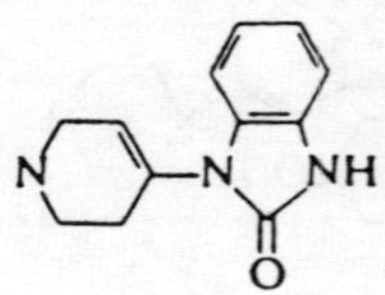	250	250	700	<1/2	2	33	33	5
		Piperazine Derivatives							
Butropipazone	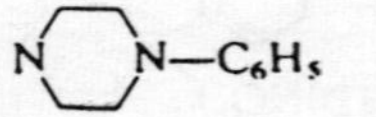	1.6	5/8	1/2	3/4	1.5	1/2	1	3

Fluanisone	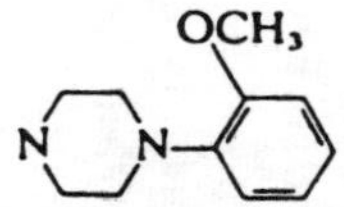	5	12	5	<$^1/_2$	2	3	3	8
Chlorpromazine		1	1	1	$^3/_4$	1.5	1	1	1

[a] Trained dogs are required to cross a hurdle when presented with an auditory warning signal to avoid shock. The efficacy of compound in producing avoidance is determined.

[b] Relative potency compared (ED_{50} mg/kg) with chlorpromazine, which is arbitrarily assigned a value of 1.

[c] Dogs administered s.c. doses of experimental compounds are challenged with an emetic dose of apomorphine at various time intervals thereafter. Inhibition of emesis, onset, and duration of effect are determined.

[d] Experimental compounds are administered s.c. to rats. The experimental procedure is the same for dogs.

[e] Ability of s.c. administered compounds to inhibit compulsory gnawing and chewing responses to i.v. dose of 10 mg/kg of amphetamine is determined.

[f] Ability of s.c. administered compounds to prevent lethality of 1.25 mg/kg(i.v.) of norepinephrine 1 hr after drug is administered is determined.

[g] The 4′-F substituent of the butyrophenone is replaced by CH_3O.

Adapted from Kaiser, C. and Setler, P. E., in *Burger's Medicinal Chemistry,* 4th ed., Part III, Wolff, M. E., Ed., John Wiley & Sons, New York, 1981, 859. With permission.

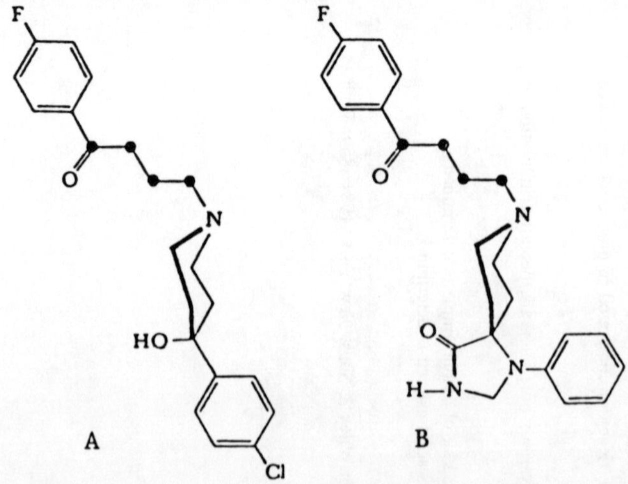

FIGURE 10. Conformation of haloperidol (A) and spiperone (B) (From Janssen, P. A., in *Psychopharmacological Agents,* Vol 3, Gordon, M., Ed., Academic Press, New York, 1974, 129. With permission.)

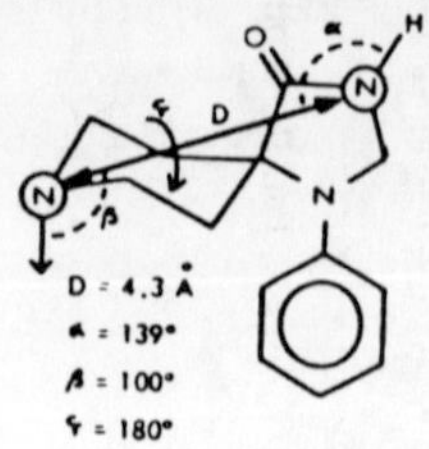

FIGURE 11. Spatial parameters in butyrophenone neuroleptics (From Koch, M. H. J., *Mol. Pharmac.*, 10, 425, 1974. By courtesy of A.S.P.E.T., copyright 1974).

It has been determined by X-ray diffraction that both the butyrophenone side chain and the phenyl ring attached to the piperidine ring of haloperidol and spiroperidol are equatorial, the piperidine ring being in the chain conformation[94] (Figure 10). The most probable conformations of the other antipsychotic butyrophenones are essentially the same or slightly different.

It was reported[95] that in order to exert high antipsychotic activity, the molecule must have a hydrogen accepting basic nitrogen in one direction and a hydrogen donor in the other, the distance between the two groups being 3.5 to 6.5 Å (Figure 11).

Diphenylbutylpiperidines

By analogy to butyrophenones, the most probable conformation of the structurally related diphenylbutylpiperidines have been predicted[94] (Figure 12).

Classification of Butyrophenones and Diphenylbutylpiperidines

The following classification is based on the nature and the substitution pattern of the basic amino group.

1. 4-Phenylpiperidine derivatives — The majority of the clinically useful butyrophenones are representatives of this class, as shown in Table 6. Penfluridol is a prominent example from the diphenylbutylpiperidines.

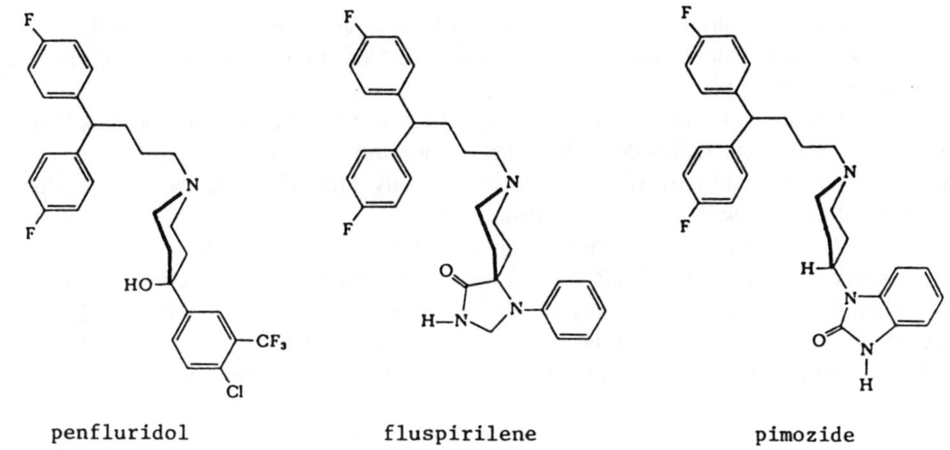

FIGURE 12. (From Janssen, P. A., in *Psychopharmacological Agents,* Vol 3, Gordon, M., Ed., Academic Press, New York, 1974, 129. With permission.)

2. Piperidine derivatives with imidazoline functions — Benperidol, droperidol, and pimozide are examples of benzimidazolinones from the butyrophenone and diphenylbutylpiperidine classes. A somewhat different type, i.e., spiroimidazolidinones, is represented by spiroperidol and fluspirilene.
3. Piperazine derivatives — Butropipazone and fluanisone are butyrophenones with 4-phenylpiperazine moieties.
4. Others — Open-chain amine derivatives with antipsychotic activity have been reviewed.[89]

Haloperidol

Pharmacokinetics

Haloperidol is rapidly absorbed after i.m. injection. In 36 fasted male volunteers, a single dose of 2 mg haloperidol led to maximum plasma levels of unchanged drug averaging 4.3 ng/mℓ, with peaks observed within 20 min after administration.[96] Terminal elimination half-life ranged from 12.8 to 35.5 hr, and averaged 20.7 ± 4.6 (SD) hr. Due to the slow elimination, a substantial accumulation of haloperidol following multiple doses was expected, e.g., two to four times higher blood levels after t.i.d. dosing compared to single dosing.[96] It was further expected that a change of dosage regimen would require 3 to 5 days before plasma levels of haloperidol had equilibrated, and before the clinical response could be evaluated.

Urinary excretion of radioactivity was maximal within the first 4 hr after oral administration of 2 mg radiolabeled haloperidol.[97] The fraction of radioactivity excreted in this period depended somewhat on the length of treatment; in 4 patients treated for 29 days, it averaged 29% of a radiolabeled dose. Fecal excretion accounted for 15% of radioactivity in 3 days, and continued in small amounts for up to 7 days.

Very little data is available on plasma levels of haloperidol after repeated administration. In two patients treated with oral dosages of 3 mg t.i.d. for 1 week, and subsequently with 6 and 9 mg for the next 2 weeks, plasma levels measured 138 and 165 ng/mℓ of haloperidol 12 hr after the last dose.[98]

Uses and Dosages

I.m. injections of haloperidol are indicated for the prompt control of agitated patients with symptoms of mania, acute delirium, or acute schizophrenic reaction. Doses of 5 to 10 mg are effective in reducing symptoms of psychomotor agitation, racing thoughts, and auditory hallucinations.[86] The onset of action occurs in 30 min. Subsequent doses, up to 80 mg in 24 hr, may be administered until control is achieved. Oral dosages should be used for the continuation of therapy.

General indications for use of haloperidol are the treatment of psychotic disorders.[99] It is effective in controlling Gilles de la Tourette's syndrome. Haloperidol is also effective in children suffering from hyperexcitability or hyperactivity. Haloperidol has been successfully used to treat patients with tardive dyskinesia.[100,101]

Initial daily dosages for adults range from 1 to 15 mg, in two or three oral doses. After a satisfactory response has been obtained, the dosage should be reduced to the lowest effective maintenance levels, often 2 to 5 mg in one dose. In patients remaining disturbed, daily doses of up to 100 mg/day have been used. Treatment of children, aged 3 to 12 years, should be initiated with 0.5 mg/day, and may be continued with dosages of 0.05 to 0.15 mg/kg/day.

Toxicity

Extrapyramidal reactions in the form of Parkinson-like symptoms occur frequently at the beginning of treatment. Less frequently these reactions include akathisia (inability to sit or stand still), dystonia (spasms in head, neck, and tongue), and other neuromuscular disturbances. Tardive dyskinesia has rarely been reported.

Drug Interactions

Haloperidol has been reported to reduce the anticoagulent effect of phenindione.[102] Antithrombotic effects of coumarins are also reduced, since their rate of metabolic transformation is increased by haloperidol.[103]

Physical-Chemical and Other Data

Haloperidol is a white to faintly yellowish, odorless, tasteless substance, m.p. 147 to 151°C. It is insoluble in water, sparingly soluble in alcohol or ethylacetate, and soluble in chloroform. A stable hydrochloride salt can be formed from the weak monovalent base (pKa 8.3).

Chemical name — 4-[4-(*p*-Chlorophenyl)-4-hydroxypiperidino]-4′-fluorobutyrophenone. $C_{21}H_{23}ClFNO_2$(375.9).

Dosage forms (U.S.) — Tablets (0.5, 1, 2, 5, and 10 mg), concentrate (2 mg/mℓ, lactate) and injection (5 mg/mℓ lactate).

Trade name — Haldol (Janssen, McNeil, U.S. only).

Haloperidol may be used as a tranquilizer in conjunction with analgesics in preparation of a patient for surgery, as an antiemetic[104] or as an anxiolytic.[105]

Benzamides

SARs

The synthesis of metoclopramide, the first example of a series of potential antipsychotic agents with a novel benzamide structure, was reported by Justin-Besançon and co-workers in 1964.[106] Although high doses of metoclopramide caused sedative, antiamphetamine, and other effects typical of antipsychotic drugs in animals, low doses were only antiemetic.[107-109] Metoclopramide has no clinical utility as an antipsychotic agent. It is used as an antiemetic drug and as an aid in GI diagnosis (intubation of small bowels and radiological examination of the stomach and small intestine). The therapeutically useful benzamides are shown in Table 7.

Sulpiride was synthesized by Justin-Besançon and co-workers in 1967.[110] Like standard phenothiazines and butyrophenones, sulpiride inhibited apomorphine-induced emesis and stereotyped behavior,[111] increased dopamine turnover in the CNS by receptor blockade,[112]

Table 7
STRUCTURES OF BENZAMIDES

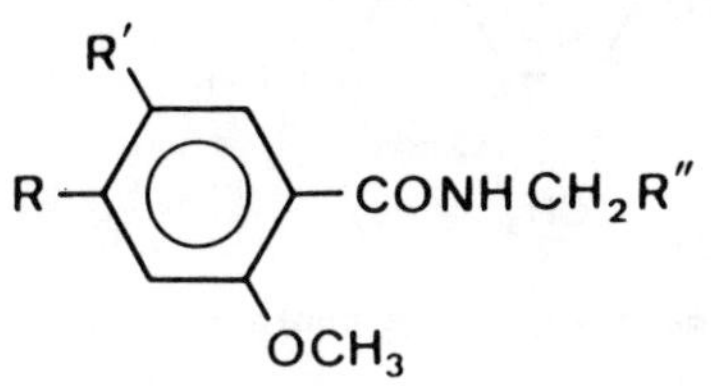

	R	R'	R"
Metoclopramide	NH_2	Cl	$CH_2N(C_2H_5)_2$
Sulpiride	H	NH_2SO_2	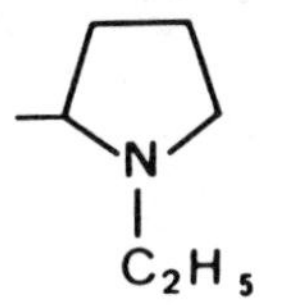
Sultopride	H	$C_2H_5SO_2$	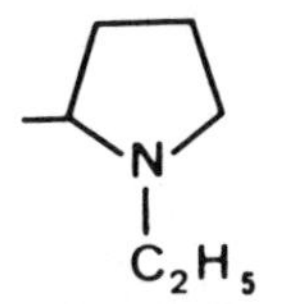
Tiapride	H	CH_3SO_2	$CH_2N(C_2H_5)_2$

and displaced ^{3}H-haloperidol from its binding site in rat striatal preparations.[113] Unlike other standard antipsychotic agents, sulpiride did not block the conditioned avoidance response in rats[114] (improved response at 100 mg/kg, reduced response at 150 mg/kg) and was only weakly cataleptogenic.[113] Sulpiride has been proposed to selectively block dopamine receptors of the D_2-type (see Mechanism of Action). Clinically, sulpiride is useful in the treatment of schizophrenia and ulcers.

Tiapride and sultopride, two additional benzamides produced by the same group of researchers,[115] are also of clinical interest. Tiapride is used in the treatment of delirium tremens, Huntington's chorea, and painful digestive dyskinesia.[116] Sultopride is an antipsychotic agent with rapid onset of action and rapid elimination.[111]

Pharmacological, biochemical, and clinical effects of benzamide drugs related to sulpiride have been reviewed recently.[111,116,117]

Stereochemical and Conformational Features of Benzamides

Pharmacological studies with sulpiride and sultopride have shown that their dopamine-antagonistic effects are caused predominantly by the (−)-enantiomers.[118,119] S-(−)-sulpiride and (−)-sultopride were 40 and 100 times more potent than their corresponding (+)-enantiomers in displacing ^{3}H-spiperone from its specific binding site in rat striatal preparations. Only the (−)-enantiomer increased the brain concentrations of dopamine metabolites (homovanillic acid, 3,4-dihydroxyphenylacetic acid) and antagonized apomorphine-induced locomotor activity in mice. Prolactin release in rats was stimulated by (−)-sulpiride to a much higher degree than by the (+)-enantiomer.

Intramolecular hydrogen bonds determine the conformations of metoclopramide and sulpiride.[120,121] These hydrogen bonds, common to both drugs, occur between the amide hydrogen and the ether oxygen, and the ammonium hydrogen and the amide oxygen (Figure 13). The resulting conformations with a folded intermediate chain represent energy minima. In contrast to metoclopramide, sulpiride can also assume conformations with extended chains,

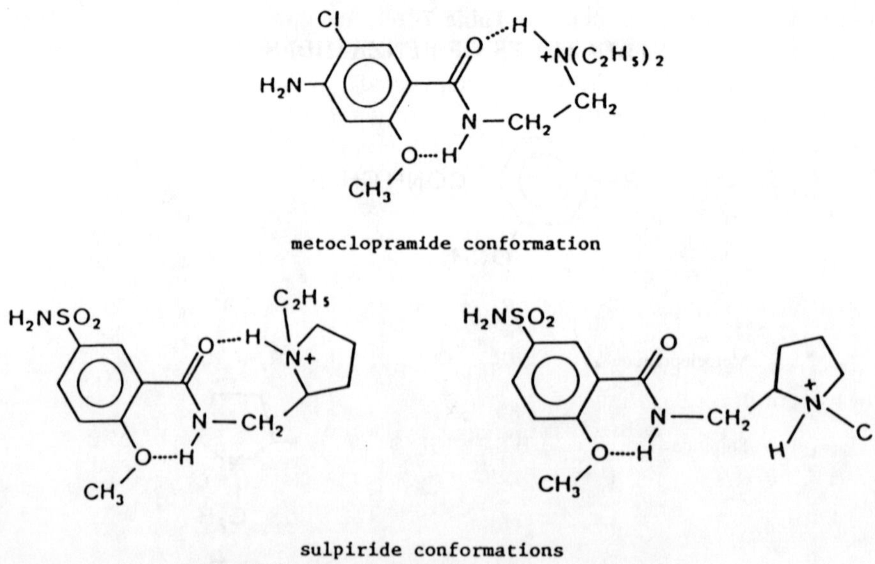

FIGURE 13. Conformations of benzamides.

since these are of only marginally higher energy (3 to 4 kCal/mol, and probably less if hydration effects are included). The distances between the center of the aromatic ring and the ammonium nitrogen are approximately 6.2 Å for the folded conformers, and 7.2 Å for the extended ones. The extended conformer of sulpiride was postulated as the active form responsible for D-2 antagonistic effects.[122]

Sulpiride

Pharmacokinetics

After administration of an oral dose of 108 mg radiolabeled sulpiride to 4 healthy male volunteers, peak plasma levels ranged from 232 to 403 ng/mℓ (mean 294 ng/mℓ) and occurred between 1 and 3 hr.[117] The plasma half-life ranged from 7.2 to 10.0 hr with a mean of 8.3 hr. The mean apparent volume of distribution was 128 ℓ and the mean plasma clearance rate was 197 mℓ/min. Urinary excretion of sulpiride accounted for 30 to 40% of the total dose[117,123-125] and was practically complete in 48 hr.[117] Fecal excretion was slower, accounting for 47 to 70% within 1 week. Sulpiride was only minimally metabolized in man: urine contained 2 to 5% of the 5-oxopyrrolidine analogue, whereas no evidence for metabolites was observed in plasma. In contrast to this finding, the 5-oxopyrrolidine analogue represented 10 to 30% of total radioactivity in urine and bile of rhesus monkeys; a second, unidentified metabolite was 10% or less.

Uses and Dosage

Sulpiride is used in the treatment of schizophrenia, GI disorders including gastric and duodenal ulcers, migraine, and vertigo. Although sulpiride is not marketed in the U.S., the world's largest market for antipsychotic drugs, it dominates three of the four next largest markets: Japan (#2), France (#4), and Argentina (#5), and it leads pharmacy sales of antipsychotics in many European countries.

The spectrum of activity of sulpiride as an antipsychotic is dominated by its strong antiautistic, antideliriant, and antihallucinatory effects.[111] In 44 psychiatric patients, oral doses of 300 to 600 mg/day during 1 month led to significant improvement in half of schizophrenic patients, and in one third of patients with affective psychoses, depressions, or neuroses, respectively.[126] Improvement was observed in a wide variety of symptoms referring to anxiety, defects in motivation and affect, agressivity, and sleep disturbances; but many of the improvements became statistically significant only after 5 to 10 days. Main side effects were sleep disturbances and extrapyramidal symptoms when the dosage was rapidly increased. In 112 long-term schizophrenic males, administration of 600 mg/day during 3 weeks led to statistically significant improvement in patients with hebephrenic and autistic schizophrenia.[127]

Fourteen schizophrenic patients were treated with 800 mg sulpiride for 1 week by i.m. injection and then orally for 5 weeks. Five of them improved markedly and could be discharged, one improved partially, 4 showed no change and 4 deteriorated.[128] Sulpiride and haloperidol were found equally effective in a double-blind trial involving 76 schizophrenic patients.[129]

The usual dosage of sulpiride for treatment of schizophrenia is 300 to 800 mg/day, by mouth or by i.m. injection. Higher doses may be required in severe psychiatric conditions; doses up to 2.6 g daily have been administered chronically.[125]

Toxicity

Sleep disturbances, overstimulation, and agitation may be associated with sulpiride. Extrapyramidal effects appear to be few and occur mainly when doses are increased rapidly. Hypertension may be aggravated by sulpiride. Sulpiride is contraindicated in patients with phaeochromocytoma.

Physical-Chemical and Other Data

Sulpiride is a white, odorless, crystalline substance, m.p. 175 to 182°C (d). It is practically insoluble in water and ether, and sparingly soluble in methanol. pKa 9.1.

Chemical name — 5-(Aminosulfonyl)-*N*-[(1-ethyl-2-pyrrolidinyl)methyl]-2-methoxybenzamide, $C_{15}H_{23}N_3O_4S$ (341.4).

Dosage form (France) — Capsules (50 mg), tablets (200 mg), injectable (100 mg), drinkable solution (500 mg, 1 g).

Trade names — Sulpiride is sold under a variety of trade names: Dogmatil (Delagrange, France), Abilit (Sumitomo, Japan), Dobren (Ravizza, Italy), Miradol (Mitsui, Japan), Paspertin (Kali Chemie, Germany). Sulpiride is not available in the U.S.

Sulpiride has also been used in the treatment of GI disorders (gastric and duodenal ulcers), migraine, and vertigo, as well as an antiemetic and antihallucinatory comedication in surgical procedures.

Dibenzo-Azepines

SARs

Several examples of dibenzo-oxazepines, dibenzo-thiazepines, dibenzo-diazepines, and dibenzo-azepines are of clinical interest (see Table 8). Loxapine, clotiapine, and metiapine are classical antipsychotics with a pharmacological profile that is typical of antipsychotic drugs. Clozapine has an unusual pharmacologic profile, since it does not cause catalepsy, does not antagonize apomorphine induced emesis, and is a weak antagonist of apomorphine-induced stereotyped behavior in rats. As previously described, clozapine produces minimal extrapyramidal symptoms and does not block dopamine receptors, although it does inhibit dopamine-sensitive adenylate-cyclase. Perlapine is devoid of antipsychotic activity but displays sedative/hypnotic properties.

Amoxapine, a loxapine analogue with a hydrogen on the piperazine nitrogen in position 4 in place of the methyl group as in loxapine, has antidepressant properties.

SARs of dibenzo-oxazepines, dibenzo-thiazepines, and dibenzo-diazepines have been described in depth by Schmutz.[130] SARs of a wide variety of tricyclic systems with 6-7-6 membered rings, especially of dibenzo-thiepines, have been reviewed by Kaiser and Setler.[89]

Table 8
STRUCTURES OF DIBENZO-AZEPINES

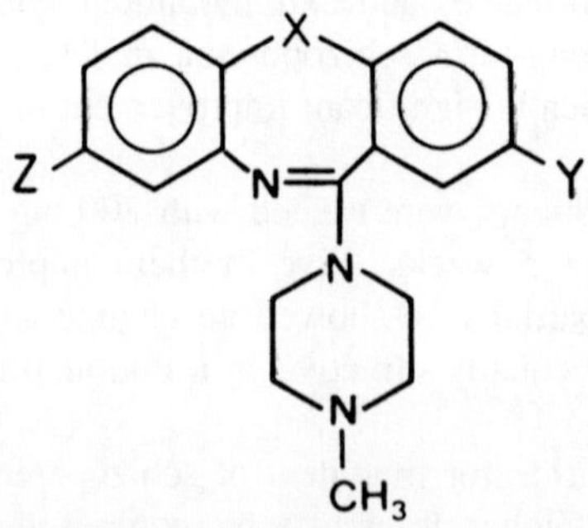

Generic name	Chemical class	X	Y	Z
Loxapine	Dibenzo-oxazepine	0	Cl	H
Clotiapine	Dibenzo-thiazepine	S	Cl	H
Metiapine	Dibenzo-thiazepine	S	CH_3	H
Clozapine	Dibenzo-diazepine	NH	H	Cl
Perlapine	Dibenzo-azepine	CH_2	H	H

Substitution on one benzene ring was found essential to neuroleptic activity.[130] Of all the chlorosubstituted compounds tested, only 2-substituted dibenzo-azepines were found to antagonize apomorphine-induced effects in rats and to cause catalepsy. A linear relationship was found between the logarithm of the dose for apomorphine antagonism and the Hammet constant σ_m: strongly electron-withdrawing substituents in 2-position led to highly potent compounds. In the oxazepine and thiazepine series, electron-withdrawing substituents in 2-position resulted in low cataleptic activity, thus yielding compounds with a desirable profile of a high ratio of apomorphine-antagonistic to cataleptic activity.

Antipsychotic effects of dibenzo-azepines were strongly dependent on the nature of the side chain: only dibasic piperazine substituents showed activity.

Stereochemical and Conformational Features

Clozapine possesses a rigid, planar tricyclic ring structure due to the considerable steric hindrance imposed by extensive hydrogen-hydrogen interactions. The conformational rigidity also found in anticholinergic drugs may explain the unusual pharmacologic profile of clozapine. Among all antipsychotic drugs, clozapine has the highest anticholinergic activity. This activity seems to be responsible for the lack of catalepsy and the minimal extrapyramidal symptoms produced by clozapine.

The nonrigid nonplaner clotiapine, metiapine, and loxapine are classical antipsychotics with a pharmacological profile that is typical of antipsychotic drugs.

The structure of clozapine, shown in Figure 14 has been determined.

Loxapine
Pharmacokinetics

Heel and co-workers have reported unpublished data from Lederle Laboratories on the pharmacokinetic properties of loxapine.[131] Absorption in man is rapid and nearly complete after oral administration. In 40 volunteers, oral doses of 25 mg in liquid concentrate or capsule form led to peak serum levels of 12.60 mg/mℓ (between 1 and 2 hr) and 9.83 ng/mℓ (between 2 and 3 hr) respectively. Excretion of radioactivity occurred predominantly in urine (56 to 70%, as glucuronide or sulfate conjugates) and to a lesser extent in the feces

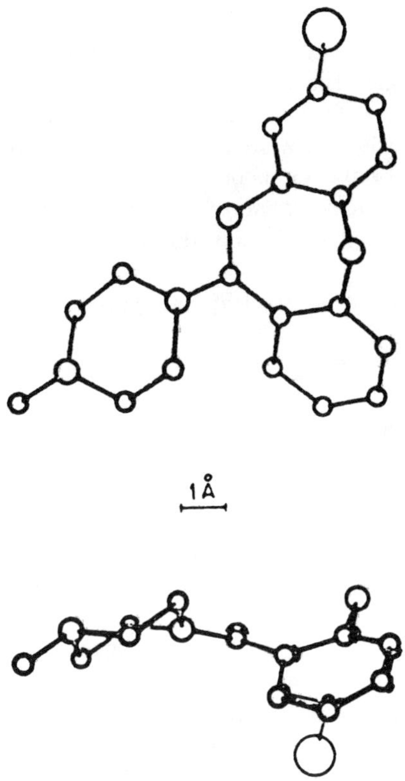

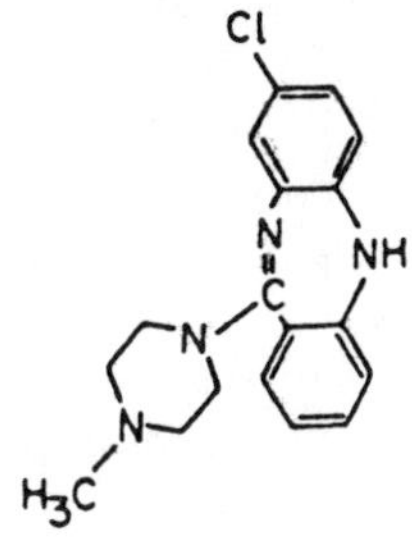

Clozapine Base

FIGURE 14. Two orthogonal views of the observed conformation of clozapine. (From Pauling, P., Ramsay, W., and Ingold, C., in *Psychotherapautic Drugs,* Part 1, Usdin, E. and Forrest, I. S., Eds., Marcel Dekker, New York, 1976, 59. With permission.)

(15 to 22%, mainly unconjugated metabolites). Loxapine was rapidly and extensively metabolized; no unchanged drug was found in either urine or feces. The major metabolite, accounting for more than 50% of the excreted radioactivity, was hydroxyloxapine. Hydroxyloxapine-*N*-oxide, loxapine-*N*-oxide, and hydroxydesmethyl loxapine were also observed in substantial quantities. The pharmacological activity of loxapine metabolites has not been studied, but its rapid metabolism suggests that one or more of the metabolites may represent the active form in man.[131]

Uses and Dosage

Loxapine is used for the treatment of schizophrenia. In open studies, loxapine therapy resulted predominantly in mild to moderate improvement of chronic schizophrenia.[132-135] Comparative trials showed that loxapine is about equieffective with chlorpromazine, haloperidol, trifluoperazine, and thiothixene, and superior to placebo, in acute or chronic schizophrenia.[168] Loxapine was superior to chlorpromazine and trifluoperazine in patients with paranoid schizophrenia.[136]

An initial oral dosage of 10 mg twice daily is recommended for the treatment of schizophrenia, but in severely disturbed patients initial dosage up to 50 mg daily may be desirable. Dosage should be increased fairly rapidly over the first 7 to 10 days until effective control of psychotic symptoms is achieved. For maintenance therapy, the dosage should be reduced to the lowest level compatible with symptom control. The usual daily dosage is 60 to 100 mg, but for many patients 20 to 60 mg may be adequate.

Toxicity

Extrapyramidal reactions (39% of patients) and sedation during initial treatment (28%) are the most common side effects of loxapine. The overall incidence of EPS caused by loxapine is similar to that caused by other phenothiazines. Rigidity (34%), tremor (16%), akathisia (15%), dystonia (14%), and hypersalivation were the predominant forms of these reactions.[131] Persistent tardive dyskinesia has not been reported with loxapine, but cannot be ruled out in light of the similarity of loxapine to other antipsychotic agents. The incidence of extrapyramidal effects appears to be doubled when daily doses of over 150 mg are administered.

Anticholinergic or cardiovascular effects (hypotension, tachycardia) occurred in 8 and 5% of patients with loxapine, respectively.

Physical-Chemical Data

Loxapine is a slightly yellow crystalline material, m.p. 109 to 110°C. Loxapine is used clinically as the succinate or hydrochloride salt.

Chemical name — 2-Chloro-11-(4-methyl-1-piperazinyl)-dibenz[b,f][1,4]oxazepine, $C_{18}H_{18}ClN_3O$), (327.81). Succinate: $C_{18}H_{18}ClN_3O{\cdot}C_4H_6O_4$ (445.9).

Dosage forms — (U.S. Loxitane® Lederle: Capsules (succinate, equivalent to 5, 10, 25, 50 mg loxapine base), oral concentrate (hydrochloride, 25 mg base/mℓ), injection (hydrochloride, 50 mg base/mℓ).

Other Agents

Antipsychotic activity is by no means restricted to the five classes previously discussed. The structures of three agents representing different chemical classes are shown below. Molindone (Moban®) is often used in the U.S. for the management of symptoms of schizophrenia. Oxypertine is utilized in Europe and Japan in the treatment of psychoses in withdrawn patients and of anxiety neuroses. Butaclamol is a pharmacological tool used in dopamine receptor binding studies with antipsychotic drugs.[49,50]

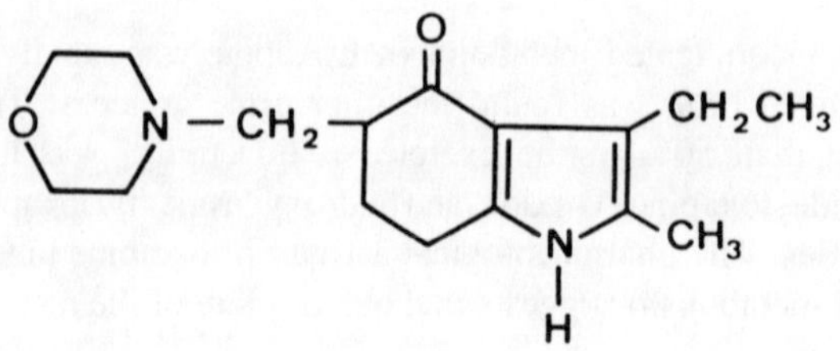

Molindone

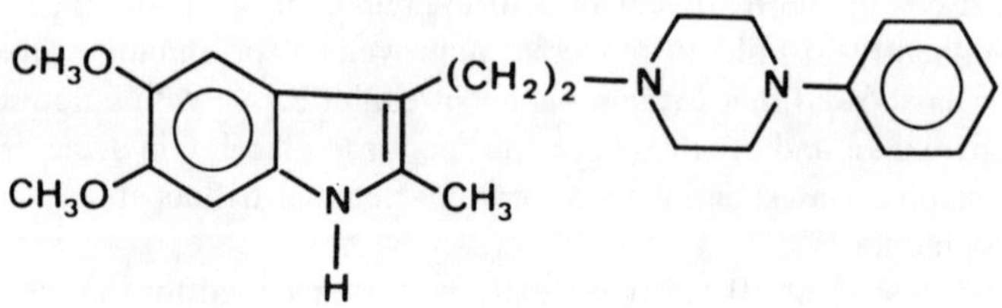

Oxypertine

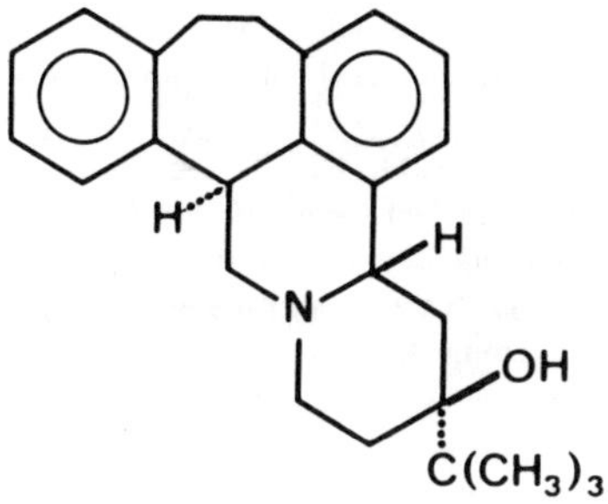

Butaclamol

REFERENCES

1. **Jarvik, M. E.,** Drugs used in the treatment of psychiatric disorders, in *The Pharmacological Basis of Therapeutics, 3rd ed.,* Goodman, L. S. and Gilman, A., Eds., Macmillan, New York, 1965, 159.
2. **Tedeschi, D. H.,** Comparative effects of neuroleptics on central and peripheral autonomic outflow, in *Neuropsychopharmacology,* Vol. 5, Int. Congr. Series #129, Brill, H. and Cole, J. O., Eds., Excerpta Medica, Amsterdam, 1967, 314.
3. **Jacobsen, E.,** The pharmacological classification of central nervous depressants, *J. Pharm. Pharmacol.,* 10, 273, 1958.
4. **Reinhard, J. F. and Vida, J. A.,** unpublished observation, 1974.
5. **Winter, C. A. and Flataker, L.,** The effect of cortisone, desoxycorticosterone, and adrenocorticotrophic hormone upon the responses of animals to analgesic drugs, *J. Pharmacol. Exp. Ther.,* 103, 93, 1951.
6. **Caviezel, R. and Baillod, A.,** Zur Differenzierung Psychotroper Substanzen Anhand Der Motilität Der Maus, *Pharm. Acta Helv.,* 33, 469, 1958.
7. **Riley, H. and Spinks, A.,** Biological assessment of tranquillisers. I, *J. Pharm. Pharmacol.,* 10, 657, 1958.
8. **Janssen, P. A. J.,** Methods of evaluating neuroleptic drugs, in *Methods in Drug Evaluation,* Mantegazza, P. and Piccinini, F., Eds., North-Holland, Amsterdam, 1966, 234.
9. **Tedeschi, R. E., Tedeschi, D. H., Mucha, A., Cook, L., Mattis, P. A., and Fellows, E. J.,** Effects of various centrally acting drugs on fighting behavior of mice, *J. Pharmacol. Exp. Ther.,* 125, 28, 1959.
10. **Randall, L. O., Schallek, W., Heise, G. A., Keith, F. A., and Bagdon, R. E.,** The psychosedative properties of methaminodiazepoxide, *J. Pharmacol. Exp. Ther.,* 129, 163, 1960.
11. **Cook, L. and Kelleher, R. T.,** Effects of drugs on behavior, in *Annual Review of Pharmacology,* Vol 3, Cutting, W. C., Ed., Annual Reviews, Palo Alto, Calif., 1963, 205.
12. **Dews, P. B. and Morse, W. H.,** Behavioral pharmacology, in *Annual Review of Pharmacology,* Cutting, W. C., Ed., Vol 1, Annual Reviews, Palo Alto, Calif., 1961 145.
13. **Riley, H. and Spinks, A.,** Biological assessment of tranquillisers. II, *J. Pharm. Pharmacol.,* 10, 721, 1958.
14. **Skinner, B. F.,** The control of human behavior, *Trans. N.Y. Acad. Sci.,* 17, 547, 1955.
15. **Sidman, M.,** Avoidance conditioning with brief shock and no exteroceptive warning signal, *Science,* 118, 157, 1953.
16. **Janssen, P. A. J. and Van Bever, W. F.,** Advances in the search for improved neuroleptic drugs, in *Current Developments in Psychopharmacology,* Vol. 2, Essman, W. E. and Valzelli, L., Eds., Spectrum Publications, New York, 1975, 165.
17. **Janssen, P. A. J., Niemegeers, C. J. E., Schellekens, K. H. L., and Lenaerts, F. M.,** Is it possible to predict the clinical effects of neuroleptic drugs (major tranquilizers) from animal data?, *Arzneimittelforsch.,* 17, 841, 1967.
18. **Carlsson, A.,** Drugs which block the storage of 5-hydroxytryptamine and related amines, in *Handbook of Experimental Pharmacology,* Vol. 19, Eichler, O. and Farah, A., Eds., Springer-Verlag, Berlin, 1965, 529.
19. **Carlsson, A.,** Antipsychotic drugs and catecholamine synapses, *Catecholamines and Schizophrenia,* Matthysse, S. W. and Kety, S., Eds., Pergamon Press, 1975, 57.
20. **Hornykiewicz, O.,** Dopamine (3-hydroxytyramine) and brain function, *Pharmacol. Rev.,* 18, 925, 1966.

21. **Carlsson, A.,** Mechanism of action of neuroleptic drugs, in *Psychopharmacology: A Generation of Progress,* Lipton, M. A., Dimascio, A., and Killam, K. F., Eds., Raven Press, N.Y., 1978, 1057.
22. **Anden, N-E., Butcher, S. G., Corrodi, H., Fuxe, K., and Ungerstedt, U.,** Receptor activity and turnover of Dopamine and noradrenaline after neuroleptics, *Eur. J. Pharmacol.,* 11, 303, 1970.
23. **Jackson, D. M., Anden, N-E., and Dahlstrom, A.,** A functional effect of dopamine in the nucleus accumbens and in some other dopamine-rich parts of the rat brain, *Psychopharmacologia,* 45, 139, 1975.
24. **Pisnenburg, A. J. J., Woodruff, G. N., and van Rossum, J. M.,** Ergometrine induced locomotor activity following intracerebral injection into nucleus accumbens, *Brain Res.,* 59, 289, 1973.
25. **Ahlenius, S. and Engel, J.,** Antagonism by haloperidol of L-dopa-induced description of a successive discrimination in rat, *J. Neural. Transm.,* 36, 43, 1975.
26. **Ahlenius, S. and Engel, J.,** Potentiation by a-methyltyrosine of the suppression of food-reinforced lever pressing behavior induced by antipsychotic drugs, *Acta. Pharmcol. Toxikol.* (KBH), 40, 115, 1977.
27. **Anden, N. E., Roos, B. E., and Werdinius, B.,** Effects of chlorpromazine, haloperidol and reserpine on the levels of phenolic acids in rabbit corpus striation, *Life Sci.,* 3, 149, 1964.
28. **Carlsson, A. and Lindquist, M.,** Effect of chlorpromazine of haloperidol on formation of 3-methoxytyramine and normetanephrine in mouse brain, *Acta Pharmakol. Toxicol. (KBH),* 20, 140, 1963.
29. **Da Prada, M. and Pletscher, A.,** Acceleration of the cerebral dopamine turnover by chlorpromazine, *Experientia,* 22, 465, 1966.
30. **Nybäck, H. and Sedvall, G.,** Further studies on the accumulation and disappearance of catecholamines formed from tyrosine-'4C in mouse brain. Effect of some phenothiazine analogues, *Eur. J. Pharmacol.,* 10, 193, 1970.
31. **Roos, B. E.,** Effects of certain tranquilizers on the level of homovanillic acid in the corpus striatum, *J. Pharm. Pharmacol.,* 17, 820, 1965.
32. **Bartholini, G., Stadler, H., Gadea-Ciria, M., and Lloyd, K. G.,** The effect of antipsychotic drugs on the release of neurotransmitters in various brain areas, in *Antipsychotic Drugs, Pharmacodynamics and Pharmacokinetics,* Sedvall, G., Ed., Pergamon Press, Oxford, 1976, 105.
33. **Carlsson, A.** Mechanism of action of neuroleptic drugs, in *Psychopharmacology: A Generation of Progress,* Lipton, M. A., Dimascio, A., and Killam, K. F., Eds., Raven Press, New York, 1978, 1059.
34. **Miller, R. J. and Hiley, C. R.,** Antimuscarinic properties of neuroleptics and drug induced Parkinsonism, *Nature (London),* 248, 596, 1974.
35. **Snyder, S. H., Greenberg, D., and Yamamura, H.,** Antischizophrenic drugs and brain cholinergic receptors, *Arch. Gen. Psychiatr.,* 31, 58, 1974.
36. **Snyder, S. H., Banerjee, S. P., Yamamura, H. I., and Greenberg, D.,** Drugs, neurotransmitters and schizophrenia, *Science,* 184, 1243, 1974.
37. **Kehr, W., Carlsson, A., Lindquist, M., Magnusson, T., and Atack, C.,** Evidence for a receptor-mediated feedback control of striatal tyrosine hydroxylase activity, *J. Pharm. Pharmacol.,* 24, 744, 1972.
38. **Christiansen, J. and Squires, R. F.,** Antagonistic effects of apomorphine and haloperidol on rat striatal synaptosomal tyrosine hydroxylase, *J. Pharm. Pharmacol.,* 26, 367, 1974.
39. **Christiansen, J. and Squires, R. F.,** Antagonistic effects of neuroleptics and apomorphine on synaptasomal tyrosine hydroxylase in vitro, *J. Pharm. Pharmacol.,* 26, 742, 1974.
40. **Farnebo, L. O. and Hamberger, B.,** Drug induced changes in the release of 3 H monamines from field stimulated rat brain slices, *Acta Physiol. Scand. Suppl.,* 371, 35, 1971.
41. **Farnebo, L. O. and Hamberger, B.** Catecholamine release and receptors in brain slices in *Frontiers in Catecholamine Research,* Usdin, E. and Snyder, S. H., Eds., Pergamon Press, New York, 1973, 589.
42. **Seeman, P. and Lee, T.,** Antipsychotic drugs: direct correlation between clinical potency and presynaptic action on dopamine neurons, *Science,* 188, 1217, 1975.
43. **Kebabian, J. W., Petzold, G. L., and Greengard, P.,** Dopamine sensitivity adenylate cyclase in caudate nucleus of rat brain, and its similarity to the "dopamine receptor", *Proc. Natl. Acad. Sci. U.S.A.,* 69, 2145, 1972.
44. **Iversen, L. L., Rogawski, M. A., and Miller, R. J.,** Comparison of the effects of neuroleptic drug on pre and post synaptic dopaminergic mechanisms in the rat striatum, *Mol. Pharmacol.,* 12, 251, 1976.
45. **Janssen, P. A. J. and Van Bever, W. F. M.,** Advances in the search for improved neuroleptic drugs, in *Current Developments in Psychopharmacology,* Vol. 2, Essman, W. B. and Valzelli, L., Eds., Spectrum, New York, 1975, 165.
46. **Creese, I. and Snyder, S. H.,** Behavioral and biochemical properties of the dopamine receptor, in *Psychopharmacology: A Generation of Progress,* Lipton, M. A., DiMascio, A., and Killam, K. F., Eds., Raven Press, New York, 1978, 377.
47. **Burt, D. R., Enna, S. J., Creese, L., and Snyder, S. H.,** Dopamine receptor binding in the corpus striatum of mammalian brain, *Proc. Natl. Acad. Sci. U.S.A.,* 72, 4655, 1975.
48. **Creese, I., Burt, D. R., and Snyder, S. H.,** Dopamine receptor binding: differentiation of agonist and antagonist states with ^{3}H-dopamine and ^{3}H-haloperidol, *Life Sci.,* 17, 993, 1975.

49. **Creese, I., Burt, D. R., and Snyder, S. H.,** Dopamine receptor binding predicts clinical and pharmacological potencies of antischizophrenic drugs, *Science*, 192, 481, 1976.
50. **Seeman, P., Lee, T., Chau-Wong, M., and Wong, K.,** Antipsychotic drug doses and neuroleptic/dopamine receptors, *Nature (London)*, 261, 717, 1976.
51. **Spano, P. F., Memo, M., Stefanini, E., Fresia, P., and Trabucchi, M.,** Detection of multiple receptors for dopamine, in *Advances in Biochemical Psychopharmacology*, Vol. 27, Pepeu, G., Kuhar, M. J., and Enna, S. J., Eds., Raven Press, New York, 1980, 243.
52. **Kebabian, J. W. and Calne, D. B.,** Multiple receptors for dopamine, *Nature (London)*, 277, 93, 1979.
53. **Kebabian, J. W. and Cote, T. E.,** Dopamine receptors and cyclic AMP: a decade of progress, *Trends Pharmacol. Sci.*, 2, 69, 1981.
54. **Seeman, P.,** Anti-schizophrenic drugs — membrane receptor sites of action, *Biochem. Pharmac.*, 26, 1741, 1977.
55. **Caron, M. G.,** Dopaminergic receptors in the anterior pituitary gland, *J. Biol. Chem.*, 253, 2244, 1978.
56. **Hyttel, J.,** Effects of neuroleptics on ^{3}haloperidol and ^{3}H-Cis(Z) — flupenthixol binding and on adenylate cyclase activity *in vitro, Life. Sci.*, 23, 551, 1978.
57. **Calne, D. B., Williams, A. C., Neopaytides, A., Plotkin, C., Nutt, J. G., and Teychenne, P. F.,** Long-term treatment of Parkinsonism with Bromocriptine, *Lancet*, 1, 735, 1978.
58. **Janssen, P. A. J.,** Structure-activity relationships (SAR) and drug design as illustrated with neuroleptic agents, in *International Encyclopedia of Pharmacology and Therapeutics*, Section 5, Vol. 1, Cavallito, C. J., Ed., Pergamon Press, 1973, New York, 1973, 34.
59. **Hollister, L. E.,** *Antipsychotics in Clinical Pharmacology of Psychotherapeutic Drugs*, Churchill Livingstone, New York, 1978, 131.
60. **Curry, S. H.,** Plasma protein binding of chlorpromazine, *J. Pharmacol.*, 22, 193, 1970.
61. **Usdin, E.,** Metabolic pathways of antipsychotic drugs, in *Psychopharmacology: A Generation of Progress*, Lipton, M. A., DiMascio, A., and Killam, K. F., Eds., Raven Press, New York, 1978, 896.
62. **Dlabac, A.,** Pharmacology of long-acting neuroleptic drugs from perathiepin series, *J. Pharmacol (Paris)*, 5 (Suppl. 2), 26, 1974.
63. **Ota, K. Y., Kurland, A. A., and Slotnick, V. B.,** Safety evaluation of penfluridol, a new long-acting oral antipsychotic agent, *J. Clin. Pharmacol.*, 14, 202, 1974.
64. **Van Praag, H. M., Schut, T., Dols, L., and Van Schilfgaarden, R.,** Controlled trail of penfluridol in acute psychosis, *Br. Med. J.*, 4, 710, 1971.
65. **Gallant, D. M., Mielke, D. H., Spirtes, M. A., Swanson, W. C., and Bost, R.,** Penfluridol: an efficacious long acting oral antipsychotic compound, *Am. J. Psychiatr.*, 131, 699, 1974.
66. **Julou, L.,** Etude expérimentale des neuroleptiques injectables à action prolongée, in *Actualités Pharmacologiques*, Vol. 25, Cheymol, J., Boissier, J. R., and Lechat, P., Eds., Masson, Paris, 1972, 23.
67. **Ginestet, D. and Julou, L.,** Long acting injectable neuroleptics, *Trends Pharmacol. Sci.*, 3, 62, 1982.
68. **Horn, A. S. and Snyder, S. H.,** Chlorpromazine and dopamine: conformational similarities that correlate with the antischizophrenic activity of phenothiazine drugs, *Proc. Natl. Acad. Sci. U.S.A.*, 68, 2325, 1971.
69. **Pauling, P., Ramsay, W., and Ingold, C.,** The significance of conformation, in *Psychotherapeutic Drugs*, Part 1, Usdin, E. and Forrest, I. S., Eds., Marcel Dekker, New York, 1976, 59.
70. **Kaiser, C. and Setler, P. E.,** Antipsychotic agents, in *Burger's Medicinal Chemistry*, 4th ed., Part III, Wolff, M. E., Ed., John Wiley & Sons, New York, 1981, 874.
71. **Janssen, P. A. J., Niemegeers, C. J. E., and Schellekens, K. H. L.,** Is it possible to predict the clinical effects of neuroleptic drugs (major tranquilizers) from animal data? I. Neuroleptic activity spectra for rats, *Arzneim.-Forsch.*, 15, 104, 1965.
72. **Charpentier, P.,** 10-Dialkylaminoalkyl-2-or 4-Chlorophenothiazines, U.S. Patent 2,645,640 (1953 to Rhone Poulenc).
73. **Delay, J., Deniker, P., and Harl, J. M.,** Utilisation en thérapeutique psychiatrique d'une phénothiazine d'action centrale élective (4560 RP), *Ann. Med. Psychol. Fr.*, 110, 112, 1952.
74. **Shader, R. I. and DiMascio, A.,** *Psychotropic Drug Side Effects*, Williams & Wilkins, Baltimore, 1970.
75. **Davis, J. M.,** Comparative doses and costs of antipsychotic medications, *Arch. Gen. Psychiatr.*, 33, 858, 1976.
76. **Sakalis, G., Traficante, L. J., Gardos, G., and Gershon, S.,** Treatment of drug resistant schizophrenics with mesoridazine, in *Phenothiazines and Structurally Related Drugs: Basic and Clinical Studies*, Usdin, E., Eckert, H., and Forrest, I. S., Eds., Elsevier, New York, 1980, 207.
77. **Gardos, G., Tecce, J. J., Hartmann, E., Bowers, P., and Cole, J. O.,** Treatment with mesoridazine and thioridazine in chronic schizophrenia, *Comp. Psychiatr.*, 19, 519, 1978.
78. **Post, M. L., Kennard, O., and Horn, A. S.,** Stereoselective blockade of the dopamine receptor and the x-ray structures of α- and β-Flupenthixol, *Nature (London)*, 256, 342, 1975.
79. **Mϕller-Nielsen, I., Pedersen, V., Nymark, M., Franck, K. F., Bocek, V., Fjalland, B., and Christensen, A. V.,** The comparative pharmacology of flupenthixol and some reference neuroleptics, *Acta Pharmacol. Toxicol.*, 33, 353, 1973.

80. **Hobbs, D. C., Welch, W. M., Short, M. J., Moody, W. A., and Van der Velde, C. D.,** Pharmacokinetics of thiothixene in man, *Clin. Pharmac. Ther.*, 16, 473, 1974.
81. **Bergling, R., Mjorndal, T., Oreland, L., Rapp, W., and Wold, S.,** Plasma levels and clinical effects of thioridazine and thiothixene, *J. Clin. Pharmacol.*, 15, 178, 1975.
82. **Sterlin, C., Ban, T. A. and Jarrold, L.,** The place of thiothixene among the thioxanthenes, *Curr. Ther. Res.*, 14, 205, 1972.
83. **Wolpert, A., Quintos, A., White, L., and Merlis, S.,** Thiothixene and chlorprothixene in behavior disorders, *Curr. Ther. Res.*, 10, 566, 1968.
84. **Kellam, A. M. P. and Jones, K. S.,** A double blind controlled trial of thiothixene and perphenazine in chronic schizophrenics, shown to require maintenance therapy, *Acta Psychiat. Scand.*, 47, 174, 1971.
85. **Belda, F. J. M. and De Haro, T.,** Clinical trial with thiothixene parenteral in agitated schizophrenic patients, *Curr. Ther. Res.*, 11, 599, 1969.
86. **Brophy, J. J.,** Psychiatric disorders, in *Current Medical Diagnosis and Treatment*, Krupp, M. A. and Chatton, M. J., Eds., Lange Medical Publications, Los Altos, 1982, 607.
87. *Physicians Desk Reference*, Medical Economics Co., Oradell, N. J., 1982, 1638.
88. **Janssen, P. A. J.,** Structure-activity relations (SAR) and drug design as illustrated with neuroleptic agents, in *Antipsychotic Drugs: Pharmacodynamics and Pharmacokinetics*, Sedvall, G., Uvnas, B., and Zotterman, G., Eds., Pergamon Press, New York, 1976, 5.
89. **Kaiser, C. and Setler, P. E.,** Antipsychotic agents, in *Burger's Medicinal Chemistry*, 4th ed., Part III, Wolff, M. E., Ed., John Wiley & Sons, New York, 1981, 859.
90. **Janssen, P. A. J., Van de Westeringh, C., Jageneau, A. H. M., Demoen, P. J. A., Hermans, B. K. F., Van Daele, G. H. P., Schellekens, K. H. L., Van der Eycken, C. A. M., and Niemegeers, C. J. E.,** Chemistry and pharmacology of CNS depressants related to 4-(4-hydroxy-4-phenylpiperidino) butyrophenone. I. Synthesis and screening data in mice, *J. Med. Pharm. Chem.*, 1, 281, 1959.
91. **Janssen, P. A. J. and Niemegeers, C. J. E.,** Analysis of the influence of haloperidol and pharmacologically related drugs on learned avoidance-escape habits of Wistar rat in a "jumping box" situation, *Arzneim.-Forsch.*, 11, 1037, 1961.
92. **Harper, N. J., Simmonds, A. B., Wakama, W. T., Hall, G. H., and Vallance, D. K.,** Some basic ketones with central nervous system depressant activity, *J. Pharm. Pharmacol.*, 18, 150, 1966.
93. **Grogan, C. H., Geschickter, C. F., Freed, M. E., and Rice, L. M.,** Spiranes. VII. Neuroleptics derived from azaspiranes, *J. Med. Chem.*, 8, 62, 1965.
94. **Janssen, P. A.,** Butyrophenones and diphenylbutylpiperidines, in *Psychopharmacological Agents*, Vol. 3, Gordon, M., Ed., Academic Press, New York, 1974, 129.
95. **Koch, M. H. J.,** The conformation of neuroleptic drugs, *Mol. Pharmac.*, 10, 425, 1974.
96. **Cressman, W. A., Bianchine, J. R., Slotnick, V. B., Johnson, P. C., and Plostnieks, J.,** Plasma level profile of haloperidol in man following intramuscular administration, *Eur. J. Clin. Pharmacol.*, 7, 99, 1974.
97. **Johnson, P. C., Jr., Charalampous, K. D., and Braun, G. A.,** Absorption and excretion of tritiated haloperidol in man (a preliminary report), *Int. J. Neuropsychiatr.*, 3, S24, 1967.
98. **Marcucci, F., Mussini, E., Airoldi, L., Fanelli, R., Frigerio, A., De Nadai, F., Bizzi, A., Rizzo, M., Morselli, P. L., and Garattini, S.,** Analytical and pharmacokinetic studies on butyrophenones, *Clin. Chim. Acta*, 34, 321, 1971.
99. *Physician's Desk Reference 36*, Medical Economics Co., Oradell, N.J., 1982, 1158.
100. **Kazamatsuri, H., Chiung-piao Chen, and Cole, J. O.,** Treatment of tardive dyskinesia. II. Short-term efficacy of dopamine-blocking agents haloperidol and thiopropazate, *Archs. Gen. Psychiatr.*, 27, 100, 1972.
101. **Kazamatsuri, H., Chiung-piao Chen, and Cole, J. O.,** Long-term treatment of tardive dyskinesia with haloperidol and tetrabenazine, *Am. J. Psychiatr.*, 130, 479, 1973.
102. **Grosshandler, S., Hanschel, E. O., and Kampine, J. P.,** Toxic reactions due to drug synergism and antagonism, *Anesth. Analg. Curr. Res.*, 47, 345, 1968.
103. **Wright, I. S.,** Recent developments in antithrombotic therapy, *Ann. Intern. Med.* 71, 823, 1969.
104. **Shields, K. G. and Ballinger, C. M.,** Antiemetic effectiveness of haloperidol in human volunteers challenged with apomorphine, *Anesth. Analg. Curr. Res.*, 50, 1017, 1971.
105. **Lord, D. J. and Kidd, C. B.,** Haloperidol versus diazepam. A double-blind crossover clinical trial, *Med. J. Aust.*, 586, 1973.
106. **Justin-Besançon, L., Laville, C., and Thominet, M.,** Metoclopramide and its homologs — introduction to their biological study, *C. R. Acad. Sci. (Paris)*, 258, 4384, 1964.
107. **Bossier, J. R., Simon, P., Lwoff, J. M., and Fichell-Pagny, J.,** Action psycholeptique du métoclopramide, *C. R. Acad. Sci. (Paris)*, 158, 1859, 1964.
108. **Laville, C.,** Effets du métoclopramide sur le systeme nerveux central, *Pathol. Biol.*, 12, 723, 1964.
109. **Laville, C.,** Effets du métoclopramide sur le systeme nerveux central, *Sem Ther.*, 40, 649, 1964.
110. **Justin-Besançon, L., Thominet, M., Laville, C., and Margarit, J.,** Constitution chimique et propriétés du sulpiride, *C. R. Acad. Sci. (Paris)*, 265, 1253, 1967.

111. **Deligné, P. and Bunodière, M.,** Neurople giques et neuroleptiques antipsychotiques. IX. Les benzamides (2eme partie), *Ann. Anesth. Franc.*, 4, 477, 1980.
112. **Tagliamonte, A., De Montis, G., Olianis, M., Vargiu, L., Corsini, G. U., and Gessa, G. L.,** Selective increase of brain dopamine synthesis by sulpiride, *J. Neurochem.*, 24, 707, 1975.
113. **Jenner, P., Elliott, P. N. C., Clow, A., Reavill, C., and Marsden, C. D.,** A comparison of *in vitro* and *in vivo* dopamine receptor antagonism produced by substituted benzamide drugs, *J. Pharm. Pharmacol.*, 30, 46, 1978.
114. **Fontaine, O., Libon, P., and Richelle, M.,** Action of a new psychotropic drug (sulpiride) on avoidance behavior in rats, *Psychopharmacologia*, 39, 309, 1974.
115. **Justin-Besançon, L., Laville, C., Margarit, J., and Thominet, M.,** Chemical composition and biological properties of substituted o-anisamides containing an alkyl sulfone group, *C. R. Acad. Sci. (Paris)*, 279, 375, 1974.
116. **Deligné, P. and Bunodière, M.,** Neuroplegiques et neuroleptiques antipsychotiques. VIII. Les benzamides (lere partie), *Ann. Anesth. Franc.*, 1, 97, 1980.
117. **Imondi, A. R., Alam, A. S., Brennan, J. J., and Hagerman, L. M.,** Metabolism of sulpiride in man and rhesus monkeys, *Arch. Int. Pharmacodyn.*, 232, 79, 1978.
118. **Jenner, P., Clow, A., Reavill, C., Theodoron, A., and Marsden, C. D.,** Stereoselective actions of substituted benzamide drugs on cerebral dopamine mechanism, *J. Pharm. Pharmacol.*, 32, 39, 1980.
119. **Hofmann, M., Jommi, G. C., Montefusco, O., Tonon, G. C., Spano, P. F., and Trabucci, M.,** Stereospecific effects of (–)sulpiride on brain dopamine metabolism and prolactin release, *J. Neurochem.*, 32, 1547, 1979.
120. **Waterbeemd, H. and Testa, B.,** PCILO and CD. Conformational study of sulpiride, a dopamine antagonist, *Helv. Chim. Acta*, 64, 2183, 1981.
121. **Pannatier, A., Anker, L., Testa, B., and Carrupt, P. A.,** A theoretical conformational study of substituted O-anisamides as models of a class of dopamine antagonists, *J. Pharm. Pharmacol.*, 33, 145, 1981.
122. **Waterbreemd, H. and Testa, B.,** Conformational studies of substituted benzamide drugs, in North American Medicinal Chemistry Symposium, Toronto, Canada, June 20 to 24, 1982, p. 103.
123. **Koishi, H., Okuda, T., and Yamashita, S.,** 5th Congr. on Therapeutic Research on New Drugs, Tokyo, 1968, 27.
124. **Kleimola, T., Leppanen, O., Kanto, J., Mantyla, R., and Syvalahti, E.,** Spectrophotofluorometric method for quantitative determination of sulpiride in human plasma and urine, *Ann. Clin. Res.*, 8, 104, 1976.
125. **Imondi, A. R., Alam, A., Brennan, J. J., and Hagerman, L. M.,** Sulpiride metabolism in rhesus monkeys and man, *Fed. Proc.*, 36, 1031, 1977.
126. **Bente, D., Koeller, D. M., Koufen, H., Pietzker, A., and Schulte, P. W.,** Klinisch-statistische Untersuchung des Wirkungsbildes von Sulpirid, einem neuen Psychopharmakon, *Arzneim.-Forsch.*, 24, 107, 1974.
127. **Eckmann, F.,** Klinische Untersuchung mit Sulpirid bei langjährig hospitalisierten schizophrenen Kranken, *Arzneim.-Forsch.*, 24, 993, 1974.
128. **Elizur, A. and Davidson, S.,** The evaluation of the anti-autistic activity of sulpiride, *Curr. Ther. Res.*, 18, 578, 1975.
129. **Cassano, G. B., Castrogiovanni, P., and Conti, L.,** Sulpiride versus haloperidol in schizophrenia: a double blind trial, *Curr. Ther. Res.*, 17, 189, 1975.
130. **Schmutz, J.,** Neuroleptic piperazinyl-dibenzo-azepines, *Arzneim.-Forsch*, 25, 712, 1975.
131. **Heel, R. C., Brogden, R. N., Speight, T. M., and Avery, G. S.,** Loxapine: a review of its pharmacological properties and therapeutic efficacy as an antipsychotic agent, *Drugs*, 15, 198, 1978.
132. **Gallant, D. M., Bishop, G., Steel, C. A., and Bishop, M. P.,** Loxapine: a six-month evaluation in severely ill schizophrenic patients, *Curr. Ther. Res.*, 15, 205, 1973.
133. **Nair, N. P. V., Decker, B. L., and Schwartz, G.,** Loxapine succinate in the treatment of chronic schizophrenia, *Curr. Ther. Res.*, 20, 802, 1976.
134. **Squitino, B. S. G., Coletty, E. B., and Lara, L. F. M.,** Treatment of chronic schizophrenia. Loxapine, a new neuroleptic agent, *A Folha Medica*, 74, 329, 1977.
135. **Wolpert, A., White, L., Dana, L., Sugerman, A. A., Arengo, A. D., Simpson, G. M., Bishop, M. P., and Gallant, D. M.,** Clinical pharmacological trial of loxapine succinate, *Clin. Pharmacol.*, 10, 175, 1970.
136. **Bishop, M. P., Simpson, G. M., Dunnett, C. W., and Kiltie, H.,** Efficacy of loxapine in the treatment of paranoid schizophrenia, *Psychopharmacology*, 51, 107, 1977.

ANTIANXIETY AGENTS

J. A. Vida and J. M. Essery

STATE OF ANXIETY

Anxiety accompanies birth and is thereafter man's constant companion. Anxiety is an unpleasant emotional state of daily living characterized by apprehensive tension, uneasiness, and distress, which indicates the presence of imminent danger and anticipation of unpleasruable events. The state of anxiety is usually associated with changes in the autonomic nervous system and behavior (feelings of tension, guilt, apprehension, loss of confidence, disturbed breathing, increased heart activity, vasomotor changes, muscular disturbances, sweating, G I upset, etc.).

Anxiety describes a multitude of states. At one end of a gradual spectrum is the objective anxiety, a normal, realistic human emotion characterized by appropriate reactions to real or threatened danger. At the other end of the spectrum are the pathologic anxiety syndromes consisting of inappropriate reactions to unreal or unknown sources of threat; the manifestations can be so severe as to interfere in a disabling fashion with everyday functioning. Anxiety can also be mixed with depression; as a result, anxiety is not always easily diagnosed.

Arbitrarily, antianxiety agents may be classified as follows: (1) benzodiazepines, (2) propanediol carbamates, (3) dipenylmethanes, (4) buspirone, (5) pyrazolopyridines, (6) triazolopyridazines, (7) purines, and (8) miscellaneous agents.

BENZODIAZEPINES*

Effects on Neurotransmitters

In general, benzodiazepines, the most important and most widely studied anxiolytic agents decrease the turnover rates of neurotransmitters. As a result, the activity of the various transmitters is reduced. Neuroleptics characteristically induce a ''dopamine deficiency syndrome'', that, in turn, causes an increase in dopamine synthesis and turnover as measured by increased accumulation of dopamine and its metabolic products. Antidepressants inhibit the uptake of norepinephrine and/or serotonin. The neuroleptics and antidepressants and other psychopharmacological agents usually affect neurotransmission in any one of the following manners:

1. Mimicking the transmitter action
2. Blocking the transmitter receptor function
3. Blocking the reuptake of the transmitter
4. Blocking the release of transmitter
5. Increasing transmitter synthesis
6. Decreasing transmitter synthesis
7. Increasing transmitter metabolism
8. Decreasing transmitter metabolism

Although benzodiazepines reduce the concentrations or data and other turnover rates of norepinephrine,[1] dopamine,[2] serotonin,[2,3,] and acetylcholine,[4] the doses required to produce these effects are 10- to 100-fold greater than the doses required to elicit anxiolytic responses in man. Unlike the neuroleptics, antidepressants and other psychopharmacological agents,

* SAR monographs on benzodiazepines may be found in the chapter entitled Nonbarbiturate Sedatives and Hypnotics.

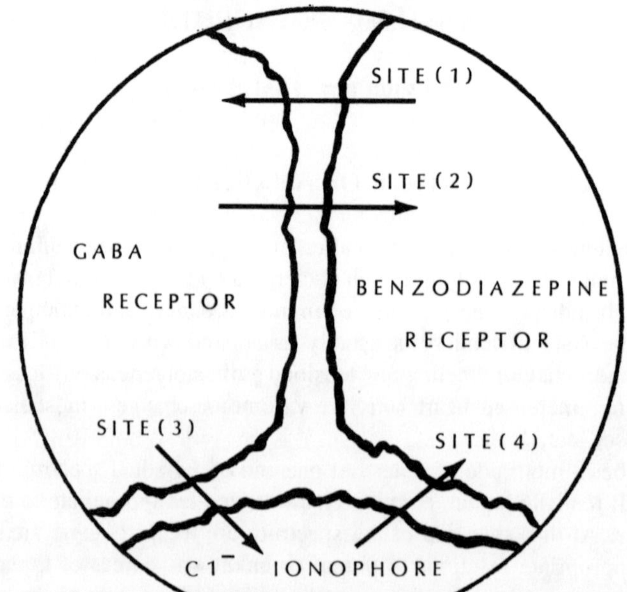

FIGURE 1. Enlarged hypothetical scheme of a postsynaptic receptor complex.

whose mechanism of action can be explained by the effect on a specific neurotransmitter, benzodiazepines could be a prototype of a second generation of psychopharmacological agents. This type of activity is ascribed to their action on a specific modulatory process[5] that physiologically regulates receptor sensitivity, in this case γ-aminobutyric acid (GABA) receptors. Benzodiazepines apparently can only maximize or minimize the receptor responses but cannot abolish them or increase them without limits.

Benzodiazepines modify the actions of the inhibitory neurotransmitter GABA. This hypothesis is supported,[5] in part, by the finding that benzodiazepines are able to potentiate the actions of GABA in several regions of the CNS and, in part, that benzodiazepines are able to protect animals against seizures induced by drugs known to act on GABA processes. Benzodiazepines are also involved in presynaptic inhibition.

It is not known as of now exactly how the benzodiazepines potentiate the actions of GABA. At least two populations of GABA receptors exist in the CNS.[6] The ratio of the two receptor populations changes in various brain areas and the receptor-binding sites are linked to chloride ionophores. The postsynaptic receptor complex may consist of the GABA receptor, benzodiazepine, receptor, and chloride ionophore shown in Figure 1.

Possible Mechanism of Benzodiazepine Action

Site 1

An increase in benzodiazepine concentration of the postsynaptic receptor complex will increase the GABA concentration

$$[\text{Bzd}] \uparrow \qquad [\text{GABA}] \uparrow$$

where the brackets denote concentrations and ↑ denote increases.

This phenomenon explains the fact that benzodiazepines increase the primary afferent depolarization in the spinal cord because they facilitate GABAergic transmission.[7] This phenomenon also explains the observation that convulsions caused by deficiency of GABAergic transmission can be blocked by benzodiazepines[8] which, according to this theory, should facilitate the GABAergic transmission. This phenomenon also explains the fact that benzodiazepines competitively antagonize the inhibition of GABA binding to synaptic membranes[9] caused by an endogenous protein[10] called GABA-modulin.[11] Also, the fact that the concentration of benzodiazepines in synaptic membranes (high occupancy of binding sites for benzodiazepines in synaptic membranes) could increase the number of high affinity GABA binding sites.[12]

Site 2

An increase in GABA concentration of the postsynaptic receptor complex will increase the affinity of the benzodiazepine receptor to bind benzodiazepine.

$$[\text{GABA}]\uparrow \qquad \text{Affinity of Bzd receptor to bind Bzd}\uparrow$$

This phenomenon explains the finding that the high-affinity binding of benzodiazepines can be promoted by drugs that stimulate GABA receptors and inhibited by drugs that block GABA receptor function.[13] This phenomenon also explains the fact that the occupancy of the high-affinity binding sites for GABA could increase the affinity of the specific binding sites for benzodiazepines.[13]

More flexible GABA agonists (dihydromuscimol, muscimol, thiomuscimol) can activate binding, whereas conformationally restricted GABA agonists THIP (4,5,6,7-tetrahydroisoxazole [5,4-c] pyridin-3-ol), isoguvacine, PSA (piperidine-4-sulfonic acid) do not activate this binding.

Both isoniazid and AOAA (amino-oxyacetic acid), block the GABA-T transaminase enzyme and therefore increase GABA concentrations and enhance benzodiazepine effects.

Conversely, a decrease in GABA concentrations at the high-affinity site of the postsynaptic receptor complex will decrease the affinity of the benzodiazepine receptor to bind benzodiazepine. This phenomenon explains the fact that the action of diazepam to abolish fear-suppressed behavior can be antagonized by blockers of GABA receptors or changed by drugs that abolish the synthesis of GABA.[14]

Site 3

The increase in GABA concentration of the postsynaptic receptor complex increases the flow of chloride ion into the postsynaptic membrane

$$[\text{GABA}]\uparrow \qquad [\text{Cl}^-]\uparrow$$

Influx of chloride into the postsynaptic receptor complex results in an outward flow of current observed as IPSP (inhibitory postsynaptic potential).

The GABA antagonists bicuculline and picrotoxin decrease the GABA concentration at the high-affinity site of the postsynaptic receptor complex and thereby block the hyperpolarization and ensuing IPSP, but do not prevent normal chloride uptake. Conversely, while benzodiazepines themselves do not produce IPSP, benzodiazepines do potentiate the effect of submaximal concentrations of GABA agonists.[11]

Site 4

The ability of barbiturates and pyrazolopyridines to increase the apparent affinity of benzodiazepines is explained by the effects on chloride channels. Both barbiturates and pyrazolopyridines increase the conductance by chloride ions.

The "presynaptic" GABA-mediated inhibition by benzodiazepines has been demonstrated

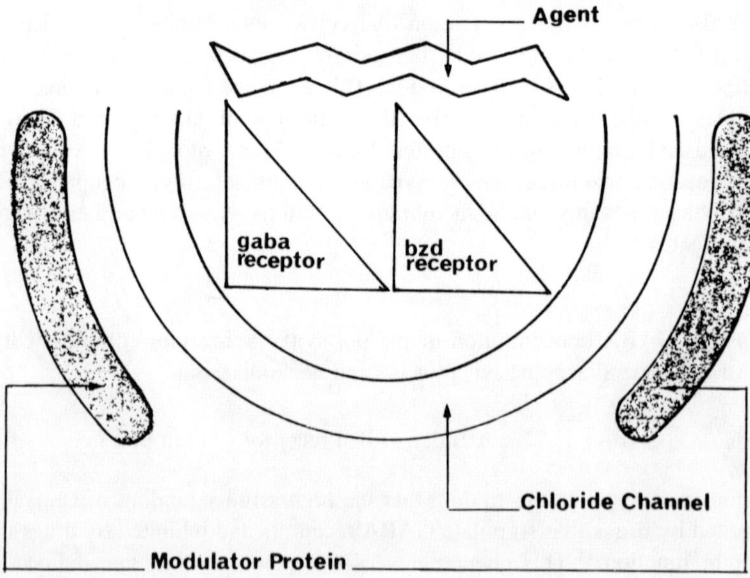

FIGURE 2. Receptor recognition site.

in several studies.[15] Benzodiazepines also interact (perhaps by mediation by GABAergic transmitters) with other CNS neurotransmitters. It has been suggested that the catecholamine antagonistic properties of benzodiazepine are responsible for the sedative activity of benzodiazepines,[16] and the 5-HT inhibitory activity of benzodiazepine is responsible for the behavioral effects of benzodiazepines, as measured by the release of punishment-suppressed behavior in the rat conflict test,[17] as measured by the high rate of bar presses during the punishment period.

None of these interactions can adequately explain the mechanism of action of benzodiazepines, however. It is clear that GABA receptor recognition sites are functionally linked to benzodiazepine recognition sites, and an increase in receptor activation for one leads to an enhancement in the receptor affinity for the other.

A Model for GABA, Benzodiazepine (Bzd), and Chloride Interaction

A model is provided to illustrate the interaction between the GABA receptor, chloride channel, and benzodiazepine receptor (Figures 2 to 9).

The receptors may undergo a variety of conformational changes within the space provided for the reception, but fill only part of the space.

The following may take place.

Resting State

At rest the benzodiazepine receptor and GABA receptor will be in a low-affinity state and the chloride channels are partially blocked. A steady state of chloride conductance against electrochemical gradients is maintained. Both the benzodiazepine and GABA receptor recognition is low. The modulator protein exerts its influence over both GABA and benzodiazepine receptors.

Benzodiazepine Administration

Due to the dynamic nature of the benzodiazepine receptor, the presence of administered benzodiazepine at the receptor site induces a conformational change in the benzodiazepine receptor, which in turn enhances GABA release and produces inhibition. The chloride channels are closed. The modulator protein loses its control on the GABA receptor.

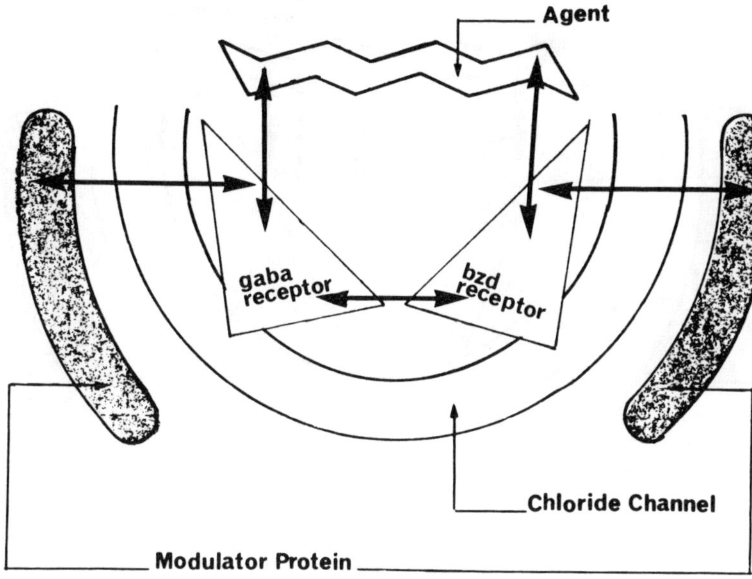

FIGURE 3. Receptor recognition site.

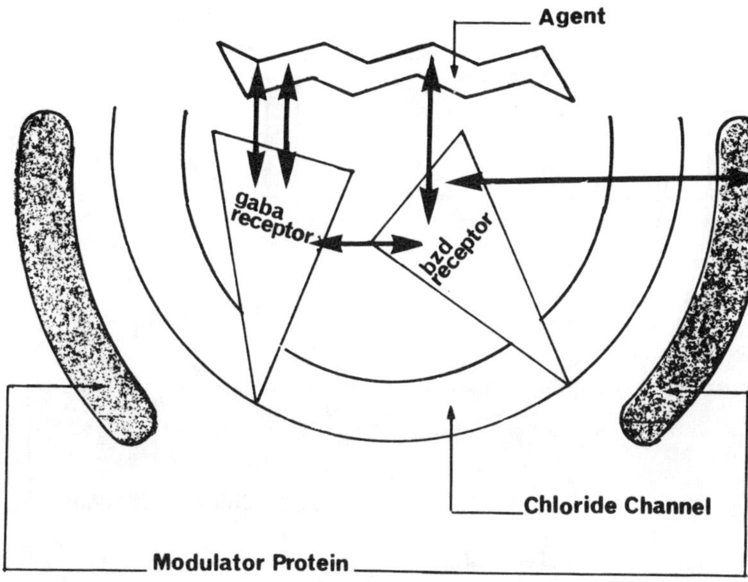

FIGURE 4. Receptor recognition site.

GABA or GABA Agonist Stimulation

Administration of a GABA agonist produces a conformational change in the GABA receptor which in turn induces a conformational change in the benzodiazepine receptor. As a result, the affinity of the benzodiazepine receptors is increased. The conformational changes open the chloride channels (usually causing hyperpolarization). The modulator protein loses its control of the benzodiazepine receptor.

Partial GABA Agonist Stimulation

Partial GABA agonists (THIP, isoguvacine, PSA, etc.) induce a moderate conformational

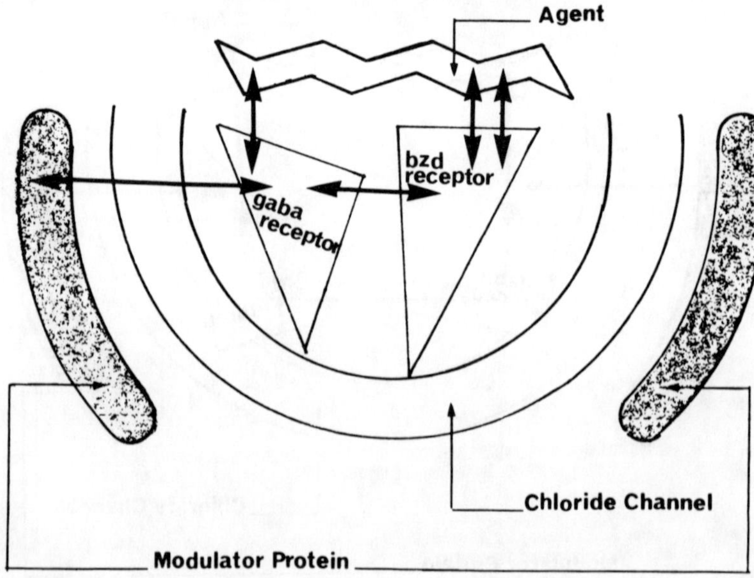

FIGURE 5. Receptor recognition site.

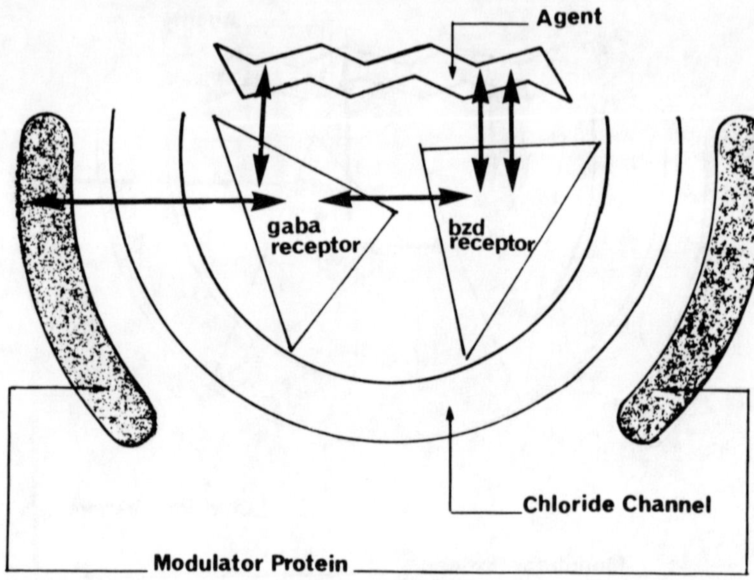

FIGURE 6. Receptor recognition site.

change in the GABA receptor so that the benzodiazepine receptors are partially activated while the chloride channels may open.

GABA-Antagonist Inhibition

Bicuculline and picrotoxin, both convulsant alkaloids, block the GABA receptor site either by direct action on the chloride channels (picrotoxin) or by indirect action (bicuculline). Bicuculline inhibits the effect of GABA agonists and GABA on benzodiazepine binding while picrotoxin does not interfere directly with the GABA-benzodiazepine interaction.

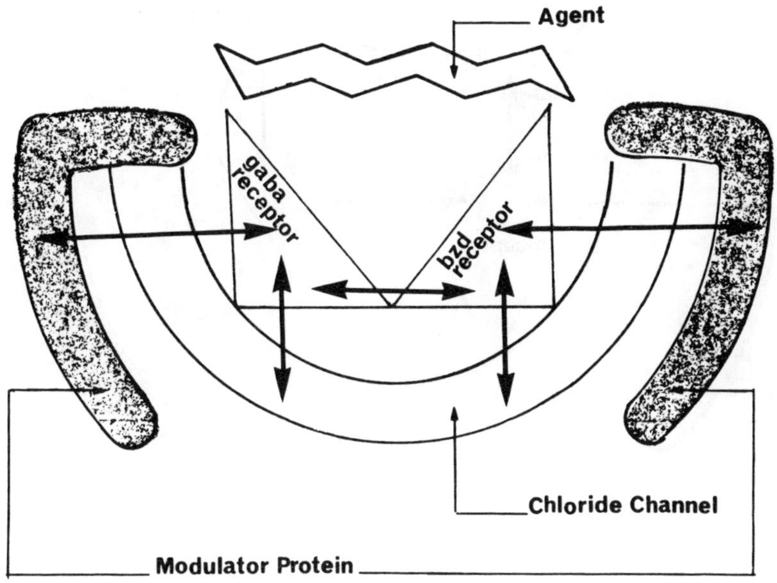

FIGURE 7. Receptor recognition site.

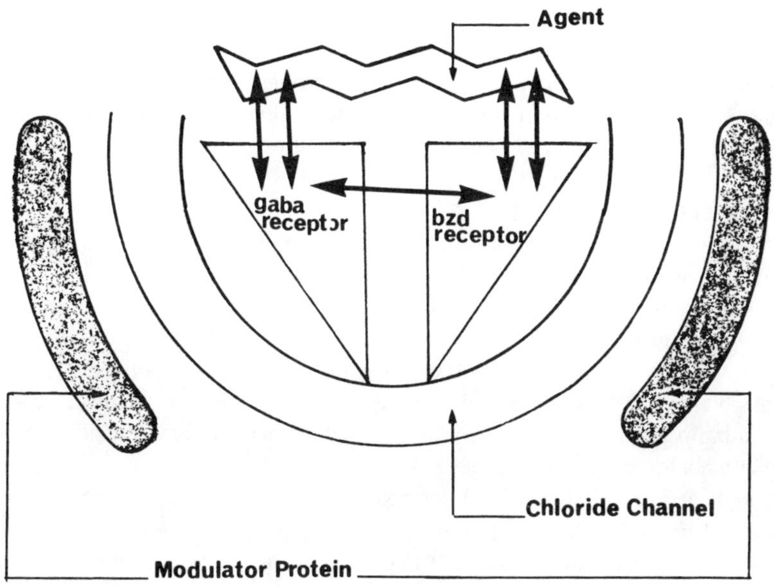

FIGURE 8. Receptor recognition site.

Simultaneous GABA and Benzodiazepine Stimulation

In the presence of benzodiazepine, release from presynaptic terminals is enhanced, which may activate the GABA modulator protein, which in turn can induce in the GABA receptor a high-affinity state. As a result, the endogenous modulator protein will lose its control on the GABA receptor. Similarly, the modulator protein will lose its control on the benzodiazepine receptor. Simultaneously there is an increase in the chloride conductance of postsynaptic membranes.

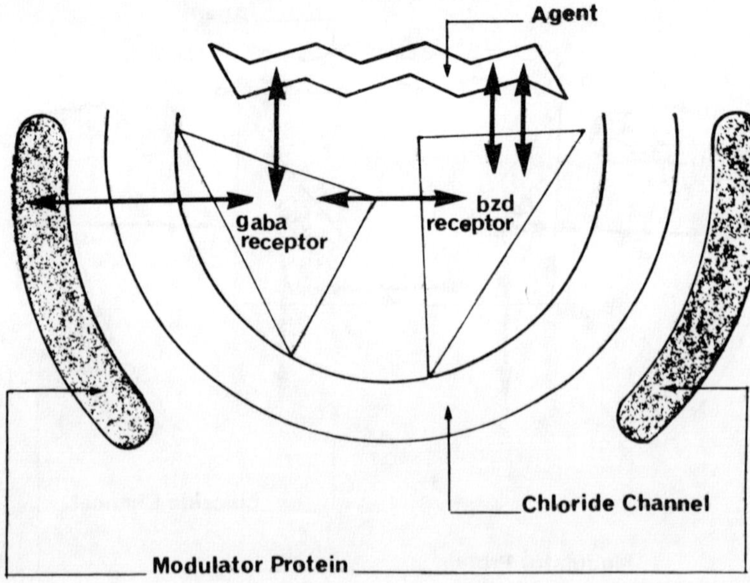

FIGURE 9. Receptor recognition site.

Barbiturate Stimulation

In the presence of barbiturates (or pyrazolopyridines) the chloride conductance is increased which produces an allosteric modification (conformational change) and the affinity of benzodiazepine receptor binding is increased. Both bicuculline and picrotoxin are capable of antagonizing barbiturate-enhanced benzodiazepine binding.

The Benzodiazepine Receptor

There is a high degree of correlation between the ability of various benzodiazepines to displace binding of radiolabeled ^{3}H-diazepam and ^{3}H-flunitrazepam from the high-affinity binding sites of mammalian brain and their activity in various behavioral, muscle relaxant, and anti convulsant tests.[15] Binding to these sites in vitro is rapid, reversible, stereo-specific, and saturable.[18,19] Since there seems to be a significant correlation between the behavioral and clinical potency of various benzodiazepines and their ability to displace ^{3}H-diazepam binding from brain preparations in vitro,[20,21] the benzodiazepine-binding site may also mediate the pharmacological actions of these drugs.

Specific high affinity binding of ^{3}H-diazepam could be obtained in the brain, kidney, liver, lung, and several cultured cells of various types. The widespread distribution of benzodiazepine receptors in the CNS results in the benzodiazepines affecting several neuronal pathways and interacting at many sites of the brain to produce their effect. Furthermore, there are multiple populations of benzodiazepine receptors in the CNS. Displacement of ^{3}H-flunitrazepam by triazolopyridazines produced negative correlation coefficients indicating regional differences between receptor subpopulations.[22]

Direct evidence for different classes of benzodiazepine-binding sites was obtained by Sieghart and Karobath[23] via photoaffinity-labeling experiments with irreversibly bound ^{3}H-flunitrazepam. In the cerebellum of the rat, the ligand bound to a single protein band which had an apparent molecular weight of 51,000 daltons (P_{51}). In other regions of the brain, in addition to P_{51}, other protein bands with apparent molecular weights of 53,000 (P_{53}), 55,000 (P_{55}), and 59,000 (P_{59}) were labeled with varying degrees of intensity. Binding of ^{3}H-flunitrazepam to these proteins was enhanced by GABA, and this stimulation was bicuculline sensitive. The topic of multiple benzodiazepine receptors is discussed further in the section on triazolopyridazines.

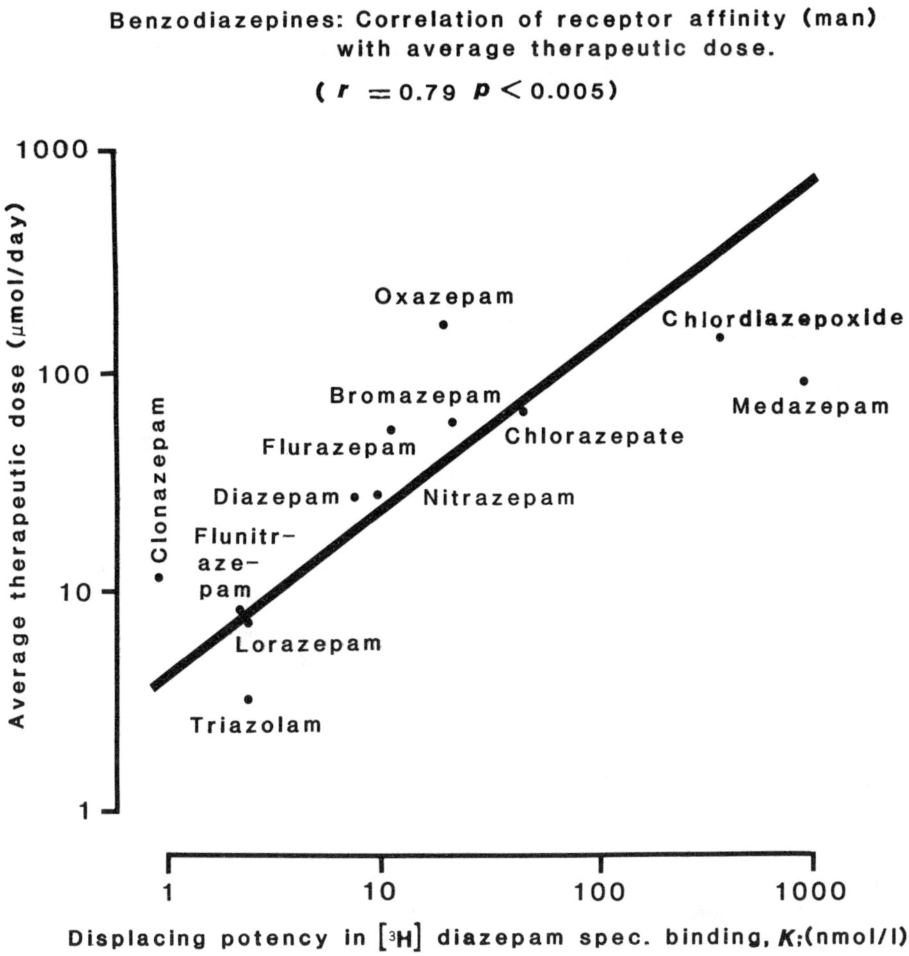

FIGURE 10. Correlation between the in vitro affinity of some benzodiazepines to the specific [^{3}H] diazepam binding sites and their average therapeutic dose.

The presence in the CNS of benzodiazepine receptors implies that there may exist natural substances which normally bind to the receptor and thereby function as endogenous ligands. Naturally occurring materials which have been proposed as endogenous ligands include hypoxanthine, inosine,[24-26] and nicotinamide.[27] The pharmacological and electrophysiological properties of these substances are to some extent compatible with a benzodiazepine-like mode of action, but they all appear to be weak inhibitors of ^{3}H-diazepam binding.

The most potent compound so far isolated is beta-carboline-3-carboxylic acid, obtained as its ethyl ester (β-CCE) from human urine.[28] This compound is an attractive candidate as an endogenous ligand because it is highly specific for benzodiazepine receptors and has an even higher affinity than diazepam.[29] However, beta-CCE is not present as such in any measurable concentration in brain in vivo. It is possible that beta-carboline-3-carboxylic acid is a component of a natural neuromodulator.

The affinity of various benzodiazepine drugs to the benzodiazepine receptor can be determined by measuring the concentration required to displace radiolabeled ^{3}H-diazepam from the specific binding sites. It was found[21] that a correlation exists between the concentrations of the benzodiazepine drugs required to displace the radioligand binding and their potency as anxiolytics in vivo (Figure 10).

Recently, it has been found that the specific high-affinity binding of labeled benzodiazepines to the synaptosomal fraction of the CNS could be inhibited by RO-15-1788 (Structure 1) an imidazodiazepine.[30]

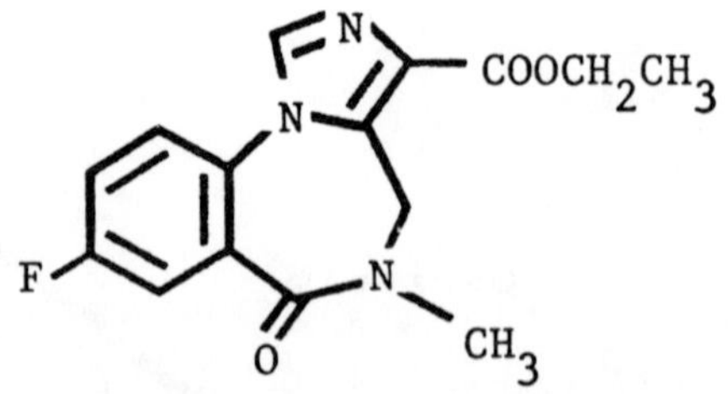

Structure of RO-15-1788

Structure 1

By inhibiting the central synaptosomal but not peripheral binding of benzodiazepines, this nontoxic compound blocked the pharmacologic effects of benzodiazepines. RO-15-1788 did not block the pharmacological effects of several other centrally acting depressants such as phenobarbital, meprobamate, ethanol, or muscimol, a GABA analog that binds to GABA receptors.[31] The benzodiazepine antagonistic properties of RO-15-1788 could be confirmed in man. It has been found[32] that benzodiazepine-induced sedation in man was very rapidly reversed by i.v. injection of RO-15-1788.

A Model for Benzodiazepine Action in Man

As previously discussed, benzodiazepines may act on a "presynaptic" site of the GABA receptor by increasing the amount of GABA released by an action potential, or by increasing the sensitivity of GABA to excitatory influences. Alternatively, benzodiazepines may act on a "postsynaptic" site by increasing the efficacy of GABA at the subsynaptic membrane.

Benzodiazepines seem to interact in a highly specific manner with the GABA neurotransmitter system. Anxiety may be due to neuronal hyperactivity particularly in the limbic system and may be reduced by potentiating the synaptic inhibitory mechanisms. Administration of a benzodiazepine may facilitate GABA transmission by binding to the GABA receptor. As a result, GABA is released into the synapse and binds to its recognition site on its receptor and may open a chloride channel. The increased permeability to chloride results in inhibition of neuronal firing, which may account for a reduction in anxiety.

Pharmacokinetics

Serious intoxication following overdosage with a benzodiazepine alone is unusual.[33] Intoxication following ingestion of a benzodiazepine together with another drug depends largely on the amount and type of the other drug taken.[33] The contribution of benzodiazepine to the severity of intoxication depends on the quantity and pharmacokinetics of the benzodiazepine.

There are definite differences among the pharmacokinetic properties of benzodiazepines. The pharmacokinetic characteristics, i.e., the rate of absorption, distribution, metabolism, and elimination will determine the drug concentrations at the receptor site of action. These factors have clinical implications in terms of dosage schedules, route of administration, drug accumulation during chronic administration, drug interactions, and drug therapy in different disease states (e.g., alcoholism, epilepsy, anxiety, etc.).

Single Dose Administration

The clinical situation determines the desirability of rapid onset of clinical effects. If the benzodiazepine is used as a hypnotic for the treatment of anxiety, or for the relief of a muscle spasm or for acute anxiety attack, then rapid onset of action is required. A rapid onset of action will be elicited if the benzodiazepine is rapidly absorbed following oral dosage. On the other hand, if the benzodiazepine is deployed with the purpose of treating a chronic condition (e.g., anxiety condition), then other effects are unwanted. For example, drowsiness or a disphoric "spaced out' condition, brought about by a fast absorption rate of a benzodiazepine is undesirable for the treatment of a chronic anxiety condition. Therefore, a fast absorption rate may be unfavorable.

Data reveal that diazepam and clorazepate are very rapidly absorbed, chlordiazepoxide, lorazepam, and flurazepam have an intermediate absorption rate, and prazepam and oxazepam are least rapidly absorbed.

Following absorption, the duration of action of a single dose of benzodiazepine depends upon the rate and extent of distribution, the rate of elimination, and the rate of active metabolite formation. It is difficult to predict which of these factors will predominate. For example, diazepam, due to its rapid absorption, could be used as a hypnotic. The use of a single dose of diazepam as a hypnotic may further be rationalized by the finding that it seems to have a short duration of action[34] in spite of the long elimination half-life, ranging from 20 to 80 hr.[35] The short duration of action is attributed to the rapid and extensive distribution from blood to fatty tissues, causing a decrease or termination of single-dose effects.

Lorazepam, on the other hand, although its absorption is much slower than that of diazepam, has a much shorter elimination half life, with an average of 15 hr.[36,37] The effective duration of a single dose of lorazepam is longer than that of diazepam, because the distribution from blood to fatty tissue is much less extensive and effective concentrations are present in the blood and brain for many hours.[38]

Multiple Dose Administration

The two factors responsible for a steady-state plasma concentration of a benzodiazepine are the rate and the extent of benzodiazepine accumulation, which in turn depend on the elimination half-life and clearance. The longer the half-life, the slower is the rate of accumulation. In general, a steady-state condition is obtained in more than 90% extent after a period of at least four times the half-life has elapsed since the start of the treatment.[39] In addition to the half-life, the extent of accumulation is also dependent upon the drug dosage and volume of distribution. The longer the half-life the larger the steady-state plasma concentration. As expected, the larger the dose, the larger the steady-state plasma concentration. However, the drug effects do not necessarily increase in direct proportion to the dose during chronic therapy. Tolerance to chronic drug administration may develop.

Pharmacokinetic Classification of Benzodiazepines

Based on the elimination half-lives of benzodiazepines and their metabolites, the following classification of benzodiazepines are proposed (Table 1, slightly modified version of Greenblatt et al.[40]).

Long-Acting Benzodiazepines

The parent drug and/or its active metabolite have a combined chemical half-life of 24 hr or longer. The common metabolites of 7-chloro 1,4-benzodiazepines are shown in Figure 11. Desmethyldiazepam (nordiazepam) with a half-life of 200 hr or more is a common metabolite of several commercially available drugs. Five of these are shown in Figure 11. In fact, two of these, clorazepate and prazepam, can be considered to be simple prodrugs of desmethyldiazepam, which is the active species. Drug accumulation of both parent drugs and active metabolites proceeds at different rates and to different extents depending on the

Table 1
PHARMACOKINETIC CLASSIFICATION OS BENZODIAZEPINE DERIVATIVES[a]

Parent compound	Clinically important active metabolites	Selected ref.
Long (effective half-life usually>24 hr)		
Chlordiazepoxide (5—30)	Desmethylchlordiazepoxide	
	Demoxepam	41
	Desmethyldiazepam (36—200)	
Diazepam (20—100)	Desmethyldiazepam (36—200)	34
		42
[Clorazepate][b]	Desmethyldiazepam (36—200)	43
[Prazepam][b]	Desmethyldiazepam (36—200)	44
Medazepam	Desmethylmedazepam	45
	Diazepam	46
	Desmethyldiazepam (36—200)	
Clobazam (12—60)	Desmethylclobazam	47
[Flurazepam][b]	Desalkylflurazepam (40—250)	Unpublished data
Intermediate — short (half-life 5—24hr)		
Nitrazepam (15—28)		48,49
Flunitrazepam (20—30)	Desmethylflunitrazepam	Unpublished data
Estazolam (10—30)	? of hydroxylated metabolite	50
Bromazepam (10—20)	? 3-hydroxybromazepam	51
Alprazolam (6—20)		Unpublished data
Lorazepam (10—20)		37,52,53
Temazepam (8—22)		54
Oxazepam (4—15)		55,56
Ultrashort (half-life<5 hr)		
Triazolam		57
Midazolam		58

[a] Approximate range of elimination half-life values is shown in parentheses when the data are available.
[b] Brackets indicate prodrugs or drug precursors that do not reach the systemic circulation as such in clinically important amounts.

From Greenblatt, D. J., Shader, R. I., Divoll, M., and Harmatz, J. S., *Br. J. Clin. Pharmacol.*, 11, 115, 1981. With permission.

half-life and clearance of each compound. The accumulation may lead to benefit, e. g., single bedtime therapy of chronic anxiety or disadvantage, e. g., unwanted drowsiness and impairment of psychomotor performance. The metabolism follows the route of N-desmethylation and hydroxylation, followed by conjugation with glucuronic acid.

Intermediate-Acting Benzodiazepines

These are characterized by a half-life of 5 to 24 hr and generally do not undergo metabolism to an active metabolite (except for flunitrazepam, estazolam, and bromazepam). The therapeutic activity exerted by these drugs is generally due to the drugs themselves and the steady-state condition is reached rapidly. The accumulation during multiple-dose therapy is not attenuated.

These benzodiazepines are used as hypnotics or for the treatment of anxiety where any residual day time effects must be avoided. If patients are prone to be poorly compliant, intermediate-acting benzodiazepines are not appropriate since, due to absent drug levels in periods of noncompliance, symptoms may recur. The pathways of 7-nitro-1,4-benzodiazepines are shown in Figure 12.

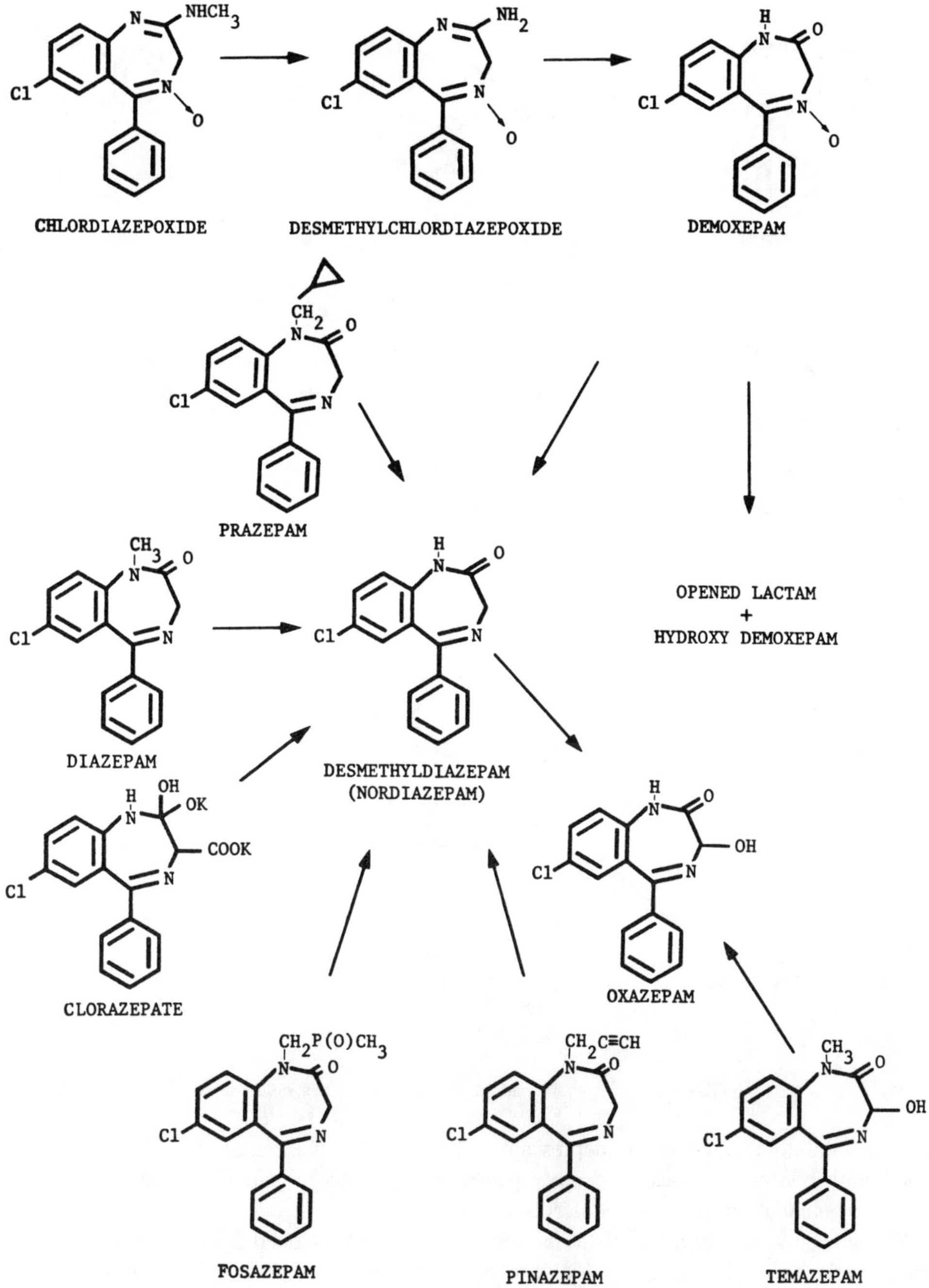

FIGURE 11. Common metabolites of 7-chloro-1,4-benzodiazepines.

Ultrashort-Acting Benzodiazepines

Triazolam and midazolam have a less than 5-hr half-life. The former is used primarily as a hypnotic while the second is being studied as a short-acting anesthetic-induction agent.

Drug Interactions

Since benzodiazepines exhibit CNS depressant effects, these effects can be enhanced by other CNS depressants such as alcohol, barbiturates, narcotics, phenothiazines, monamine

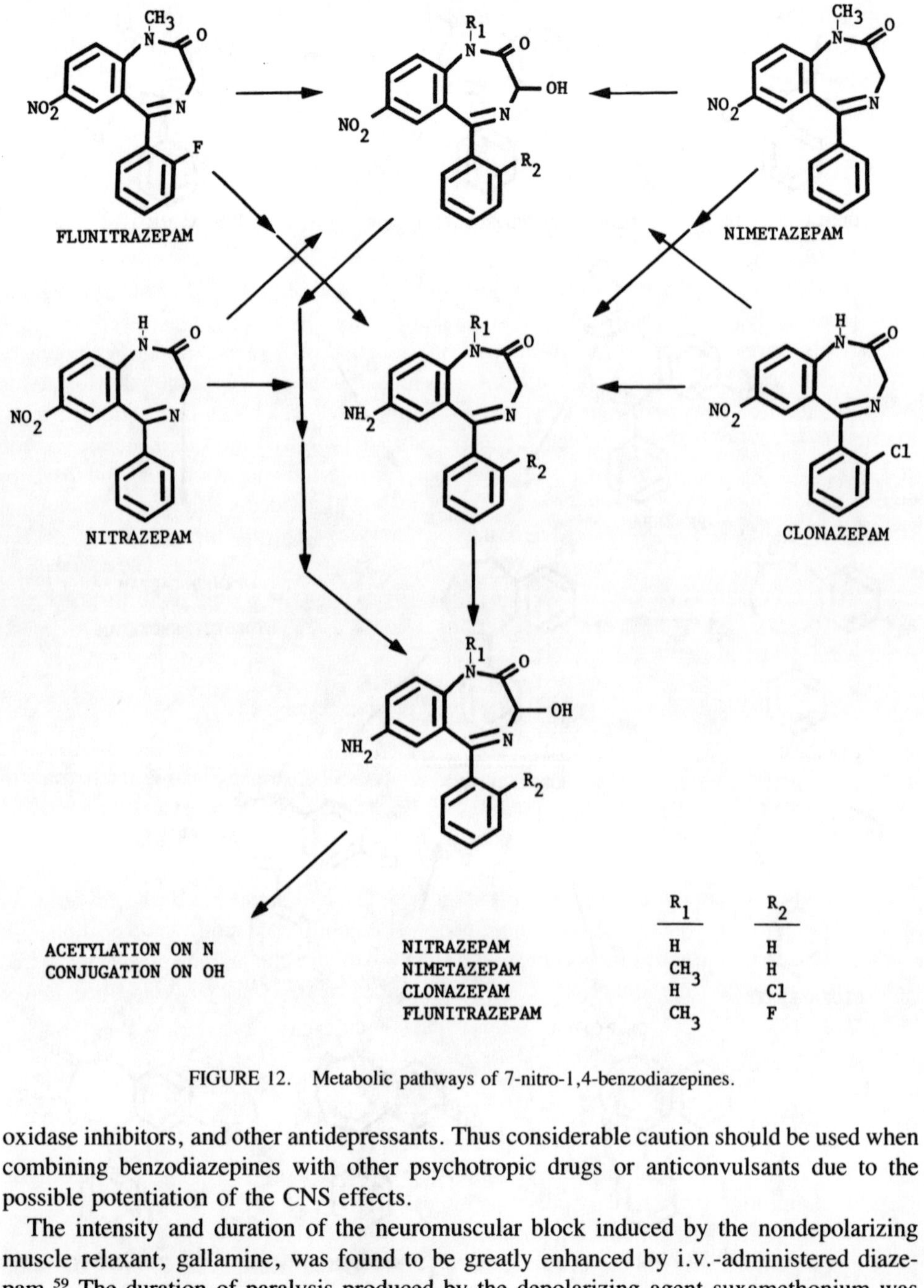

FIGURE 12. Metabolic pathways of 7-nitro-1,4-benzodiazepines.

oxidase inhibitors, and other antidepressants. Thus considerable caution should be used when combining benzodiazepines with other psychotropic drugs or anticonvulsants due to the possible potentiation of the CNS effects.

The intensity and duration of the neuromuscular block induced by the nondepolarizing muscle relaxant, gallamine, was found to be greatly enhanced by i.v.-administered diazepam.[59] The duration of paralysis produced by the depolarizing agent suxamethonium was reduced by i.v. diazepam.[59]

While the long-term use of low-dose estrogen-containing oral contraceptives did not alter the volume of distribution of diazepam, a significant impairment of total metabolic clearance occurred, resulting in a marked prolongation of the elimination half-life of diazepam.[60] This study employed diazepam administered i.v., but it can be reasonably assumed that the same pharmacokinetic interaction will occur with diazepam given orally since the drug is rapidly absorbed and 100% bioavailable by this route of administration.

Interaction between diazepam and cimetidine is of clinical importance since these are two of the most widely used drugs, and many patients receive them at the same time. Cimetidine significantly impairs the elimination of diazepam, resulting in an enhancement of sedation in patients.[61] Cimetidine has also been shown to enhance the absorption of diazepam and lorazepam.[62]

Important marketed benzodiazepines appear in Table 2.

PROPANEDIOL CARBAMATES

The introduction of meprobamate into therapy in 1954[63] began the era of modern anxiolytic agents. Up until that time, patients were treated with sedatives such as barbiturates which, although they were effective in small doses, presented severe problems of overdosage and psychic and physical dependence. Meprobamate resulted from an extensive effort (1200 compounds were synthesized and evaluated over a period of 8 years) to improve upon the properties of the 1,2-propanediol mephensin, a muscle relaxant whose duration of action was too short for it to be useful clinically. Meprobamate, which can be considered to be the first true anxiolytic, became very popular because it allowed mild to moderate tension and anxiety to be controlled in patients without causing too much sedation. Of perhaps greater significance, the clinical success of meprobamate provided a stimulus for medicinal chemists to search for improved anxiolytics, and this led to the discovery of the benzodiazepines.

Of the 1,3-propanediols synthesized, meprobamate is the most widely used. Its properties are described in the chapter entitled ''Drugs Used to Treat Neuromuscular Disorders: Antiparkinsonism Agents and Skeletal Muscle Relaxants''.

DIPHENYLMETHANES

While the major activity of the antihistamine drugs is that of antagonizing the actions of histamine on various peripheral structures,these compounds also show a variety of actions, particularly in the CNS. Such actions are unwanted side effects in some cases, but in other situations the CNS effects may contribute to the principal use of a compound.

In the diphenylmethane series of antihistaminics, diphenhydramine has the useful property of suppressing motion sickness as a consequence of central depression. Another diphenylmethane, hydroxyzine, has found clinical use as an anxiolytic although its precise locus and mechanism of action are unknown. The properties of hydroxyzine are described in the monograph section.

BUSPIRONE

Buspirone, 8-{4[4(2-pyrimidinyl)-1-piperazinyl]butyl}-8- azaspiro [4,5] decane-7,9-dione hydrochloride (Structure 2), was first synthesized by Wu et al.[64] in 1972. It is an anxiolytic agent with a unique structure and pharmacological profile.

Inhibition of the rhesus monkey's normal aggressive response to prodding with a pole has been correlated with anxiolytic drug action.[65] The activity of buspirone in this pole prod test led[66] to a clinical study in outpatients with a primary diagnosis of anxiety neurosis. In this double-blind study carried out by Goldberg and Finnerty,[67] buspirone was compared with diazepam for safety and efficacy. From the study, in which 18 patients received buspirone, 20 received diazepam and 18 placebo over a 4-week period, it was concluded that buspirone was as effective an antianxiety agent as diazepam. It was also concluded that side effects due to buspirone (light-headedness and dizziness) are certainly no greater, and may be considerably less, than those due to diazepam (dry mouth, depression, drowsiness, and confusion). The anxioselective profile of buspirone has since been verified in several additional clinical studies.

Table 2
IMPORTANT MARKETED BENZODIAZEPINES

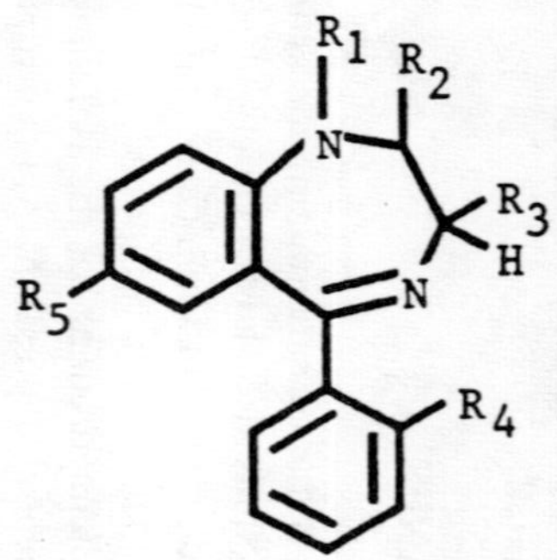

Name	R_1	R_2	R_3	R_4	R_5	Principal use
Bromazepam	–H	=O	–H	Skeletal =N–	–Br	Anxiolytic
Camazepam	CH_3	=O	$OCON(CH_3)_2$	–H	–Cl	Anxiolytic
Clonazepam	H	=O	–H	–Cl	$-NO_2$	Anticonvulsant
Clorazepate	H	–OH –OK	–COOK	–H	–Cl	Sedative anxiolytic
Chlordesmethyl diazepam	H	=O	–H	–Cl	–Cl	Sedative
Demethyldiazepam (nordiazepam)	–H	=O	–H	–H	–Cl	Sedative anxiolytic
Diazepam	$-CH_3$	=O	–H	–H	–Cl	Anxiolytic
Doxefazepam	$-(CH_2)_2OH$	=O	–OH	–F	–Cl	Sedative hypnotic
Ethyl loflazepate	–H	=O	$-COOC_2H_5$	–F	–Cl	Anxiolytic
Flurazepam	$-(CH_2)_2N(C_2H_5)_2$	=O	–H	–F	–Cl	Sedative hypnotic
Flutoprazepam	$-CH_2-\triangleleft$	=O	–H	–F	–Cl	Anxiolytic
Fosazepam	$CH_2P(O)(CH_3)_2$	=O	–H	–H	–Cl	Sedative
Halazepam	$-CH_2CF_3$	=O	–H	–H	–Cl	Anxiolytic
Lorazepam	–H	=O	–OH	–Cl	–Cl	Anxiolytic sedative

Lormetazepam	$-CH_3$	$=O$	$-OH$	$-Cl$	$-Cl$	Hypnotic
Medazepam	$-CH_3$	$=H,H$	$-H$	$-H$	$-Cl$	Anxiolytic
Metaclazepam	$-CH_3$	$-CH_2OCH_3,H$	$-H$	$-Cl$	$-Br$	Anxiolytic
Nimetazepam	$-CH_3$	$=O$	$-H$	$-H$	$-NO_2$	Hypnotic
Nitrazepam	$-H$	$=O$	$-H$	$-H$	$-NO_2$	Hypnotic
Oxazepam	$-H$	$=O$	$-OH$	$-H$	$-Cl$	Anxiolytic
Pinazepam	$-CH_2C{\equiv}CH$	$=O$	$-H$	$-H$	$-Cl$	Anxiolytic
Prazepam	$-CH_2-\triangleleft$	$=O$	$-H$	$-H$	$-Cl$	Anxiolytic
Quazepam	$-CH_2CF_3$	$=S$	$-H$	$-F$	$-Cl$	Sedative hypnotic
Temazepam	$-CH_3$	$=O$	$-OH$	$-H$	$-Cl$	Sedative hypnotic
Flunitrazepam	$-CH_3$	$=O$	$-H$	$-F$	$-NO_2$	Sedative hypnotic

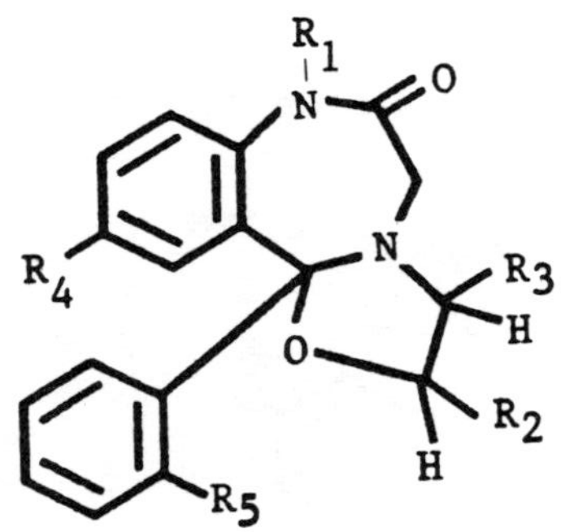

Cloxazolam	$-H$	$-H$	$-H$	$-Cl$	$-Cl$	Anxiolytic
Flutazolam	$-CH_2CH_2OH$	$-H$	$-H$	$-Cl$	$-F$	Anxiolytic
Haloxazolam	$-H$	$-H$	$-H$	$-Br$	$-F$	Hypnotic sedative
Mexazolam	$-H$	$-H$	$-CH_3$	$-Cl$	$-Cl$	Anxiolytic
Oxazolam	$-H$	$-CH_3$	$-H$	$-Cl$	$-H$	Anxiolytic

Table 2 (continued)
IMPORTANT MARKETED BENZODIAZEPINES

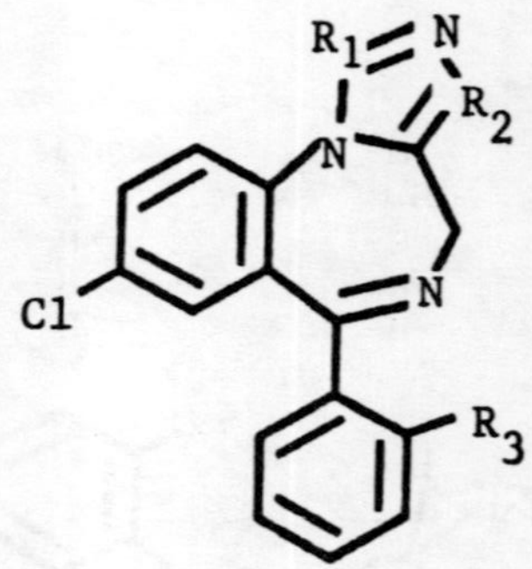

Name	R_1	R_2	R_3	Principal use
Alprazolam	CH_3–C	N	–H	Anxiolytic
Estazolam	CH	N	–H	Sedative hypnotic
Midazolam	CH_3–C	CH	–F	Hypnotic
Triazolam	CH_3–C	N	–Cl	Sedative anxiolytic
Adinazolam	$(CH_3)_2NCH_2$–C	N	–H	Sedative anxiolytic

Loprazolam

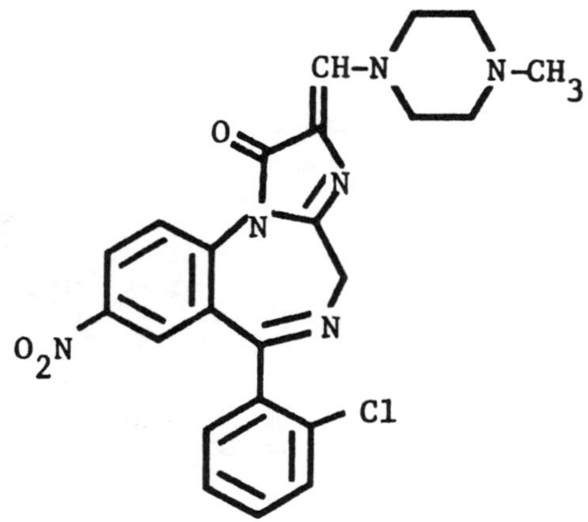

Sedative
Hypnotic

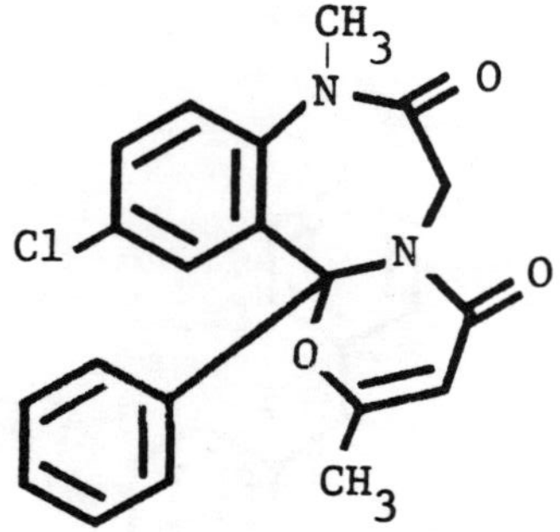

Ketazolam
Anxiolytic

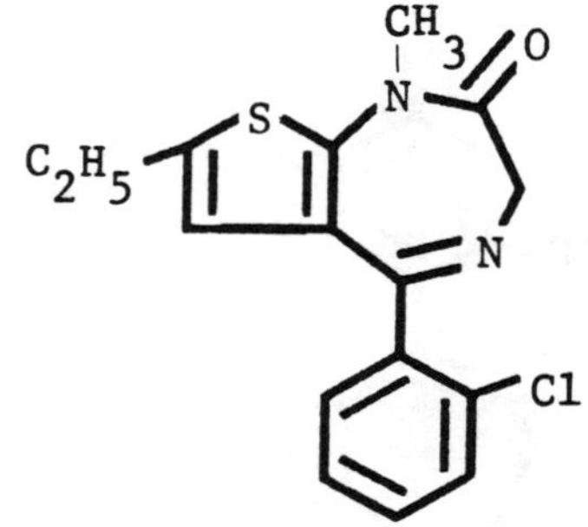

Clotiazepam
Anxiolytic

Table 2 (continued)
IMPORTANT MARKETED BENZODIAZEPINES

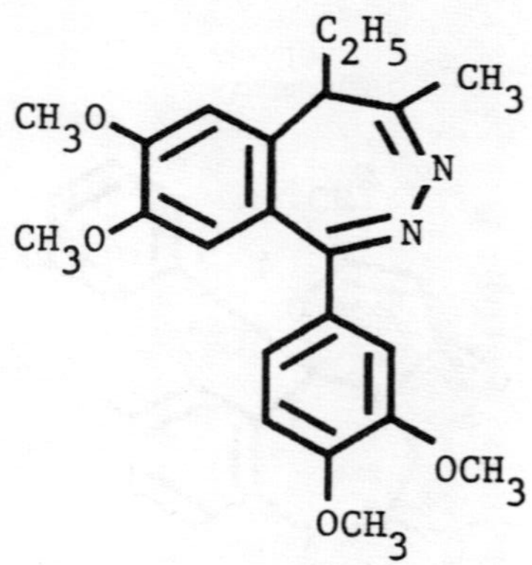

Tofisopam
Anxiolytic

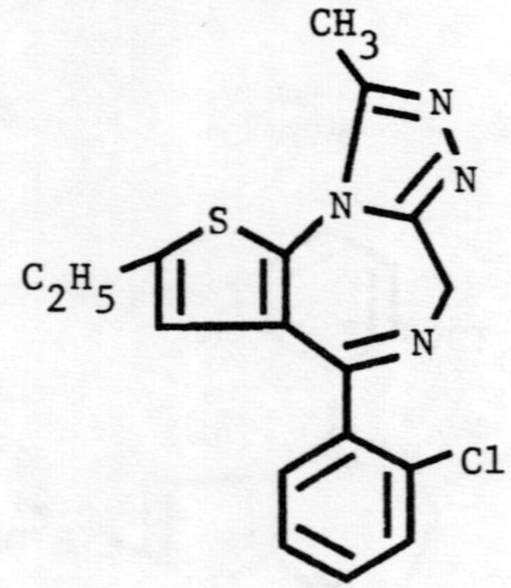

Etizolam
Anxiolytic

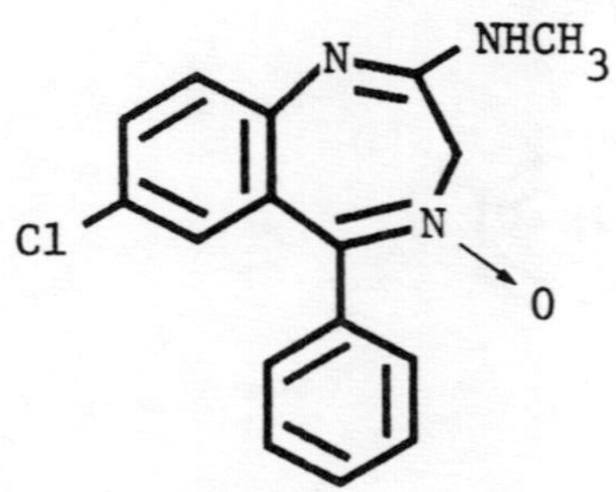

Chlordiazepoxide
Anxiolytic

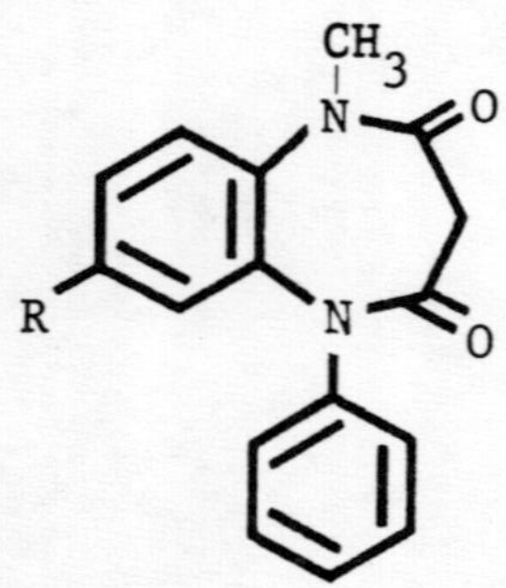

R=Cl=Clobazam
R=CF_3=Triflubazam
Anxiolytics

Pharmacological studies in animals have differentiated buspirone from the benzodiazepines, and the differences have been summarized by Stanton et al.[68] Buspirone attenuated conflict behavior in rats and monkeys at doses similar to benzodiazepines, and in contrast to diazepam, effective doses of buspirone did not produce ataxia.[69-71] However, buspirone does not appear to inhibit or stimulate [^{3}H]-benzodiazepine receptor binding in vitro, and it does not appear to affect the stimulation of such binding by GABA or halide ions. Buspirone does not interfere with the uptake or binding of GABA.[72] Buspirone was found to have no effect on chemically induced convulsions such as convulsions induced by pentylenetetrazol, bicuculline, strychinine, or picrotoxin in rodents,[73] and may lower the electroconvulsive threshold in rats. In contrast, diazepam has been shown to be active against all of these convulsant agents. Buspirone, unlike most anxiolytics, does not show sedative properties in animals and it does not potentiate the effects of CNS depressants such as barbiturates and alcohol.[73] In contrast with diazepam, buspirone does not induce muscle relaxation. Based on the results of animal studies, it appears unlikely that buspirone will suffer from abuse potential or physical dependance liabilities.[73]

Presently, the biochemical mechanism(s) by which buspirone exerts its anxiolytic effect is not fully elucidated. Buspirone has no effect on biogenic amine uptake, nor on in vitro binding to [^{3}H]-ligand labeled sites for the following receptors: α_1,α_2,β-adrenergic, muscarinic cholinergic, glutamate, glycine, H_1- and H_2-histamine, Type I opiate, and Type 1 and Type 2 serotonergic.[68]

Buspirone has, however, demonstrated high-affinity binding to central dopamine receptors using [^{3}H]-spiperone (dopamine antagonist) and [^{3}H]-N-propylapomorphine (dopamine agonist) ligands.[74] Quantitative EEG studies, rotational behavior, catalepsy reversal, decreased striatal l-dopa levels and the perfused isolated rabbit ear artery preparation indicate that buspirone has similar properties to dopamine agonists.[73] However, at higher doses effects similar to dopamine antagonists on stereotypy and levels of striatal dihydroxyphenylacetic acid occur.[73] Benzodiazepines facilitate GABAergic neurotransmission, which in turn may activate dopaminergic systems.[75] The anxioselective action of buspirone has been hypothesized to occur through differential interaction at the multiple dopamine receptors which occur in different locations within the CNS[76] Anxioselective agents, such as buspirone, may act preferentially at subclasses of dopamine receptors in the mesocortical and possibly mesolimbic pathways. Buspirone may preferentially affect those systems responsible for anxiolytic/anticonflict actions without appreciably affecting systems responsible for sedative, muscle relaxant, and anticonvulsant effects. In contrast, the less selective benzodiazepines affect all of these systems.

Buspirone is absorbed rapidly from the GI tract, and is subject to extensive first-pass metabolism which is presumed to occur mainly in the liver. Metabolites arise from oxidation of the azaspirodecanedione moiety, the butylene side chain, and the aromatic ring.[77] A major metabolite is the 5-hydroxypyrimidinyl compound (Structure 3), and the corresponding glucuronide. The plasma half-life of (Structure 3) is about 5 hr, while that of buspirone is 2.5 hr.[78]

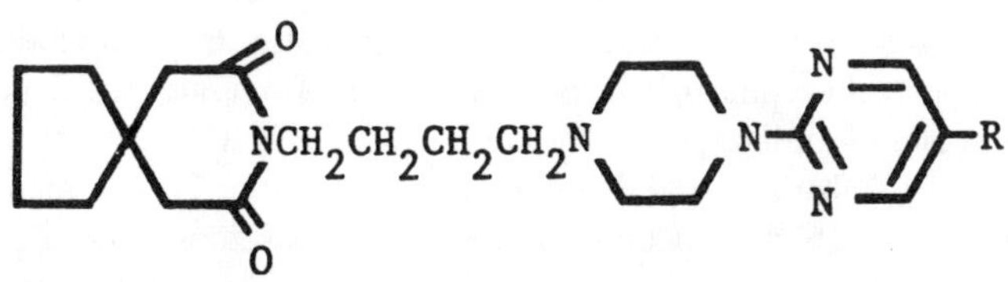

Structure 2, R = H
Structure 3, R = –OH

PYRAZOLOPYRIDINES

A new compound, SQ 20,009 (Structure 4) was reported[79] to be about as potent as diazepam and more potent than chlordiazepoxide in a behavioral test designed to predict the anxiety-reducing effect of drugs.[80] Chemically, SQ 20,009 is 1-ethyl-4 (isopropylidine-hydrazino)-1H-pyrazolo[3,4-b]pyridine-5-carboxylic acid ethyl ester hydrochloride, and it was later named etazolate. Since etazolate, diazepam, and chlordiazepoxide are potent inhibitors of cyclic AMP phosphodiesterase activity in the brain, and cyclic AMP activity has been implicated in the behavioral test, it was speculated that the anxiolytic properties of these compounds may be mediated by cyclic AMP in the brain.

An analog, SQ 65,396 (cartazolate, Structure 5) 4-butylamino-1-ethyl-1H-pyrazolo [3,4-b]pyridine-5-carboxylic acid ethyl ester, was reported to be an effective anxiolytic in human trials.[81] This compound was found to "increase" the binding of ^{3}H-diazepam by increasing the affinity of the substrate for its binding site rather than changing the number of binding sites. Both etazolate and cartazolate were shown to increase Na^{+} -independent ^{3}H-GABA binding to rat cerebral cortex membranes in vitro.[82] Furthermore, SQ 20,009 was shown to increase the apparent sensitivity of the GABA recognition site which is part of the GABA-benzodiazepine receptor complex.[83] Moreover, this effect was enhanced by chloride ions. In the absence of GABA, SQ 20,009 stimulates benzodiazepine binding by increasing the affinity of the benzodiazepine-binding sites, this effect being chloride ion dependent.[84,85] It was concluded[83] that the actions of etazolate are not related to the inhibitory effects on cyclic AMP phosphodiesterase in the brain.

A closely related pyrazolopyridine, tracazolate, ICI 136,753 (Structure 6) was reported to have anxiolytic properties in animals, but did not show significant muscle relaxant or anticonvulsant effects.[86]

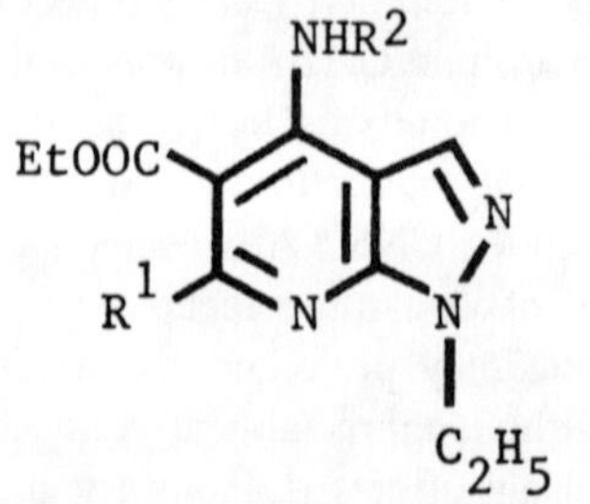

Structure 4, $R^1 = H$, $R^2 = N = C(CH_3)_2$
Structure 5, $R^1 = H$, $R^2 = C_4H_9$
Structure 6, $R^1 = CH_3$, $R^2 = C_4H_9$

TRIAZOLOPYRIDAZINES

As mentioned in the discussion on the benzodiazepine receptor, significant correlations have been obtained between the in vitro and in vivo ability of benzodiazepines to displace brain-specific ^{3}H-diazepam binding and the potency of these drugs as anxiolytic agents in vivo. The first nonbenzodiazepine to selectively displace ^{3}H-binding (with a potency between those of diazepam and chlordiazepoxide) was CL 218,872.[87] This compound, 3-methyl-6-[3-(trifluoromethyl)]-1,2,4-triazolo [4,3-b]pyridazine (Structure 7), is a member of a novel group of chemicals[88] with highly specific actions on benzodiazepine receptors. In addition, CL 218,872 afforded protection against convulsions induced by pentylenetetrazole and was active in conflict procedures employing shock-induced suppression of feeding behavior in rats. These procedures have been correlated with the anxiolytic activity of benzodiazepines.[89] However, unlike the benzodiazepines, CL 218,872 was only weakly active in tests designed to measure the muscular incoordination and sedation produced by benzodiazepines.[87,90] It was suggested[87] that there may be more than one class of benzodiazepine receptor in the brain. Type I may be involved in the mediation of anxiolytic actions, while another class (Type II) may be responsible for the production of side effects. This hypothesis supposes that the benzodiazepines have almost equal affinity for both receptor types, while CL 218,872 has a greater affinity for Type I than Type II and thus exhibits anxiolytic actions without side effects.

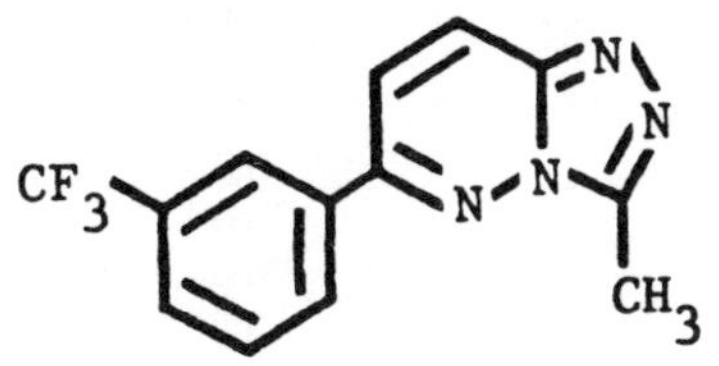

Structure 7

More recently it has been shown that two types of benzodiazepine receptors can be physically separated.[91] The Type I receptors are more highly concentrated in the cerebellum and corpus striatum and show higher affinity for CL218,872 and beta-carboline esters. The Type II receptors are more concentrated in the hippocampus. Several benzodiazepines were found not to differentiate between these receptor types.

Fujimoto et al.[92] have described experiments leading to a classification of various ligands based on the effect of modulators on their affinities for the benzodiazeine receptor found in rat cerebellar and hippocampal membranes. As shown in Table 3, the first class (diazepam type) showed an affinity which increased with GABA, chloride ion, or a combination of these. The affinity of the second class (CL 218,872 type) increased only with GABA, while only chloride ion increased the affinity of the third type of ligand (beta-carboline type). The fourth (inactive) type, consisting of nicotinamide and pentylenetetrazole, was unaffected by GABA, chloride ion, or a combination.

PURINES

It is highly probable that endogenous ligands for benzodiazepine receptors exist in the brain and several attempts have been made to isolate such substances. The purine derivatives inosine and hypoxanthine have been proposed as putative endogenous ligands.[24,26] However, while some biochemical experiments appear to implicate these purines as ligands, results of pharmacological experiments suggest that purines are not likely to be involved.[93]

Skolnick et al.[94] found that a derivative of adenosine, EMD 28422, N^6-[2-(4-chlorophenyl)-bicyclo[2.2.2.]octyl-(3)-adenosine (Structure 8), had a spectrum of pharmacological effects similar to the benzodiazepines. It increased punished responding in a conflict situation in rats, decreased spontaneous motor activity in mice, and protected mice from pentylenetetrazole- and caffeine-induced seizures. In contrast to the benzodiazepines, however, EMD 28422 "increased" the binding or ^{3}H-diazepam to benzodiazepine receptors in vitro[95] and in vivo.[93] This change was due to an increase in the number of benzodiazepine receptors rather than an increase in the affinity of the radiolabeled ligand for the receptor. Neither of the components of EMD 28422, adenosine, and the amine 2-(4-chlorophenyl) bicyclo[2.2.2]octylamine, showed the pharmacological effects of diazepam.

Table 3
EFFECTS OF 10 μM GABA AND/OR 150 mM Cl⁻ ION ON THE AFFINITY OF VARIOUS LIGANDS FOR THE BDZ RECEPTOR IN THE CEREBELLUM DETERMINED AT 0°C

Compound (N)	K_i value under standard conditions (nM)	Relative binding potency (K_i value of the control/K_i value)			
		Control	NaCl	GABA	NaCl + GABA
Diazepam type					
Diazepam (7)	12 ± 1	100	131 ± 5[a]	173 ± 12[a]	266 ± 36[d]
Nitrazepam (4)	22 ± 3	100	133 ± 4[a]	177 ± 13[a]	250 ± 26[d]
Estazolam (4)	19 ± 1	100	149 ± 11[a]	163 ± 7[a]	283 ± 8[d]
Medazepam (4)	1,900 ± 100	100	140 ± 6[a]	144 ± 7[a]	202 ± 12[d]
CL218872 type					
Oxazepam (4)	36 ± 1	100	109 ± 3	199 ± 10[a]	229 ± 9[b]
Lorazepam (4)	2.3 ± 0.2	100	104 ± 3	139 ± 6[a]	158 ± 6[b]
CL218872 (7)	86 ± 9	100	107 ± 4	230 ± 8[a]	226 ± 19[b]
Inosine (6)	950,000 ± 40,000	100	104 ± 3	128 ± 9[a]	114 ± 8
β-Carboline type					
Harmane (4)	19,000 ± 2,000	100	186 ± 16[a]	109 ± 10	179 ± 19[c]
Harmine (4)	110,000 ± 9,000	100	188 ± 17[a]	106 ± 14	197 ± 15[c]
Harmaline (4)	440,000 ± 30,000	100	171 ± 9[a]	125 ± 9	177 ± 7[c]
β-CCE (7)	1.2 ± 0.1	100	126 ± 4[a]	101 ± 6	91 ± 5[b]
Inactive type					
Nicotinamide (4)	2,400,000 ± 200,000	100	112 ± 4	105 ± 3	109 ± 3
PTZ (4)	2,500,000 ± 200,000	100	100 ± 2	109 ± 4	118 ± 9

[a] Significantly different from control, $p < 0.05$.
[b] Significantly different from NaCl (150 mM) alone, $p < 0.05$.
[c] Significantly different from GABA (10 μM) alone, $p < 0.05$.
[d] Significantly different from either NaCl or GABA alone, $p < 0.05$.
Others are not significantly different. Each value is the mean ± SEM.

Used by kind permission of Life Sciences.

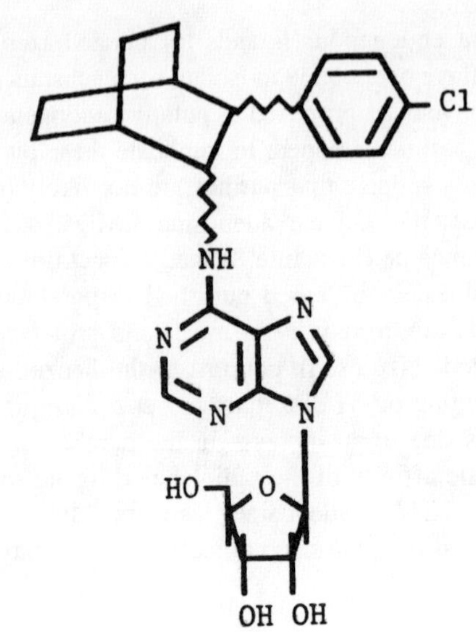

Structure 8

MISCELLANEOUS AGENTS

Zopiclone

Zopiclone, RP 27267, a pyrrolo-[3,4-b]-pyrazine derivative (Structure 9) represents another new chemical type which possesses some of the characteristic pharmacological properties of the benzodiazepines.[96] It was found to bind to the benzodiazepine receptor with a potency comparable to that of diazepam and nitrazepam in vitro and chlordiazepoxide in vivo.[97] Zopiclone had a profile typical of the benzodiazepines in the rat variable interval-fixed ratio conflict test, with a potency equivalent to that of diazepam.[93]

In clinical trials, however, zopiclone exhibited only mild anxiolytic activity, and it is being developed as a sleep inducer having similar properties to nitrazepam.

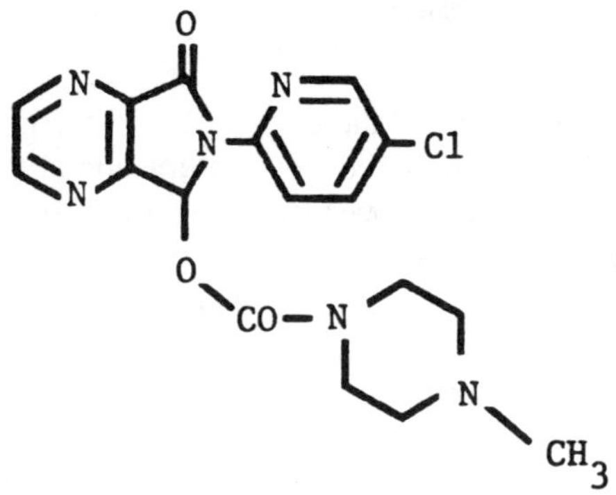

Structure 9

Pyrazolo[4,3-c]Quinolin-3-Ones

A series of 2-arylpyrazolo[4,3-]quinolin-3-ones (Structures 10 a to c) having very high affinity for benzodiazepine receptors has been reported.[98] Of these compounds, Structure 10a, compounds, like Ro 15-1788 (Structure 1), showed potent benzodiazepine antagonist activity. The chloro derivative, Structure 10b had activity similar to diazepam in an anti-conflict behavioral test and in the anticonvulsant test, while Structure 10c behaved as a benzodiazepine agonist at low doses and an antagonist at higher doses.

Structure 10a designated CGS 8216, inhibited ^{3}H-flunitrazepam (^{3}H-FLU) binding to rat synaptosomal membranes in vitro and prevented the in vivo labeling of benzodiazepine receptors by ^{3}H-FLU when given orally to mice.[99] The regional distribution of ^{3}H-CGS 8216 binding sites in the brain was similar to that of ^{3}H-FLU. It was also found that GABA did not enhance the binding of ^{3}H-CGS 8216, and this supports the suggestion that the effects of GABA on the binding of various ligands may distinguish between benzodiazepine agonists and antagonists.[100] When ^{3}H-diazepam is used as a radioligand in binding studies, it has to be used at low temperatures (0 to 4°C). The high affinity of ^{3}H-CGS 8216 at 37°C should make it a useful research tool for studies of in vivo labeling of benzodiazepine receptors.[99]

PHARMACOLOGIC EVALUATION OF ANXIOLYTICS

Tranquilizers, when administered to animals, exert a multiplicity of pharmacologic effects. The relative importance of these pharmacologic effects in deciding potential tranquilizing action in man has been reviewed.[101,102]

A variety of tests are used today for the evaluation of anxiolytics.

Tests for Selective Depression of the CNS

Muscle relaxation and paralysis — These include measurement of the loss of the righting

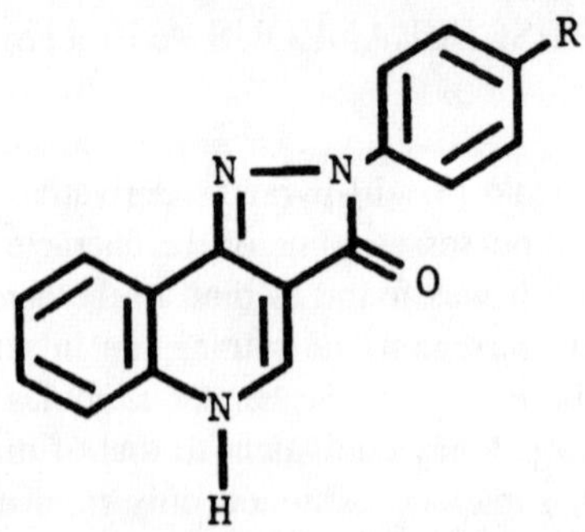

Structure 10a, R=H
Structure 10b, R=Cl
Structure 10c, R=OCH_3

reflex,[103] performance on the rotarod,[104] the traction test,[105] the inclined screen test,[106] and the horizontal griddle test[107] in rodents. The gross observation of the relaxed position of the body and the hind legs when cats are suspended by the scruff of the neck is also used.[108]

Sedative/hypnotic activity — Potentiation of hexobarbital sleeping time[63] or other barbiturates[109] administered either in hynotic or subhypnotic doses measures general CNS depression as opposed to specific tranquilizing property. Another procedure measures the effect of tranquilizers on chlorprothixene anesthesia.[110]

Locomotor activity — The effects of tranquilizers on the spontaneous activity of small animals has been measured in a photocell activity cage[111] and in jiggle cages.[112]

Anticonvulsant activity — These tests measure the ability of tranquilizers to antagonize artificially induced convulsions by chemical or physical means. For chemical convulsions, Metrazole (pentylenetetrazole),[113] for physically induced convulsions the maximal electroshock[114] and minimal electroshock[115] tests are most often used. Other methods have extensively been reviewed.[116]

Tests for Neurophysiological Effects

These tests make extensive use of implanted electrodes and other electrophysiological techniques. The primary purpose of these tests is to help localize the site of action and shed light on the mechanism of action. Clues for the mechanism of action are obtained by comparing the effects of experimental compounds with those of standard agents in areas of the brain. Most often tests include effects on reflexes,[63,117,118] electrical activity determinations in the cerebral cortex,[119] subcortical areas,[119] reticular activating system,[120] limbic system,[121] and other areas.[122]

Behavioral Tests

Open-field test and hole-board test — The open-field test[123] has been used for the measurement of the behavioral activity of small animals for the determination of emotion or fear.[124] The number of times the animal reared, groomed, licked, or defecated in a well-lighted open-field is taken as a measure of behavior, emotion, and fear. Tranquilizers increase the area covered by the animal without any of the previously described events. The hole-board test[125] is based on responses of small animals to an unfamiliar environment by measuring the number of times the animal dips its head into holes in a time period. Tranquilizers increase the number of holes examined.

Spontaneous hostility — Tranquilizers reduce fighting that occurs when a new mouse is introduced into a group of mice used to living together,[126] and fighting in rabbits.[127] Tranquilizers also exert taming effects in monkeys[128] and hamsters.[129]

Experimentally induced aggressiveness — In the isolation-induced aggressiveness test,[130] mice that are aggressive by a 3-week isolation will attack a nonisolated mouse placed in the cage. Aggressiveness can be induced in pairs of mice by footshock[131] that is manifested by fighting between the mice. Aggressiveness can also be induced with apomorphine[132] and lesions in the septal area of rats.[133] Tranquilizers decrease the level of aggressiveness.

Conflict tests — The most widely used behavioral model is termed the conflict test, so named because behavioral responses of the animal are both rewarded and punished. This type of test measures the ability of an anxiolytic to disinhibit behaviors which are readily suppressed by punishment. Examples of such a test are the approach-avoidance (conflict) paradigm of Geller and Seifter,[134] and the shock-induced suppression of drinking behavior test.[80] Such procedures are very selective for anxiolytics, and the clinical efficacy of the benzodiazepines has been positively correlated with their activity in these tests.[135]

The use of numerous other tests has been reviewed.[89,136]

HYDROXYZINE

The effects of hydroxyzine are not apparently exerted by an effect on metabolism in the brain. It is not a cortical depressant but its action may be a result of suppression of activity in certain regions of the subcortical region of CNS.

SARS

Hydroxyzine (Structure 11) is related chemically to the antihistaminic drugs of the diphenylmethane type. It is the most useful anxiolytic agent among the many diphenylmethane compounds having a piperazine substituent. Slight variations in the structure of hydroxyzine cause great changes in pharmacological properties.

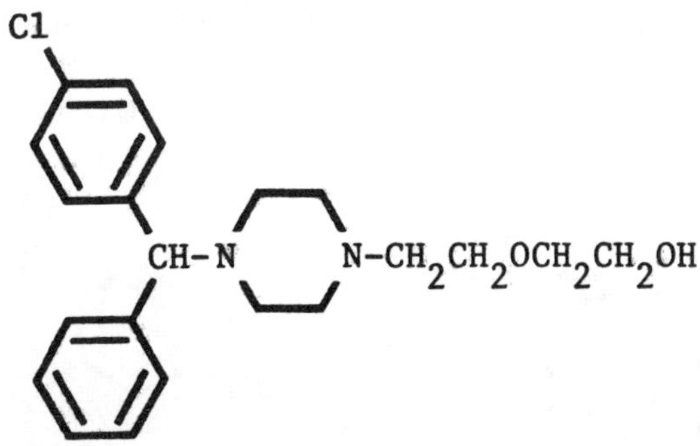

Structure 11

Pharmacokinetics

Hydroxyzine is rapidly absorbed from the GI tract and clinical effects are usually observed within 30 min following oral administration.

Uses and Dosages

Hydroxyzine is used for the symptomatic relief of anxiety and tension associated with psychoneuroses, nervous tension, and emotional stress. It has also been used as an antiemetic, for the management of symptoms of alcohol withdrawal, and in the management of certain allergic conditions (urticaria and dermatoses).

For the treatment of anxiety and tension, the usual dose for adults is 50 to 100 mg given orally four times daily. For children over 6 years of age the range is 50 to 100 mg in divided doses.

Toxicities

Side effects associated with hydroxyzine therapy are usually mild and transitory in nature. The commonest are drowsiness and dry mouth. Hydroxyzine exhibited teratogenic effects in rodents and rabbits and should not be administered to pregnant women.

Drug Interactions

Hydroxyzine potentiates the activity of such CNS depressants as narcotics, nonnarcotic analgesics, barbiturates, and alcohol.

Physical-Chemical and Other Data

Hydroxyzine hydrochloride (Atarax) is 2-[2-[4-[(4-chlorophenyl) phenylmethyl]1-piperazinyl]ethoxy]ethanol dihydrochloride $C_{21}H_{27}ClN_2O_2 \cdot 2HCL$. White odorless powder with a bitter taste; melting point 196 to 204°C. Solubility in water, 1 part in 1.

Hydroxyzine pamoate (Vistaril) is 2-[2-[4-[(4-chlorophenyl) phenylmethyl]-1-piperazinyl]ethoxy]ethanol·4,4′-methylenebis[3-hydroxy-2-naphthalene-carboxylic acid]. $C_{21}H_{27}ClN_2O_2 \cdot C_{23}H_{16}O_6$. Pale yellow, almost odorless powder. Insoluble in water; soluble to the extent of 1 part in 700 in ethyl alcohol and 1 part in 10 in dimethylformamide.

Chlordiazepoxide (Librium®)

Chlordiazepoxide was the first 1,4-benzodiazepine to be marketed, in 1960. Shortly after its introduction into medical practice, it became the most often prescribed drug until 1969, when it was overtaken by diazepam (Valium®), another member of the 1,4-benzodiazepine class.

SARS

Originally, based on analogous synthetic reactions described in the literature, a heptoxidiazine structure was assigned to chlordiazepoxide. Additional chemical reactions showed that the assigned structure was incorrect and a new structure was soon devised, representing the correct structure (Structures 12 and 13).

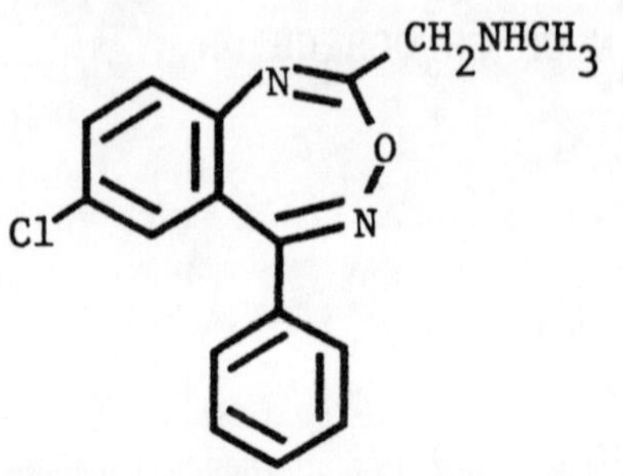

Heptoxdiazine structure

Structure 12

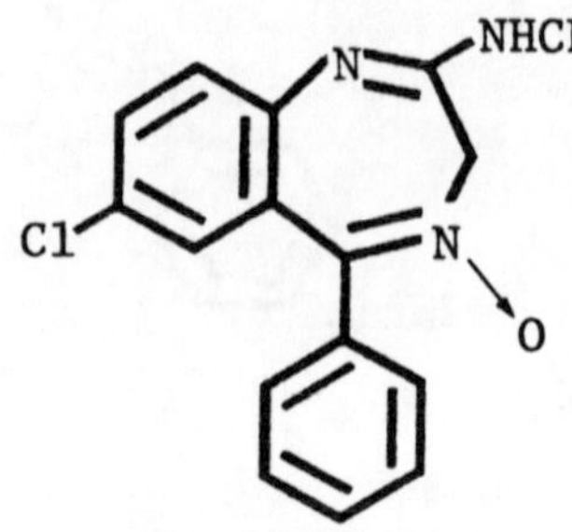

Chlordiazepoxide

Structure 13

Following the discovery of the striking pharmacological effects of chlordiazepoxide by Sternbach and Randall at Roche, a large number of derivatives have been prepared. Chlordiazepoxide turned out to be one of the most potent members of this class of compounds. None of the other analogues provided a compound with superior overall biological activity.

The following SAR generalizations can be made:

1. Increasing the size of the substituent at carbon-2 beyond methyl decreases the potency.

2. Replacing the phenyl group at carbon-5 by other substituents decreases the potency.

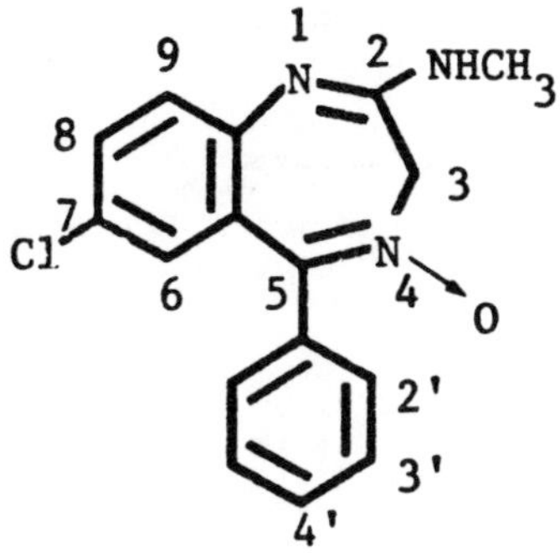

Structure 14

3. The N-oxide oxygen is not essential for pharmacologic activity.
4. The basic substituent attached to carbon 2 is not essential for pharmacologic activity.

Alternatively, rigorous SAR studies revealed that the following features are indispensable for retaining pharmacologic activity:

1. The presence of the 1,4-benzodiazepine ring system
2. The presence of the chlorine at the 7-position
3. The presence of the phenyl group at the 5-position

Due to the fact that chlordiazepoxide was hygroscopic, bitter, unstable, and not suitable for making syrup or elixir formulations for pediatric and geriatric use, a number of chemical studies were initiated on chlordiazepoxide. Upon standing in acidic solutions, chlordiazepoxide is rapidly converted into Structure 15 which has the same pharmacologic activity as chlordiazepoxide. Removal of the N-oxide function is also easily achieved and Compound 13 is perhaps even more potent than chlordiazepoxide. Interestingly Compound 12 on treatment with acetic anhydride undergoes the Polonovsky rearrangement to form 3-acetyloxazepam, which on mild hydrolysis yields oxazepam.

A general route to the synthesis of chloridazepoxide consists of acylation of an appropriate oxime followed by treatment of the chloromethyl quinazoline 3-oxide with a primary amine. In the last step the quinazoline-3-oxide rearranges to a 1,4-benzodiazepine.

Pharmacokinetics

Chlordiazepoxide is readily absorbed from the GI tract. Its peak time is 0.5 to 4 hr with a peak level of 350 to 2050 ng/mℓ. The elimination half-life after a single oral dose is about 24 hr. It is excreted in the urine and the bile. Only a small amount of drug is excreted as the unchanged drug. Several metabolites have been identified as outlined below. After i.m. injection, absorption of chlordiazepoxide is slow and erratic, resulting in peak concentrations that are usually lower than after oral administration.

Chlordiazepoxide is first metabolized to desmethylchlordiazepoxide and demoxepam, both of which are pharmacologically active. The latter is partly converted to nordiazepam, an active metabolite, and partly to a ring cleaved lactam and hydroxydemoxepam which are pharmacologically less active or inactive.

Chlordiazepoxide has an apparent volume of distribution (V_d) of 0.26 to 0.58 ℓ/kg and a 94 to 97% protein binding.

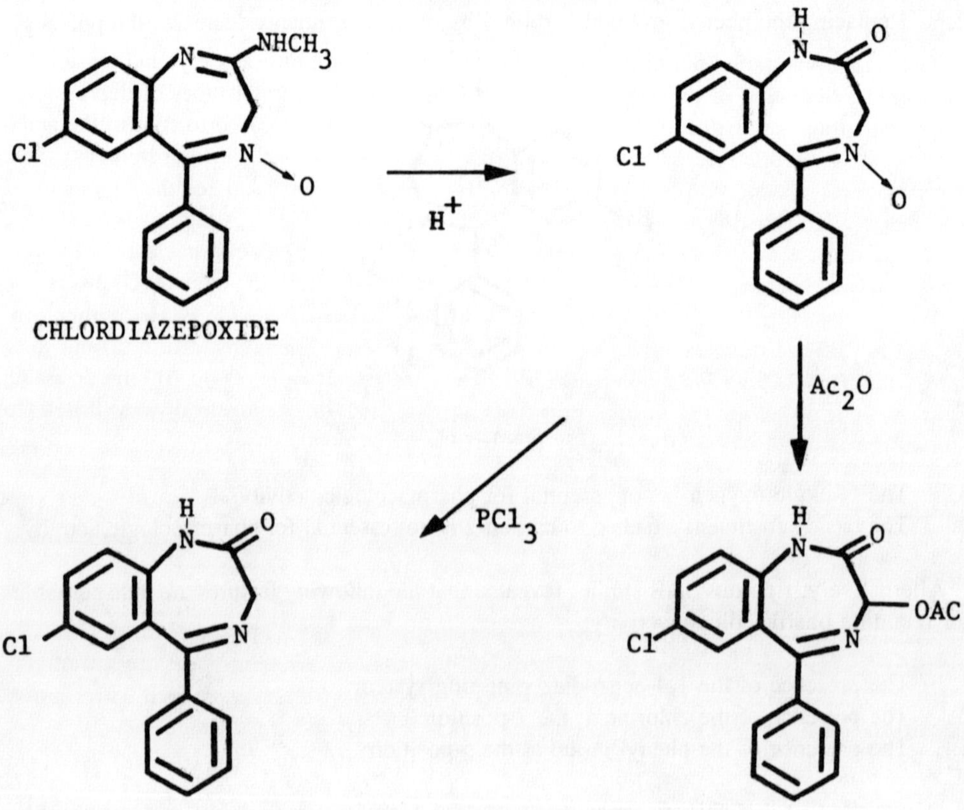

Structure 15

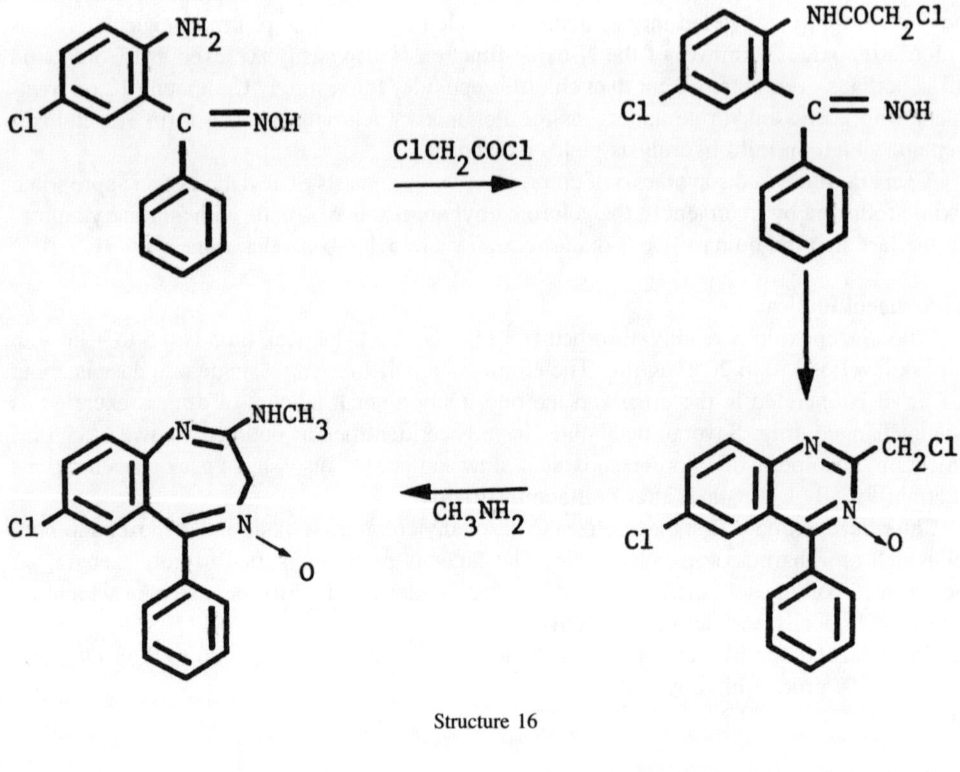

Structure 16

Uses and Dosage

Chlordiazepoxide is used in the treatment of neurotic conditions, anxiety, and tension but not for the treatment of psychotic patients. It is also used in certain types of depressions, largely anxious depression or endogenous depression. It has been used to control the acute symptoms of the alcohol withdrawal syndrome (e.g., agitation or delirium tremens). Due to its modest muscle relaxant effects, chlordiazepoxide is also used for the treatment of musculo-skeletal disorders associated with emotional disturbances. Additionally it is used as a preanesthetic sedative. Chlordiazepoxide has a synergistic effect on some anticholinergic drugs such as clidinium and is used as a combination product for the treatment of GI disorders.

As a preanesthetic drug due to its sedative effects, chlordiazepoxide is used either i.m. in 50- to 100-mg doses 1 hr before operation, or the drug is employed in a 50-mg dose orally the night before and given again 2 hr before the operation in 50- to 100-mg dose. As an anxiolytic in adults the normal oral daily dosage is 15 to 30 mg in divided doses. In children, the usual daily dosage orally is 5 to 20 mg in divided doses.

Toxicities

Drowsiness, dizziness, sleepiness, fatigue, constipation, apathy, and ataxia are the most frequently seen side effects. Indigestion, diarrhea, skin rashes, impotency, headache, and nausea have occasionally been reported.

The effects of chlordiazepoxide may be accentuated by other CNS depressants especially barbiturates and narcotic analgesics. There is a pronounced increase of the alcohol effects when coupled with a dose of chlordiazepoxide. Chlordiazepoxide must be used with caution in patients with renal or hepatic impairment and in patients using anticoagulants.

Physical-Chemical and Other Data

Chlordiazepoxide hydrochloride.Methaminodiazepoxide. 7-chloro-2-methylamino-5-phenyl-3*H*-1, 4-benzodiazepine 4-oxide, $C_{16}H_{14}ClN_3O$. HC1 Molecular weight = 336.2, pKa 4.76.

It is a white crystalline substance with a bitter taste. It is soluble in 10-fold volume of water and 40-fold volume of ethanol and is insoluble in chloroform and ether.

Chlordiazepoxide hydrochloride is available in 5-mg capsules, in 5-, 10-, and 25-mg tablets, and as a 100-mg injectable.

Chlordiazepoxide hydrochloride should be protected from light.

OXAZEPAM (SERAX®)

Soon after the discovery of chloridazepoxide by Sternbach, its chemical reactions were extensively investigated by chemists at Wyeth and Roche. Bell and Childress at Wyeth discovered the outstanding biologic activity of oxazepam which could be obtained from chlordiazepoxide by acetylation, hydrolysis, treatment of the product with acetic anhydride followed by mild hydrolysis of the 3-acetoxy derivative shown in Structure 17.

Alternatively, oxazepam can be prepared by any of the following sequences, which are general reactions for a large number of 1,4-benzodiazepines in Structure 18.

SARs

A great number of analogues have been synthesized both at Wyeth and Roche and examined for biological effects. The following generalizations can be made:

1. An electron-withdrawing substituent (chloro, bromo, nitro, or trifluoromethyl) at position 7 is required. Displacement of the electron withdrawing substituent by an electron-donating group (methyl or methoxy) reduces the potency. Substitution at carbons 6, 8, or 9 decreases potency.

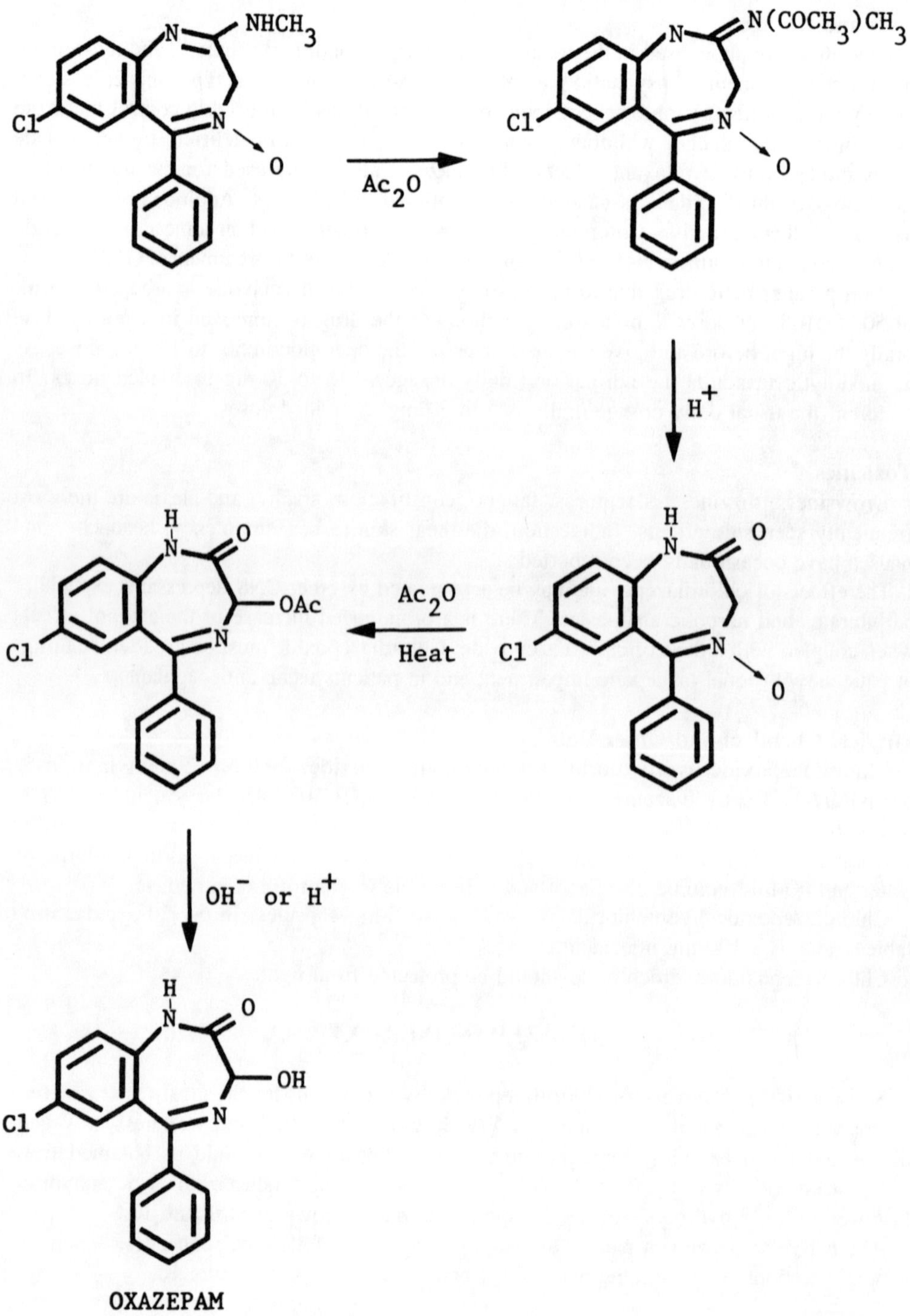

Structure 17

2. A methyl substituent at position 1 increases the potency. Certain other substituents at position 1 (diethylaminoethyl, cyclopropylmethyl) also have a favorable effect on the activity. Fusion of an additional ring to positions 1 and 2 often produces compounds of superior activities. A carbonyl or an imine function or a precursor leading to any of these functions at position 2 is required for good activity. The double bond between positions 4 and 5 is required.

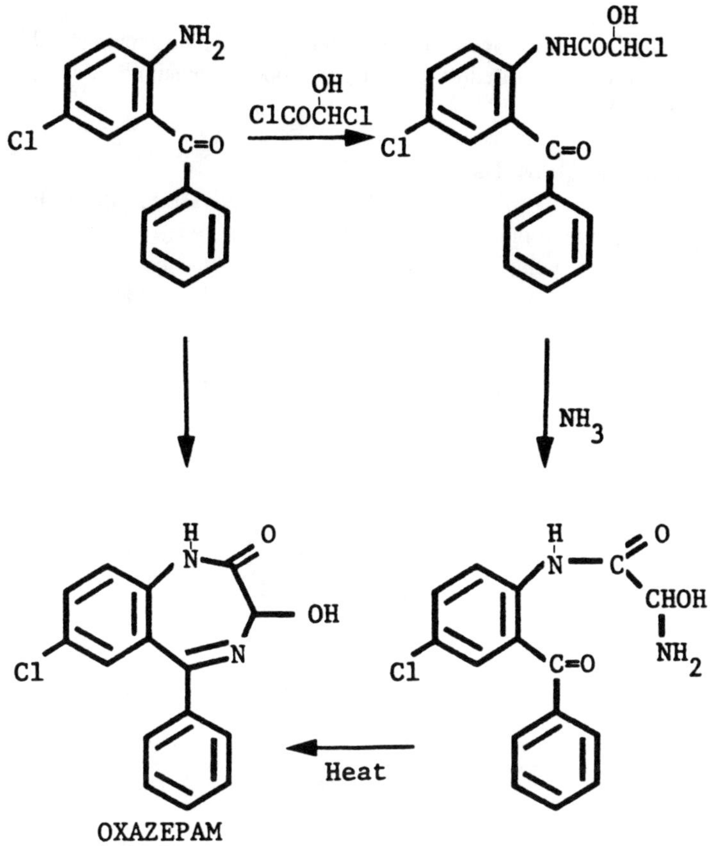

Structure 18

3. A phenyl group at position 5 is required. Electron-withdrawing (especially halogens) substituents at position 2 increase the potency. Any substitution at position 4 decreases potency.

Pharmacokinetics

Oxazepam is absorbed slowly from the GI tract. Its peak time is between 1 to 4 hr with a peak level of 120 to 190 ng/mℓ. Its elimination half-life after a single dose is 4 to 13 hr. It is excreted in the urine mainly as the *O*-glucoronide conjugate and in the feces in the unchanged form.

Oxazepam has no active metabolite and it is directly conjugated with glucuronic acid. Oxazepam has an apparent volume of distribution (V_d) of 0.6 ℓ/kg and a 87 to 90% protein binding.

Uses and Dosage

Oxazepam is used in the treatment of neurotic conditions accompanied by anxiety and tension. It is not used in the treatment of psychotic conditions. Oxazepam has been found to be effective for sleep induction in higher dosages than those used for the relief of anxiety. Due to its kinetic profile, the absence of active metabolites, and the shorter elimination half-life, oxazepam is an attractive benzodiazepine for use in routine practice.

Oxazepam may be useful in the treatment of the acute symptoms (agitation and delirium tremens) of alcohol withdrawal syndrome.

The usual dose for the treatment of anxiety conditions is 45 to 120 mg daily in divided doses. In severe conditions 90 to 180 mg is used, also in divided doses.

Toxicities

Toxicities are similar to those seen with chlordiazepoxide. Oxazepam should be used with caution in patients who are required to drive a car or operate machinery, since it may reduce the person's mental alertness.

Physical-Chemical and Other Data

Oxazepam. 7-chloro-2,3-dihydro-3-hydroxy-5-phenyl-1*H*-1,4-benzodiazepin-2-one. $C_{15}H_{11}ClN_2O_2$. Molecular weight = 286.7, pKa 1.62.

It is an off-white to pale yellow odorless substance. It is almost insoluble in water, soluble in alcohol, chloroform, and dioxan. A suspension in water has a pH of 4.8 to 7.

Oxazepam is available in 10-, 15-, and 30-mg capsules and 10- and 15-mg tablets.

DIAZEPAM (VALIUM®) (STRUCTURE 19)

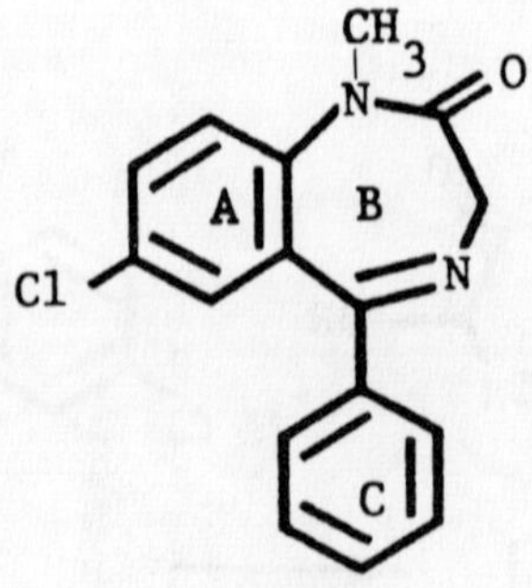

DIAZEPAM

Structure 19

SARs

After the discovery of chlordiazepoxide a great number of simple benzodiazepinones without the N-oxide function were synthesized. While researchers at Wyeth concentrated on oxazepam and related analogues, Sternbach and co-workers at Roche prepared benzodiazepinones bearing various substituents at different positions in ring A and substituents in position 1 of ring B. Among these analogues, the 7-chloro-1,3-dihydro-1-methyl-5-phenyl-2*H*-1,4-benzodiazepin-2-one derivative, called diazepam, was several times more potent than chlordiazepoxide, shown:

COMPARISON OF PHARMACOLOGIC ACTIVITIES[a]

	Inclined screen	Foot shock	Cat	Pentylene tetrazole	Electroshock MAX	Electroshock MIN
Chlorodiazepoxide	100[a]	40	2	18	92	150
Diazepam	30	10	0.2	1.4	6.4	64

[a] The numbers denote the dose (mg/kg) of orally administered drug to elicit the same effect.

Diazepam was extensively investigated and marketed in 1963 under the trademark of Valium®.

Diazepam soon became the most successful anxiolytic drug as well as one of the most successful drugs ever among all prescription drugs. Due to its success many other companies became involved in the investigation of benzodiazepines. Several thousands of analogues have been prepared and more than 40 different 1,4-benzodiazepine derivatives have been marketed worldwide.

Pharmacokinetics

Diazepam is readily absorbed from the GI tract. Peak concentrations of 175 to 400 ng/mℓ are obtained in about 1.5 to 2 hr after administration of 10 mg diazepam orally. Its elimination half-life after multiple oral doses is 20 to 70 hr. Diazepam is extensively metabolized and its active metabolites include nordiazepam, oxazepam, and temazepam. Diazepam and its metabolites are excreted in the urine and a small amount is excreted in the feces. Metabolites have been found in the urine and in the plasma 14 days after a 10-mg dose. Due to its very long elimination time, diazepam belongs to the long-acting benzodiazepines. Its apparent volume of distribution (V_d) is 1 to 2 ℓ/kg and it is 97 to 99% protein bound. In older patients, its apparent volume of distribution is increased and its elimination half-life is longer. After i.m. injection, absorption of diazepam is slow and erratic, usually producing lower plasma concentrations than those after oral administration.

Uses and Dosage

The use of diazepam is very similar to that of chlordiazepoxide. It is used primarily in the treatment of neurotic conditions accompanied by anxiety. It is also used in the treatment of skeletal-muscle spasms as a muscle relaxant and in severe status epilepticus as an injectable anticonvulsant. The usual dose for the treatment of anxiety states is 5 to 15 mg in divided doses and 15 to 30 mg in severe anxiety states, also in divided doses. In muscle spasms the usual dose is 2 to 15 mg daily in divided doses, while as an anticonvulsant it is used intravenously in a 10-mg dose. To prevent recurrence of epileptic seizures, 100- to 200-mg doses are administered over a 24- to 48-hr period as an i.v. infusion.

Toxicities

The toxicities of diazepam are very similar to those produced by chlordiazepoxide. In addition, dependence of the barbiturate/alcohol type may occur. Diazepam reduces mental alertness and therefore adversely affects the ability of patients to drive or operate machinery. Its effects are enhanced by alcohol. In severely neurotic patients diazepam may produce a paradoxical, stimulating effect instead of sedation.

Physical-Chemical and Other Data

Diazepam. 7-chloro-2,3-dihydro-1-methyl-5-phenyl-1*H*-1,4-benzodiazepin-2-one. $C_{16}H_{13}ClN_2O$. Molecular weight = 284.7, pKa 3.17.

A white or off-white, odorless, crystalline powder with a bitter after-taste. Very slightly soluble in water. Soluble in alcohol and chloroform.

Available in 2- and 5-mg capsules, 2-, 5-, and 10-mg tablets, 2-mℓ and 4-mℓ ampoules of an injection containing 5 mg/mℓ and as a syrup containing 2 mg in 5 mℓ. A once daily 15-mg diazepam controlled release capsule has been approved by the FDA for the USA; Roche has demonstrated that the 15-mg Valrelease capsule is equivalent to the thrice daily administration of 5-mg oral Valium®, tablets.

PRAZEPAM (VERSTRAN®, CENTRAX®)

Prazepam is a prodrug of desmethyldiazepam (nordiazepam) since it is nearly completely metabolized in the liver. However oral administration of prazepam yields desmethyldiazepam gradually almost resembling a sustained-release kinetic profile. During multiple dose therapy with prazepam, desmethyldiazepam accumulates in sufficient quantities and at a proper rate to make it possible to use prazepam as a once-daily anxiolytic.

Pharmacokinetics

Prazepam is slowly absorbed from the GI tract. Prazepam is extensively metabolized to desmethyldiazepam as shown in Structure 20.

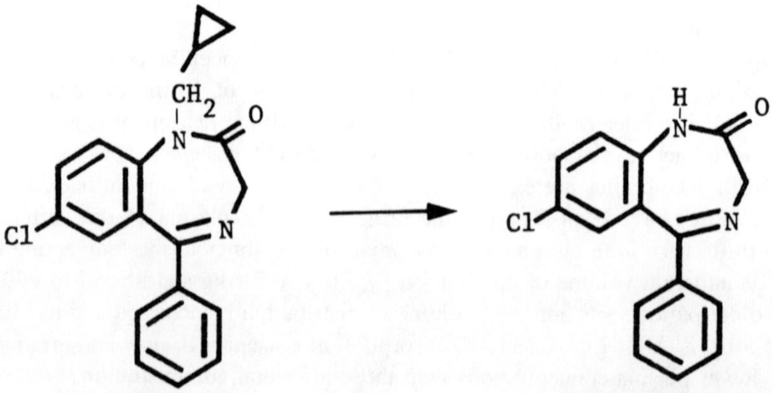

PRAZEPAM

DESMETHYLDIAZEPAM (NORDIAZEPAM)

Structure 20

Desmethyldiazepam is the only active unconjugated benzodiazepine present in the blood as a result of prazepam administration. The blood levels of prazepam are negligible.

Uses and Doses

Since prazepam is extensively metabolized to desmethyldiazepam, which in turn is responsible for the clinical activity of prazepam, the clinical efficacy of prazepam is similar to other long-acting benzodiazepines. Prazepam is valuable as a once-daily anxiolytic; its clinical efficacy has been established in controlled clinical studies. Aside from its pharmacokinetic advantage, prazepam does not have a superiority over other long-acting benzodiazepines.

Toxicities

The slow generation of desmethyldiazepam may be advantageous in patients that cannot tolerate the general CNS depressant effects associated with benzodiazepine therapy.

Since desmethyldiazepam is responsible for the activity of prazepam, the side effects observed as a result of prazepam administration are also similar to those seen with other long-acting benzodiazepines.

Physical-Chemical and Other Data

Prazepam. 7-chloro-1-cyclopropylmethyl-5-phenyl-2,3-dihydro-1*H*-1,4-benzodiazepine-2-one. $C_{19}H_{17}ClN_2O$. Molecular weight = 324.81.

A colorless, odorless crystalline powder. Available in 10-mg tablets.

LORAZEPAM (ATIVAN®)

Lorazepam is very similar both in chemical structure and in properties to oxazepam (Structure 21).

Like oxazepam, lorazepam does not undergo oxidative metabolism and therefore does not utilize hepatic microsomal enxymes. Like oxazepam, lorazepam is a short-acting benzodiazepine and has no active metabolites.

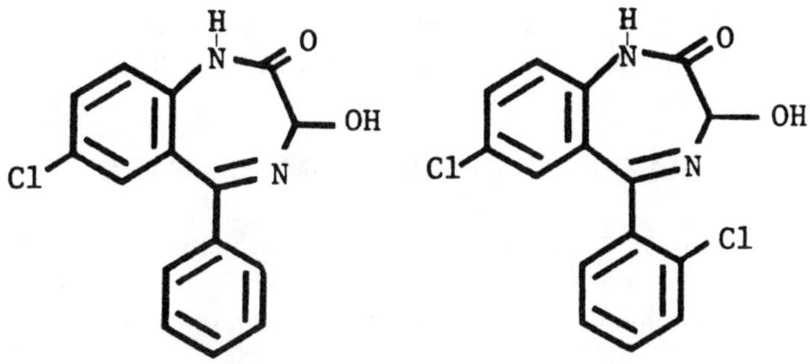

OXAZEPAM LOREZAPAM

Structure 21

Pharmacokinetics

Lorazepam is absorbed from the GI tract. It reaches a peak level of 17 to 20 ng/mℓ in a peak time of 1 to 2 hr in a fasting subject after oral dosing. More than 80% of the oral dose reaches the systemic circulation. Intramuscularly administered lorazepam is rapidly and completely absorbed with a mean peak time of 1.5 hr. Lorazepam has an elimination half-life of 12 hr after both single and multiple oral doses. The apparent volume of distribution (V_d) of lorazepam is 0.7 to 1ℓ/kg and it is 85% protein bound.

Lorazepam has no active metabolite and is excreted after conjugation with glucuronic acid. Multiple dose lorazepam administration produces steady-state conditions within 3 days. If lorazepam therapy is discontinued, lorazepam is cleared from the system within 4 days after the final dose as opposed to long-acting benzodiazepines, which require as much as 2 weeks to be completely eliminated.

Uses and Dosage

The uses of lorazepam are similar to those of diazepam and oxazepam. It is used in the treatment of anxiety tensions and as a premedication to anesthesia.

Lorazepam has a great advantage over chlordiazepoxide, diazepam, and other benzodiazepines that are available for parenteral administration. While the absorption of both chlordiazepoxide and diazepam after parenteral administration is erratic, lorazepam is completely absorbed with a peak time of 1.5 hr. Furthermore, serum or plasma concentrations decline very rapidly after an i.v. bolus injection of diazepam, due to its rapid and extensive distribution which causes a rapid decline of the drug effects of diazepam (anticonvulsant, sedative, or amnestic effect). Lorazepam, on the other hand, is distributed less rapidly and to a lesser extent prolonging the clinical effects. As a result, despite its shorter elimination half-life, lorazepam may elicit longer effective clinical activity than diazepam, due to its less rapid and less extensive tissue distribution. Lorazepam, therefore, may be preferred to diazepam as a preanesthetic medication.

It should be pointed out, however, that a consistent clinical advantage of lorazepam over other longer acting benzodiazepines has not been demonstrated.

The usual dosage is 2 to 3 mg/day in divided doses.

Toxicities

If they occur, toxicities are usually seen at the beginning of therapy. Most frequent side effects include sedation, dizziness, weakness, and unsteadiness. Less frequent side effects are disorientation, depression, nausea, change in appetite, headache, sleep disturbance, agitation, etc.

Physical-Chemical and Other Data

Lorazepam. 7-Chloro-5-(2′-chlorophenyl)-2,3-dihydro-3-hydroxy-1*H*-1,4-benzodiazepine-2-one. $C_{15}H_{10}Cl_2N_2O_2$. Molecular weight = 321.2. A white, odorless substance. Practically insoluble in water, soluble in alcohol, acetone, and glacial acetic acid. Available in 0.5-, 1-, and 2-mg tablets.

CLORAZEPATE DIPOTASSIUM (TRANXENE®)

Clorazepate is a prodrug of nordiazepam since it converts promptly to nordiazepam in the acidic medium of the stomach and probably elsewhere in the body. Only nordiazepam appears in the systemic circulation after oral administration of clorazepate (Structure 22).

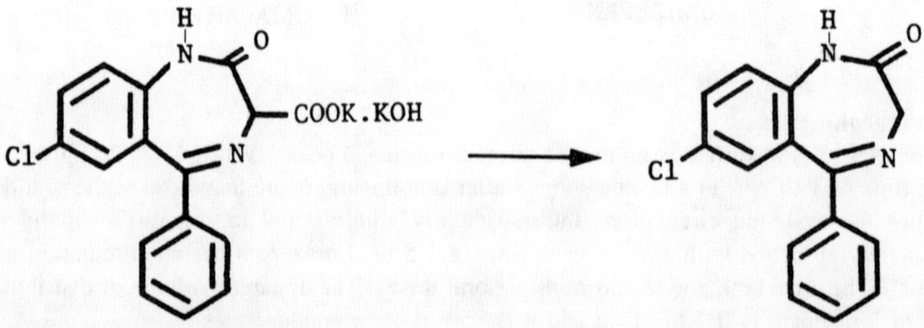

Structure 22

Pharmacokinetics

Clorazepate is absorbed from the GI tract and converted by the acid of the stomach to nordiazepam (desmethyldiazepam), which is the major active metabolite of diazepam. Since nordiazepam has a prolonged half-life, once a day dosing with clorazepate is often adequate.

After oral administration nordiazepam is rapidly formed even in the presence of antacids. It has been shown that achlorhydria does not influence the pattern of appearance of nordiazepam in the blood.

Intramuscularly injected clorazepate produces a clorazepate peak plasma concentration within 2 hr and is subsequently coverted to nordiazepam with a half-life of 1 to 2 hr. The elimination half-life of nordiazepam is 7 to 10 days. Nordiazepam has an apparent volume of distribution of 0.95 to 1.25 ℓ/kg and is 97.5% protein bound.

Uses and Doses

Clorazepate is used in the treatment of anxiety, often as a single daily dose of 15 mg. The clinical efficacy of clorazepate is similar to that of other long-acting benzodiazepines. Its antianxiety effectiveness has been shown to be equivalent to diazepam in controlled clinical studies.

The usual dose is 15 to 60 mg daily, in divided doses. The unit dose is 15 mg.

Toxicities

The toxicities are very similar to those observed with diazepam.

Physical-Chemical and Other Data

Clorazepate dipotassium. Dipotassium 7-chloro-2,3-dihydro-2,2-dihydroxy-5-phenyl-1*H*-1,4-benzodiazepine-3-carboxylate. $C_{16}H_{11}ClK_2N_2O_4$. Molecular weight = 408.9. A fine, odorless, light yellow substance. Very soluble in water. Insoluble in organic solvents. Available in 15-mg tablets.

CLOBAZAM (FRISIUM®, URBANYL®)

This anxiolytic agent has a 1,5-benzodiazepine structure as opposed to other benzodiazepines which have a 1,4-benzodiazepine structure.

Clobazam was synthesized by researchers at Maestretti in Milan in 1969 based on the hypothesis that replacement of the 4-5 imine function by the carbonyl group at position 4 should not diminish anxiolytic activity.

The pharmacologic evaluation of clobazam revealed that there was a wide separation between doses of the drug which would induce a tranquilizing effect and doses required to produce ataxia. In a variety of pharmacologic models clobazam produced less disruption of motor coordination than the 1,4-benzodiazepines at doses which produce equivalent tranquilizing effects.

Clinical trials indicate that twice the doses of clobazam are required to produce a comparable antianxiety effect to that produced by diazepam.

Clobazam appears to cause less objectively measured sedation and psychomotor impairment in spite of the fact that subjective measurement of drowsiness appears to be of comparable frequency to that produced by diazepam. Clobazam has minimal muscle relaxant and hypnotic activity.

Pharmacokinetics

Clobazam is virtually completely absorbed from the GI tract. Peak levels of 250 to 430 ng/mℓ are obtained 1 to 4 hr after the administration of 20 mg clobazam. After a single oral dose the elimination half-life is 9 to 30 hr. Clobazam is present in the serum mainly as the parent compound and is metabolized to desmethylclobazam, 4′-hydroxy-clobazam and 4′-hydroxy-desmethylclobazam. The main metabolite, desmethylclobazam (or norclobazam) has an elimination half-life of 36 to 46 hr but is considerably less potent than clobazam. The formation of phenyl-hydroxylated metabolites is unique to clobazam and is not observed among the 1,4-benzodiazepines (Structure 23).

Clobazam resembles the long-acting 1,4-benzodiazepines in its pharmacokinetic properties, due to its long overall elimination half-life. It is 87 to 90% protein bound.

Uses and Doses

Clobazam is an effective antianxiety agent in the treatment of neurotic conditions associated with anxiety. It is not useful for the treatment of psychotic conditions. Clobazam may be useful in the anxiety condition in patients who experience pyschomotor impairment with 1,4-benzodiazepines. However the pharmacodynamic profile of clobazam appears to be qualitatively similar to the 1,4-benzodiazepines producing reduction of anxiety and drowsiness as a subjectively reported side effect. Unlike 1,4-benzodiazepines, however, clobazam does not produce muscle relaxant activity.

The usual adult dose is 20 to 30 mg daily orally in divided doses. In severely neurotic patients up to 80 mg daily has been administered orally, also in divided doses.

Toxicities

Clobazam is generally well tolerated. Treatment of anxious patients with 30- to 80-mg daily doses appears to produce as much drowsiness or hangover as that produced with 15 to 40 mg diazepam under comparable study conditions. Other reported side effects include dizziness, weakness, lightheadedness, weight gain, orthostatic hypotension, syncope, headache, dry mouth, incoordination, and irritability.

Interaction between clobazam and drugs that affect the CNS (including alcohol) has been observed, leading, in general, to an enhancement of drug effects.

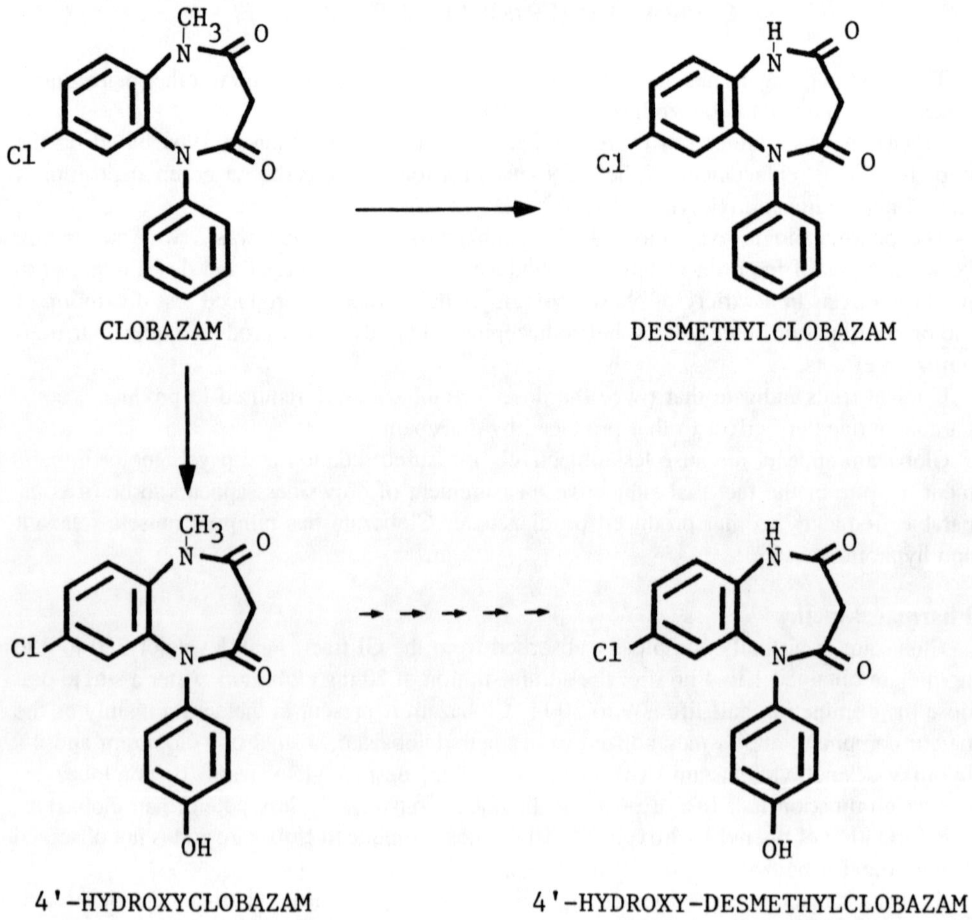

Structure 23

Physical-Chemical and Other Data

Clobazam. 7-chloro-1-methyl-5-phenyl-1*H*-1,5-benzodiazepine-2,4(3*H*,5*H*)-dione. $C_{16}H_{13}ClN_2O_2$. Molecular weight = 300.7. Clobazam is available in 10- and 20-mg tablets.

BROMAZEPAM (LEXOTANIL®)

SARs

Although it was initially believed that replacement of the phenyl substituent by other substitutents in position 5 of the B ring of 1,4-benzodiazepines decreases anxiolytic potency, it was later demonstrated that a heterocyclic aromatic ring could replace the phenyl ring. Bromazepam is the first marketed 1,4-benzodiazepine that does not have a 5-phenyl substituent. Instead, bromazepam has a 5-pyridyl substituent (Structure 24).

Pharmacokinetics

Like other 1,4-benzodiazepines, bromazepam is readily absorbed from the GI tract. Two principal metabolites have been identified, as shown in Structure 25.

The major urinary metabolites in man are 3-hydroxybromazepam accounting for 15 to 30% of the dose and 2-(2-amino-5-bromo-3-hydroxybenzoyl) pyridine accounting for 5 to 25% of the dose. Both of these metabolites are excreted as glucuronic acid or sulfate conjugates.

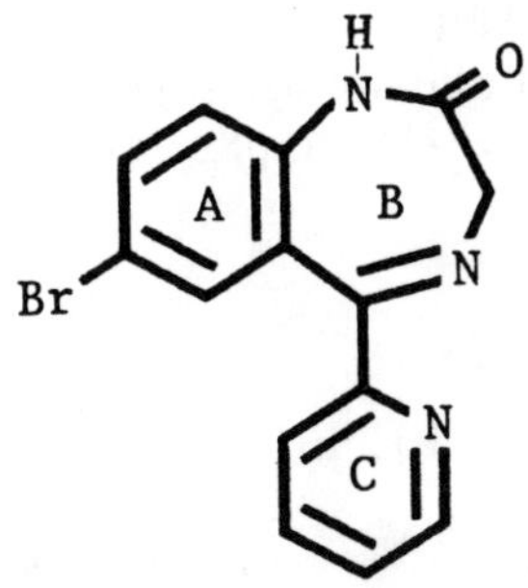

BROMAZEPAM

Structure 24

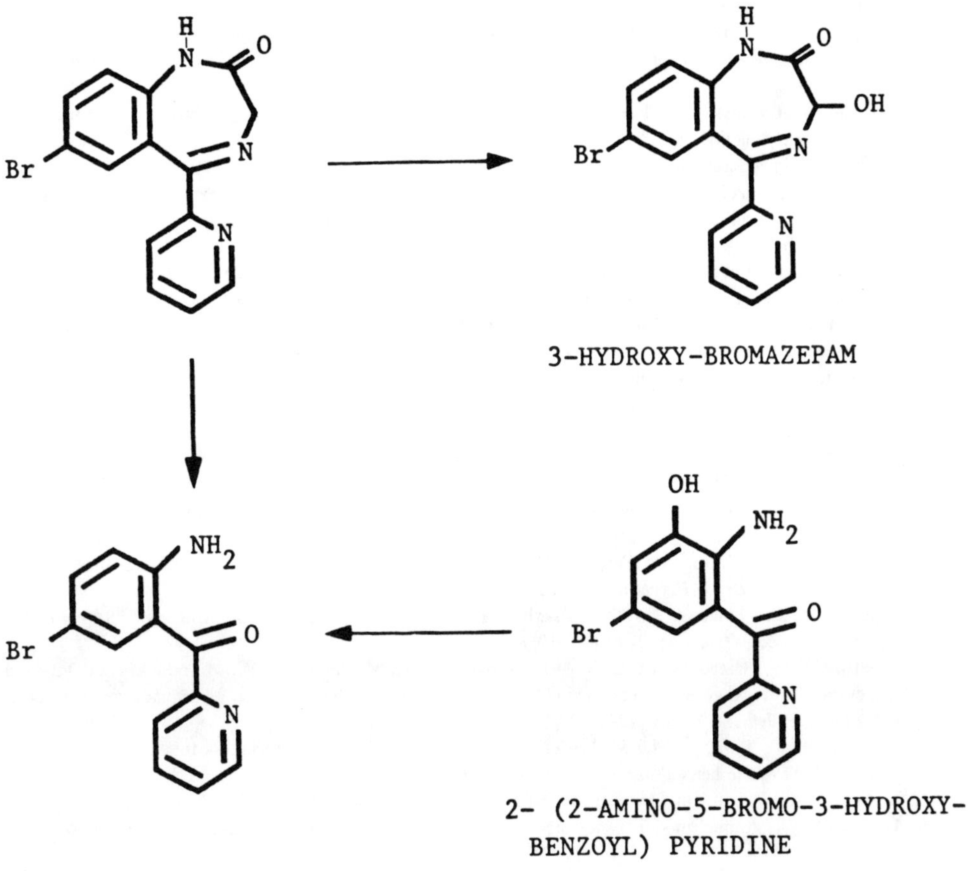

Structure 25

Uses and Doses

The uses of bromazepam are similar to those of other 1,4-benzodiazepines. The usual oral daily dose is 12 to 36 mg in divided doses.

Toxicities

Bromazepam is well tolerated. No evidence of increased tolerance has been observed. The side effects are similar to those seen with other benzodiazepines.

Physical-Chemical and Other Data

Bromazepam. 7-Bromo-2,3-dihydro-5-(2-pyridyl)-1*H*-1,4-benzodiazepine-2-one. $C_{14}H_{10}BrN_3O$. Molecular weight = 316.16, pKa 2.9. Available in 6-mg tablets.

REFERENCES

1. **Taylor, K. M. and Laverty, R.,** The interaction of chlordiazepoxide, diazepam and nitrazepam with catecholamines and nitrazepam with catecholamines and histamine in regions of rat brain, in *The Benzodiazepines,* Garattini, S., Mussini, F., and Randall, L. O., Eds., Raven Press, New York, 1973, 191.
2. **Lidbrink, P., Corrodi, H., Fuxe, K., and Olson, L.,** The effects of benzodiazepines, meprobamate and barbiturates on central monoamine neurons, in *The Benzodiazepines,* Garattini, S., Mussini, E., and Randall, L. O., Eds., Raven Press, New York, 1973, 203.
3. **Stein, L., Wise, C. D., and Belluzi, J. D.,** Effect of benzodiazepines on central serotoninergic mechanism, in *Mechanisms of Action of Benzodiazepines,* Costa, E. and Greengard, P., Eds., Raven Press, New York, 1975, 1.
4. **Consolo, S., Ladinsky, H., Peri, G., and Garattini, S.,** Effect of diazepam on mouse whole brain and brain area acetylcholine and choline levels, *Eur. J. Pharmacol.,* 27, 266, 1974.
5. **Costa, E.,** Benzodiazepines and neurotransmitters, *Arzneim. Forsch/Drug Res.,* 30, 858, 1980.
6. **Guidotti, A., Baraldi, M., Schwartz, J. P., and Costa, E.,** Molecular mechanisms regulating the interactions between the benzodiazepines and GABA receptors in central nervous system, *Pharmacol. Biochem. Behav.,* 10, 803, 1979.
7. **Schmidt, R. F., Vogel, M. E., and Zimmermann, M.,** Die wirkung von Diazepam auf die prasynaptische hemmung und andere ruckenmarks reflexe, *Naunyn — Schmiedebergs Arch. Pharmacol.,* 258, 69, 1967.
8. **Costa, E., Guidotti, A., Mao, C. C., and Surda, A.,** New concepts on the mechanism of action of benzodiazepines, *Life Sci.,* 17, 167, 1975.
9. **Guidotti, A., Toffano, G., and Costa, E.,** An endogenous protein modulates the affinity of GABA and benzodiazepines in rat brain, *Nature (London),* 275, 533, 1978.
10. **Toffano, G., Guidotti, A., and Costa, E.,** Purification of an endogenous protein inhibitor of the high affinity binding of γ-aminobutyric acid to synaptic membranes of rat brain, *Proc. Natl. Acad. Sci. U.S.A.,* 75, 4024, 1978.
11. **Baraldi, M., Guidotti, A., Schwartz, J. P., and Costa, E.,** GABA receptors in clonal cell lines: a model for study of benzodiazepine action at molecular level, *Science,* 205, 821, 1979.
12. **Costa, E., Guidotti, A., and Toffano, G.,** Molecular mechanisms mediating the action of diazepam on GABA receptors, *Br. J. Psychiatr.,* 133, 239, 1978.
13. **Tallman, J. F., Thomas, J. W., Gallagher, D. W.,** GABA-ergic modulation of benzodiazepine site sensitivity, *Nature (London),* 274, 383, 1978.
14. **Zakusov, V. V., Ostrovskaya, R. V., Kozhechkin, S. N., Markovic, V. V., Molodavkin, G. M., and Voronina, T. A.,** Further evidence for GABA-ergic mechanisms in the action of benzodiazepines, *Arch. Int. Pharmacodyn.,* 229, 313, 1977.
15. **Tallman, J. F., Paul, S. M., Skolnick, P., and Gallagher, D. W.,** Receptors for the age of anxiety: pharmacology of the benzodiazepines, *Science,* 207, 274, 1980.
16. **Stein, L.,** Behavioral neurochemistry of benzodiazepines, *Arzneim. Forsch./Drug Res.,* 30, 868, 1980.
17. **Iversen, S. D.,** Animal models of anxiety and benzodiazepine actions, *Arzneim. Forsch./Drug Res.,* 30, 862, 1980.
18. **Braestrup, C., Albrechtsen, R., and Squires, R. F.,** High densities of benzodiazepine receptors in human cortical areas, *Nature (London),* 269, 702, 1977.
19. **Mohler, H. and Okada, T.,** Benzodiazepine receptor: demonstration in the central nervous system, *Science,* 198, 849, 1977.
20. **Braestrup, C. and Squires, R.,** Brain specific benzodiazepine receptors, *Br. J. Psychiatr.,* 133, 249, 1978.
21. **Mohler, H. and Okada, T.,** The benzodiazepine receptor in normal and pathological human brain, *Br. J. Psychiat.,* 133, 261, 1978.
22. **Williams, E., Rice, K., Paul, S., and Skolnick, P.,** Heterogeneity of benzodiazepine receptors in the central nervous system demonstrated with kenazepine, an alkylating benzodiazepine, *J. Neurochem.,* 35, 519, 1980.
23. **Sieghart, W. and Karobath, M.,** Molecular heterogeneity of benzodiazepine receptors, *Nature (London),* 286, 285, 1980.

24. **Skolnick, P., Marangos, P. J., Syapin, P., Goodwin, F. K., and Paul, S. M.,** CNS benzodiazepine receptors: physiological studies and putative endogenous ligands, *Pharmacol. Biochem. Behav.*, 10, 815, 1979.
25. **Skolnick, P., Marangos, P. J., Goodwin, F. K., Edwards, M., and Paul, S. M.,** Identification of inosine and hypoxanthine as endogenous inhibitors of ^{3}H-diazepam binding in the central nervous system, *Life Sci.*, 23, 1473, 1978.
26. **Asano, T. and Spector, S.,** Identification of inosine and hypoxanthine as endogenous ligands for the brain benzodiazepine receptor binding sites, *Proc. Natl. Acad. Sci. U.S.A.*, 76, 977, 1979.
27. **Mohler, H., Polc, P., Cumin, R., Pieri, L., and Kettler, R.,** Nicotinamide is a brain constituent with benzodiazepine-like actions, *Nature (London)*, 278, 563, 1979.
28. **Braestrup, C., Nielsen, M., and Olsen, C. E.,** Urinary and brain betacarboline-3-carboxylates as potent inhibitors of brain benzodiazepine receptors, *Proc. Natl. Acad. Sci. U.S.A.*, 77, 2288, 1980.
29. **Braestrup, C. and Nielsen, M.,** Searching for endogenous benzodiazepine receptor ligands, *Trends Pharm. Sci.*, 1, 424, 1980.
30. **Hunkeler, W., Mohler, H., Pieri, L., Polc, P., Bonetti, E. P., Cumin, R., Schaffner, R., and Haefely, W.,** Selective antagonists of benzodiazepines, *Nature (London)*, 290, 514, 1981.
31. **Gavish, M. and Snyder, S. H.,** γ-Aminobutyric acid and benzodiazepine receptors: copurification and characterization, *Proc. Natl. Acad. Sci. U.S.A.*, 78, 1939, 1981.
32. **Darragh, A., Lambe, A. L., Brick, I., and Downie, W. W.,** Reversal of benzodiazepine-induced sedation by intravenous RO 15-1788, *Lancet*, 1042, 1981.
33. **Greenblatt, D. J., Allen, M. D., Noel, B. J., and Shader, R. I.,** Acute overdosage with benzodiazepine derivatives, *Clin. Pharm. Ther.*, 21, 497, 1977.
34. **Mandelli, M. Tognoni, G., and Garattini, S.,** Clinical pharmacokinetics of diazepam, *Clin. Pharmacokinet.*, 3, 72, 1978.
35. **Greenblatt, D. J., Allan, M. D., Harmatz, J. S., and Shader, R. I.,** Diazepam disposition determinants, *Clin. Pharm. Ther.*, 27, 301, 1980.
36. **Greenblatt, D. J., Shader, R. I., Franke, K., Maclaughlin, D. S., Harmatz, J. S., Allen, M. D., Werner, A., and Woo, E.,** Pharmacokinetics and bioavailability of intravenous, intramuscular and oral lorazepam in humans, *J. Pharm. Sci.*, 68, 57, 1979.
37. **Greenblatt, D. J., Allen, M. D., Lockniskar, A., Harmatz, J. S., and Shader, R. I.,** Lorazepam kinetics in the elderly, *Clin. Pharm. Ther.*, 26, 103, 1979.
38. **George, K. A. and Dundee, J. W.,** Relative amnesic actions of diazepam, flunitrazepam and lorazepam in man, *Br. J. Clin. Pharm.*, 4, 45, 1977.
39. **Greenblatt, D. J. and Koch-Weser, J.,** Clinical pharmacokinetics, *N. Engl. J. Med.*, 293, 702, 1975.
40. **Greenblatt, D. J., Shader, R. I., Divoll, M., and Harmatz, J. S.,** Benzodiazepines: a summary of pharmacokinetic properties, *Br. J. Clin. Pharm.*, 11, 11S, 1981.
41. **Greenblatt, D. J., Shader, R. I., Macleod, S. M., and Sellers, E. M.,** Clinical pharmacokinetics, chlordiazepoxide, *Clin. Pharmacokinet.*, 3, 381, 1978.
42. **Greenblatt, D. J., Allen, M. D., Harmatz, J. S., and Shader, R. I.,** Diazepam disposition determinants, *Clin. Pharm. Ther.*, 27, 301, 1980.
43. **Greenblatt, D. J. and Shader, R. I.,** Pharmacokinetic understanding of anti-anxiety drug therapy, *South. Med. J.*, 71 (Suppl. 2), 2, 1978.
44. **Allen, M. D., Greenblatt, D. J., Harmatz, J. S., and Shader, R. I.,** Single-dose kinetics of prazepam, a precursor of desmethyldiazepam, *J. Clin. Pharm.*, 19, 445, 1979.
45. **Viukari, M. and Linnoila, M.,** Serum medazepam, diazepam, and N-desmethyldiazepam levels after single and multiple oral doses of medazepam, *Ann. Clin. Res.*, 9, 284, 1979.
46. **Desilva, J. A. F. and Puglisi, C. V.,** Determination of medazepam (Norbrium), diazepam (Valium) and their major biotransformation products in blood and urine by electron-capture gas-liquid chromatography, *Anal. Chem.*, 42, 1725, 1970.
47. **Hanks, G. W., Lader, M. H., and Lawson, D. H., Eds,** Clobazam, *Br. J. Clin. Pharm.*, 7, 7S, 1979.
48. **Breimer, D. D., Bracht, H., and Deboer, A. G.,** Plasma level profile of nitrazepam (Mogadon) following oral administration *Br. J. Clin. Pharm.*, 4, 709, 1977.
49. **Kangas, L., Hsalo, E., Kanto, J., Lehtinen, V., Pynnonen, S., Ruikka, I., Salminen, J., Sillanppa, M., and Syvalahil, E.,** Human pharmacokinetics of nitrazepam: effect of age and disease, *Eur. J. Clin. Pharm.*, 15, 163, 1979.
50. **Allen, M. D., Greenblatt, D. J., and Arnold, J. D.,** Single and multiple dose kinetics of estazolam, a trazolobenzodiazepine, *Psychopharmacology*, 66, 267, 1979.
51. **Kaplan, S. A., Jack, M. I., Weinfeld, R. E., Glover, W., Weissman, L., and Cotler, S.,** Biopharmaceutical and clinical pharmacokinetic profile of bromazepam, *J. Pharmacokinet. Biopharm.*, 4, 1, 1976.
52. **Greenblatt, D. J., Shader, R. I., Franke, K., Maclaughin, D. S., Harmatz, J. S., Allen, M. D., Werner, A., and Woo, E.,** Pharmacokinetics and bioavailability of intravenous, intramuscular and oral lorazepam in humans, *J. Pharm. Sci.*, 68, 57, 1979.

53. **Huffman, D. H., Harmatz, J. S., and Shader, R. I.,** Single and multiple-dose kinetics of oral lorazepam in humans: the predictability of accumulation, *J. Pharmacokinet. Biopharm.*, 7, 159, 1979.
54. **Lader, M. H., Makin, F. J. B., and Nicholson, A. N., Eds.,** Temazepam and related 1,4-benzodiazepines, *Br. J. Clin. Pharm.*, 8, 7S, 1979.
55. **Shull, H. J., Wilkinson, G. R., Johnson, R., and Schenker, S.,** Normal disposition of oxazepam in acute viral hepatitis and cirrhosis, *Ann. Intern. Med.*, 84, 420, 1976.
56. **Chinnery, R. and Sundwall, A., Eds.,** Pharmacodynamic, pharmacokinetic, and clinical aspects on oxazepam and related benzodiazepines, *Acta Psychiatr. Scand.*, 274, (Suppl.), 1, 1978.
57. **Metzler, C. M., Ko, H., Royer, M. E., Veldkamp, W., and Linet, O. I.,** Bioavailability and pharmacokinetics of orally administered triazolam in normal subjects, *Clin. Pharm. Ther.*, 21, 111, 1977.
58. **Brown, C. R., Sarnquist, F. H., Canup, C. A., and Pedley, T. A.,** Clinical, electroencephalographic, and pharmacokinetic studies of a water-soluble benzodiazepine, midazolam maleate, *Anesthesiology*, 50, 467, 1979.
59. **Feldman, S. A. and Crawley, B. E.,** Interaction of diazepam with the muscle-relaxant drugs, *Br. Med J.*, 1, 336, 1970.
60. **Abernethy, D. R., Greenblatt, D. J., Divoli, M., Arendt, R., Ochs, H. R., and Shader, R. I.,** Impairment of diazepam metabolism by low-dose estrogen-containing oral-contraceptive steroids, *N. Engl. J. Med.*, 306, 791, 1982.
61. **Klotz, U. and Reimann, I.,** Delayed clearance of diazepam due to cimetidine, *N. Engl. J. Med.*, 302, 1012, 1980.
62. **McGowan, W. A. W. and Dundee, J. W.,** The effects of intravenous cimetidine on the absorption of orally administered diazepam and lorazepam, *Br. J. Clin. Pharm.*, 14, 207, 1982.
63. **Berger, F.,** The pharmacological properties of 2-methyl-2-N-propyl-1,3- propanediol dicarbamate (MILTOWN), a new interneuronal blocking agent, *J. Pharmacol. Exp. Ther.*, 112, 413, 1954.
64. **Wu, J., Rayburn, J., Allen L., Ferguson, H., and Kissel, J.,** Psychosedative agents. II. 8-(4-substituted 1-piperazinylalkyl)-8-azaspiro-[4,5]decano-7,9-diones, *J. Med. Chem.*, 15, 477, 1972.
65. **Plotnikoff, A.,** Clorazepate dipotassium, Tranxene®: antianxiety and antidepressant activity, *Res. Comm. Chem. Pathol. Pharm.*, 5, 128, 1973.
66. **Tompkins, E. C., Clemento, A. J., Taylor, D. P., and Perhach, J. L.,** Inhibition of agressive behaviour in rhesus monkeys by buspirone, *Res. Comm. Psychol. Psychiat. Behav.*, 5, 337, 1980.
67. **Goldbert, H. L. and Finnerty, R. J.,** The comparative efficacy of buspirone and diazepam in the treatment of anxiety, *Am. J. Psychiatr.*, 136, 1184, 1979.
68. **Stanton, H. C., Taylor, D. P., and Riblet, L. A.,** Buspirone — an anxioselective drug with dopaminergic action, in *The Neurobiology of the Nucleus Accumbens*, Chronister, R. B. and Defrance, J. F., Eds., Haer Institute, Brunswick, Maine 1981, 316.
69. **Hartman, R. J. and Geller, I.,** Effects of buspirone on conflict behaviour of laboratory rats and monkeys, *Proc. W. Pharmacol. Soc.*, 24, 179, 1981.
70. **Hartman, R. J. and Geller, I.,** Effects of buspirone and diazepam on experimentally induced conflict in cynomolgus monkeys, *Fed. Proc.*, 42, 1637, 1982.
71. **Hyslop, D. K., Becker, J. A., Eison, M. S., Riblet, L. A., and Taylor, D. P.,** Effects of buspirone, a novel anxioselective drug, on L-DOPA levels, *Abstr. Biomed. Appl. LCEC Voltammetry*, No. 68, 1982.
72. **Riblet, L. A., Allen, L., Hyslop, D. K., Taylor, D. P., and Wilderman, R.,** Pharmacologic activity of buspirone, a novel non-benzodiazepine antianxiety agent, *Fed. Proc.*, 39, 752, 1980.
73. **Riblet, L. A., Taylor, D. P., Eison, M. S., and Stanton, H. C.,** Buspirone: pharmacology and neurochemistry, *Pharmacol. Biochem. Behav.*, in press.
74. **Taylor, D. P., Hyslop, D. K., and Riblet, L. A.,** Buspirone: a model for anxioselective drug action, *Soc. Neurosci. Abstr.*, 6, 791, 1980.
75. **Fuxe, K., Andersson, K., Ögren, S-O, Perez de La Mora, M., Schwarz, R., Hökfelt, T., Eineroth, P., Gustafsson, J-A., and Skett, P.,** GABA neurons and their interaction with monamine neurons. An anatomical, pharmacological and functional analysis, in *GABA Neurotransmitters*, Krogsgard-Larsen, P., Scheel-Kruger, J., and Kofod, H., Eds., Academic Press, New York, 1979, 74.
76. **Taylor, D. P., Riblet, L. A., Stanton, H. C., Eison, A. S., Eison, M. S., and Temple, D. L.,** Dopamine and antianxiety activity, in press.
77. **Temple, D. L., Yevich, J. P., and New, J. S.,** Buspirone. Chemical profile of a new class of anxioselective drugs, *Pharmacol. Biochem. Behav.*, in press.
78. **Gammans, R. E., Mayol, R. F., and Labudde, J. A.,** *Fed. Proc.*, in press.
79. **Beer, B., Chasin, M., Clody, D. E., Vogel, J. R., and Horovitz, Z. P.,** Cyclic adenosine monophosphate phosphodiesterase in brain: effect on anxiety, *Science*, 176, 428, 1972.
80. **Vogel, J. R., Beer, B., and Clody, D. E.,** A simple and reliable conflict procedure for testing antianxiety agents, *Psychopharmacologia*, 21, 1, 1971.
81. **Beer, B., Klepner, C. A., Lippa, A. S., and Squires, R. F.,** Enhancement of ^{3}H-diazepam binding by SQ 65,396: a novel anti-anxiety agent, *Pharmacol. Biochem. Behav.*, 9, 849, 1978.

82. **Placheta, P. and Karobath, M.,** In vitro modulation by SQ 20,009 and SQ 65,396 of GABA receptor binding in rat CNS membranes, *Eur. J. Pharmacol.*, 62, 225, 1980.
83. **Supavilai, P. and Karobath, M.,** Interaction of SQ 20,009 and GABA-like drugs as modulators of benzodiazepine receptor binding, *Eur. J. Pharmacol.*, 62, 229, 1980.
84. **Williams, M. and Risley, E. A.,** Enhancement of the binding of ^{3}H-diazepam to rat brain membrane *in vitro* by SQ 20,009, a novel anxiolytic, γ-aminobutyric acid (GABA) and muscimol, *Life Sci.*, 24, 833, 1979.
85. **Supavilai, P. and Karobath, M.,** Stimulation of benzodiazepine receptor binding by SQ 20,009 is chloride dependent and picrotoxin sensitive, *Eur. J. Pharmacol.*, 60, 111, 1979.
86. **Patel, J. B. and Malick, J. B.,** ICI 136,753: a novel benzodiazepine, non-sedative anxiolytic, *Pharmacologist,* 33(Abstr.), 260, 554, 1980.
87. **Lippa, A. S., Coupet, J., Greenblatt, E. N., Klepner, C. A., and Beer, B.,** A synthetic non-benzodiazepine ligand for benzodiazepine receptors: a probe for investigating neuronal substrates of anxiety, *Pharmacol. Biochem. Behav.*, 11, 99, 1979.
88. **Albright, J. D., Trust, R. I., and Moran, D. B.,** The synthesis and anxiolytic activity of 6-substituted-phenyl-1,2,4,-triazolo[4,3,-b]-pyridazines and 6- and 7-substituted-1,2,4,-triazolo[4,3-b]-1,2,4-triazines, paper presented at 177th Natl. Meet. Am. Chem. Soc., Honolulu, Hawaii, 1979.
89. **Lippa, A. S., Nash, P. A., and Greenblatt, E. N.,** Preclinical neuropsychopharmacological testing procedures for anxiolytic drugs, in *Anxiolytics, Industrial Pharmacology,* Vol. 3, Fielding, S. and Lal, H., Eds., Futura Publishing, Mount Kisco, N.Y., 1979, 41.
90. **Klepner, C. A., Lippa, A. S., Benson, D. I., Sano, M. C., and Beer, B.,** Resolution of biochemically and pharmacologically distinct benzodiazepine receptors, *Pharmacol. Biochem. Behav.*, 11, 457, 1979.
91. **Lo, M. M. S., Strittmatter, S. M., and Snyder, S. H.,** Physical separation and characterization of two types of benzodiazepine receptors, *Proc. Natl. Acad. Sci. U.S.A.*, 79, 680, 1982.
92. **Fujimoto, M., Hirai, K., and Okabayashi, T.,** Comparison of the effects of GABA and chloride ion on the affinities of ligands for the benzodiazepine receptor, *Life Sci.*, 30, 51, 1982.
93. **Sepinwall, J. and Cook, L.,** Mechanism of action of the benzodiazepines: behavioral aspect, *Fed. Proc.*, 39, 3024, 1980.
94. **Skolnick, P., Lock, K-L., Paugh, B., Marangos, P. J., Windsor, R., and Paul, S.,** Pharmacologic and behavioral effects of EMD 28422: a novel purine which enhances ^{3}H-diazepam binding to brain benzodiazepine receptors, *Pharmacol. Biochem. Behav.*, 12, 685, 1980.
95. **Skolnick, P., Lock, K-L., Paul, S. M., Marangos, P. J., Jonas, R., and Irmscher, K.,** Increased benzodiazepine receptor number elicited *in vitro* by a novel purine, EMD 28422, *Eur. J. Pharmacol.*, 67, 179, 1980.
96. **Bardone, M. C., Ducrot, R., Garret, C., and Julou, L.,** Benzodiazepine-like central effects of R.P. 27267, a dihydro-7-oxo-5H-pyrrolo-[3,4-b]pyrazine derivative, in *7th Int. Congr. Pharmacol. London,* Pergamon Press, New York, 1978, 743.
97. **Blanchard, J. C., Boireau, A., Garret, C., and Julou, L.,** In vitro and in vivo inhibition by zopiclone of benzodiazepine binding to rodent brain receptors, *Life Sci.*, 24, 2417, 1979.
98. **Yokoyama, N., Ritter, B., and Neubert, A. D.,** 2-Arylpyrazolo [4,3-c]quinolin-3-ones: novel agonist, partial agonist and antagonist of benzodiazepines, *J. Med. Chem.*, 25, 337, 1982.
99. **Czernik, A. J., Petrack, B., Kalinsky, H. J., Psychoyos, S., Cash, W. D., Tsai, C., Rinehart, R. K., Granat, F. R., Lovell, R. A., Brundish, D. E., and Wade, R.,** CGS 8216: receptor binding characteristics of a potent benzodiazepine antagonist, *Life Sci.*, 30, 363, 1982.
100. **Ehlert, F. J., Roeske, W. R., and Yamamura, H. I.,** Multiple benzodiazepine receptors and their regulation by γ-aminobutyric acid, *Life Sci.*, 29, 235, 1981.
101. **Gordon, M.,** Chemistry and pharmacology of 1,4-benzodiazepines, in *Psychopharmacological Agents,* Academic Press, New York, 1974, 175.
102. **de Angelis, L.,** Animal techniques for evaluating benzodiazepine drugs, Meth. Find. Exp./Clin. Pharmacol, 1 (3), 129, 1979.
103. **Berger, F. M. and Bradley, W.,** The pharmacological properties of α:β dihydroxy-(2-methylphenoxy)-propane (Myanesin), *Br. J. Pharmacol.* 1, 265, 1946.
104. **Dunham, N. W. and Miya, T. S.,** A note on a simple apparatus for detecting neurological deficit in rats and mice, *J. Am. Pharm. Assoc.*, 46, 208, 1957.
105. **Boissier, J. R. and Simon, P.,** L'Utilization du test de la traction test du (Julou-Courvoisier) pour l'etude des psycholeptiques, *Therapie,* 15, 1170, 1960.
106. **Graham, R. C. B., Lu, F. C., and Allmark, M. G.,** Combined effect of tranquilizing drugs and alcohol on rats, *Fed. Proc.*, 16, 302, 1957.
107. **Kondziella, W.,** Eine neue Methode zur messung der Muskularen relaxation bei weissen Mausen, *Arch. Int. Pharmacodyn.*, 152, 277, 1964.
108. **Randall, L. O., Schallek, W., Heiss, G. A., Keith, E. F., and Bagdon, R. E.,** The pyschosedative properties of methaminodiazepoxide, *J. Pharmacol. Exp. Ther.*, 129, 163, 1960.

109. **Dervinis, A., Glassman, J. M., and Seifter, J.,** The effects of meprobamate on barbiturate response in mice, *J. Pharmacol. Exp. Ther.*, 122, 17A, 1958.
110. **Randall, L. O., Schalleck, W., Scheckel, C. L., Stefko, P.L., Banziger, R. F., Pool, W., and Moe, R. A.,** Pharmacological studies on flurazepam hydrochloride (R05-6901), a new psychotropic agent of the benzodiazepine class, *Arch. Int. Pharmacodyn.*, 178, 216, 1969.
111. **Dews, P. B.,** The measurement of the influence of drugs on voluntary activity in mice, *Br. J. Pharmacol.*, 8, 46, 1953.
112. **Caviezel, R. and Baillod, A.,** Zur differenzieurung psychotroper subtanzen anhand der motilitat der maus, *Pharm. Acta Helv.*, 33, 469, 1958.
113. **Everett, G. M. and Richards, R. K.,** Comparative anticonvulsant action of 3,5,5-trimethyl-oxazolidine-2,4-dione (Tridione), dilantin and phenobarbital, *J. Pharmacol. Exp. Ther.*, 81, 402, 1944.
114. **Toman, J. E. P., Swinyard, E. A., and Goodman, L. S.,** Properties of maximal seizures, and their alteration by anticonvulsant drugs and other agents, *J. Neurophysiol.*, 9, 231, 1946.
115. **Swinyard, E. A., Brown, W. C., and Goodman, L. S.,** Comparative assays of antiepileptic drugs in mice and rats, *J. Pharmacol. Exp. Ther.*, 106, 319, 1952.
116. **Reinhard, J. F. and Reinhard, J. F., Jr.,** Experimental evaluation of anticonvulsants, in *Anticonvulsants*, Vida, J. A., Ed., Academic Press, New York, 1977, chap. 3.
117. **Berger, F. M.,** The mode of action of myanesin, *Br. J. Pharmacol.*, 2, 241, 1947.
118. **Domino, E. F.,** Centrally acting muscle relaxants, in *Evaluation of Drug Activities: Pharmacometrics*, Laurence, D. R. and Bacharach, A. L., Eds., Academic Press, New York, 1964, 313.
119. **Randall, L. O., Heise, G. A., Schallek, W., Bagdon, R. E., Banziger, R., Boris, A., Moe, R. A., and Abrams, W. A.,** Pharmacological and clinical studies on valium, a new psychotherapeutic agent for the benzodiazepine class, *Curr. Ther. Res.*, 3, 405, 1961.
120. **French, J. D., Verzeano, M., and Magoun, H. W.,** An extralemniscal sensory system in the brain, *Arch. Neurol. Psychiatr.*, 69, 505, 1953.
121. **Kletzkin, M. and Berger, F. M.,** Effect of meprobamate on limbic system of the brain, *Proc. Soc. Exp. Biol. Med.*, 100, 681, 1959.
122. **Oswald, I.,** *Sleeping and Waking, Physiology and Psychology*, Elsevier, Amsterdam, 1962, 4.
123. **Hall, C. S.,** Emotional behavior in the rat. I. Defecation and urination as measures of individual differences in emotionality, *J. Comp. Psychol.*, 18, 385, 1934.
124. **Archer, I.,** Test for emotionality in rats and mice: a review, *Anim. Behav.*, 21, 205, 1973.
125. **Boissier, J. R. and Simon, P.,** La reaction de'exploration chez la souris, *Therapie*, 17, 1225, 1962.
126. **Oelkers, H. A.,** Zur pharmakologie des meprobamats, *.Arzneimittelforschung*, 11, 530, 1957.
127. **Wolf, A. and Von Haxthausen, E. F.,** Zur analyse der wirkung einiger zentral-sedativer substanzen, *Arzneimittelforschung*, 1. 50, 1960.
128. **Heise, G. A. and Boff, E.,** Taming action of chlordiazepoxide, *Fed. Proc.*, 20, 393, 1961.
129. **Kreiskott, H. and Vater, W.,** Verhaltensstudien am goldhamster unter dem einfluss zentral wirksamer substanzen, *Naunyn Schmiedeberg Arch. Pharmacol.*, 236, 100, 1959.
130. **Yen, C. Y., Stanger, R. L., and Millman, N.,** Ataractic suppression of isolated-induced aggressive behavior, *Arch. Int. Pharmacodyn.*, 123, 179, 1959.
131. **Tedeschi, R. E., Tedeschi, D. H., Mucha, A., Cook, L., Mattis, T., and Fellows, E. J.,** Effects of various centrally-acting drugs on fighting behavior of mice, *J. Pharmacol. Exp. Ther.*, 125, 28, 1959.
132. **Senault, G.,** Comportement de'aggressivite intra-specifique induit par l'apomorphine chez le rat, *Psychopharmacologia*, 18, 271, 1970.
133. **Brady, J. V. and Nauta, W. J. H.,** Subcortical mechanism in emotional behavior: affective changes following septal forebrain lesions in the albino rat, *J. Comp. Physiol. Psychol.*, 46, 339, 1953.
134. **Geller, I. and Seifter, J.,** The effects of meprobamate, barbiturates, d-amphetamine and promazine on experimentally induced conflict in the rat, *Psychopharmacologia*, 1, 482, 1960.
135. **Geller, I., Kulak, J. T., and Seifter, J.,** The effects of chlordiazepoxide and chlorpromazine on a punishment discrimination, *Psychopharmacologia*, 3, 374, 1962.
136. **Berger, F. M.,** Evaluation of tranquilizers and techniques and their meaning, in *Methods in Drug Evaluation*, Mantegazza, P. and Piccinini, F., Eds., North-Holland, Amsterdam, 1966, 218.

DRUGS USED TO TREAT NEUROMUSCULAR DISORDERS; ANTIPARKINSONISM AGENTS AND SKELETAL MUSCLE RELAXANTS

John L. Neumeyer and Norman R. Boisse

Drugs described in this chapter have in common the ability to reduce muscle tone by virtue of their action on the CNS. These drugs however have distinctly different mechanisms of action, therapeutic uses, and chemical properties. One group includes those drugs useful in treating Parkinson's disease and related syndromes. The other group of drugs, the centrally acting muscle relaxants, depress neuronal synapses controlling muscle tone and are used for treating acute muscle spasms, tetanus, and some orthopedic conditions.

DRUGS USED FOR PARKINSON'S DISEASE

For more than a century, drugs with central anticholinergic properties, used empirically and with limited success, constituted the major approach to the pharmacotherapy of Parkinson's disease (paralysis agitans) first described by James Parkinson in 1817. The clinical usefulness of the belladonna alkaloids was believed to be due to their central anticholinergic properties, and based on this reasoning, the supposition of increased cerebral cholinergic activity in parkinsonism was made. The application of this concept led to the introduction of several synthetic anticholinergic drugs for the treatment of Parkinson's disease (Table 1). Such therapy was often supplemented with antihistamines. However, it was not until sophisticated biochemical assays and histochemical techniques were developed, that the pathophysiology of Parkinson's disease was understood. The discovery that large doses of levodopa produce dramatic improvement in the symptoms of this disease led to the use of this drug and the combination of levodopa and decarboxylase inhibitors in therapy.

Although the anticholinergic agents are still useful for parkinsonism they are less effective than levodopa. The chemistry and mechanism of action of the anticholinergics and antihistamines are discussed in greater detail in separate chapters in this volume. This chapter will discuss primarily the pathophysiology of Parkinson's disease and the events which led to the use of dopamine agonists for the treatment of this disease.

Pathophysiology of Parkinson's Disease

In addition to its many other functions, the extrapyramidal system maintains posture, muscle tone, and modulates voluntary movement. In patients with parkinsonism it has been observed that dopamine levels in the striatum are depleted or severely reduced. Dopamine is believed to act as a neurotransmitter at certain striatal synapses which are concerned with mediating inhibition in the nigrostriatal pathway. The discovery by Ehringer and Hornykiewicz[1] that there was a gross depletion of striatal dopamine in the brains of patients with idiopathic and postencephalitic parkinsonism examined at postmortem led to the present hypotheses regarding the role of dopamine in the pathogenesis of parkinsonism. It is currently believed that acetylcholine is an excitatory transmitter for neurons running between the substantia nigra, the pallidum, and the striatum and that in parkinsonism there is an imbalance between these two transmitter systems in the direction of cholinergic dominance (see Figure 1). In parkinsonism the dopaminergic input to the corpus striatum is deficient, whereas the cholinergic input remains unchanged. The resulting imbalance explains the apparent increased effect of acetylcholine in parkinsonism and now justifies the former empirical use of anticholinergic drugs, although their usefulness is now limited.

Symptoms of Parkinson's disease may be induced by drugs that cause a relative CNS dopamine-acetylcholine imbalance by one of the following mechanisms: (1) the depletion of dopamine from intraneuronal stores; (2) by rendering the dopamine receptor less accessible

Table 1
DRUGS USED FOR PARKINSONISM

Class and nonproprietary name	Chemical structure	Trade name	Single oral dose (mg) (initial)
	Naturally occurring anticholinergic alkaloids		
Atropine	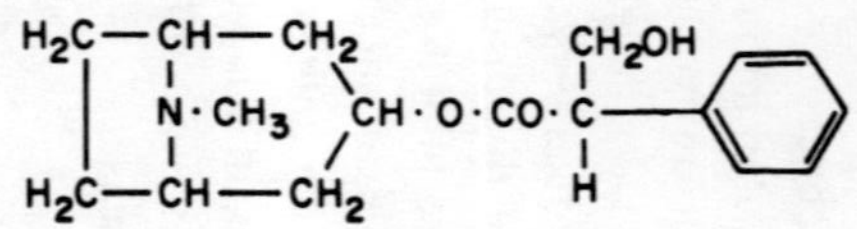		
Scopolamine	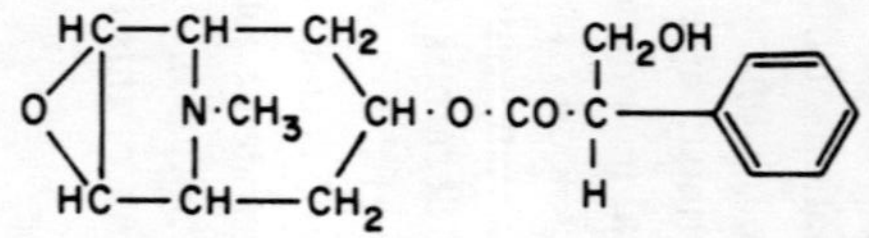		
	Synthetic anticholinergic agents		
Benztropine mesylate	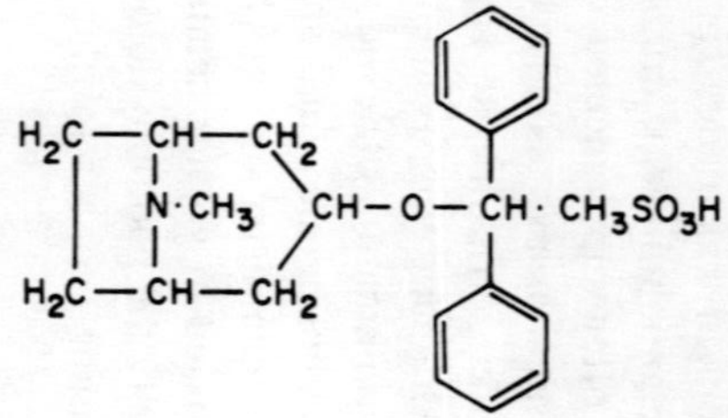	Cogentin	0.5—1.0

Trihexyphenidyl hydrochloride	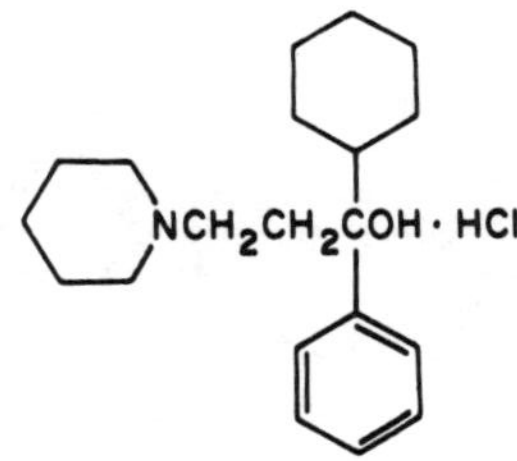	Artane Pipanol Tremin	1.0—2.0
Biperiden hydrochloride	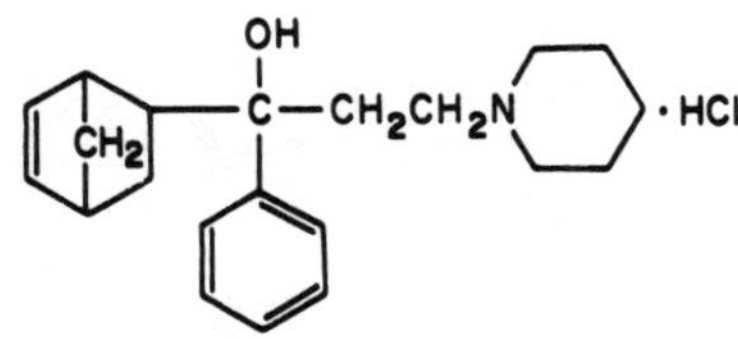	Akineton	1.0—2.0
Cycrimine hydrochloride	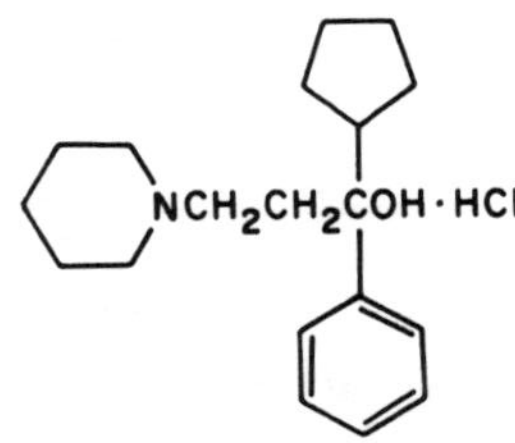	Pagitane	1.25—5.0
Procyclidine hydrochloride	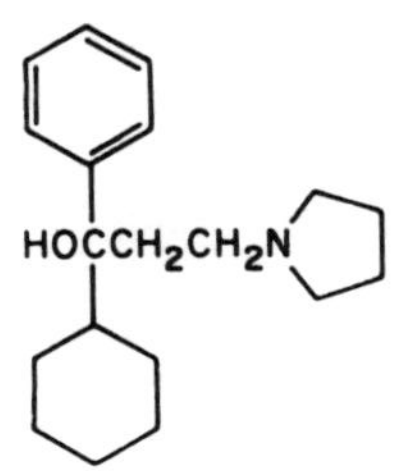	Kemadrin	2.0—2.5

Table 1 (continued)
DRUGS USED FOR PARKINSONISM

Class and nonproprietary name	Chemical structure	Trade name	Single oral dose (mg) (initial)
	Antihistamines		
Chlorphenoxamine hydrochloride	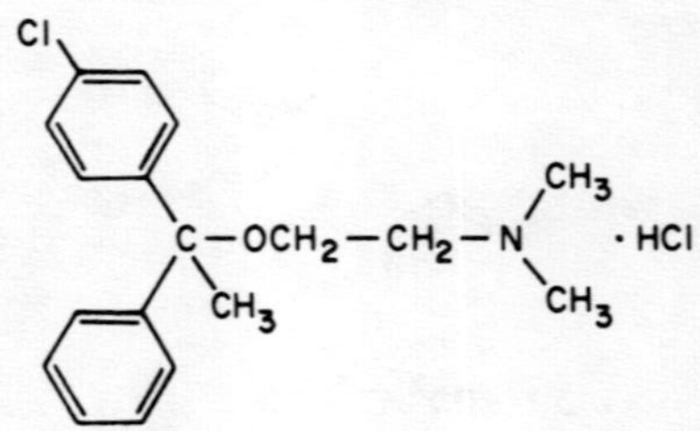	Phenoxene	50.0
Orphenadrine hydrochloride	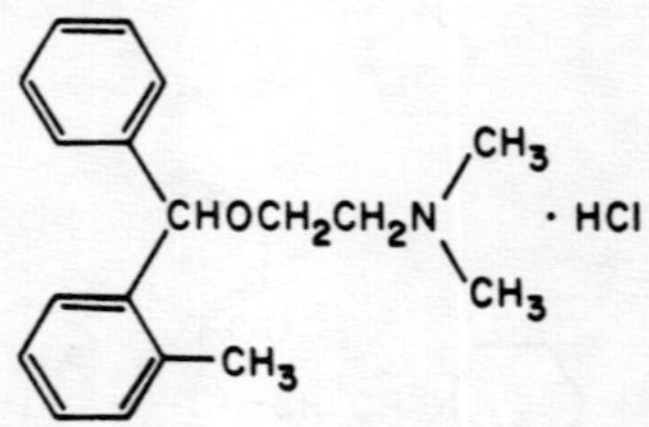	Disipal	50.0
Diphenhydramine	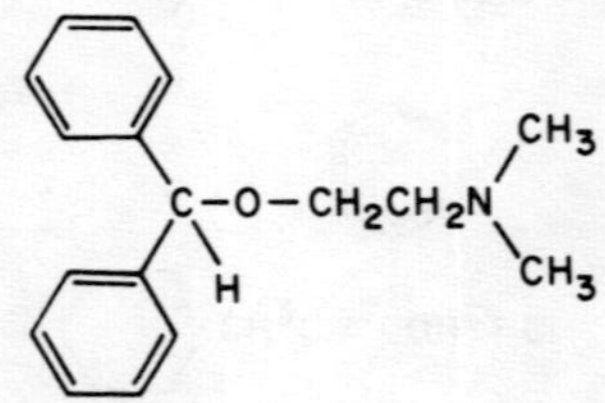		

	Phenothiazines (antihistamines)		
Ethopropazine hydrochloride	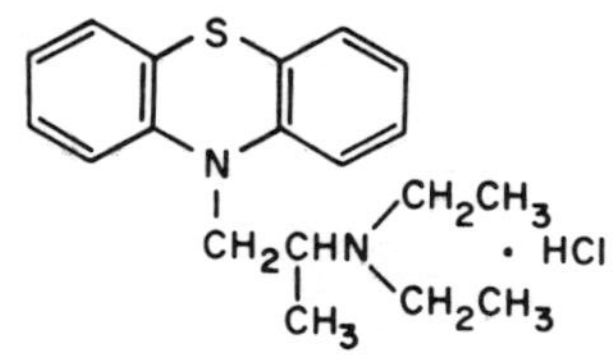	Parsidol	50.0
Promethazine	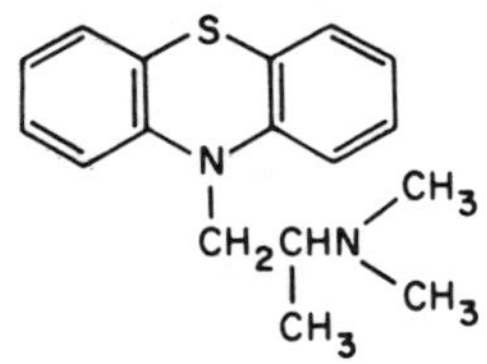	Phenergan	25.0—50.0
	Dopamine agonists		
Levodopa	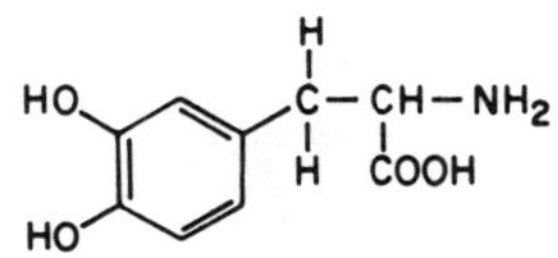	Bendopa Dopar Larodopa	0.5—1.0 g
	Decarboxylase inhibitors		
Carbidopa (Mk-486)	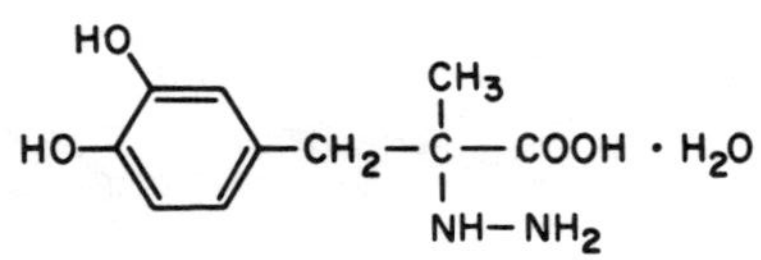		10—25

Table 1 (continued)
DRUGS USED FOR PARKINSONISM

Class and nonproprietary name	Chemical structure	Trade name	Single oral dose (mg) (initial)
Benserazide hydrochloride (Ro 4-4602)	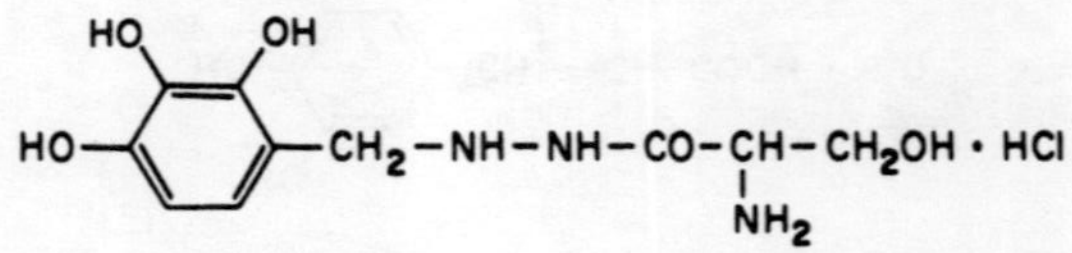		
	Dopamine-releasing agents		
Amantadine hydrochloride	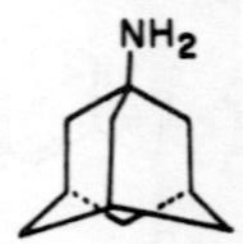	Symmetrel	100

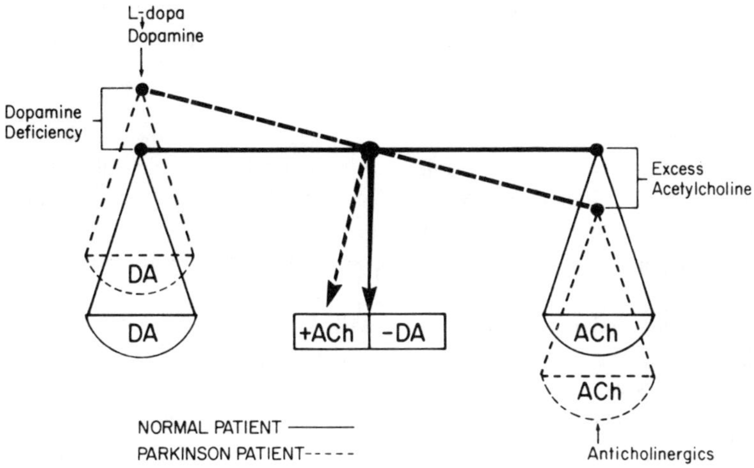

FIGURE 1. Schematic representation of the imbalance between the excitatory neurotransmitter acetylcholine, and the inhibitory neurotransmitter dopamine, in the basal ganglia of Parkinsonian patients.

to dopamine; or (3) by increasing acetylcholine levels. Reserpine is an example of a drug in the first category. Parkinsonian symptoms induced by such drugs can be reversed with exogenous levodopa. Among the drugs that render the dopaminergic receptor less accessible to dopamine are the neuroleptic agents (see chapter entitled "Antipsychotic Agents") such as butyrophenones, thioxanthines, dibenzodiazepines, and many of the phenothiazines. It is believed that these drugs selectively block the dopamine receptors preventing the interaction between dopamine and its receptor. Drugs which increase acetylcholine levels such as the cholinesterase inhibitors, physostigmine, and the cholinergic agent carbachol aggravate parkinsonism.

Parkinson's disease, independent of specific etiology, usually appears insidiously in the latter decades of life and produces a slowly increasing disability in movement. The disease is usually characterized by four major clinical features: tremor, bradykinesia, rigidity, and a disturbance of posture.[2]

It has been well accepted that gradually decreasing levels of dopamine observed in Parkinson's disease and in advancing age contributed to the accelerated process of cell aging. Other factors which have been recognized as contributing to cell aging are anoxia, viruses, and the accumulation of some toxic trace metals such as manganese.

Anticholinergic Therapy

For many years drug therapy of parkinsonism depended primarily on the limited efficacy of the belladonna alkaloids. With the newer synthetic structural analogues of such alkaloids (Table 1) attempts were made to increase the central anticholinergic effects as well as to reduce their undesirable peripheral effects. Unfortunately the side effects of these agents are troublesome, the most severe being their action on the CNS which include delusions, hallucinations, somnolence, ataxia, and dysarthria. The peripheral effects such as dry mouth, blurred vision, constipation, urinary retention, and tachycardia are also disturbing deterrents to their use.

In general anticholinergic drugs rarely produced more than 20% improvement and, despite continued use, the symptoms of the disease continue to progress. The most important present-day usage of the anticholinergic agents is their concomitant administration with L-dopa where in addition to diminishing the cholinergic striatal effects, anticholinergics may inhibit the reuptake and storage of dopamine in the striatum.[3] The antihistamines, particularly those with central anticholinergic effects, are generally better tolerated in the elderly and may produce slightly greater relief from tremor. With the advent of L-dopa therapy this mode of treatment is rarely used today.

Dopaminergic Therapy

Several approaches may be used to alleviate the striatal dopamine deficiency syndrome. These are (1) augmentation of the synthesis of brain dopamine, (2) direct stimulation of dopamine receptors, (3) stimulation of dopamine release from presynaptic sites, (4) decreasing reuptake of dopamine at presynaptic sites, and (5) decreasing dopamine catabolism.

Augmentation of Dopamine Synthesis — Levodopa Therapy

The initial report describing dramatic symptomatic improvement of parkinsonian patients given high oral doses of racemic dopa by Cotzias et al.,[4] in 1967, was followed by more clinical trials that confirmed the efficacy and safety of the *levo* isomer. The effectiveness of levodopa required penetration of the drug into the CNS and its subsequent enzymatic decarboxylation to dopamine. Dopamine does not cross the blood-brain barrier probably due to its basicity (pKa 8.9 (OH), 10.6 (NH_2)),[5] since it exists primarily in its protonated form under physiological conditions. However, the precursor amino acid which is less basic (pKa 2.3 (COOH); 8.7 (NH_2) 9.7 (OH); 13.4 (OH))[5] can penetrate the CNS.

Levodopa is formed from L-tyrosine as an intermediary metabolite in the biosynthesis of catecholamines (Figure 2). Dopamine is formed from levodopa by the cytoplasmic enzyme, aromatic L-amino acid decarboxylase. The effects observed following systemic administration of levodopa have been attributed to its catabolites dopamine, norepinephrine, and epinephrine acting at various sites in the periphery and in the brain. The principal metabolic pathways for levodopa are shown in Figure 2. A small amount is methylated to 3-*O*-methyldopa, which accumulates in the CNS due to its long half-life. Most is converted to dopamine, small amounts of which are metabolized to norepinephrine and epinephrine. Metabolism of dopamine proceeds rapidly to the principal excretion products 3,4-dihydroxyphenylacetic acid (DOPAC) and 3-methoxy-4-hydroxyphenylacetic acid (homovanillic acid, HVA). The first step which appears to be the main metabolic pathway is the decarboxylation of levodopa by the pyridoxine dependent enzyme, aromatic-L-amino acid decarboxylase (dopa decarboxylase) to form dopamine. In man, dopa decarboxylase activity is greater in the liver, heart, lungs, and kidneys than in the brain.[2] Therefore, ingested levodopa is converted to dopamine in the periphery and in preference to the brain. It is thought that, in man, levodopa will thus enter the brain only when administered in dosages high enough to overcome losses due to peripheral metabolism (3 to 6 g daily). Inhibition of peripheral decarboxylase markedly increases the proportion of levodopa that crosses the blood-brain barrier. This has been accomplished by the co-administration of such decarboxylase inhibitors as carbidopa[6] and benserazide,[7] agents incapable of crossing the blood-brain barrier.

The actions of dopamine are complex because the amine acts on several different receptors located both centrally and peripherally. In the periphery, β-adrenergic stimulation enhances heart rate, α-adrenoreceptor stimulation cause vasoconstriction, and dopaminergic receptor stimulation causes renal and mesenteric vasodilation[8] and delayed gastric emptying. One of the most common side effects of levodopa therapy is gastric upset with nausea and vomiting. This appears to be the result of direct GI irritation as well as stimulation of the chemoreceptor trigger zone (CTZ) in the medulla. The CTZ, which lies outside the blood-brain barrier, also has dopamine receptors. One of the results of combining levodopa with a peripheral decarboxylase inhibitor such as carbidopa, (Sinemet®) is a significant decrease in the incidence and severity of nausea and vomiting as well as diminished effects on the cardiovascular system.

Because combination therapy permits a 75 to 80% reduction of the dosage of levodopa,

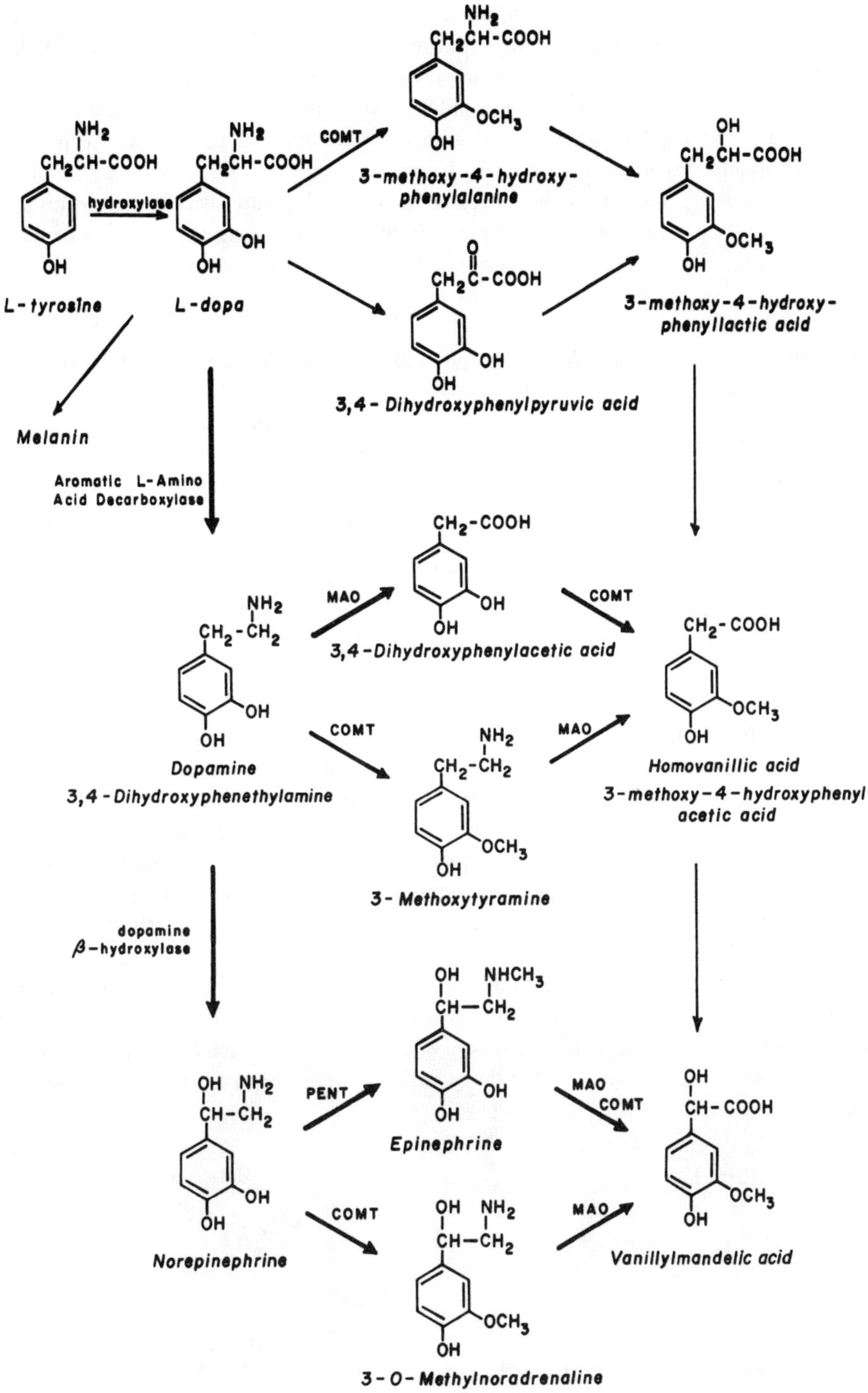

FIGURE 2. Important pathways in the biosynthesis and metabolism of levodopa. Major pathways are shown by heavy arrows. COMT, catechol-*O*-methyltransferase: MAO, monoamine oxidase; PENT, phenylethanolamine *N*-methyl transferase.

the risk of side effects is reduced. Pyridoxine (a co-enzyme for dopa decarboxylase) reverses the therapeutic effects if concomitantly administered with levodopa by increasing the decarboxylase activity which results in more levodopa being converted to dopamine in the periphery and consequently less being available for penetration into the CNS. When peripheral dopa decarboxylation is blocked with an agent such as carbidopa, the pyridoxine effect on peripheral levodopa metabolism is negligible.

Parkinsonian patients not previously treated with levodopa are usually started on a combination therapy with Sinemet®, which is available most often in a fixed ratio of one part cardipoda and ten parts levodopa, either 10/100 mg or 25/250 mg.

Physical Chemical Data

Levodopa, Bendopa, Lavodopa, levopa, Dopar, $C_9H_{11}NO_4$, (−) -3- (3,4-dihydroxyphenyl)-L-alanine. Colorless to white, odorless and tasteless crystals (m.p. 276 to 278°C dec.) $[\alpha]_D^{13}$ −13.1°C. Readily soluble indilute HCl and formic acid, pKa 2.3 (carboxyl), 8.7 (amine), 9.7, and 13.4 (phenol). See structure in Table 1.

Carbidopa, Lodosyn, (component of Sinemet®), $C_{10}H_{14}N_2O_4 \cdot H_2O$, (−)-L-α-hydrazino-3,4-dihydroxy-α-methylhydrocinnamic acid monohydrate. Crystals from hot water (m.p. 203 to 205°C,(dec)). Structure in Table 1.

Benserazide hydrochloride — R04-4602 (combination with levodopa, *madopar*), $C_{10}H_{16}Cl N_3O_5$,*N*-(DL-seryl)-N′-(2,3,4-trihydroxybenzyl) hydrazone. White crystalline powder (m.p. 146 to 148°C), soluble in water. See structure in Table 1.

Monoamine Oxidase and Dopamine β-Hydroxylase Inhibitors

Another means of increasing brain dopamine levels in the basal ganglia has been suggested. By the administration of monoamine oxidase (MAO) inhibitors which can inhibit the intracerebral metabolic degradation of dopamine, the dose of levodopa can be reduced. Deprenyl, in contrast to the nonspecific MAO inhibitors such as phenelzine and isocarboxazid, is a selective inhibitor of MAO-B, the isoenzyme that predominates in certain regions of the CNS.[9] The administration of deprenyl does not result in the profound and potentially lethal potentiation of the effects of catecholamines when administered concurrently with a centrally active amine such as tyramine.[2] Deprenyl is not currently available in the U.S.

Fusaric acid, a potent inhibitor of the enzyme dopamine β-hydroxylase which mediates the final step in norepinephrine synthesis, has also been tried in Parkinson's disease in an attempt to increase central dopamine levels. Even though this seemed like a rational approach to therapy, fusaric acid failed to alter the severity of parkinsonism in otherwise untreated patients.[10]

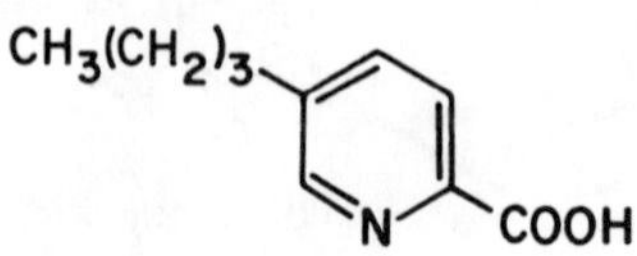

Fusaric acid
5-butyl-2-pyridine carboxylic acid

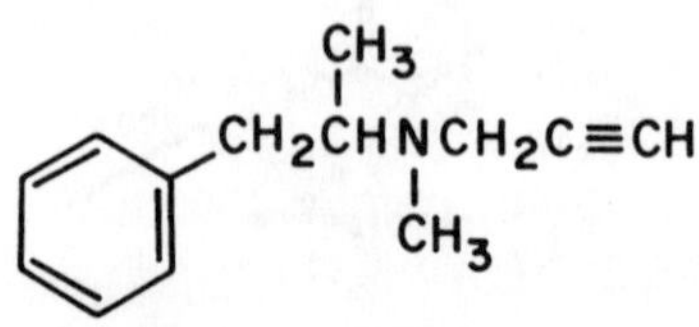

Deprenyl
(R)-(−)-N,α-dimethyl-N-prop-2-ynylphenethylamine

Direct Stimulation of Dopamine Receptors

Drugs that act by directly stimulating the dopamine receptors are not yet marketed in the U.S., although several are currently undergoing clinical trials. The potential advantage of such drugs compared to levodopa is that their effect is independent of striatonigral degeneration, whereas levodopa depends on the ability of the remaining neurons to decarboxylate levodopa to dopamine.

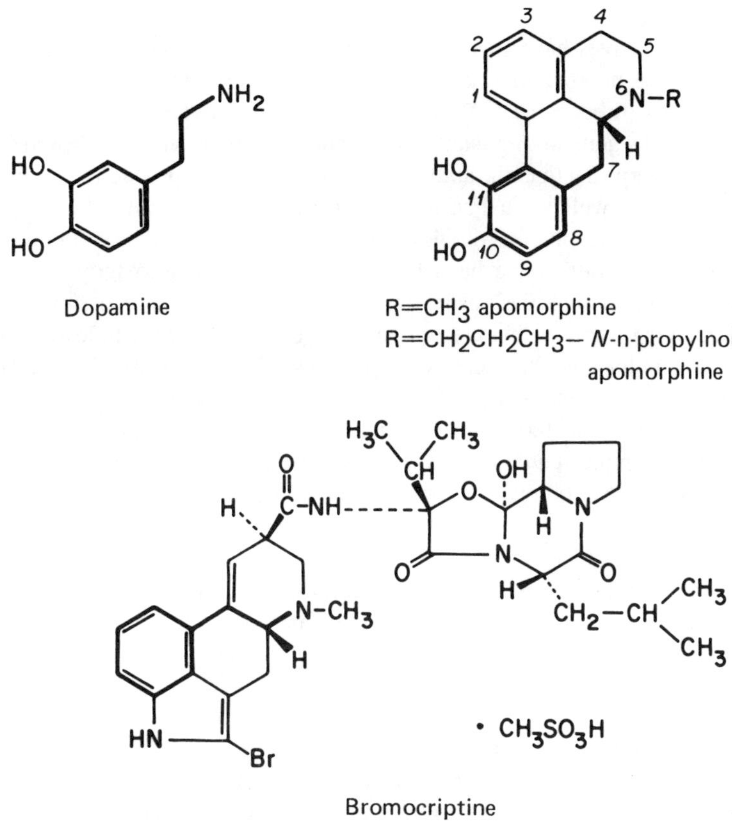

FIGURE 3. Structural relationship between dopamine, apomorphine, and the ergoline bromocriptine.

Among those drugs that act predominantly at the dopamine receptor are the aporphines (R)-apomorphine and its homolog (R)-*N*-*n*-propylnorapomorphine, and the ergolines, lergotrile, lisuride, and bromocriptine. Both the aporphines and the ergolines are structurally related to dopamine as shown in Figure 3.

Aporphines

A variety of compounds in which the catechol ring and the amino group of dopamine are held in rigid conformation can be prepared in order to assess more accurately the structural and conformational requirements of the dopamine receptor. The aporphine alkaloid, apomorphine, obtained by the acid catalyzed rearrangement of morphine and recognized for years as a powerful centrally acting emetic by its action on the medullary CTZ was found to produce effects similar to dopamine. It is evident that apomorphine can pass through the blood-brain barrier whereas dopamine cannot. Of a variety of other apomorphine analogues which have been examined, *N*-*n*-propylnorapomorphine (NPA) (Figure 3) was found to be more potent as the dopamine agonist. Cotzias[12] has reported beneficial effects with NPA in the treatment of Parkinson's disease. NPA is available in the U.S. for experimental use only.

Physical Chemical Data

Apomorphine hydrochloride $C_{17}H_{17}NO_2 \cdot HCl$ ((R)-(−)10,11-dihydroxyaporphine hydrochloride. Small, off-white crystals that decompose and turn green on exposure to light and air. One gram disolves in 50 mℓ water; pKa = 8.92 (acid), 7.0 (base). See structure in Figure 3.

Ergolines

Several ergot derivatives have shown dopaminergic activity in animal models of parkinsonism and mimic the neuroendocrinological effects of dopamine to decrease the secretion of prolactin and growth hormone.[13] Bromocriptine is available for clinical use in the U.S. for the short-term management of galactorrhea (prolonged lactation), and experimentally for the treatment of hyperprolactinemia and acromegaly (excessive production of growth hormone in the adult), as well as in Parkinson's disease.[14] While studies with lergotrile in parkinsonism demonstrate pharmacological properties similar to those of bromocriptine, further clinical investigations have been halted because of hepatotoxicity ascribed to lergotrile.[2] In many patients with Parkinson's disease, high doses of bromocriptine (50 to 100 mg) elicit therapeutic responses that are equivalent to those obtained with levodopa. In man, bromocriptine is rapidly absorbed and concentrations of the drug usually peak in plasma within 2 hr., therapeutic concentrations for the treatment of Parkinson's disease persist three to four times longer than do those of levodopa (given without a decarboxylase inhibitor). The dose of bromocriptine required to control parkinsonism (20 to 150 mg daily) is considerably larger than that required to suppress galactorrhea.[2]

Physical-Chemical Data

Bromocriptine mesylate — parlodel, $C_{32}H_{40}BrN_5O_5 \cdot CH_3SO_3H$, (2-Bromoergocryptine monomethane sulfonate) m.p. 192 to 196°C (dec). For structure, see Figure 3.

Stimulation of Dopamine Release

Amantadine

Agents which can release dopamine from neuronal storage sites should be beneficial in the treatment of parkinsonism. Currently amantadine (also used as an antiviral agent) is the only such drug which has clinically significant antiparkinsonian effects which are enhanced in the presence of levodopa. Amantadine appears to increase the release of dopamine and cause the inhibition of the membrane pump (resulting in the enhanced accumulation of brain dopamine). Thus patients who respond to amantadine (indicating that their dopamine receptors are still capable of responding to dopamine) will generally respond to treatment with levodopa. The rapid onset of optimal benefits from amantadine, within 2 weeks, affords an advantage over levodopa. Amantadine also produces fewer side effects than with levodopa or the anticholinergic drugs. Amantadine for parkinsonism is usually given orally, 100 to 300 mg daily, and is readily absorbed from the GI tract.

Amantadine is a primary amine with a pKa of 10.8. The amino group is more basic than the amino group of dopamine (pKa 10.6). Since most of the drug is in the protonated form at physiological pH, how does it pass through the blood-brain barrier to exert its effect on dopaminergic neurons? A feasible explanation would be that the cage-like structure of amantadine not only increases its lipophilicity but also precludes its catabolism by oxidative enzymes making more of the drug available for penetration into the CNS, even though it may be in minute amounts. Metabolism studies have shown that indeed amantadine is excreted in the urine in its unchanged form.

Physical Chemical Data

Amantadine — Symmetrel, $C_{10}H_{17}N \cdot HCl$ (1-Adamantanamine Hydrochloride). Crystalline compound (dec. 360°C), soluble in water (1:20) and alcohol, pKa 10.8; structure, see Table 1.

SKELETAL MUSCLE RELAXANTS

The skeletal muscle relaxant drugs are used to relieve muscular spasticity, a condition

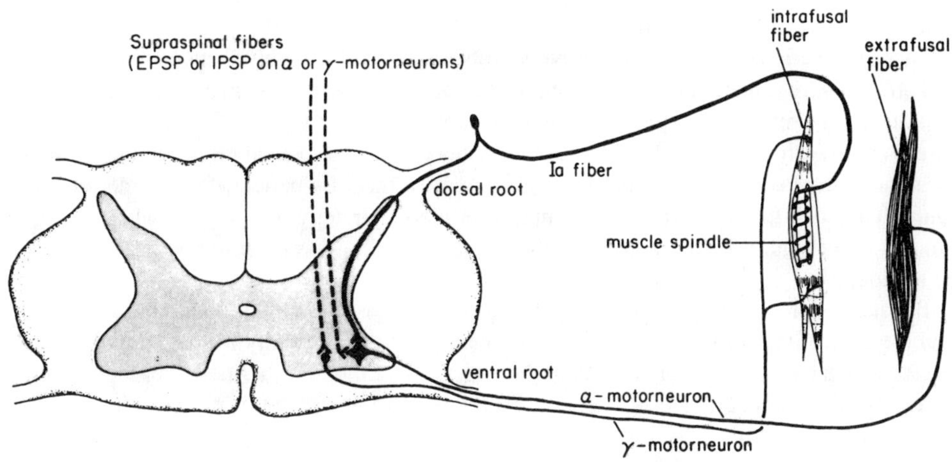

FIGURE 4. Physiology of muscle tone and spasticity.

characterized by exaggerated resting tone of a muscle. Muscle hypertonus is usually accompanied by increased resistance to passive stretch. Spasticity may result from musculoskeletal trauma or from spinal cord or brain lesions or diseases. Regardless of the cause, the mechanisms essential for the expression of spasticity are contained in the spinal cord.

Physiology of Muscle Tone and Spasticity

Skeletal muscle tone is due to the asynchronous activation of individual motor units. It arises from the activation of the alpha motorneurons in the spinal cord which in turn initiates a twitch contraction of only few individual fibers which it innervates. Accordingly, the frequency of action potentials in motorneurons at rest determines the level of muscle tone. If the nerve innervating a skeletal muscle is cut (axotomy), the neural drive for muscle tone is lost and the muscle becomes flaccid to touch and paralyzed. In spasticity, the muscles are hypertonic due to an abnormally high frequency of activation of α-motorneurons, the final common pathway for the expression of motor function (Sherrington).

As shown in Figure 4, the factors which normally influence muscle tone and cause spasticity are the factors which determine the frequency of action potentials in α-motorneurons. The dendrites and soma of the α-motorneurons are continuously bombarded by excitatory and inhibitory synaptic influences (EPSPs and IPSPs) which summate spatically and temporally. Alpha motorneurons are excited by monosynaptic connections with sensory nerves (Ia afferents) which arise from the muscle spindle or annulospiral nerve endings of intrafusal muscle fibers. The sensitivity of the muscle spindle is under the control of the γ-motorneurons which innervate striated poles of the intrafusal fiber. The more active the γ-motorneurons are, the more the muscle spindles are stretched and, accordingly, the more sensitive they are for activation by passive stretch. The patellar tendon reflex is essentially a test of excitability of this reflex arc to a passive stretch (tendon tap). The decerebrate preparation is an animal model of spacticity that is caused by overactivation of the γ-motorneurons. Decerebration, achieved by a midcollicular lesion removes a tonic inhibitory supraspinal influence on γ-motorneurons resulting in net overactivation. This form of spasticity is due to reflex activation of α-motorneurons by Ia afferent fibers arising from the muscle spindles and can be relieved by cutting dorsal roots and leaving ventral roots which carry motorneurons intact. Normally both α- and γ-motorneurons are continuously bombarded by excitatory and inhibitory spinal and supra-spinal influences. Any change in these inputs which results in greater net activity would promote hypertonicity or spacticity. Conversely, skeletal muscle relaxant drugs with a central action can be expected to influence the excitability of α-motorneurons directly or indirectly via synaptic inputs.[15-18]

Clinical Evaluation of Spasticity

There is a serious lack of quantitative methodology to objectively demonstrate neurophysiological improvement in the drug treatment of spastic states in man.[17] Simple tests of muscle tone or reflex latency have been impossible or useless in man. So far, global clinical assessments such as the number of painful spasms per day have been more useful. However, even the combined subjective impressions of improvement by patient, family and physician cannot establish that the drug is working. A valuable strategy to establish whether a drug produces any benefit is to gradually withdraw the drug; however, abrupt withdrawal should be avoided.[16]

Because of the enormous diversity of neurological disorders that culminate in spasticity and the subjectivity of many of the measurements, there is presently inadequate information to establish the superiority of any one of the antispastic drugs. Furthermore, spasm frequently co-exists with pain. Drug efficacy may be related to both skeletal muscle relaxation and analgesia.[19]

Animal Pharmacologic Screens

Traditional methods used to evaluate and characterize effective skeletal muscle relaxants are not likely to lead to the development of better drugs. Most of the available tests were developed based on the pharmacologic activity of mephenesin[20] and may be useful for comparative purposes. However, animal models of spasticity representative of the diverse neurological disorders associated with in man are needed.

Standard testing includes the determination of lethality (LD_{50}) and muscle paralysis (PD_{50}) in mice and rats. Effective skeletal muscle relaxants produce abolition of skeletal muscle tone (paralysis or loss of righting) without impairment of consciousness. However, the validity of this test has been questioned. Loss of the righting reflex in rodents is not a suitable criterion for identifying muscle relaxants because it is not directly relevant to the endpoint sought in clinical use. Paralytic potency is evaluated in rats and mice; the relative duration of paralytic activity is usually evaluated in dogs.[21]

Since mephenesin, the prototype interneuronal blocking agent, protects against strychnine-induced seizures and lethality, several of the muscle relaxants have been evaluated for this activity in rats and mice. Since strychnine blocks glycine receptors and glycine is the major inhibitory transmitter of the spinal cord, antistrychnine activity also implies selectivity of action for the spinal cord. However, antistrychnine activity does not directly reflect interneuronal blocking activity. In addition, meprobamate is much more effective against pentylenetetrazole-induced seizures than against strychnine seizures, suggesting a greater supraspinal component activity which is consistent with meprobamate's proven effectiveness as a sedative-hypnotic.

Interneuronal blocking activity is evaluated by measuring the inhibition of polysynaptic (flexor reflex) and monosynaptic (knee-jerk) reflexes in the spinal cat. However, such selectivity of action is not requisite for effective skeletal muscle relaxant activity.

Mechanisms of Action

The antispastic drugs are diverse both in their chemical structure and in their site and mechanism of action. The first drug recognized to exhibit antispastic activity was Antodyne or 3-phenoxy-1,2-propanediol. In guinea pigs and rabbits, Antodyne produced prolonged paralysis without impairing consciousness. Antodyne was introduced into clinical medicine in 1910 as an analgesic and antipyretic. However, the duration of its skeletal muscle relaxant effect was too short lived to be clinically useful. In 1943, SAR studies of a series of simple glyceryl ethers related to Antodyne by Berger and Bradley identified the most active compound which underwent pharmaceutical development in 1946, mephenesin or 3-*0*-toloxy-1,2-propanediol.[22] Pharmacological studies revealed that mephenesin selectivity depressed polysynaptic while sparing monosynaptic spinal cord reflexes. Because of its safety and selective action on the spinal cord, mephenesin became the first widely prescribed centrally acting skeletal muscle relaxant. Accordingly, 'mephenesin'' is the prototype for the ''interneuronal blocking type'' of muscle relaxant. Several other compounds with mephenesin-like pharmacologic profiles, some longer lasting, were developed and marketed as muscle relaxants. These include carisoprodol, chlorphenesin carbamate, chlorzoxaone, metaxalone, methocarbamol, and orphenadrine. The general pharmacology and neuropharmacology of propanediol carbamates have been recently reviewed.[23]

The classification of muscle relaxants as interneuronal blockers is suggestive of an understanding of the mechanism of action of these drugs. However, depression of neural activity by these drugs is not limited to interneurons. Furthermore, some compounds block interneuron activity (polysynaptic spinal reflexes) without exhibiting muscle relaxant activity. Therefore, interneuronal blocking activity is neither a necessary nor sufficient condition for skeletal muscle relaxant activity.[24]

The pharmacology of the mephenesin-like muscle relaxants is remarkably similar to that of the sedative-hypnotics. Indeed, the only apparent difference is the greater selectivity for the spinal cord and consequently, lesser sedation for antispastics. Both classes produce a reversible nonspecific depression of the CNS.[25]

The second group of antispastic drugs to be developed are the benzodiazepines, in particular, ''diazepam.'' Diazepam exerts its skeletal muscle relaxant effect by binding to specific receptors which enhance the efficacy of GABA receptor binding and consequently synaptic inhibition by opening chloride channels in nerves. Diazepam probably has been supraspinal and spinal sites of antispastic action. Spinal actions are sufficient to relieve spasticity in patients with complete spinal cord transections. The ability of diazepam to enhance GABA-mediated inhibition in the spinal cord of the cat is well documented.[17,26] Sedation may accompany the antispastic effect of benzodiazepines. The molecular mechanism of action of benzodiazepines is more fully discussed elsewhere in this volume. Like the interneuronal blockers, diazepam selectively depresses poly over monosynaptic spinal reflexes. Diazepam is an effective drug in the treatment of spasticity as well as nonspastic disorders like tetanus and stiff man syndrome.

In an attempt to elevate gamma aminobutyric acid (GABA) activity in the brain more directly, the GABA structural analogue ''baclofen'' [4-amino-3-(4-chlorophenyl)butanoic acid] was developed. Unlike GABA, baclofen is lipophilic and will penetrate the blood-brain barrier. However, baclofen fails to meet important criteria to be GABA-mimetic. Although it depresses neural activity, its effects cannot be blocked by the GABA receptor antagonist bicuculline. Furthermore, it has no effect on inhibitory postsynaptic potentials. The most important site of action by baclofen is the primary afferent fiber terminal where baclofen blocks the release of excitatory neurotransmitter. In contrast to the interneuronal blockers, baclofen depresses both monosynaptic and polysynaptic activity in the spinal cord. Baclofen hyperpolarizes primary afferent terminals in contrast to GABA which depolarizes them.[27] Like diazepam, baclofen probably acts primarily in the spinal cord as it is an effective antispastic drug in complete spinal transections. Baclofen also shows analgesic activity in the whole animal which may also enhance relief. Baclofen is effective in the treatment of spasticity in paraplegic or quadriplegic patients with multiple sclerosis or traumatic lesions to the cord.[17,26]

With the development of ''dantrolene'' in 1974 no longer are all muscle relaxants centrally acting. Dantrolene, a peripherally acting agent, relaxes skeletal muscle by suppressing the release of calcium by the sarcoplasmic reticulum. By appropriate adjustment of dose, dantrolene will reduce spasticity by a graded reduction in the force of muscular contraction. Dantrolene has no other pharmacologically significant site of action in the stretch reflex arc. It does, however, have other CNS depressant effects like euphoria, light-headedness, dizziness, and drowsiness early in treatment, but tolerance develops for these effects since they subside with time. No tolerance develops to the antispastic action of dantrolene. Interestingly, dantrolene is also a valuable drug to alleviate the signs of malignant hyperthermia. This is a rare genetically determined condition which can be precipitated by administration of neuromuscular blocking drugs and inhalation anesthetics. Muscular contraction and hyperthermia are apparently precipitated by release of excess calcium ions from the sarcoplasmic reticulum. Dantrolene is an effective antispastic drug, but because it weakens skeletal muscles, it is more likely to diminish voluntary motor activity compared to the centrally acting muscles relaxants.[16,17]

Interestingly, phenothiazines have been shown to be effective in alleviating the spasticity and rigidity developed in the decerebrate cat. Apparently, phenothiazines exert their antispastic effect by blocking tonically active noradrenergic fibers which excite γ-motorneurons. Unfortunately, too much sedation accompanies the antispastic effect of phenothiazines. Raines and co-workers demonstrated that the capacity of chloropromazine to abolish decerebrate rigidity could be potentiated by phenytoin.[28] A clinical trial of phenytoin-chlorpromazine in spastic patients has shown that the combination is more effective than either drug alone and the incidence of lethargy and somnolence was reduced.[29] This combination, however, is usually attempted in patients for whom the other drug choices are unacceptable.

Doses

The usual muscle relaxant dose of the major drugs in use are shown in Table 2.

Toxicities

For some patients spasticity may provide a functional compensation for serious muscle weakness. In such patients the administration of skeletal muscle relaxant drugs may so compromise the contraction of antigravity muscles as to further impair mobility.[15-17] All skeletal muscle relaxants, even the peripherally acting dantrolene, have a tendency to depress the CNS as evidenced by drowsiness, light-headedness, and confusion.[26] Tolerance usually develops to sedation. Additive CNS depression may occur if these drugs are taken concurrently with alcohol, anxiolytics, or sedative-hypnotics.

Withdrawal reactions may occur if skeletal muscle relaxants are abruptly terminated after chronic treatment. Diazepam withdrawal seizures and baclofen withdrawal hallucinations have been described.[26] Even in the absence of new drug withdrawal phenomena, spasticity may show rebound aggravation.

Dantrolene is potentially hepatotoxic and may cause a fatal hepatitis. Accordingly, it is important to monitor liver function in patients treated with dantrolene. Diarrhea is a complaint that might be avoided by gradual dose escalation, but prohibits use of this drug in some patients.[17,24]

Chemical Classes of Skeletal Muscle Relaxants

Classificaiton of drugs according to their site or mechanism of action is impractical, partially because of the incomplete knowledge of the pathophysiology of spasticity/spasm and because the drugs acting as muscle relaxants have several sites of action. Drugs in this section are classified whenever possible according to their structural relationships.

α-Glycerol Ethers and Derivatives

Many compounds containing the α-glycerol ether structure have been prepared in the search for the ideal muscle relaxant. Substitution of a small alkyl, alkoxy, or halogen group in the benzene nucleus of antodyne generally increases its activity. The o-substituted compounds are usually more potent than the corresponding meta or para isomers. On the other hand, substitution at the benzene ring by an amino, hydroxy, hydroxymethyl, carboxy,

Table 2
SKELETAL MUSCLE RELAXANTS

Class and nonproprietary name	Chemical structure	Trade name	Single dose
	Glycerol monoethers and derivatives		
Mephenesin	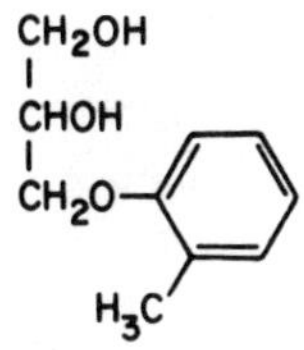	Tolserol	1—2 g (oral)
Chlorophenesin carbamate	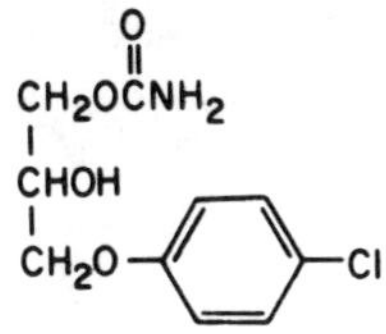	Maolate	800 mg (oral)
Methocarbamol	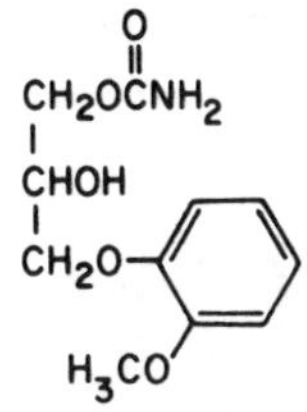	Robaxin	1—2 g (oral), 500 mg i.m., 1—3g (i.v), (slowly)
	Substituted alkanediols and derivatives		
Carisoprodol	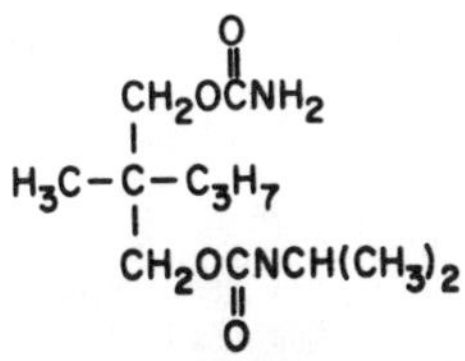	Rela Soma	250—350 mg
	Oxazolidinones		
Metazalone	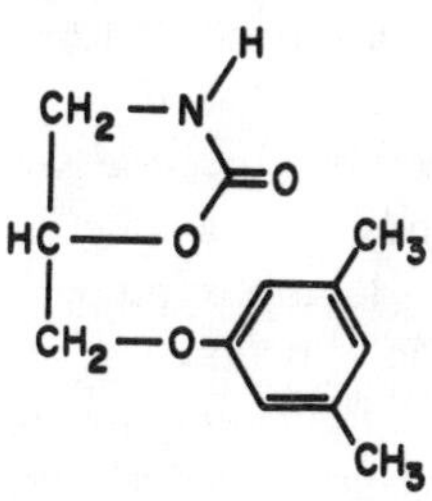	Skelaxin	800 mg
	Benzazoles		
Chlorzoxazone	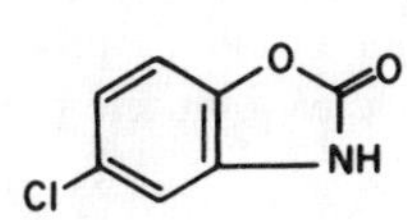	Paraflex	250—750 mg

Table 2 (continued)
SKELETAL MUSCLE RELAXANTS

Class and nonproprietary name	Chemical structure	Trade name	Single dose
	Miscellaneous		
Orphenadrine citrate[a]	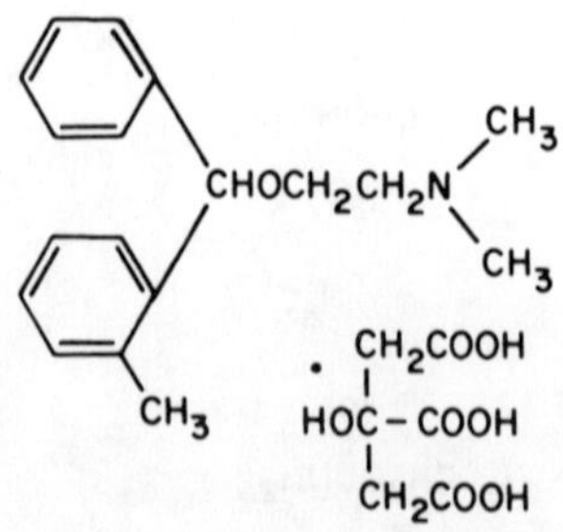	Norflex	100 mg (oral), 60 mg i.m. or i.v.
Dantrolene	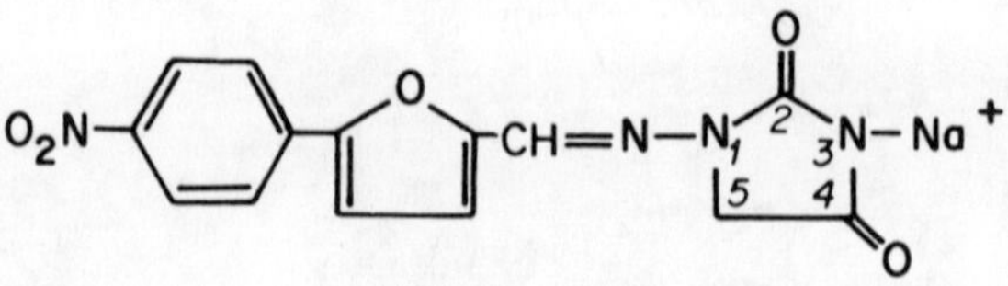	Dantrium	25—100 mg
Cyclobenzaprine hydrochloride	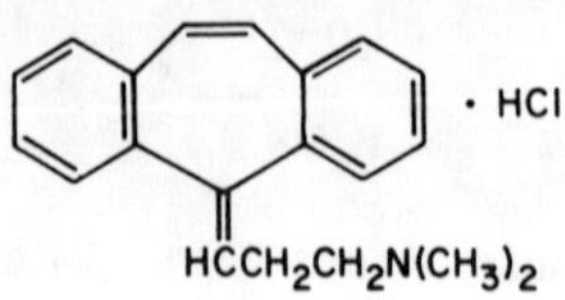	Flexeril	10 mg
Baclofen	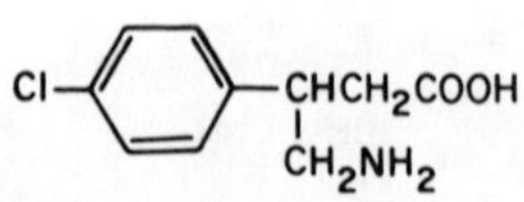	Lioresal	50—20 mg

[a] This product has been removed from the U.S. market.

carbomethoxy, carbamyl, acetamido, acetyl, or acetyloxime group generally decreases or abolishes the activity of antodyne.

In order to elicit muscle relaxant activity, the aryl group in the structure of glycerol ether derivatives is not necessary. Alkyl-substituted glycerol ethers also possess muscle relaxant activity, although it is generally less than that of the aryl ethers.

Most glycerol diethers and triethers are much less potent than the corresponding monofunctional ethers. Several compounds have a long duration of activity. The most active compound is 3-ethoxy-2-hydroxy-1-isopropoxypropane, which is less potent than mephenesin but has a much longer duration of action.

Since many diethers and triethers of glycerol have some muscle relaxant activity, it may be concluded that the free hydroxyl groups of glycerol are not necessary for the muscle relaxant activity. This is further substantiated by the finding that many carbamate and ether derivatives of glycerol are also potent muscle relaxants, with a prolonged duration of action.

It has been reported that the major metabolic product of mephenesin is β—(*o*-toloxy)-lactic acid. This compound is inactive as a muscle relaxant.

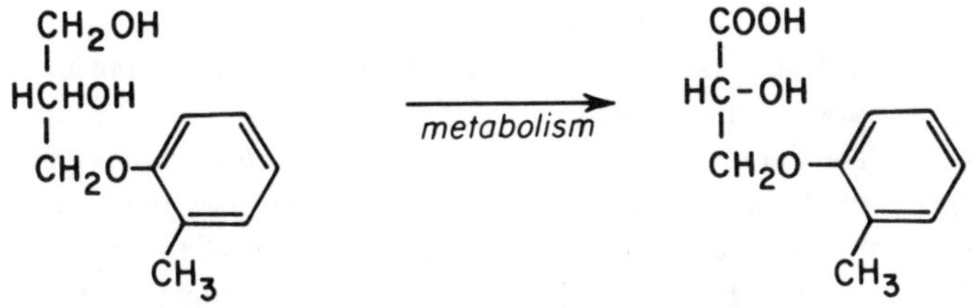

Mephenesin β-(O-toloxy)-lactic acid

In order to eliminate its formation, therefore, many compounds that have a protecting group on the α-hydroxyl function have been synthesized.

Protection of the α-hydroxyl function in the form of an ether derivative, as previously shown with diethers and triethers, usually produces compounds with a longer duration of action than the parent α-monoether. Carbamate and ester derivatives of the parent α-monoether also have longer activity. Although only half as active as mephenesin, a half ester, mephenesin succinate, has a much longer duration of action.

Similarly, an oral dose of the α-carbamate of mephenesin, 3-(*o*-toloxy)-2-hydroxypropyl carbamate, yields a higher and longer-lasting concentration of the drug in the blood of dogs than does mephenesin.

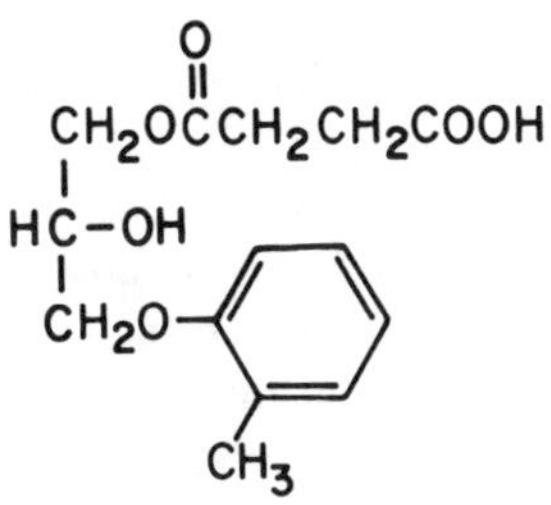

Mephenesin succinate

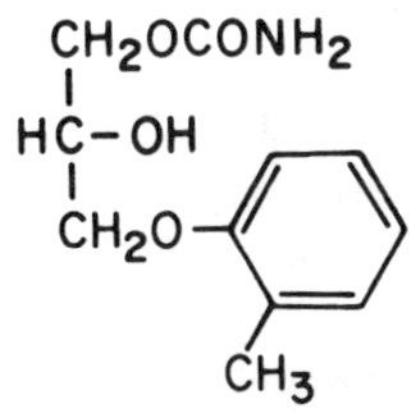

Mephenesin carbamate

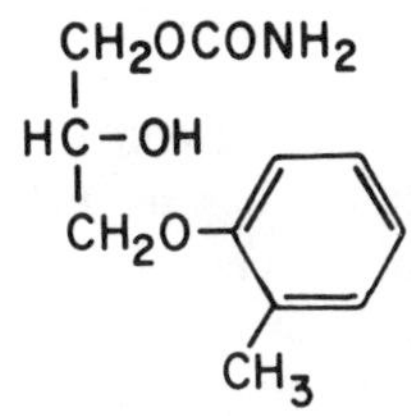

Methocarbamol

Several other carbamates, including chlorphenesin carbamate, methocarbamol, and 3-(α,α,α-trifluoro-*m*-toloxy)-2-hydroxypropyl carbamate, have a greater potency and longer duration of action than mephenesin.

''Methocarbamol'' has largely replaced mephenesin and its carbamate as a muscle relaxant, because the duration of action is significantly longer than that of mephenesin carbamate. Methocarbamol, however, produces a significant sedative effect and is more apt to produce toxic side effects (such as dizziness, drowsiness, fainting, hypotension, vertigo, and thrombophlebitis). In severe cases of spasticity or in postoperative conditions when oral administration is not feasible, perenteral administration is indicated. When given intravenously, it is effective as an adjunct in treating spinal spasticity, spinal root spasms, and musculoskeletal spasms, which it may reduce for brief periods.

Physical Chemical Data

Mephenesin, Tolserol, $C_{10}H_{14}O_3$, 3-(*o*-Methylphenoxy)-1,2-propanediol, crystals (m.p. 70 to 72° C) with a bitter taste. Freely soluble in alcohol, propylene glycol, and chloroform. At 20°C one part dissolves in 85 parts water. See structure in Table 2.

Chlorphenesin carbamate, Maolate, $C_{10}H_{12}ClNO_4$, 3-(*p*-chlorophenoxy)-1,2-propanediol 1-carbamate. Crystals (m.p. 89 to 91°C), readily soluble in dioxane; almost insoluble in water. See structures in Table 2.

Substituted Alkanediols and Derivatives

Many substituted 1,3-alkanediols and their derivatives are potent muscle relaxant drugs. Compared to mephenesin, ''meprobamate,'' 2-methyl-2-*n*-propyl-1,3-propanediol dicarbamate, is less potent. The duration of action, however, is about eight times that of mephenesin. Both compounds have a similar margin of safety. Meprobamate is more extensively used as an antianxiety agent or as a daytime sedative than as a muscle relaxant.

Carisoprodol, *N*-isopropyl-2-methyl-2-*n*-propanediol dicarbamate is more potent and less toxic than mephenesin. Interestingly, it is ineffective as a strychnine antagonist in experimental animals.

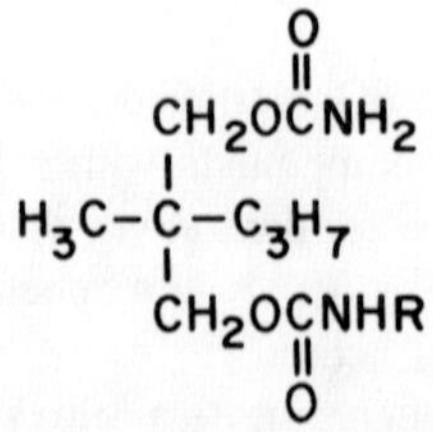

Meprobamate	R=H
Carisoprodol	R=$CH(CH_3)_2$
Tybamate	R=$(CH_2)_3CH_3$

Like mephenesin, meprobamate is a useful muscle relaxant in several dyskinetic disorders, particularly in patients with painful muscle spasms associated with arthritis and orthopedic disorders. Unlike mephenesin, the most important use of meprobamate was in the treatment of moderately severe anxiety. The use of meprobamate as an anxiolytic is discussed in the chapter entitled ''Antianxiety Agents''.

Carisoprodol is a powerful muscle relaxant with a rapid onset and long duration of action. It also has an analgesic effect. Carisoprodol is a safe drug that is useful for the treatment of muscle spasm and pain and stiffness associated with inflammatory, traumatic, and degenerative muscle conditions. Tybamate, the N-butyl substituted analogue of meprobamate was introduced in 1965 for the treatment of psychoneurotic anxiety and tension states. It is not currently used as a muscle relaxant and no longer marketed.

Physical Chemical Data

Carisprodol, Rela, Soma, $C_{12}H_{24}N_2O_4$, 2-methyl-2-propyl-1,3-propanediol carbamate 150-propylcarbamate, crystals (m.p. 92 to 93° C), slightly soluble in water (30 mg in 100 mℓ at 25° C). Stable in dilute acids and alkalies. See structure in Table 2.

Oxazolidinones

The oxazolidinone, metaxalone, has a long duration of action. It can be considered a cyclic carbamate (oxazolidinone) of an alkanediol such as methocarbamol. Metaxalone has limited utility as an adjunct in temporary relief of skeletal muscle spasm of local origin.

Physical Chemical Data

Metaxalone, Skelaxin$C_{12}H_{15}NO_3$, 5-(3,5-dimethyl phenoxymethyl)-2-oxazolidinone. Crystals from ethyl acetone m.p. 191 to 191.5° C. Sparingly soluble in water; soluble in methanol, ethanol, and isopropyl alcohol. See structure in Table 2.

Benzazoles

Several benzazoles with the following general structure have good muscle relaxant activity. Benzimidazole produces muscle relaxation and decreases voluntary movement for several hours. 2-aminobenzothiazole is also a potent muscle relaxant, with a wide margin of safety. Two benzoxazole derivatives, 5-chloro-2-benzoxazolinone (chlorzoxazone; Paraflex) and 5-chloro-2-aminobenzaxazole (zoxazolamine) display good muscle relaxant potencies.

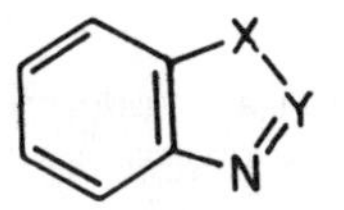

Benzimidazole: X = NH, Y = CH
Benzothiazole: X = S, Y = CH
Benzozazole: X = O, Y = CH
Benzotriazole: X = NH, Y = N

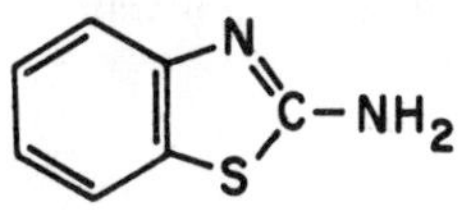

2-Aminobenzothiazole

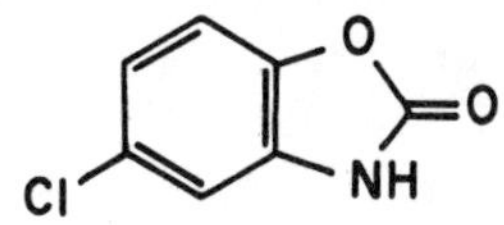

Chlorzoxazone

Chlorzoxazone is effective as an adjunct to rest, physical therapy, and other appropriate measures in treating the discomfort associated with acute, painful musculoskeletal conditions. Benzotriazole is also a potent muscle relaxant with a wide margin of safety.

Physical Chemical Data

Chlorzoxazone, Paraflex $C_7H_4ClNO_2$, 5-chlorobenzoxazolidone, crystals from acetone, m.p. 191 to 191.5°C. Sparingly soluble in water, soluble in methanol, ethanol, and isopropyl alcohol. See structure in Table 2.

Orphenadrine Citrate

This is an analogue of the antihistamine, diphenhydramine. Orphenadrine has about the same actions and degree of effectiveness as other centrally acting skeletal muscle relaxants. It also displays anticholinergic activity. It is administered orally and via injection (i.m. or i.v.). The hydrochloride salt is used in Parkinson's disease. (Disipal) See Table 1.

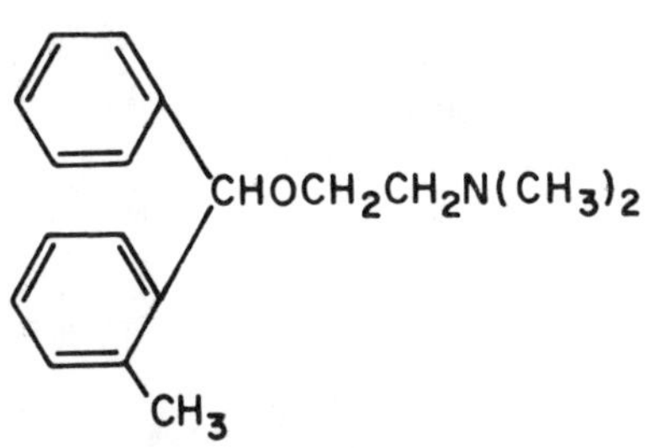

Orphenadrine

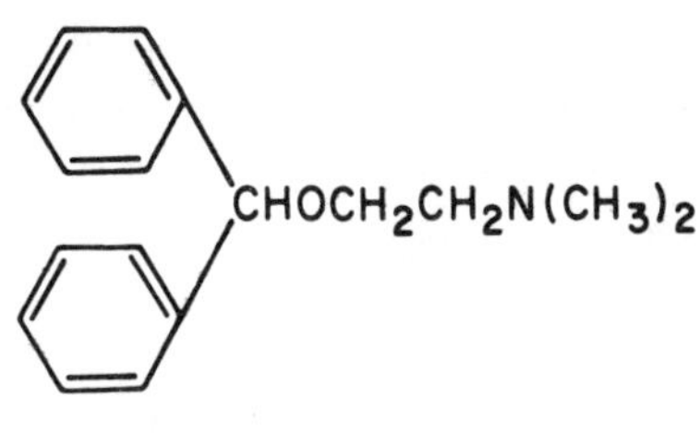

Diphenhydramine

Physical Chemical Data

Orphenadrine citrate, Norflex, $C_{24}H_{31}NO_8$, *N,N*-Dimethyl-2-[(2-methylphenyl) phenylmethoxy] ethanamine citrate. See complete structure in Table 2.

Cyclobenzaprine

Closely related to the tricyclic antidepressant both in its structure and pharmacological action, cyclobenzaprine is difficult to categorize in comparison with other available skeletal muscle relaxants.[33] It is clear that this drug acts neither at the neuromuscular junction nor directly on skeletal muscle. Little information is available that clarifies the mechanism by which this agent modifies muscle spasm.

Physical Chemical Data

Cyclobenzaprine hydrochloride, Flexeril, $C_{20}H_{21}N{\cdot}HCl$, 3-(5-H-Dibenzo [a,d] cycloheptene-5-ylidene)-*N,N*-dimethyl-1-propanamine hydrochloride; crystal m.p. 216 to 218° C, soluble in water. This drug is no longer available in the U.S. due to toxicity.

Baclofen

Baclofen is a derivative of the neurotransmitter gamma-aminobutyric acid (GABA) is a recently developed drug for the treatment of spasticity in patients with multiple sclerosis. Although structurally related to GABA, baclofen acts by hyperpolarizing primary afferent nerve terminals. The drug is absorbed rapidly after oral administration and has a half-life in plasma of about 3 to 4 hr.

Physical Chemical Data

Baclofen, Lioresal, $C_{10}H_{12}ClNO_2$ β-(Aminomethyl)-*p*-chlorohydrocinnamic acid; 4-amino-3-(4-chlorophenyl)butanoic acid, crystals m.p. 179 to 181° C. See structure in Table 2.

Dantrolene

Dantrolene was introduced in 1974 and represents a unique class of skeletal muscle relaxants in that it exerts its effects by a direct action on skeletal muscle.[30] It does not affect the membrane potential or electrical excitability of skeletal muscle, nor does it alter neuromuscular transmission.[31,32] Dantrolene sodium (Dantrium) is slowly absorbed from the GI tract, but dose-related blood levels are obtained. It is slowly metabolized by the liver, to give the 5-hydroxy and acetamido metabolites, but the unchanged drug is also found in the urine.

Physical Chemical Data

Dantrolene sodium, Dantrium $C_{14}H_9N_4O_5Na$; 1-[(5-(p-nitrophenyl) furfurylidene) amino] hydantoin. See Table 2 for structure.

ACKNOWLEDGMENTS

The authors wish to thank Carl Regillo for help in preparing the manuscript, Noralie Barnett for the art work, and Roger Avelino for typing this chapter. The partial support of BRSG Grant S07 05830-03 awarded by the Biomedical Research Support Grant Program, Division of Research Resources, National Institute of Health is greatfully acknowledged.

REFERENCES

1. **Ehringer, H. and Hornykiewicz, O.,** Verteilung von Noradrenalin and dopamin (3-hydroxytyramin) im Gehirn des Menschen and ihr Verhalten bei Erkrankungen des extrapyramidalen Systems, *Klen. Wohenschr.*, 38, 1236, 1960.
2. **Bianchine, J. R.,** in the *Pharmachological Basis of Therapeutics,* 6th ed., Gilman, A. G., Goodman, L. S., and Gilman, A., Eds., Macmillan, New York, 1980, 475.
3. **Yahr, M. D.,** Overview of present day treatment of Parkinson's disease, *J. Neural. Transm.*, 43, 227, 1978.
4. **Cotzias, G. C., Van Woert, M. H., and Schiffer, L. M.,** Aromatic amino acids and modification of parkinsonism, *N. Engl. J. Med.*, 276, 374, 1967.
5. **Nagatsu, T.,** in *Biochemistry of Catecholamines,* University Park Press, 1973, 289.
6. **Bianchine, J. R. and Shaw, G. M.,** Clinical pharmacokinetics of levodopa in Parkinson's disease, *Clin. Pharmacokinet.*, 1, 313, 1976.
7. **Palfreyman, M. G., Danzin, C., Bey, P., Jung, M. J., Riberbeau-Gayon, G., Aubry, M., Vevert, J. P., and Sjoerdsmar, A.,** Difluoromethyl dopa, a new enzyme-activated irreversible inhibitor of aromatic L-amino acid decarboxylase, *J. Neurochem.*, 31, 927, 1978.
8. **Goldberg, L. I., Volkman, P. H., and Kohli, J. D.,** A comparison of the vascular dopamine receptor with other dopamine receptors, *Annu. Rev. Pharmacol. Toxicol.*, 18, 57, 1978.
9. **Knoll, J.,** The possible mechanism of action of (−) deprenyl in Parkinson's disease, *J. Neural. Transm.*, 43, 177, 1978.

10. **Barbeau, A.,** in *Psychopharmacology: A Generation of Progress,* Lipton, M. A., Di Mascio, A., and Killam, K. F., Eds., Raven Press, New York, 1978, 771.
11. **Neumeyer, J. L., Lal, S., and Baldessarini, R. J.,** in *Apomorphine and Other Dopaminomimetics,* Vol. 1, Gessa, G. L. and Corsini, G. V., Eds., Raven Press, New York, 1981, 1.
12. **Cotzias, G. D., Papavasiliou, P. S., Tolosa, E. S., Mendez, J. S., and Bell-Midura, M.,** Treatment of parkinsonism with aporphines: possible role of growth hormone, *N. Engl. J. Med.,* 294, 567, 1967.
13. **Calne, D. B.,** Developments in the pharmacology and therapeutics of parkinsonism, *Ann. Nuerol.,* 1, 111, 1977.
14. **Keller, H. H. and Daprada, M.,** Central dopamine agonistic activity and microsomal biotransformation of lisuride, lergotrile and bromocriptine, *Life Sci.,* 24, 1211, 1979.
15. **Petersen, E.,** *Spasticity — Mechanism, Measurement and Management,* Charles C. Thomas, Springfield, Ill., 1969.
16. **Young, R. R. and Delwaide, P. J.,** Drug therapy: spasticity (first of two parts), *N. Engl. J. Med.,* 304 (1), 28, 1981.
17. **Davidoff, R. A.,** Pharmacology of spasticity, *Neurology (Minneapolis),* 28, 46, 1978.
18. **Dimitrijefic, M. R. and Sherwood, A. M.,** Spasticity: medical and surgical treatment, *Neurology,* 30 (2), 19, 1980.
19. **Elenbaas, J. K.,** Drug therapy reviews: centrally acting oral skeletal muscle relaxants, *Am.J. Hosp. Pharm.,* 37, 1313, 1980.
20. **Crossland, J.,** *Lewis's Pharmacology,* 4th ed., Livingston, Edinburgh, 1970, 343.
21. **Roszkowski, A. P.,** A pharmacological comparison of therapeutically useful centrally acting skeletal muscle relaxants, *J. Pharmacol. Exp. Ther.,* 129, 75, 1960.
22. **Berger, F. M. and Bradley, W.,** The pharmacological properties of α: a dihydroxy -v-(2-methylphenoxy) propane (Myanesin) *Br. J. Pharmacol. Chemother.,* 1, 265, 1946.
23. **Haefely, W., Schaffner, R., Polc, P., and Pieri, L.,** General pharmacology and neuropharmacology of propanediol carbomates, in *Handbook of Experimental Pharmacology,* Vol. 55/II, Hoffmeister, E. and Stille, G., Eds., Springer-Verlag, Berlin, 1981, 263.
24. **Bianchine, J. R.,** Drugs for Parkinson's disease; centrally acting muscle relaxants, in *Goodman and Gilman's The Pharmacological Basis of Therapeutics,* 6th ed., Gilman, A. G., Goodman, L. S., and Gilman, A., Eds., Macmillan, New York, 1980, 488.
25. **Smith, C. M.,** Relaxants of skeletal muscle, in *Physiological Pharmacology, Vol. 2, Part B,* Root, W. S. and Hofman, F. G., Eds., Academic Press, New York, 1965, 2-96.
26. **Young, R. A. and Delwaide, P. J.,** Drug therapy: spasticity. II, *N. Engl. J. Med.,* 304 (2), 96, 1981.
27. **Davidoff, R. A.,** Effects of baclofen on synaptic activity in the spinal cord, in *Spasticity: Disordered Motor Control,* Feldman, R. G., Young, R. R., and Koella, W. P., Eds., Yearbook Medical Publishers, Chicago, Ill., 1980.
28. **Anderson, R. J. and Raines, A.,** Suppression of decerebrate rigidity by phenytoin and chloropromazine, *Neurology,* 26, 858, 1976.
29. **Cohan, S. L., Raines, A., Panagakos, J., and Ormitage, P.,** Phenytoin and chloropromazine in the treatment of spasticity, *Arch. Neurol.,* 37, 360, 1980.
30. **Snyder, H. R., Jr., Davis, C. S., Bickerton, R. K., and Halliday, R. P.,** 1-((5-Arylfuerfurylidine)aminohydantoins. A new class of muscle relaxants, *J. Med. Chem.,* 10, 807, 1967.
31. **Davidoff, R. A. and Sears, E. S.,** The effects of lioresal on synaptic activity in the isolated spinal cord, *Neurology (Minneapolis),* 24, 957, 1974.
32. **Davidoff, R. A.,** Pharmacology of spasticity, *Neurology (Minneapolis),* 28, 46, 1978.
33. **Ashby, P., Burke, D., and Rao, S.,** Assessment of cyclobenzaprine in the treatment of spasticity, *J. Neurol. Neurosurg. Psychiatr.,* 35, 599, 1972.

NARCOTIC ANALGESICS, ANTAGONISTS, AND ANTITUSSIVE AGENTS

Michael A. Eissenstat and William F. Michne

INTRODUCTION

Historical

Pain is a universal experience of man which knows no bounds of time or intensity. The subject of pain and its relief is found among the writings of the most ancient of civilizations. A brief but interesting history of pain relief was given by Tainter in 1948.[1] He divided the evolution of analgesic drugs into four major eras. The first of these begins with primitive man in the darkness of superstition and religious mysticism. Pain was viewed as being caused by demons which could be appeased or frightened away by wearing special charms supplied by the medicine man or priest. Gradually, ancient man learned to use natural plant preparations such as poppy, mandragora, hemp, and henbane, whose narcotic properties gave comfort to all sufferers. By the end of the 18th century they remained the principal drugs used for the alleviation of pain, and their preparations had been essentially unchanged for some 2000 years. The second major era in the evolution of analgesics began very early in the 19th century when a relatively uneducated apothecary's assistant named Sertürner isolated the active constituent of opium and proceeded to study its effects in animals and in man. Sertürner named his new substance "morphine" after Morpheus, the Greek god of dreams. Morphine is the first alkaloid ever to be isolated; this work ushered in a most prolific scientific period characterized by the extraction and purification of most of the important alkaloids from their natural plant sources and their wide applications in medicine which have largely persisted to the present day. The third major era in the evolution of analgesics actually began in 1828 when the German chemist Wohler reported his synthesis of urea from inorganic materials. This discovery shattered the theory that a "vital force" present in living material was necessary for the formation of organic compounds and began the period of the first successful analgesic drugs made by chemical synthesis. For the most part these drugs are relatively simple chemically, such as aspirin and aminopyrine, but represented the most potent agents of synthetic origin until recent times. The fourth and present era of analgesic drug evolution has a nebulous beginning early in the 20th century. The embryonic sciences of organic chemistry and pharmacology were developing rapidly and quantitative evaluations of progressively more complex synthetic molecules were becoming commonplace. The total synthesis of morphine,[2] the chemical and pharmacological characterization of a great variety of opioid agonists and antagonists,[3-6] the demonstration of opiate receptors in nerve tissue,[7] and the isolation of endogenous opioid peptides[8] are among the triumphs of this period.

Animal Models in Analgesic Testing

The evaluation of a new chemical compound as an analgesic using animal models is an extremely complex task. Nothing is known about the perception of pain by animals and the situation is made even less clear when attempts are made to relate test results to humans, taking into account the variability and subjectivity of individuals. Nevertheless, as a result of a fairly large number of analgesics having been studied in man, pharmacologists have been able to devise tests for analgesics in animals and correlate their data with human clinical observations. Those testing methods become widely used which show a high correlation with human data, are reproducible, simple to run, and are sufficiently sensitive to detect both weak and potent analgesics. However, comparisons of data from two laboratories using the same test should be approached with caution owing to variations in actual methodology and/or data handling. With this in mind the following brief description of the most commonly encountered tests in the analgesic medicinal chemistry literature is given.

Writhing tests — There are a number of procedures in which administration of a chemical nociceptive agent to a rodent results in a characteristic response known as a writhe which is blocked by analgesic drugs. The most common chemical irritants used are acetic acid,[9] acetylcholine,[10] and phenylquinone.[11] These tests have the advantages of simplicity, sensitivity, and reproducibility, and the correlation with human analgesic data is generally good. The disadvantage of these tests is that activity is shown by some other classes of drugs such as anticholinergics.[12,13] Such false positives can only be determined by subsequent ancillary testing. Nevertheless, the writhing tests are frequently used as high-capacity primary analgesic screens.

Thermal stimulus tests — Application of a thermal stimulus, usually to a rodent, of sufficient intensity and duration causes an escape attempt on the part of the animal within a reproducible period of time, referred to as the latency. Analgesic drugs extend the latency for the escape reaction. In the tail flick procedure,[14] the rodent's tail is exposed to a radiant heat or hot water stimulus; the response is tail withdrawal. In the hot plate procedure[15] the animal is placed on a heated plate; the response includes jumping and paw licking. These assays are quite good for narcotics, and less dependable for narcotic antagonist analgesics and nonnarcotics. Not surprisingly, sedatives and muscle relaxants may show up as false positives.

Dependence testing — For narcotic and narcotic antagonist analgesics it is of interest to determine a drug's physical dependence profile. This is most commonly done in monkeys[16] or rats.[17] In the monkey model animals are maintained on morphine injections and then withdrawn. A single dose of test compound is substituted for morphine. The drug either suppresses, exacerbates, or has no effect on withdrawal symptoms. In the rat model morphine is administered by continuous infusion followed by substitution of the test drug and observation of the withdrawal symptoms. Additionally, a test compound may be studied in either model for its ability to produce dependence itself.

In vitro models — The use of isolated tissues has the advantage of being free from complications associated with absorption, distribution, and metabolism. Further, experimental parameters may be precisely controlled and very little test compound is used. There are three commonly used in vitro assays for opiates. The guinea pig ileum[18] and mouse vas deferens[19] preparations are used to measure the agonist and antagonist effects of opiates. Taken together the two tests can give some indication of a test drug's selectivity for the subsets of opiate receptors.[20] Opiate receptor binding[21] measures a compound's affinity for the opiate receptor, and in some cases a distinction between agonists and antagonists can be made based on the differential effect of sodium ion.[22]

This chapter is organized by the major structural classes of the narcotic analgesics. Within each class one or more compounds of therapeutic importance are discussed in detail. Treatment of the synthetic chemistry of these substances is minimal; the interested reader is referred to the monograph edited by de Stevens,[23] and the reviews by Casy,[5,6] Archer and Harris,[3] and Archer and Michne.[4] More specific chemistry reviews are noted in the text as appropriate. Since the monographs on individual drugs may derive much of their data from monographs in cited works, reference to these works will only be made at the first appropriate point for each drug.

Mechanism of Action

The mechanism of action of the opioids is not clearly understood. The subject has been dealt with extensively in the literature,[24-31] and an in-depth discussion of this topic is beyond the scope of this review. Only a brief discussion will be presented here.

The relatively recent discoveries of opioid receptors and endogenous opioids have stimulated considerable research in the opioid area. It appears that binding of an exogenous or endogenous ligand to an opioid receptor initiates a series of events that ultimately lead to analgesia. Interaction of opioids with these receptors is stereospecific, saturable, and naloxone reversible. The distribution of enkephalins correlates well with that of the receptors. Some of the proposed consequences of the interaction of the opioid with its receptor are discussed below.

The transmembrane transport of Ca^{2+} has been proposed as a critical step in neuronal function. The effect of opioids on this process may be integral to their action. Morphine acutely depletes brain Ca^{2+} levels and prevents uptake of Ca^{2+} into brain slices and synaptosomes.[32] Mice or rats apparently adapt to long-term morphine administration by increasing the basal levels of synaptosomal Ca^{2+}. This causes more morphine to be required than originally to reach subnormal levels of Ca^{2+}, thus explaining tolerance. The increased levels of Ca^{2+} in addicted animals may explain the withdrawal syndrome seen upon cessation of opiate administration. Administration of Ca^{2+} antagonizes morphine analgesia. La^{3+}, which inhibits Ca^{2+} binding and transport, enhances morphine analgesia and is itself analgesic at high doses. Animals tolerant to morphine have been shown to be cross-tolerant to La^{3+}.

Changes in cyclic AMP levels have also been postulated to be important in the mechanism of opiate action.[33,34] The neuronal effects of both Ca^{2+} and cyclic AMP may be mediated by their effects on phosphorylation of synaptic membrane proteins. Since cyclic AMP stimulates the phosphorylation of certain synaptic membrane proteins in the presence of Ca^{2+} and others in its absence, the phosphorylation mechanism may be dependent on Ca^{2+} levels. Opioids appear to produce a selective and naloxone-reversible inhibition of adenylate cyclase in certain tissues, resulting in decreased levels of cyclic AMP and presumably reduced levels of neuronal transmission. On chronic administration of opiates cyclic AMP levels increase, thus leading to an analogous explanation of tolerance and withdrawal as was made for Ca^{2+} *vide supra*.

The distributions of enkephalins and substance P in the CNS show considerable overlap. Depletion of substance P has been associated with antinociception. The release of substance P like immunoreactivity has been shown to have some Ca^{2+} dependence. Substance P has thus been postulated to be one of the transmitters of nociceptive information. In certain brain tissues substance P release is inhibited stereospecifically by opioids and this effect is reversed by naloxone.[35] This inhibition may be a major contributor to the mechanism of analgesic action of the opioids.

Among the other putative neurotransmitters which may be involved in opioid analgesia are acetylcholine,[36] catecholamines,[37] and serotonin.[38] Thus the mechanism of opioid analgesia is quite complex, with a number of putative neurotransmitters thought to be involved. However, although various associations between analgesia and these neurotransmitters have been made, it is not known which if any are actually required in the mechanism of analgesia.

HEXACYCLICS

Beginning in the late 1950s Bentley and co-workers described the synthesis of a variety of Diels-Alder adducts of thebaine and the further transformation of these compounds into potent narcotic analgesics and antagonists. This work has been extensively reviewed.[4,39-42]

The parent Structure 1 (R=H) of the hexacyclics (oripavines) has 40 times the potency of morphine as an antinociceptive agent. Some structure-activity relationships (SARs) of derivatives of Structure 1 have been established. Compounds with R=H are generally 10 to 50 times as potent as those with R=Me. Alkyl substitution at C_{16} decreases potency. Saturated alkyl substitution at $C_{8\alpha}$ may give increased antinociceptive potency but unsaturated groups (e.g., CO_2Et) at that position may have the opposite effect.

Interest has been particularly strong in Structure 2 which incorporates a hydroxy-substituted alkyl side chain at $C_{7\alpha}$. In this series antinociceptive activity generally decreases when R^3

is larger than methyl and potent narcotic antagonists may be generated when R^3 is allyl or cyclopropylmethyl. Reduction of the 6,14-etheno bridge or removal of the 6-methoxy group has little effect on potency.[43] For straight chain alkyl substitutents at R^1, when R^2=methyl, antinociceptive potency peaks when R^1=propyl or butyl. Antagonist activity is not found in Structure 2 when R^1 is larger than ethyl. It should be noted that a hydroxy-substituted side chain is not required for potent antinociceptive activity since Structure 3 has 1000 times the potency of morphine.

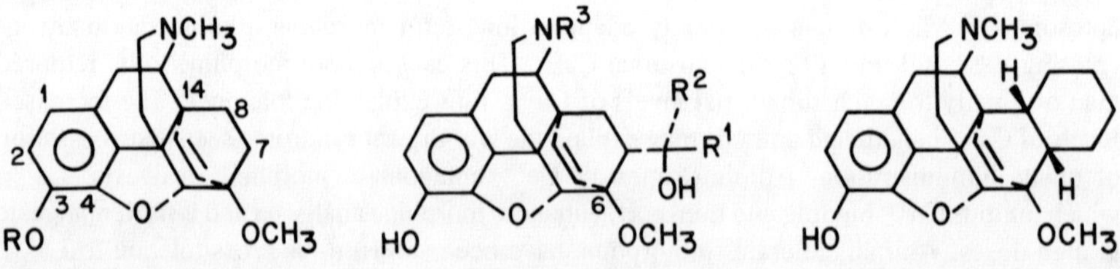

Structure 1 Structure 2 Structure 3

Several derivatives of Structure 2 have been marketed. Etorphine (Structure 2, R^1=,Pr, R^2=R^3=Me), an analgesic 1000 times as potent as morphine, has been used in combination with a neuroleptic for immobilization of large animals. Diprenorphine (Structure 4), a narcotic antagonist 100 times as potent as nalorphine, has been used to revive animals under the influence of etorphine. Buprenorphine (Structure 5) is a potent mixed agonist-antagonist analgesic used in man.

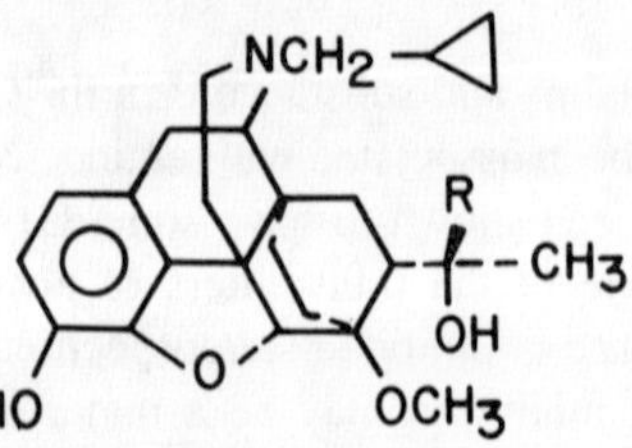

Structure 4; R=Me

Structure 5; R=t-Bu

Buprenorphine Hydrochloride

Pharmacokinetics

There is no direct relationship between plasma levels and pharmacologic effect for this drug. Peak plasma levels are 18.6 μg/ℓ 2 min after 0.3 mg i.v., 3.6 μg/ℓ 5 min after 0.3 mg i.m., or 2 hr after 0.3 mg p.o., and 0.74 μg/ℓ 30 to 300 min after 0.4 mg sublingual;[44,45] 96% of the circulating drug is protein bound and some drug is detectable in the blood after 1 day. Analgesia lasts for 9 hr following sublingual administration and 6 to 8 hr following i.m. administration. Most of the drug is excreted unchanged in the feces and some appears in the urine as N-dealkylated and conjugated metabolites.

Uses and Dosage[28,46-48]

Buprenorphine is a potent mixed agonist-antagonist analgesic indicated for the relief of moderate to severe pain. The compound produces analgesia of similar quality to that of morphine, but slower in onset and longer in duration. It is about 30 times as potent as morphine as an analgesic and equipotent to naloxone as a narcotic antagonist. As had been observed in earlier mixed agonist-antagonists such as pentazocine the compound has lower abuse potential than classical opiates. Withdrawal symptoms are relatively mild and delayed in onset; the drug has been used clinically in the treatment of narcotic addiction. It may be the drug of choice for patients with any cardiac condition since clinical trials showed no change in hemodynamic parameters.[49] Buprenorphine has been used for acute postoperative pain and its long duration of action implies possible use in the treatment of chronic pain. It suppresses some aspects of stress response and is effective as an analgesic supplement to anesthesia.[50]

Buprenorphine may be administered i.m. (0.2 to 0.6 mg), by slow i.v. administration of the same dose, or sublingually (0.4 to 0.8 mg) every 6 to 8 hr as needed.

Toxicity

The most commonly observed side effects are sedation, nausea, vomiting, dizziness, sweating, mild euphoria, and respiratory depression. Peak respiratory depression occurs 3 hr after i.m. injection and is milder than that of morphine but not readily reversed by narcotic antagonists. Cardiovascular effects are qualitatively similar to morphine. Peak miosis occurs 6 hr after i.m. injection of a recommended dose. Psychotomimetic effects are not observed.

Warnings — The drug should be used with caution in patients with respiratory conditions or impaired liver function. It may precipitate withdrawal in narcotic addicts and can antagonize the actions of narcotic analgesics previously administered. It should be used cautiously in patients who have recently taken MAO inhibitors. Patients should be warned against attempting potentially hazardous tasks after taking the drug. Use during pregnancy is not recommended.

Physical - Chemical and Other Data

The structure of buprenorphine (Structure 5) may be found in the SAR discussion.

[5α,7α(S)]-17-(Cyclopropylmethyl)-α-(1,1-dimethylethyl)-4,5-epoxy-18,19-dihydro-3-hydroxy-6-methoxy-α-methyl-6,14-ethenomorphinan-7-methanol hydrochloride; $C_{29}H_{41}NO_4 \cdot HCl$; Temgesic (Reckitt and Colman); Buprenex (Norwich Eaton).

PENTACYCLICS

The correct structure of morphine (Structure 6) the major pentacyclic compound isolated from the opium poppy, was proposed by Gulland and Robinson[51] in 1925 and was confirmed via total synthesis by Gates and Tschudi[2] in 1952. Some of the many morphine derivatives which retain the pentacyclic ring structure are shown in Table 1.[28,52] However, in general these compounds showed little or no therapeutic advantage over morphine for the treatment of moderate to severe pain. Although more potent analgesics were found, the side effects generally were correspondingly severe. The synthesis of heroin (Structure 7) in the 1880s by acetylation of the 3- and 6-OH's of morphine provided an analgesic considerably more potent than morphine, but which also had severe addiction liablity.

Codeine (Structure 8), the 3-methoxy analogue of morphine also isolated from the opium poppy, is considerably less potent than morphine as an analgesic but is very useful nevertheless, in part due to its oral activity, decreased side effects, and its antitussive effect. In general, the potency of 3-alkoxy analogues of morphine decreases as the size of the alkoxy group increases.[53] However the presence of a 3–OH is not required for potent analgesic activity. Recently, workers at NIH have found that although removal of the 3–OH to provide 3-desoxymorphine lowered antinociceptive potency, when other changes were also made, such as hydrogenation of the 7-8 double bond, the difference in activity between the parent compound and the 3-desoxy derivative was small.[54] In the codeine series, the 1-position of the aromatic ring has been substituted by chloro, bromo, acetyl, and fluoro, but only the last retained activity.[27]

Table 1
MORPHINE DERIVATIVES[28,52]

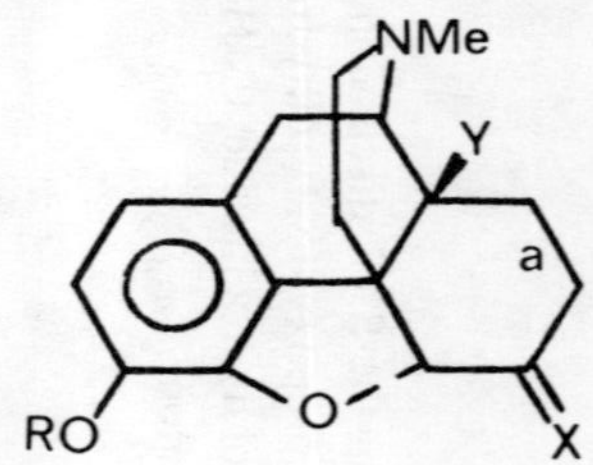

Name	R	Double bond a	X	Y	Relative potency (morphine = 1)	Uses and comments
Morphine	H	Yes	α-OH	H	1	Analgesic
Codeine	Me	Yes	α-OH	H	0.15	Antitussive, mild analgesic
Ethylmorphine	Et	Yes	α-OH	H	0.1	Opthalmology (chemotic)
Heroin	Ac	Yes	α-OAc	H	4	Potent analgesic with severe addiction liability, prohibited in U.S., common drug of abuse
Hydromorphone, dihydromorphinone	H	No	O	H	5	Introduced in 1926, less sedative, shorter acting than morphine
Hydrocodone, dihydrocodeinone	Me	No	O	H	0.7	Analgesic, antitussive, orally effective, marketed in U.S. only with additional ingredients
Oxymorphone	H	No	O	OH	10	Analgesic, introduced in 1959
Oxycodone	Me	No	O	OH	1	Analgesic, antitussive, orally effective, marketed in U.S. only with additional ingredients
Dihydrocodeine	Me	No	α-OH	H	0.3	Analgesic, faster acting than morphine, marketed in U.S. only with additional ingredients, may cause hypotension

Table 2
MORPHINE-DERIVED ANTAGONISTS

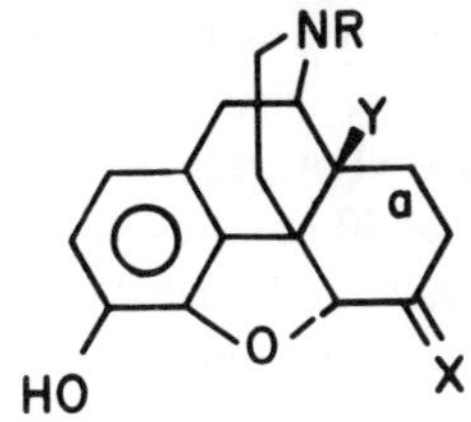

Name	R	X	Double bond a	Y	Uses and comments
Nalorphine	$CH_2CH=CH_2$	α-OH	Yes	H	First narcotic antagonist used in medicine, antidote for narcotic overdosage, potent analgesic
N-Propylnormorphine	$CH_2CH_2CH_3$	α-OH	Yes	H	Potent antagonist with less analgesic activity than nalorphine
Naloxone	$CH_2CH=CH_2$	O	No	OH	Pure narcotic antagonist, >10 times nalorphine, short acting, few if any side effects
Naltrexone	CH_2 	O	No	OH	Potent antagonist 17 times nalorphine, little agonist activity, longer acting than naloxone
Nalbuphine	CH_2 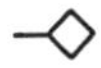	α-OH	No	OH	Mixed agonist-antagonist, as potent as morphine as agonist, 1/4 nalorphine as antagonist

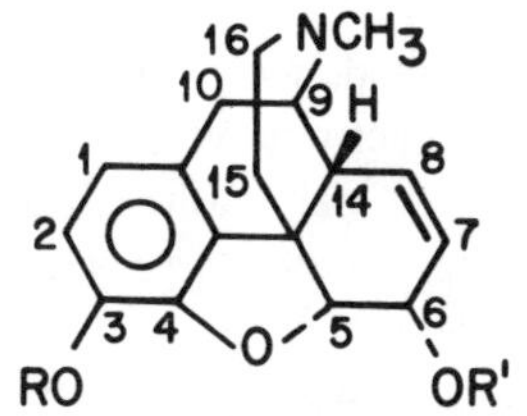

Structure 6; R=H R′=H

Structure 7; R=Ac R′-Ac

Structure 8; R=Me R′=H

Much of the synthetic activity on pentacyclics has involved varying the *N*-alkyl substituent. Normorphine, the N-demethylated derivative of morphine, is a weaker antinociceptive agent. Quaternization of the nitrogen considerably attenuates analgesic activity presumably because the compound cannot readily penetrate the blood-brain barrier. Antinociceptive activity is decreased as the *N*-alkyl group is increased in size to ethyl but is restored for larger substituents such as amyl and hexyl, and is increased further for substituents such as phenethyl and cyclohexylethyl.[27] However, of more interest than this increase in potency was the discovery that substituents slightly larger than ethyl could generate narcotic antagonists.[3,4,55,56] The first report of a narcotic antagonist, *N*-allylnorcodeine, was made in 1915 by Pohl.[57] *N*-allylnormorphine (nalorphine) was synthesized in the early 1940s and shown to be a potent narcotic antagonist. Among the N-substituents which have also been found to provide potent antagonists are propyl, isobutyl, methallyl, cyclopropylmethyl, and cyclobutylmethyl. Some narcotic antagonists based on the pentacyclic ring system are shown in Table 2.

A great deal of chemistry has been done on the C-ring of morphine.[27,53] The antinociceptive activity of the pentacyclics tolerates considerable variation at C_6. Inverting the 6–OH causes little change in activity. Replacing the 6–OH by H or oxidizing it to a carbonyl decreases antinociceptive activity in the morphine series but increases activity in the dihydromorphine series. Dihydromorphine is approximately as potent as morphine. Etherification or acylation of the 6–OH also gives antinociceptive agents more potent than morphine. Replacement of the 6–OH in dihydromorphine by methyl or methylene increases activity and an azide at the 6-position provides an analgesic about 50 times as potent as morphine in man. Compounds containing a C_6–C_7 double bond such as Structure 9 can also be quite potent. Methyl substitution at C_5 of dihydromorphinone provides the potent analgesic metopon (Structure 10). Cyclopropanation but not epoxidation of the olefin in codeinone increases antinociceptive activity.[58] In the dihydrocodeinone series small alkyl groups at $C_{8\beta}$ may increase potency but large alkyl groups are detrimental.[59]

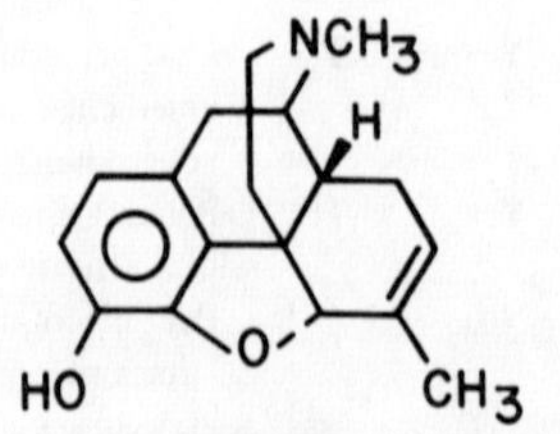

Structure 9

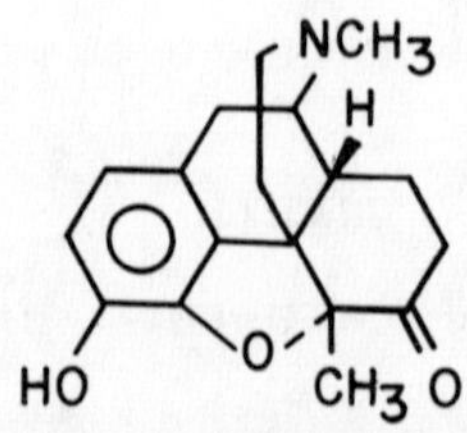

Structure 10

Significant enhancement of potency is often attained by substitution at the $C_{14\beta}$ position, particularly in the presence of a carbonyl at C_6. Most of the compounds studied have a hydroxy group at $C_{14\beta}$ but other substituents at that position which also enhance activity include acetoxy, methyl, ethyl, halo, mercapto, nitro, and amino.[60,61] 14β-Hydroxymorphinone (Structure 11) is approximately 10 times as potent as morphine. A major advance in the area of narcotic antagonists was made when allyl and cyclopropylmethyl groups known to imbue normorphine with antagonist activity were placed on noroxymorphone. The potent antagonists naloxone (Structure 12) and naltrexone (Structure 13) were thus generated. *N*-Cyclobutylmethylation of dihydronoroxymorphone provided nalbuphine (Structure 14), which is a clinically useful mixed agonist-antagonist analgesic. Many of the derivations previously described for the narcotic agonists have also been done for naloxone or naltrexone. Other derivatives which have been made include naloxazone (Structure 15) and naltrexazone (Structure 16),[62] which have increased duration of action, and the irreversible antagonists β-chlornaltrexamine (CNA, Structure 17) and β-funaltrexamine (FNA, Structure 18).[63]

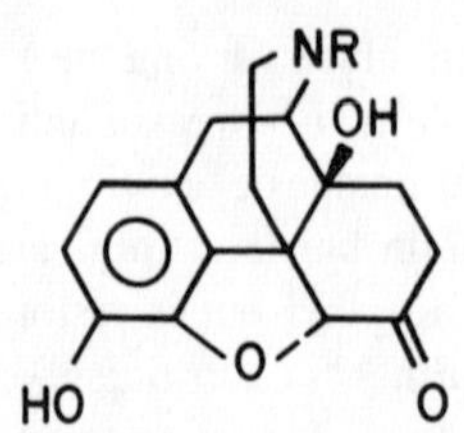

Structure 11; $R{=}CH_3$

Structure 12; $R{=}CH_2CH{=}CH_2$

Structure 13; $R{=}CH_2$ —◁

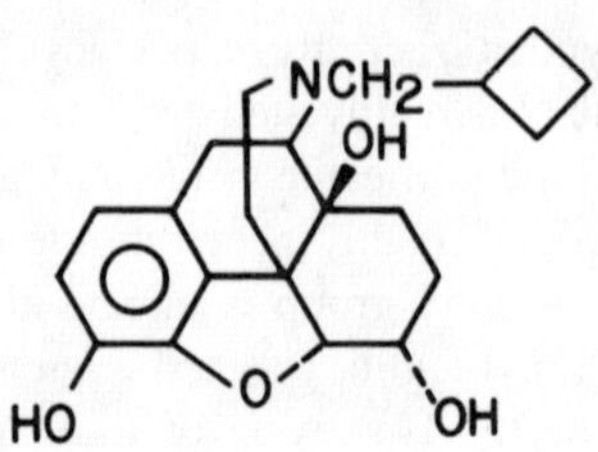

Structure 14

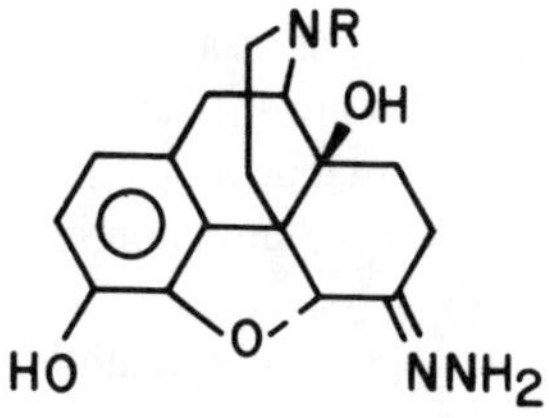

Structure 15; R=$CH_2CH{=}CH_2$

Structure 16; R=CH_2 —◁

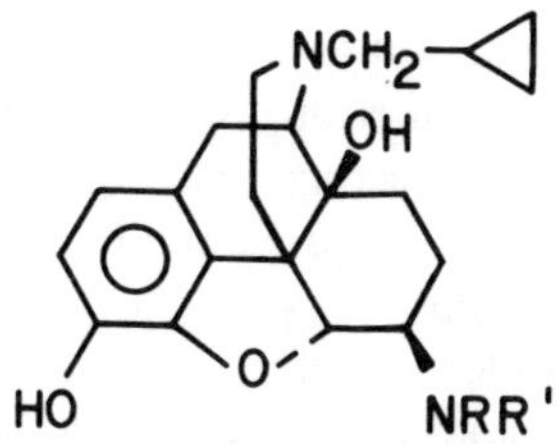

Structure 17; R=R′=CH_2CH_2Cl

Structure 18; R=H, R′=C(O)C=CCO_2Me

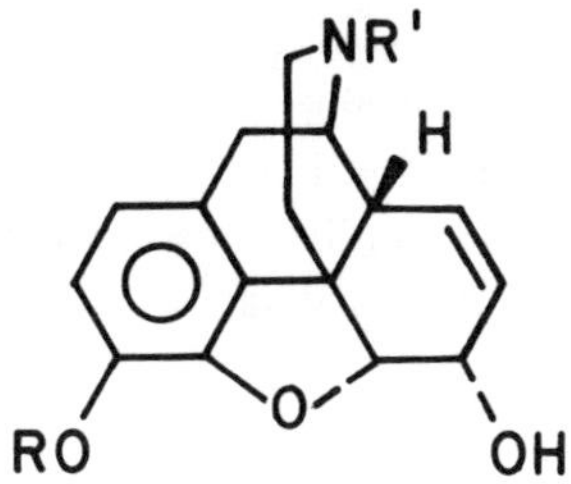

R=H	R′=Me	10%
R=glu	R′=Me	75%
R=H	R′=H	1%
R=glu	R′=H	4%

FIGURE 1. Metabolites of morphine.

All of the pentacyclics which have been discussed are related to the naturally occurring (−) isomer of morphine. The (+) isomer has been synthesized and is inactive as an antinociceptive agent.[53] Another variation on the morphine ring system, removing the ether ring, also gives potent analgesics which will be described in the section on tetracyclics.

Morphine

Pharmacokinetics[28,64-66]

Although morphine is readily absorbed from the GI tract, nasal mucosa, and lung, it is most effective when administered parenterally. An oral dose is partially metabolized in the liver and the peak plasma level is only 1/5 to 1/3 that found after parenteral administration. Peak plasma concentrations are 60μg/ℓ after 10 mg i.v. and 70 μg/ℓ 10 to 20 min after 10 mg i.m. The plasma half-life of the drug is 2 to 3 hr. Maximum analgesia occurs 20 min after an i.v. dose, 60 to 90 min after s.c., and 30 to 60 min after i.m. administration and lasts 4 to 7 hr. One third of the drug in the serum is protein bound, and free morphine leaves the blood and accumulates in parenchymatous tissues including the kidney, lung, liver, and spleen as well as in the skeletal muscle. The brain/plasma ratio of the drug is about 1; 90% of a therapeutic dose is excreted within 24 hr, 7 to 10% in the feces, and the rest in the urine: 10% as free drug, 75% as the 3-glucuronide, and 5% as normorphine or its conjugates (Figure 1). Small amounts of O-methylated metabolites have also been detected.

Uses and Dosage[67-70]

Morphine is a narcotic analgesic indicated for relief of moderate to severe pain. In addition it may be used as an adjunct to anesthesia, a nonspecific antidiarrheal agent, or as an antitussive. It not only decreases physiologically painful stimuli but also alters the psychological response to these stimuli in such a way that the individual is no longer disturbed by the pain. It also suppresses anxiety and apprehension. Although it is an effective antitussive it is not generally used as such because of erratic absorption after its oral administration and its dependence and abuse liabilities. The use of morphine as an antitussive is indicated when a cough is associated with severe pain and milder antitussives are ineffective. Morphine is an effective and fast-acting nonspecific antidiarrheal agent which acts by enhancing tone in long segments of longitudinal muscle and inhibiting propulsive contraction of both circular and longitudinal muscle.

Morphine is more reliable parenterally than orally and is generally administered by this route. Small to moderate doses should be used for constant dull pain and moderate to large doses for intermittent sharp pain of traumatic or visceral origin. Atropine sulfate may be used in combination with morphine to counteract the spasmodic effects of the latter. All doses should be administered every 4 hr as needed: i.v. 2.5 to 15 mg in 4 to 5 mℓ water slowly over 4 to 5 min, s.c. or i.m. 5 to 20 mg, pediatric s.c. 0.1 to 0.2 mg/kg with a maximum of 15 mg/dose. As an antitussive it is administered p.o. 2 to 4 mg, 3 to 4 times daily. Morphine sulfate is incompatible with i.v. fluids that contain aminophylline, sulfisoxazole diolamine, meperidine hydrochloride, and the sodium salts of amobarbital, chlorothiazide, phenytoin, heparin, methicillin, nitrofurantoin, novobiocin, phenobarbital, bicarbonate, iodide, sulfadiazine, and thiopental.

Availability

Morphine sulfate injection (various) — 8 and 10 mg/mℓ in 1-mℓ ampules and vials, 1- and 2-mℓ disposable syringes. 15 mg/mℓ as above as well as in 20-mℓ ampules and vials.

(Winthrop) 2, 4, 8, 10, and 15 mg/mℓ, 1 mℓ in 2-mℓ syringes. (Wyeth) 2 and 4 mg/mℓ in 1-mℓ syringes.

Tablets (hypodermic) (Lilly) 10 mg in 100s, 15 and 30 mg in 100s and 500s.

Oral solution (Philips Roxane) 2 mg/mℓ of 10% ethanol in 500-mℓ bottles.

With Atropine Sulfate — Injection (Lilly) 15 mg with 0.4 mg atropine sulfate per milliliter in 10-mℓ ampules.

(Beecham) 16 mg with 0.4 mg atropine sulfate per milliliter in 30-mℓ vials.

Tablets (hypodermic) (Lilly) 15 mg with 0.4 mg atropine sulfate in 100s.

Toxicity

Cerebral

Change in mood and attitude, disruption of thought processes, euphoria, drowsiness, and apathy may be observed.

Medullar

There is a dose-dependent depression of all phases of respiratory activity (rate, minute volume, and tidal exchange) caused by a reduction in responsiveness of the brain stem to CO_2 tension. A similar effect occurs during sleep and the effects are additive. The decreased respiratory volume is due mainly to slowed breathing and may remain below normal for 4 to 5 hr following medication. Maximal respiratory depression occurs 7 min after i.v., up to 30 min after i.m., and up to 90 min after s.c. doses. The increase in CO_2 tension in the blood caused by the respiratory depression causes increased cerebral blood flow and intracranial pressure.

Stimulation of the chemoreceptor trigger zone causes nausea in 40% and vomiting in 15% of ambulatory patients, generally about 10 min after injection. These effects are unusual in recumbent patients. Some patients never vomit after morphine treatment and others always do, although after a therapeutic dose subsequent doses of morphine or other emetics rarely induce vomiting. The antitussive effect of morphine is in part due to direct action on the cough center in the medulla.

Pituitary

Morphine suppresses secretion of luteinizing hormones, leading to decreased plasma concentration of testosterone and decreased sexual drive, sperm motility and ejaculate volume in males.

Other CNS Effects

Pupillary constriction (miosis) is a commonly observed side effect of opiates to which addicts do not generally become tolerant. The effect is probably due to excitatory action on the autonomic segment of the oculomotor nerve. Following morphine administration decreases in olfactory and auditory acuity, cutaneous perception, and sense of muscle tone may be observed. Lightheadedness, dizziness, and sweating are not uncommon and the extremities may feel heavy and the body warm. The face may itch, the mouth become dry, and the person become lethargic.

GI, Biliary, and Urinary Tracts

There is an increase in the activity and tone of the smooth muscles of the GI, biliary, and urinary tracts causing delayed digestion, constipation, urinary retention, biliary spasms, and an increase in biliary pressure up to 10 to 15 times normal.

Cardiovascular

Vasodilatation and peripheral pooling can cause orthostatic hypotension, lessened muscular activity, sleep, decreased metabolism, and respiratory depression.

Bronchial

Morphine administration can cause histamine release which stimulates the bronchial muscles and can precipitate an asthmatic attack.

Tolerance, Dependence, and Abuse Liability

Tolerance may develop to the effects of opiates repeatedly administered over a period of time, with larger doses being required to provide the same analgesic effect. Psychic and physical dependence which may develop, when combined with the euphorigenic effects of morphine, can result in excessive use of the drug and addiction.

Overdosage

The major toxic effect of morphine overdosage is respiratory depression which can be treated by ventilation and administration of a narcotic antagonist. Symptoms of morphine overdosage include coma, pinpoint pupils, respiratory depression, and sometimes shock, decreased body temperature, and pulmonary edema. Extremely high doses may produce convulsions.

Warnings

The use of morphine is contraindicated in patients with bronchial asthma, respiratory depression, and idiosyncrasy or hypersensitivity to the drug (as in pateints with myxedema, Addison's disease, or hepatic cirrhosis). It should be used with caution in infants, the aged, or the debilitated as well as in patients with increased intracranial pressure, head injuries, cerebral edema, or delirium tremens. It should also be used with caution in patients with possible respiratory problems, depleted blood volume, or some other deficiency in the ability to maintain blood pressure, convulsive disorders, cardiac arrythmias, chronic ulcerative colitis, impaired hepatic or renal function, toxic psychoses, prostatic hypertrophy, hypothyroidism, or urethral stricture. Since morphine crosses the placental barrier it should be used during pregnancy only if the benefits outweigh the potential risks to the mother and fetus. Newborn infants whose mothers have been using opiates should be monitored closely for signs of respiratory depression. Because of its sedative effects, patients should be warned against attempting potentially dangerous tasks after taking the drug.

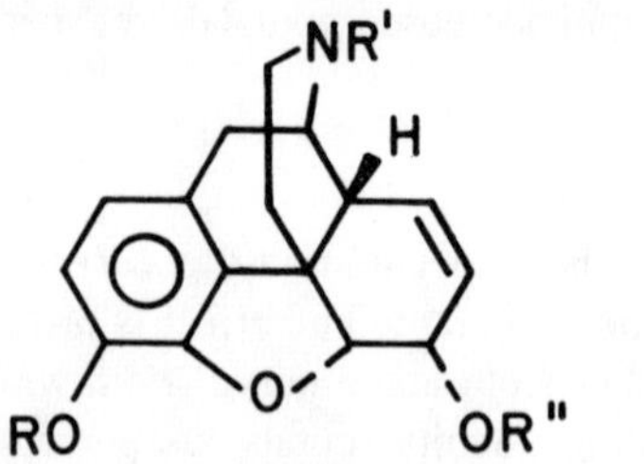

R=Me	R′=Me	R″=H	5 to 17%
R=Me	R′=Me	R″=sugar	32 to 46%
R=Me	R′=H	R″=H	Trace
R=Me	R′=H	R″=sugar	10 to 21%
R=H	R′=Me	R″=H	Trace
R=H	R′=Me	R″=sugar	5 to 13%

FIGURE 2. Metabolites of codeine.

Drug Interactions

Alcohol, antihistamines, muscle relaxants, tranquilizers, antidepressants, sedative-hypnotics, phenothiazines or other anesthetics, and other narcotic analgesics may interact with morphine to intensify overlapping actions. Phenothiazines and certain anesthetics may decrease an individual's ability to maintain blood pressure. Other classes mentioned above can intensify respiratory depression and anticholinergic effects. Anticholinergics can partially reverse biliary spasms produced by opiates, but are additive to the GI and urinary effects and can lead to severe constipation and urinary retention. MAO inhibitors may intensify the effects of morphine by decreasing its rate of metabolism.

Physical - Chemical and Other Data

The structure of morphine may be found in the pharmacokinetics section (Figure 1, R=H, R′=Me).

Base[71-74] — (5α,6α)-7,8-Didehydro-4,5-epoxy-17-methylmorphinan-3,6-diol; $C_{17}H_{19}NO_3$; air stable white rhombic prisms, fine needles, or crystalline powder; m.p. ~250°C (dec); solubility — 1 g in 5000 mℓ water, 400 mℓ boiling water, 210 mℓ ethanol, 1200 mℓ chloroform, 125 mℓ glycerol, 6500 mℓ ether; insoluble in benzene.

Sulfate — $(C_{17}H_{19}NO_3)_2 \cdot H_2SO_4 \cdot 5\ H_2O$; odorless while crystals or crystalline powder which gradually loses water of hydration in air, m.p. ~250°C (dec); α (2% solution in H_2O) −107 to −109.5°; solubility — 1 g in 16 mℓ water, 1 mℓ water at 80°C, 570 mℓ ethanol, 240 mℓ ethanol at 60°C; insoluble in chloroform or ether.

Hydrochloride — $C_{17}H_{19}NO_3 \cdot HC1 \cdot 3H_2O$; odorless, light-sensitive white crystalline powder or needle-like crystals which gradually loses water of hydration in air; α (2% solution in H_2O) −110 to −115° (based on dry salt); pKa (20°C) 8.0 (acid), 9.9 (base); solubility — 1g in 18 mℓ water (pH 5), 100 mℓ ethanol, 10 mℓ boiling ethanol, 10 mℓ glycerol; insoluble in chloroform or ether.

Tartrate — $(C_{17}H_{19}NO_3)_2 \cdot C_4H_6O_6 \cdot 3H_2O$; white crystalline powder; solubility — 1 g in 11 mℓ water, 1000 mℓ ethanol; insoluble in chloroform or ether.

Acetate — white powder; solubility — 1 g in 2.5 mℓ water, 100 mℓ ethanol, 5 mℓ glycerol; insoluble in ether.

Codeine

Pharmacokinetics[75,76]

Codeine is rapidly absorbed following either parenteral or oral administration. Peak plasma levels are 0.03 mg/ℓ 2 hr after 15 mg base p.o. and 0.11 mg/ℓ 1 hr after 60 mg sulfate p.o. The plasma half-life is 3.5 hr after oral administration and maximum analgesia occurs after

1 hr. After parenteral administration maximum analgesia occurs in 30 to 60 min and lasts 3 to 4 hr. 24 hr after a 20-mg oral dose of base the urine contains 5 to 17% free codeine, 32 to 46% codeine conjugates, a trace of norcodeine, 10 to 21% norcodeine conjugates, a trace of morphine, and 5 to 13% morphine conjugates (Figure 2).

Uses and Dosage[28,67,77]

Codeine is a weaker analgesic than morphine and is indicated for the relief of mild to moderate pain and also for use as an antitussive; 64 mg of codeine is less effective than 10 mg morphine at relieving severe pain. Although it is less addicting than other opiates, it should only be used in cases when nonaddictive analgesics are ineffective or are inadvisable for some other reason. Codeine is useful for inducing sleep in the presence of mild pain and has less psychologic effect than morphine.

Codeine is $^{2}/_{3}$ as potent orally as parenterally and may be administered by either route. It is generally given orally for the treatment of mild pain or as an antitussive. For more severe pain it may be used s.c., i.m., or by slow i.v. administration. Codeine is often used in combination with nonopiate analgesics to increase the degree of pain relief.

Doses should be administered every 4 hr as needed. As an analgesic: 30 (15 to 60) mg s.c., p.o., or i.m.; pediatric 0.5 mg/kg p.o. As an antitussive: 5 to 10 mg p.o.; pediatric 0.2 mg/kg.

Codeine phosphate injection is incompatible with aminophylline, ammonium chloride, sulfisoxazole diolamine and the sodium salts of amobarbital, chlorothiazide, heparin, methicillin, nitrofurantoin, novobiocin, pentobarbital, phenobarbital, bicarbonate, iodide, sulfadiazine, and thiopental.

Availability[78]

Codeine phosphate — Injection (various) 30 mg/mℓ in 1- and 2-mℓ ampules and 1-mℓ syringes, 60 mg/mℓ in 1-mℓ ampules. (Wyeth) 60 mg/mℓ in 1-mℓ syringes. (Winthrop) 60 mg/mℓ 1 mℓ in 2-mℓ syringes. Hypodermic tablets (Lilly) 15, 30, and 60 mg in 100s and 500s.

Codeine Sulfate — Tablets (various) 15 mg in 25s, 100s, 500s, and 1000s, 30 and 60 mg in 25s, 100s and 1000s. Hypodermic tablets (Lilly) 15, 30, and 60 mg in 100s and 500s.

Combinations — Also available are various combinations of codeine phosphate with aspirin, phenacetin, caffeine, salicylamide, and butalbital.

Toxicity

Codeine has generally milder side effects than morphine. Optimal doses produce less sedation, respiratory depression, nausea, emesis, constipation, diarrhea, mood modification, and urinary and pupillary effects than morphine, although nausea, vomiting, dizziness, and drowsiness are still common. Tolerance to codeine develops more slowly than to morphine and the drug has a lower abuse liability, but some cases of addiction have been reported. Excessive doses may cause restlessness, exhilaration, convulsions, narcosis, respiratory depression, hypotension, miosis, and tachycardia.

Warnings — Codeine is contraindicated in patients who are hypersensitive to it. Indiscriminate use may precipitate severe respiratory insufficiency due to increased viscosity of bronchial secretions and suppression of the cough reflex. It may be habit forming and appropriate precautions against abuse should be taken. Because it may cause drowsiness and dizziness patients taking the drug are advised against attempting potentially dangerous tasks.

Drug interactions — As an opiate codeine may interact with the same drugs as does morphine, but to a lesser extent. Among the compounds which may enhance its CNS depressant effects are phenothiazines, MAO inhibitors, tricyclic antidepressants, and other narcotics.

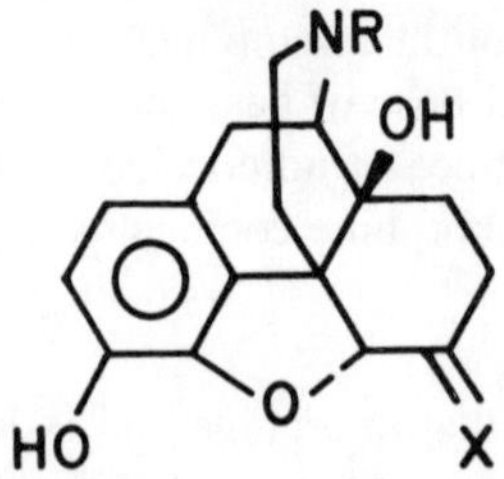

R=CH$_2$–◇ X=α-OH
R=H X=α-OH
R=CH$_2$–◇ X=O

FIGURE 3. Metabolites of nalbuphine.

Physical - Chemical and Other Data

The structure of codeine may be found in the pharmacokinetics section (Figure 2, R=R′=Me, R″=H).

Base:[69,79-83] (5α,6α)-7,8-Didehydro-4,5-epoxy-3-methoxy-17-methylmorphinan-6-ol monohydrate; $C_{18}H_{21}NO_3 \cdot H_2O$; light sensitive, white crystals or powder which effloresces slowly in dry air, m.p. (anhydrous) with a 2° range between 154 and 158°C; α (2% solution in EtOH) −142 to −146°; solubility — 1 g in 120 mℓ water (pH 9.8), 1 g in 20 mℓ boiling water, 2 mℓ ethanol, 0.5 mℓ chloroform, 50 mℓ ether, 20 mℓ benzene; also soluble in amyl alcohol, carbon disulfide, and methanol.

Phosphate: $C_{18}H_{21}NO_3 \cdot H_3PO_4 \cdot 1/2\ H_2O$; light sensitive, odorless, white crystalline powder or needle shaped crystals which effloresces in dry air; α (2% solution in H_2O) −98 to −102°; pk_a 8.2; solubility — 1 g in 2.5 mℓ water, 0.5 mℓ boiling water, 325 mℓ ethanol, 125 mℓ boiling ethanol; slightly soluble in ether or chloroform.

Sulfate: $C_{18}H_{21}NO_3 \cdot H_2SO_4 \cdot 3H_2O$; light sensitive white crystals or powder which effloresces in dry air; α (2% solution in H_2O) −112.5 to −115°; solubility — 1 g in 30 mℓ water, 6.5 mℓ water at 80°C, 1300 mℓ ethanol; insoluble in chloroform or ether.

Hydrochloride: $C_{18}H_{21}NO_3 \cdot HCl \cdot 2H_2O$; m.p. ~280°C(dec); solubility — 1 g in 30 mℓ water, 100 mℓ ethanol, 80 mℓ chloroform.

Nalbuphine Hydrochloride

Pharmacokinetics[68,84-86]

Nalbuphine is only 1/5 as potent orally as parenterally. Peak plasma levels of 0.048 mg/ℓ are observed 30 min after an i.m. dose of 10 mg. The plasma half-life of the drug is 5 hr. Analgesia is observed within 15 min of i.m. or s.c. administration and lasts for 3 to 6 hr. Following i.v. administration analgesia is observed within 2 to 3 min with the peak effect at 30 min and the duration of action 2.5 hr. The principal mode of elimination is via fecal excretion; 7% of the administered drug is found in the urine as nalbuphine and conjugates, N-dealkylated material and conjugates, and 6-oxidized material and conjugates (Figure 3).

Uses and Dosage[28,87]

Nalbuphine is a mixed narcotic agonist-antagonist analgesic, and thus is generally compared to pentazocine, the first member of this class to be successfully marketed. It is indicated for the relief of moderate to severe pain, for use in preoperative analgesia, as a supplement to surgical anesthesia, and for obstetrical analgesia during labor. As an analgesic it is approximately equipotent with morphine and 3 times as potent as pentazocine and as an antagonist 1/4 times as potent as nalorphine and 10 times as potent as pentazocine. As is generally observed with mixed narcotic agonist-antagonists the abuse potential and respiratory depression are less than for morphine. It also causes less nausea and vomiting than morphine.

Nalbuphine has been reported to exhibit low incidences of psychotomimetic effects and few adverse cardiac effects in patients with coronary artery disease.

Nalbuphine (10 mg) may be administered s.c., i.m., or i.v. every 3 to 6 hr as needed. The maximum recommended single dose in nontolerant patients is 20 mg. The injection is incompatible with diazepam or promethazine hydrochloride.

Availability

Hydrochloride injection — Nubain (Endo) 10 mg/mℓ with 0.2% of a 9:1 mixture of methyl and propyl parabens in 1- and 2-mℓ ampules and 10-mℓ vials.

Toxicity

The most commonly observed side effects following a therapeutic dose of nalbuphine are sedation 36%, sweating 9%, nausea and vomiting 6%, dizziness and vertigo 5%, dry mouth 4%, and headache 3%. A ceiling effect for respiratory depression is seen so that as the dosage is increased past 30 mg no further respiratory depression is observed. GI effects are mild and abuse potential is low. In contrast to butorphanol it did not increase cardiac work in patients with stable coronary artery disease. At high doses it is identified more as a barbiturate than as an opiate and may be dysphoric.

Warnings — Nalbuphine is contraindicated in patients hypersensitive to the drug. It does have abuse potential, although lower than that of morphine, and should be used cautiously in emotionally unstable patients or those with a history of drug abuse. It should also be used with care in patients whose respiratory function is compromised, or in those with renal or hepatic dysfunction, head injury or increased intracranial pressure, in patients undergoing biliary surgery or with myocardial infarction who exhibit nausea and vomiting, in pregnant patients, and in women delivering premature infants. There is insufficient data to warrant its use in patients under 18 years old. Because of its sedative effects, patients should be warned against attempting potentially hazardous tasks after taking the drug. Nalbuphine may precipitate withdrawal in narcotic dependent subjects.

Drug interactions — As is generally observed with narcotic analgesics its effects may be additive to those of other CNS depressants: tranquilizers such as phenothiazines, narcotic analgesics, sedatives, hypnotics, and general anesthetics.

Physical - Chemical and Other Data

The structure of nalbuphine may be found in the pharmacokinetics section (Figure 3, R=cyclobutylmethyl, X=$\alpha-$OH). (5α, 6α)-17-(Cyclobutylmethyl)-4,5-epoxymorphinan-3,6,14-triol hydrochloride; $C_{21}H_{27}NO_4 \cdot HCl$; Nubain (Endo); white to slightly off-white powder; pKa 8.71 (acid), 9.96 (base); solubility — soluble in water, slightly soluble in ethanol.

Naloxone

Pharmacokinetics[4,68,88-91]

Naloxone is only 1/50 as active orally as parenterally. Peak plasma levels of 0.01 mg/ℓ were observed 2 min after administration of 0.4 mg i.v. and the serum half-life of the drug was 64 min after the initial distribution phase. Its antagonist action began within 2 min. After s.c. or i.m. administration of this dose the antagonist action began within 5 min and lasted 1 to 4 hr; 50% of administered naloxone is excreted in the urine within 24 hr as naloxone and its conjugates, dihydronaloxone and conjugates, N-dealkylated material and conjugates, and aromatic alkoxylated and hydroxylated drug and conjugates (Figure 4).

Uses and Dosage[28,67,69,92]

Naloxone is a pure narcotic antagonist seven times more potent than nalorphine. It is indicated for complete or partial reversal of narcotic depression, including respiratory depres-

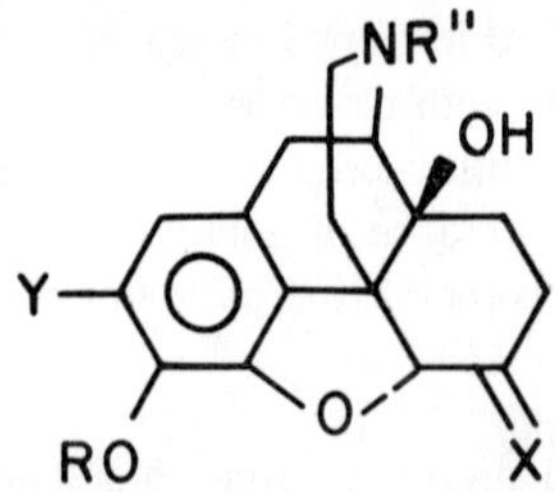

R=H	Y=H	R″=CH_2CH=CH_2	X=O
R=H	Y=H	R″=CH_2CH=CH_2	X=OH,H
R=H	Y=H	R″=H	X=O
R=Me	Y=OH	R″=CH_2CH=CH_2	X=O
R=H	Y=OMe	R″=CH_2CH=CH_2	X=O

FIGURE 4. Metabolites of naloxone.

sion, and diagnosis of suspected acute opiate overdose or narcotic addition. Naloxone is essentially inactive as an analgesic, but when it is administered prior to a narcotic it can block the latter's action. For example, 1 mg of naloxone administered i.v. blocks the effects of 25 mg of heroin. It will also reverse the actions of a narcotic when administered after the narcotic and will rapidly precipitate a severe withdrawal syndrome when given to an individual physically dependent on narcotics. No tolerance to its antagonist effects or physical or psychological dependence develops on long-term administration. When co-administered with an opiate it can, in low doses, alleviate postoperative respiratory depression while still allowing significant analgesia. Large (10 to 15 mg) doses are required to reverse the dysphoric and psychotomimetic effects of mixed agonist-antagonist analgesics. The drug is short acting and repeated administration may be required to sustain its effects. It is being investigated for its effects on appetite regulation,[93] psychiatric disorders,[94] postmenopausal flushing,[95] alcohol intoxication,[96] narcotic addiction,[97] and shock.[98,99]

Because naloxone is relatively short acting, the patient should be observed constantly to be sure that the effect of the drug has not worn off. For treatment of narcotic overdose 0.4 mg is administered i.v., i.m., or s.c. at 2 to 3 min intervals until the desired response is achieved. If after 2 to 3 doses no response is seen, narcotic overdosage is probably not the problem. After the desired response is achieved repeat doses may be given at 1 to 2 hr intervals as needed. The pediatric dose is 0.01 mg/kg administered as above. For reversal of postoperative respiratory depression 0.1 to 0.2 mg naloxone is administered i.v. at 2 to 3 min intervals until the desired effect is achieved. The dose may be repeated at 1 to 2 hr intervals as needed. The pediatric dose is 0.01 mg/kg administered in the same fashion.

Naloxone hydrochloride injection should not be mixed with preparations containing bisulfite, metabisulfite, long-chain or high-molecular-weight anions, or any alkaline solution.

Availability

Naloxone hydrochloride injection — Narcan (Endo) 0.4 mg/mℓ in 1-mℓ ampules and 10 mℓ vials. Neonatal — 0.02 mg/mℓ in 2-mℓ ampules.

Toxicity

Very few side effects have been observed with naloxone. At the large dose of 24 mg only slight drowsiness was observed. Nausea and vomiting may also occur.

Warnings — Naloxone is contraindicated in patients who are hypersensitive to it. It should be administered cautiously to patients dependent on narcotics since withdrawal symptoms may be severe. Abrupt reversal of narcotic depression may produce nausea, vomiting, sweating, tachycardia, increased blood pressure, trembling, and rarely, heart failure and pulmonary edema. Safety in children under 12 or during pregnancy has not been established. The drug should be administered cautiously in patients with cardiac irritability.

Physical - Chemical and Other Data

The structure of naloxone may be found in the pharmacokinetics section (Figure 4, R=Y=H, R″=CH_2CH=CH_2,X=O)

Base:[100] 5α-4,5-Epoxy-3,14-dihydroxy-17-(2-propenyl)-morphinan-6-one; $C_{19}H_{21}NO_4$; m.p. 177 to 180°C, solubility — soluble in chloroform, insoluble in ether.

Hydrochloride: $C_{19}H_{21}NO_4 \cdot HCl$; Narcan (Endo); white to off-white crystalline powder; m.p. 200 to 205°C; pKa 7.94 (acid); solubility — soluble in water, slightly soluble in ethanol, insoluble in ether or chloroform.

TETRACYCLICS

In the late 1940s Grewe reported the first synthesis of a tetracyclic compound lacking only the ether bridge of the morphine family.[101] Grewe as well as Schneider and co-workers soon applied this chemistry to a variety of such morphinans, and significant analgesic activity was found among these compounds.[52] The compounds were synthesized as racemates, and upon resolution it was found that most of the analgesic activity resided in the *l*-isomers, which had the same absolute stereochemistry as morphine. The *d*-isomers generally had little or no analgesic activity but often retained the antitussive activity of the opiates. For example levorphanol (Structure 19) has been used clinically and has an analgesic activity four times that of morphine, whereas the methyl ether of its enantiomer, dextromethorphan (Structure 20), is the most commonly used antitussive agent on the market. Somewhat surprisingly, the unnatural B,C-*trans* ring structure of the isomorphinans often retained significant analgesic activity which resided predominantly in the *l*-isomer. Indeed Compound 21 is a more potent analgesic than levorphanol (Structure 19).

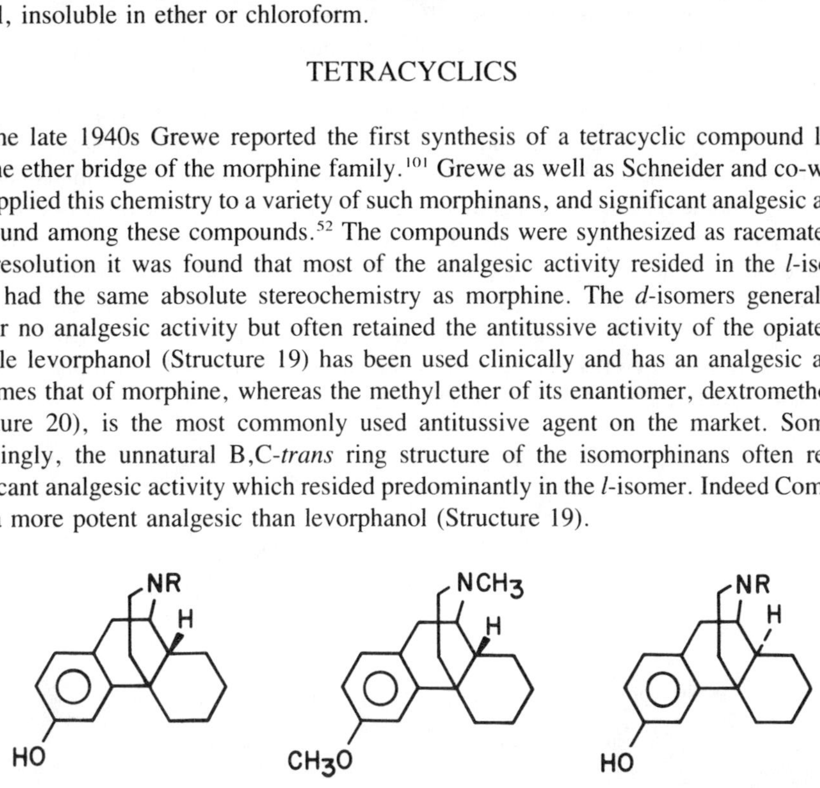

Structure 19; (−); R=Me   Structure 20; (+)   Structure 21; (−); R=Me

Many variations have been made in the A-ring of the morphinans.[52,102] The ordering of A-ring substituted *N*-methylmorphinans in terms of antinociceptive potency is 3–OH>3,4–$(OMe)_2$>3–OAc≥3–OMe≈4–OMe≥4–OH≥4–OAc>H and 4–OH>1–OH>2–OH>2–OMe. Even the unsubstituted A-ring derivative is one fifth as potent as morphine, but in the isomorphinan series loss of the A-ring hydroxy group eliminates antinociceptive activity. Introduction of a 2-methyl substituent onto levorphanol (Structure 19) also abolishes activity.

When changes are made in the C-ring as well as the A-ring the ordering in terms of potency may change.[102,103] For the 6-keto derivatives the ordering is 4–OMe≈3,4–$(OMe)_2$>4–OAc>4–OH. Besides being considerably more potent than the 4–OH derivative the 4–OMe derivative 22 has about six times the potency of morphine. Compound 23, which contains a 2–OH substituent has been studied clinically as a mixed agonist-antagonist analgesic.[104] Thus, it is clear that an A-ring 3-hydroxy group is not a prerequisite for analgesic activity.

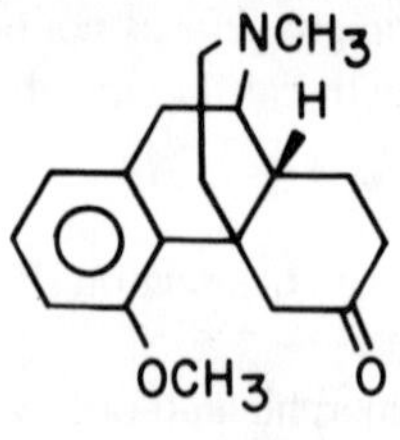

Structure 22

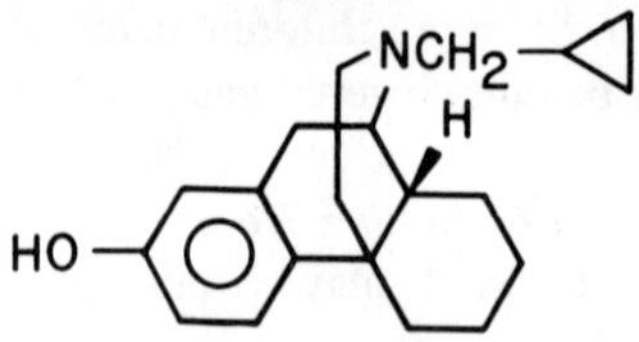

Structure 23; (−)

Substituents on the nitrogen of the morphinans generally had parallel effects on activity to the same variation in the pentacyclics.[4,52,53,55] *N*-Methyl compounds were generally more potent antinociceptive agents than *N*-H or *N*-ethyl, but about equipotent to *N*-amyl or *N*-hexyl and less so than *N*-phenethyl or related compounds. As for the pentacyclics, a number of groups intermediate in size between ethyl and amyl provided some degree of narcotic antagonism. *N*-Allyl substitution generated levallorphan (Structure 24), an antagonist twice as potent as nalorphine. *N*-Cyclopropylmethyl substitution provided the mixed agonist-antagonist cyclorphan (Structure 25) and in the B,C-*trans* series the antagonist isocyclorphan (Structure 26). The potency enhancing effect of a 14-hydroxy substituent also parallelled the pentacyclics.[105] In this way the potent antagonist oxilorphan (Structure 27) and the clinically useful mixed agonist-antagonist butorphanol (Structure 28) were generated. Workers at Bristol have shown that compounds possessing a heterocyclic C-ring can also be useful analgesics. One of the compounds generated in this way is the potent agonist-antagonist proxorphan (Structure 29).[106]

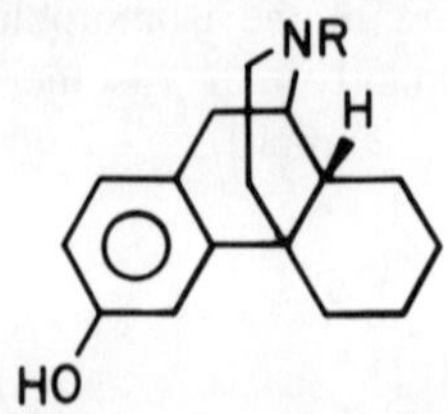

Structure 24; R=$CH_2CH{=}CH_2$

Structure 25; R=CH_2 —◁

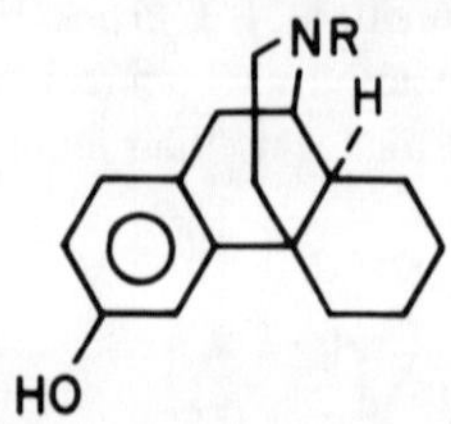

Structure 26; R=CH_2 —◁

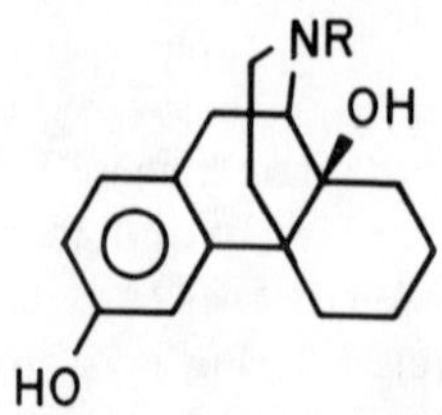

Structure 27; R=CH_2 —◁

Structure 28; R=CH_2 —◇

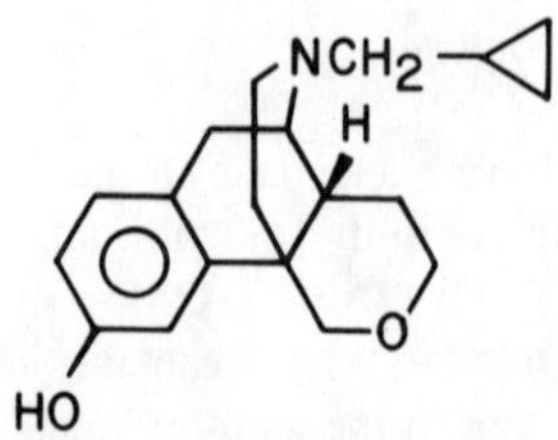

Structure 29

A few variations in the D-ring of the morphinans provided interesting antinociceptive agents. The 9-aza-compound (Structure 30) is a mixed agonist-antagonist[107] The homomorphinan (Structure 31) is as potent an antinociceptive agent as morphine,[108] but the five-membered D-ring analogue is inactive.[109]

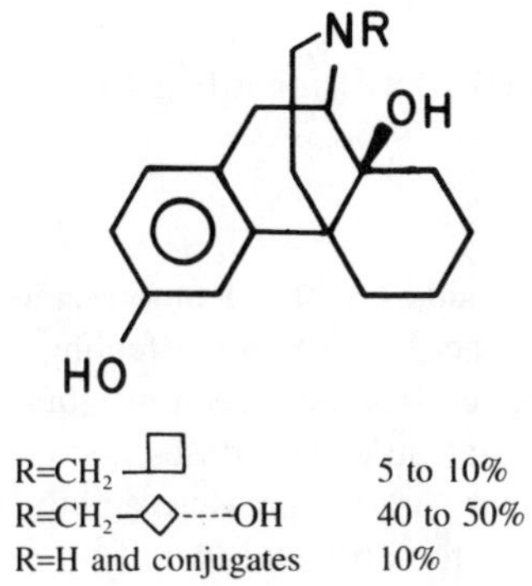

FIGURE 5. Metabolites of butorphanol.

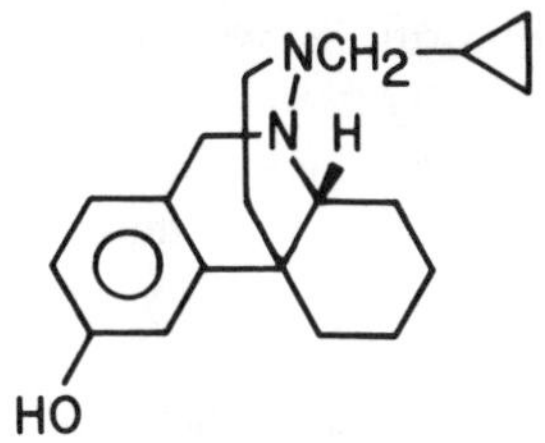

Structure 30; (−)

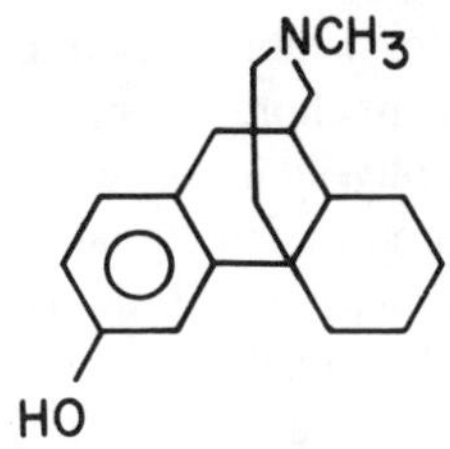

Structure 31

Butorphanol Tartrate

Pharmacokinetics[68,110-112]

Butorphanol is completely absorbed from the GI tract and i.m. injection sites, but because of extensive first pass metabolism by the liver only 17% of an oral dose reaches the circulatory system unchanged. Peak plasma levels of 0.7μg/ℓ 1 to 1.5 hr after 8 mg p.o., 2.2 μg/ℓ 30 to 60 min after 2 mg i.m., and 1.5 μg/ℓ virtually immediately after 1 mg i.v. are observed; 83% of the drug in the plasma is protein bound. The plasma half-life of the drug is about 3 hr. The drug concentrates in excretory organs and lipid-rich tissues. Onset of analgesia is observed within 30 min of an i.m. dose and 1 min of an i.v. dose. Peak analgesic effect is seen 1 hr after an i.m. dose, less than 30 min after an i.v. dose, and 1 to 1.5 hr after an oral dose, and analgesia lasts 3 to 4 hr, 2 to 4 hr, and 5 to 6 hr, respectively; 11 to 14% of a parenteral dose of butorphanol appears in the feces and 62 to 75% in the urine: 5 to 10% as butorphanol, 40 to 50% as hydroxybutorphanol, and 10% as N-dealkylated material and conjugates (Figure 5).

Uses and Dosage[4,28,69,113-115]

Butorphanol is a mixed narcotic agonist-antagonist analgesic which is 3 to 8 times as potent as morphine as an analgesic and 1/40 as potent as naloxone as an antagonist. Butorphanol is indicated for the relief of moderate to severe pain, particularly postoperatively, as well as for preoperative or preanesthetic medication, a supplement to balanced anesthesia, and relief of prepartum pain. Its onset, peak, and duration of action are similar to morphine's, but its withdrawal syndrome is slower in onset and milder than that of the latter drug. The increase in respiratory depression produced by butorphanol at high doses is less pronounced than that produced by morphine. The drug has provided good to excellent pain relief for up to 8 months in cancer patients and patients with orthopedic or neural problems. It is also a potent antitussive.

Butorphanol may be administered every 3 to 4 hr as needed in 2 mg (1 to 4 mg) i.m. or 1 mg (0.5 to 2 mg) i.v. doses. The injection is incompatible with atropine sulfate, hydroxyzine hydrochloride, or promethazine hydrochloride.

Availability

Tartrate injection — Stadol (Bristol) 1 mg/mℓ in 1 mℓ vials, 2 mg/mℓ in 1-, 2-,and 10-mℓ vials and 1-mℓ syringes

Toxicity

The most commonly observed side effects of butorphanol treatment are sedation 40%, sweating 6%, nausea 6%, headache 3%, a floating feeling 3%, dizziness and vertigo 2%, and lethargy 2%. Psychotomimetic effects are seen infrequently. Although the intensity of respiratory depression reaches a plateau at high doses, its duration increases with increasing dose. The drug increases pulmonary arterial pressure and the work of the heart. Postaddicts are indifferent to or mildly dislike the drug and identify it more frequently as a barbiturate than as an opiate. Overdosage of butorphanol may produce respiratory depression and cardiovascular and CNS effects. It should be reversed in the usual way by treatment with naloxone, ventilation, and appropriate support measures.

Warnings — Butorphanol is contraindicated in patients hypersensitive to the drug. It is generally not the drug of choice for patients with heart problems. It should not be administered to patients who are physically dependent on other narcotics since it may precipitate withdrawal. Administration to emotionally unstable individuals or those prone to drug abuse should be closely monitored. It should be used cautiously in patients with head injury, respiratory problems, or renal or hepatic dysfunction. Safe use in children or nursing mothers, or prior to biliary tract surgery has not been established. Because of its sedative effects, patients should be warned against potentially hazardous tasks after taking the drug.

Drug interactions — As is generally observed for narcotic analgesics, the CNS depressant effects are additive to such classes as general anesthetics, phenothiazines or other tranquilizers, sedatives, hypnotics, or alcohol. Concomitant administration with pancuronium may cause an increase in conjunctival effects.

Physical - Chemical and Other Data

The structure of butorphanol is given in the pharmacokinetics section (Figure 5, R=cyclobutylmethyl). (−)-17-(Cyclobutylmethyl)morphinan-3,14-diol[S-(R*,R*)]2,3-dihydroxybutanedioate (1:1) (salt); $C_{21}H_{29}NO_2 \cdot C_4H_6O_6$; Stadol (Bristol); white crystalline powder; pKa 8.6 (acid); solubility — slightly soluble in water, insoluble in ethanol.

Dextromethorphan Hydrobromide

Pharmacokinetics[116,117]

Dextromethorphan is orally active and is metabolized very rapidly with <2μg/ℓ free drug being present in the plasma. However, its antitussive effect persists for up to 12 hr leading to speculation that a metabolite may be responsible for its action. Indeed its major metabolite, dextrorphan, exhibits antitussive activity in dogs. Up to 56% of a 30-mg oral dose is excreted in the urine within 6 hr and excretion is essentially complete after 48 hr. Less than 1% unchanged drug, 46% O-demethylation to dextrorphan and conjugates, 34% N,O-demethylation and conjugates, and a small amount of N-demethylated material are found (Figure 6).

Uses and Dosage[28,68,77,118,119]

Dextromethorphan is a nonnarcotic antitussive agent 0.5 times as active as codeine and which is sold over the counter. It is indicated for the treatment of chronic unproductive cough. It is devoid of analgesic properties and has very few side effects.

Dextromethorphan should be administered orally: 10 to 20 mg every 4 hr or 30 mg every 6 to 8 hr. Dosage should be cut in half for children 6 to 12 and in one quarter and only under medical supervision for children under 6. The drug is incompatible with penicillins,

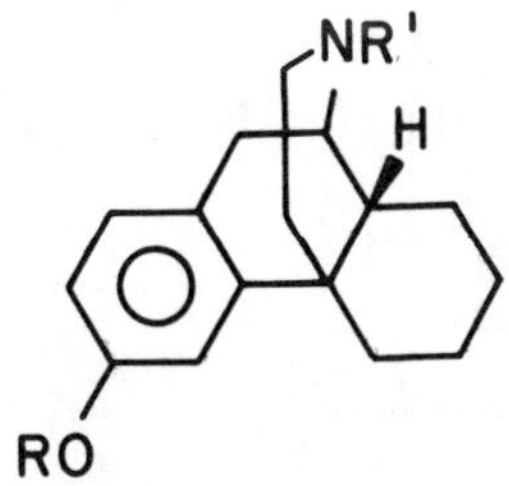

R,R′=Me	<1%
R=H,R′=Me	46%
R=H,R′=H	34%

FIGURE 6. Metabolites of dextromethorphan.

tetracyclines, salicylates, phenobarbital sodium, hydriodic acid, and sodium or potassium iodide.

Availability

Capsules — Stop-Koff (Stan Labs) 15 mg + 3 mg benzocaine in 30s.

Lozenges — Chloraseptic DM Couth Control Lozenges (Eaton) 10 mg with phenol and sodium phenolate in 10s. Spec-T Sore Throat/Cough Suppressant Lozenges (Squibb) 10 mg with 10 mg benzocaine in 10s. Formula 44 Cough Control Discs (Vick Chemical) 5 mg with 1.25 mg benzocaine and menthol, anethole, and peppermint oil in 24s. Sucrets® Cough Control Lozenges (Beecham) 7.5 mg with 3.75 mg benzocaine in 24s. Hold 4 Hour Adult Formula Lozenges (Beecham) 7.5 mg with 3.75 mg benzocaine in 10s.

Syrups and Liquids — Romilar CF 8 Hour Cough Formula Syrup (Block) 3 mg/mℓ in 20% ethanol in 52.5, 88, and 150 mℓ. Romilar Children's Cough Syrup (Block) 0.5 mg/mℓ in 90 and 180 mℓ. Benylin DM Cough Syrup (Parke-Davis), Symptom 1 Liquid (Parke-Davis) 2 mg/mℓ in 5% ethanol in 120 mℓ. Pertussin 8 Hour Cough Formula Syrup (Chesebrough-Pond's) 1.5 mg/mℓ in 95% ethanol in 90 and 180 mℓ. Congesprin Syrup (Bristol-Myers) 1 mg/mℓ in 90 mℓ. Robitussin-DM® Syrup (A. H. Robins) 3 mg/mℓ with 20 mg/mℓ guaifenesin in 120, 240, 480, and 3840 mℓ. Bayer Cough Syrup for Children (Glenbrook) 1.5 mg/mℓ with 1.8 mg/mℓ phenylpropanolamine hydrochloride in 90 mℓ.

Toxicity[120]

Slight drowsiness or GI upset may be observed after taking the drug. It causes little or no CNS depression and has no addiction liability. No tolerance develops to its effects. Large overdosage may cause stupor and ataxia which are rapidly reversed after vomiting.

Warnings — The drug is contraindicated in patients hypersensitive to it and should not be administered to patients taking MAO inhibitors or to children under 6 except under medical supervision.

Physical - Chemical and Other Data[121-124]

The structure of dextromethorphan may be found in the pharmacokinetics section (Figure 6, R=R′=Me). (9α, 13α, 14α)-3-Methoxy-17-methylmorphinan hydrobromide monohydrate; $C_{18}H_{25}NO \cdot HBr \cdot H_2O$; off-white to white bitter tasting crystalline powder with faint odor; mp ~126°C(dec); $\alpha(CHCl_3)$ +64.3 to +66.3°; α(2% solution in 1 *M* HCl) +28 to +30°; pKa 8.3; solubility — 1 g in 65 mℓ water (pH of 5.2 to 6.5), 1 g in 10 mℓ ethanol; soluble in chloroform and insoluble in ether.

TRICYCLICS

The synthesis and biological evaluation of tricyclic analogues of morphine has been

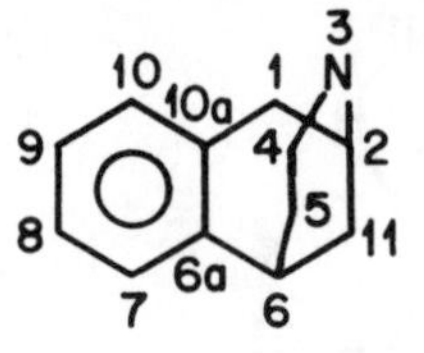

2,6-Methano-3-benzazocine

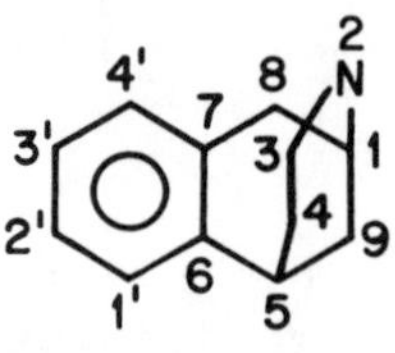

6,7-Benzomorphan

FIGURE 7.

pursued very strongly since the late 1950s,[52] in part due to the synthetic accessibility of these compounds officially known as 2,6-methano-3-benzazocines, and formerly referred to as 6,7-benzomorphans. The numbering system of the benzazocines shall be used here, but that of the benzomorphans is also shown in Figure 7.

Much of the work on these compounds has been done with 6,11α-dimethyl benzazocines. In this series substituents at the 8-position generally show the following order of antinociceptive activity: OAc>OH>OMe>H> halo.[52,125] Also fairly well studied is the 6,11α-diethyl series for which the order of activity for 8-substituents is OAc≈OH≈H>NO_2>NH_2. 8–OH substituted compounds, as in the previously discussed more complex ring systems, are thus among the most potent, and most of the variations which have been made on the benzazocines have retained this substituent. However, benzazocines also differ from the higher order ring systems in that the (+)-isomers have been reported to retain significant analgesic activity in certain cases, although most of the activity resides as expected in the (−)-isomers.

In general, N-substitution of the benzazocines gave similar results to those previously discussed for more complex structures.[52,55,126] The variation in antinociceptive activity as R is varied in Compound 32 is Et,Pr,Bu≪Me(metazocine) and amyl<phenethyl (phenazocine) and 2-thienylethyl.

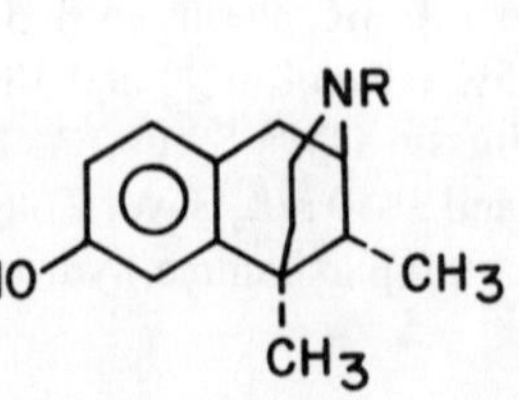

Structure 32

Phenazocine was marketed as an analgesic significantly more potent than morphine, but it retained significant addiction liability. Significant narcotic antagonist activity was observed for Compound 32 when R=allyl (SKF 10,047), propyl, propargyl, cyclopropylmethyl (cyclazocine), cyclopropylethyl, cyclobutylmethyl, cyclopentylmethyl, furanylmethyl, and other groups of similar bulk.[4,127] Cyclazocine is an effective analgesic in man as well as a potent narcotic antagonist. Although pentazocine (Compound 32, R=$CH_2CH{=}CMe_2$) is less potent than cyclazocine both as an analgesic and an antagonist, its milder spectrum of side effects allowed it to be marketed as the first of the agonist-antagonist analgesics. These compounds in general have less severe side effects than morphine and in particular much lower addiction liability.

Although most of the compounds so far described were 6,11α-dimethyl substituted, other ligands have also been attached at these positions.[52] For 6-methyl substitution the antinociceptive potency order for 11-substituents is β-Et≈β-Me>α-Et>α-Me>H. For 6-ethyl substitution the order for 11-substituents is β-Me>β-Et>H>α-Et≈α-Me. For 6-propyl substitution the order for 11-substituents is β-Pr>H>α-Me>α-Pr. Thus, it can be seen that

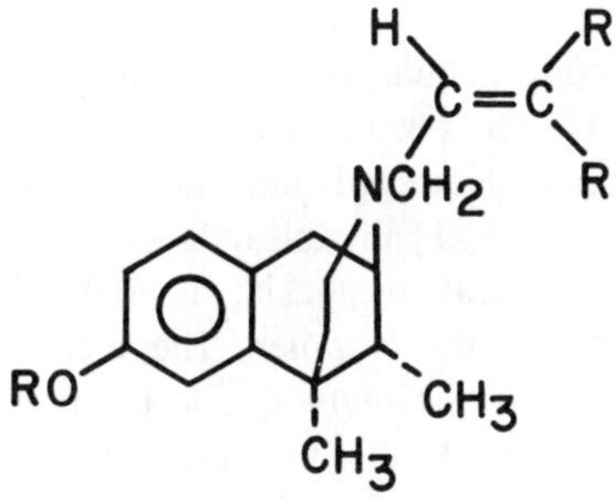

R=H or conjugates

R'=CH_3	R"=CH_3	16%
R'–CH_3	R'=CH_2OH	11%
R'=CO_2H	R"=CH_3	39%

FIGURE 8. Metabolites of pentazocine.

as in the isomorphinans (*vide supra*), 11β substituted compounds may be more potent than the ''morphine-like'' 11α-compounds and often considerably more potent than morphine.

Introduction of an 11β–OH onto the structure of pentazocine or cyclazocine increased antagonist and decreased agonist potency.[128] Dimethyl substitution at the 11-position has been reported to give more potent and longer acting compounds.[129] A particularly interesting result was noted when a substituted alkyl chain was attached at the 11β-position. Not only could antinociceptive potency be increased, but also narcotic antagonist activity could be obtained in *N*-methyl substituted compounds.[130]

Many other variations have been made on the benzazocine structure. Replacement of the aromatic ring by heteroaromatics generally decreased activity.[131] The B-ring of the hydronapthalene system has been increased or decreased in size in several ways, but no clearcut advantage has been demonstrated.[132,133] Several compounds have been synthesized which appear to act primarily at a receptor (κ) different from the classical morphine (μ) receptor. Two such compounds are ketazocine (Structure 33) and bremazocine (Structure 34).[134,135] Whether such compounds have advantages over classical analgesics remains to be seen.

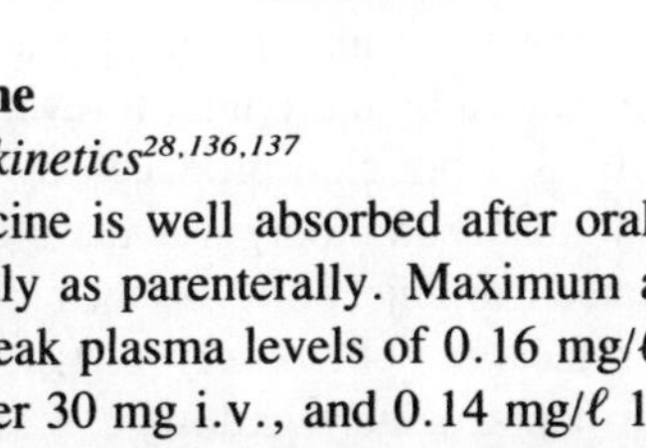

Structure 33

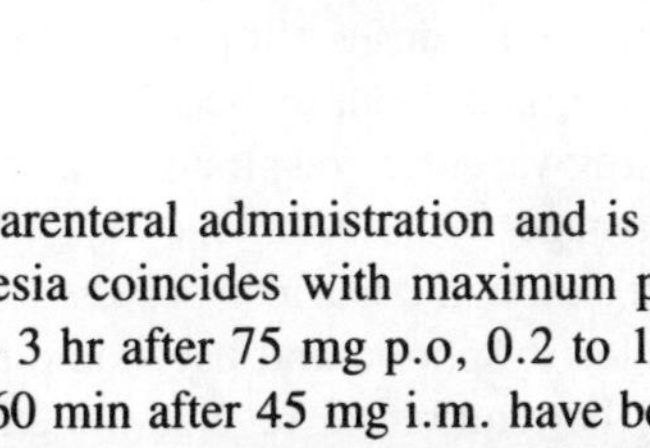

Structure 34

Pentazocine

Pharmacokinetics[28,136,137]

Pentazocine is well absorbed after oral or parenteral administration and is about 1/3 as potent orally as parenterally. Maximum analgesia coincides with maximum plasma levels of drug. Peak plasma levels of 0.16 mg/ℓ 2 to 3 hr after 75 mg p.o, 0.2 to 1.0 mg/ℓ 5 to 10 min after 30 mg i.v., and 0.14 mg/ℓ 15 to 60 min after 45 mg i.m. have been detected. The brain to plasma ratio of the drug is about 4. The plasma half-life of pentazocine is 2 to 3 hr after a p.o. or i.m. dose. Analgesia is usually experienced within 15 to 30 min after p.o., 15 to 20 minutes after i.m. or s.c., and 2 to 3 min after i.v. administration, and generally lasts for at least 3 hr. It is excreted mainly in the urine and 24 hr after a 56.5 mg dose 3% is excreted unchanged, 13% as conjugates, 11% hydroxylated on the *N*-alkyl chain and conjugates, and 39% as a carboxylic acid side chain and conjugates (Figure 8).

Uses and Dosage[67,69,138]

Pentazocine is a mixed narcotic agonist/antagonist analgesic having 1/3 the activity of morphine as an analgesic and 1/50 that of nalorphine as an antagonist. Orally it is approximately as potent an analgesic as codeine. It is indicated for the relief of moderate to severe pain, particularly postoperative pain and pain during labor, as a preoperative or preanesthetic medication, and for support of surgical anesthesia. Its analgesic and respiratory depressant activities are due predominantly to the *l*-isomer. The increase in respiratory depression produced at high doses is less than that of morphine because of a ceiling effect, so that increasing doses of pentazocine do not further increase respiratory depression. It also causes less elevation of biliary pressure. Pentazocine has a lower abuse potential than morphine but cases of psychological and physical dependence have been reported.

Pentazocine may be administered by p.o., i.v., i.m., or s.c. routes. Orally it is generally administered as the hydrochloride 50 mg every 3 to 4 hr with maximum single and daily doses in nontolerants of 100 and 600 mg, respectively. Parenterally it is usually administered as the lactate 30 mg (20 to 60 mg) every 3 to 4 hr with maximum single doses of 30 mg i.v. and 60 mg i.m. or s.c. and a maximum daily dose of 360 mg. For obstetrics, recommended doses are 30 mg i.m. during labor or 20 mg i.v. after contractions become regular, every 2 to 3 hours as needed. Solutions of pentazocine are incompatible with barbiturate salts. When antiinflammatory or antipyretic action is desired pentazocine may be co-administered with aspirin.

Availability

Oral (Hydrochloride)

Tablets — Talwin (Winthrop) 50 mg in 100s.

Injection (Lactate)

Talwin (Winthrop) 30 mg/mℓ in 1-, 1.5-, and 2-mℓ ampules and 10-mℓ vials, 30, 45, and 60 mg/mℓ, 1 mℓ in 2-mℓ syringes.

Combinations (Hydrochloride)

Tablets — Talwin Compound Caplets (Winthrop) 12.5 mg with 325 mg aspirin in 100s. Talacen (Winthrop) 25 mg with 650 mg APAP.

Toxicity[68,139]

The most commonly observed side effects of pentazocine treatment are sedation, nausea, vomiting, dizziness, light headedness, sweating, euphoria, and headache. Less frequently observed are constipation, dysphoria, weakness, disturbed dreams, insomnia, syncope, visual blurring and focusing difficulty, depression, flushing, rash, hypotension, and tachycardia. Intravenously, the drug increases systemic and pulmonary arterial pressure and vascular resistance in patients with coronary artery disease. It may increase blood pressure and heart rate especially at high doses. In patients tolerant to opiates it may precipitate a mild withdrawal syndrome, and it can reduce the analgesia produced by morphine. It reverses incompletely the cardiovascular, respiratory, and behavioral depression produced by opiates. It may cause damage to tissue at the injection site, most commonly via the s.c. route of administration although it has also occurred in repeated i.m. injections. As the recommended dose is exceeded dysphoric and psychotomimetic effects may be observed. Large doses of naloxone may be required to reverse the effects of pentazocine overdosage. Recently pentazocine has been abused most frequently in combination with the antihistamine tripelennamine.

Warnings — Pentazocine is contraindicated in patients hypersensitive to the drug. It should be used very cautiously in heart patients, those with head injury or preexisting increase in intracranial pressure, patients with severe respiratory problems, women delivering premature infants and in patients with impaired renal or hepatic function. It should also be used cautiously in patients dependent on narcotics or prone to drug dependence or abuse, and in those who are emotionally unstable or who have experienced psychotomimetic effects with the drug previously. It should be administered with care in patients prone to seizures. As with all opiates it should be avoided if possible during pregnancy. Safe use in children under 12 has not been established. To reduce the chance of tissue damage, the injection site should be rotated. Because of the dizziness and sedation experienced when using the drug, patients should be cautioned against attempting potentially hazardous tasks following its administration.

Drug interactions — The actions of pentazocine are additive to, or may potentiate, the effects of other CNS depressants such as anesthetics, sedatives, antihistamines, tranquilizers, and alcohol.

Physical - Chemical and Other Data

The structure of pentazocine is shown in the pharmacokinetics section (Figure 8, R=H, R′=R″=CH_3).

Base:[140] (2α,6α,11R*)-1,2,3,4,5,6-Hexahydro-6,11-dimethyl-3-(3-methyl-2-butenyl)-2,6-methano-3-benzazocin-8-ol; $C_{19}H_{27}NO$; Talwin (Winthrop); pale tan to white odorless crystalline solid stable to light and air; m.p. between 147 and 158°C; solubility — 1 g in 11 mℓ ethanol, 2 mℓ chloroform, 42 mℓ ether; insoluble in water, slightly soluble in benzene and ethyl acetate, and soluble in acetone.

Hydrochloride:[141] $C_{19}H_{27}NO \cdot HCl$; Polymorphic-one form m.p. ~254°C, another at 218°C; solubility — 1 g in 30 mℓ water, 16 mℓ ethanol, 4 mℓ chloroform; insoluble in ether.

Lactate:[142,143] $C_{19}H_{27}NO \cdot C_3H_6O_3$; pK_a 8.5 (acid), 10 (base); solubility — 1 g in 25 mℓ water, 12 mℓ ethanol, 25 mℓ chloroform, and 1500 mℓ ether.

ARYLPIPERIDINES AND RELATED COMPOUNDS

The first totally synthetic morphine-like analgesic agents retained only two rings from the pentacyclic morphine structure. The syntheses of these compounds began in the late 1930s with the discovery of the now widely used analgesic meperidine (Structure 35, R=Me), first reported by Eisleb in 1939.[144,145] Although this compound was synthesized in Germany as a potential antispasmodic, it was soon discovered to possess potent analgesic activity. Despite the many analogues which have been generated in the subsequent years,[146] meperidine remains the most widely accepted compound of this series for the treatment of moderate to severe pain.

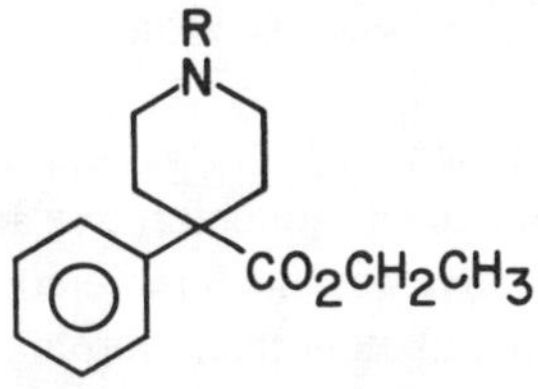

Structure 35

The *N*-methyl group of meperidine was replaced by a large variety of groups in the hope of improving the efficacy within the series.[52] Although analgesic potency many times that of meperidine could be obtained, side effects generally increased correspondingly. As in the more complex structures previously described, the potency order for N-substitution in Structure 35 is H (normeperidine) and lower alkyl <Me (meperidine). As the size of the alkyl group is increased further, potency returns and may be increased.[147] Thus Structure 35 (R=$CH_3(CH_2)_5$) has seven times the potency of meperidine. Anileridine (Structure 35

$R{=}CH_2CH_2{-}4NH_2C_6H_4$) and piminodine (Structure 35, $R{=}(CH_2)_3NH\phi$) are potent analgesics which were marketed at one time. In a departure from previously discussed classes, N-substitution of Structure 35 by allyl or cyclopropylmethyl did not provide narcotic antagonist activity. However antagonists may be obtained when other changes are also made in the molecule (vide infra).

Several variations have been made in the ester portion of the meperidine molecule.[146] The isopropyl ester properidine (Structure 36, $X{=}CO_2iPr$) is more potent, but the methyl and propyl esters less so, than meperidine.

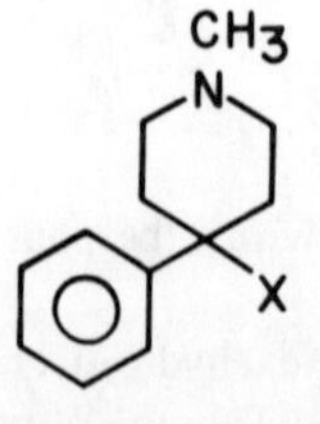

Structure 36

Reversal of the ester group also usually improved potency (Structure 36, $X{=}O_2CEt$). Keto-containing derivatives of Structure 36 (X=C(O)Et or C(O)Pr) are also quite active. Alkoxy or alkyl substitution at this position generally lowered activity.

Substitution on the piperidine ring, especially at C_3 has led to greater potency.[148,149] Methyl substitution at C_3 of the reversed ester of meperidine provided α- and β-prodine, both potent antinociceptive agents, with the former (Structure 37) having had some clinical utility as a fast-acting analgesic of short duration. Halo substituents or groups larger than allyl at C_3 generally decrease potency. Dimethyl substitution on the piperidine carbons provided the clinically useful, long-acting analgesic trimeperidine (Structure 38).

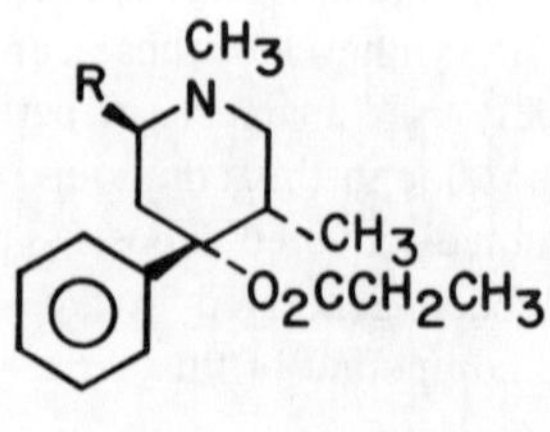

Structure 37; R=H

Structure 38; R=Me

The presence of a hydroxy substituent on the aromatic ring may greatly alter the SAR of the arylpiperidines.[145] In some compounds antinociceptive acitivity is decreased and in others increased and narcotic antagonist activity is much more common. Ketobemidone (Structure 39), a clinically active analgesic more potent than morphine, is an example of a compound which is significantly more potent than the analog lacking the *meta*-hydroxy substituent. An example of an antagonist in this series is Compound 40, which is 1/3 as potent an antagonist as nalorphine. A more striking example of the unique effect a *meta*-hydroxy substituent has on this series is the fact that Compound 41 is an antagonist of naloxone-like potency and considerably more potent an antagonist than ester 42.

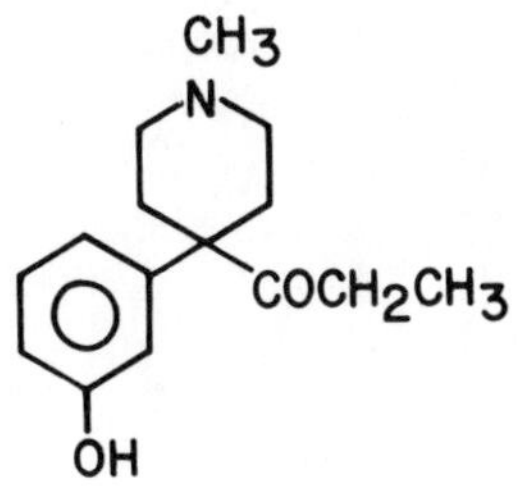

Structure 39

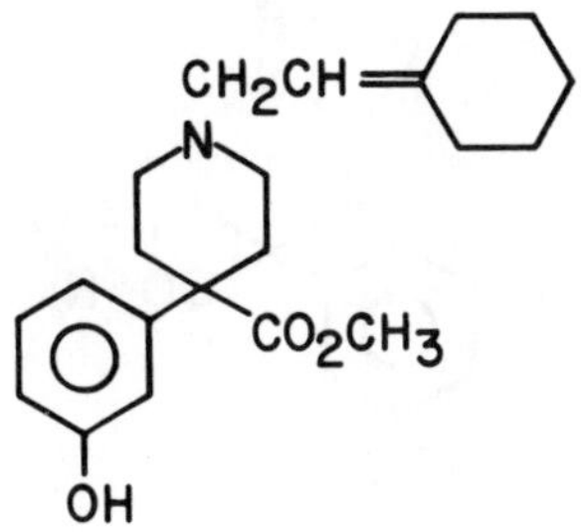

Structure 40

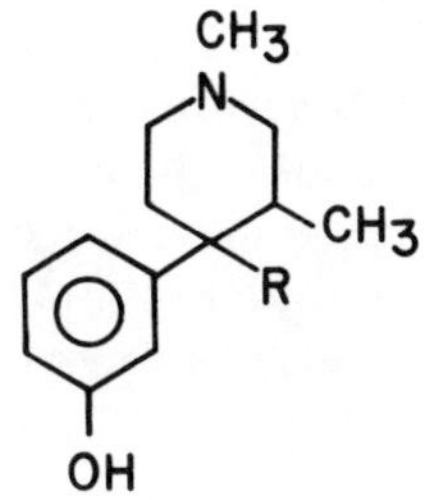

Structure 41; R=Me

Structure 42; R=O_2CEt

The 4-phenyl substituent of the arylpiperidines may be replaced by certain heteroaromatic rings without significant loss of antinociceptive activity. A much more important discovery was made by Janssen in the early 1960s, who found that the 4-phenyl and -carboxy substituents could be replaced by an anilido substituent to generate extremely potent analgesics, the most notable of which, fentanyl (Structure 43), is a short-acting analgesic 50 times as potent as meperidine in man which has found common use in surgical procedures. Some related compounds containing a second substituent at C_4 of the piperidine or methyl substitution at C_3 or on the *N*-alkyl side chain are even more potent.[150-152]

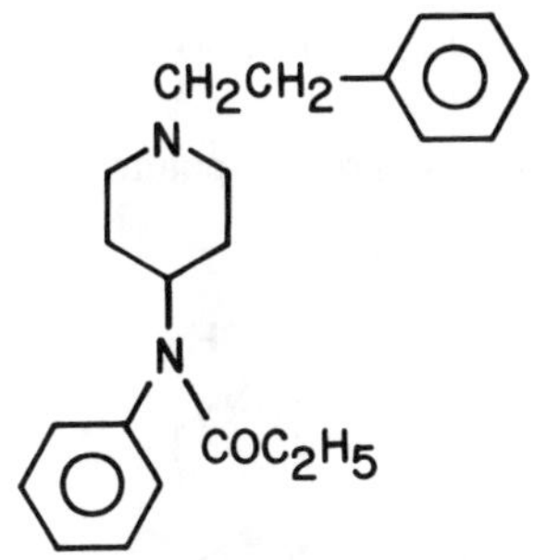

Structure 43

Expansion or contraction of the heterocyclic ring of the arylpiperidines has in some cases provided compounds with significant analgesic and/or antagonist activity. Some clinically active compounds generated in this way include prodilidine (Structure 44), ethoheptazine (Structure 45), and the mixed agonist-antagonists profadol (Structure 46) and meptazinol (Structure 47).[52,153,154]

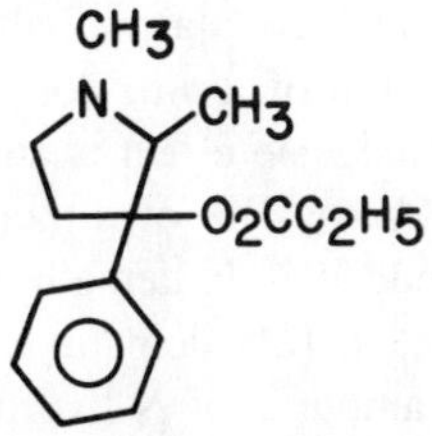

Structure 44

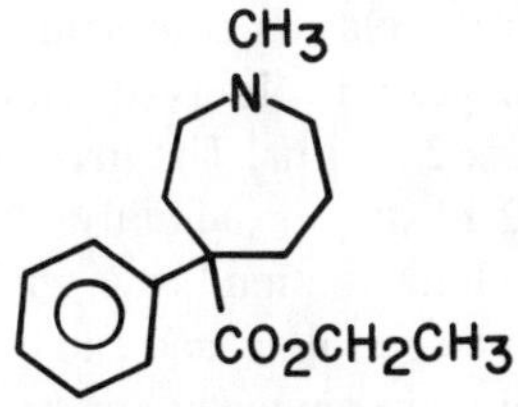

Structure 45

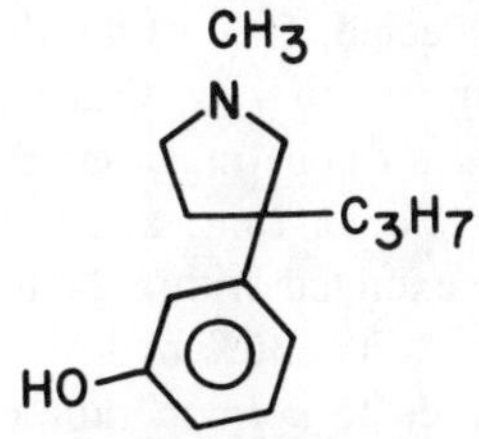

Structure 46

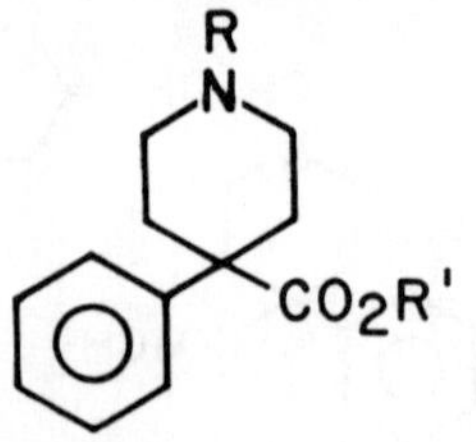

R–Me	R′=Et	10%
R=H	R′=Et	10%
R=Me	R′=H and conjugates	36%
R=H	R′=H and conjugates	18%

FIGURE 9. Metabolites of meperidine.

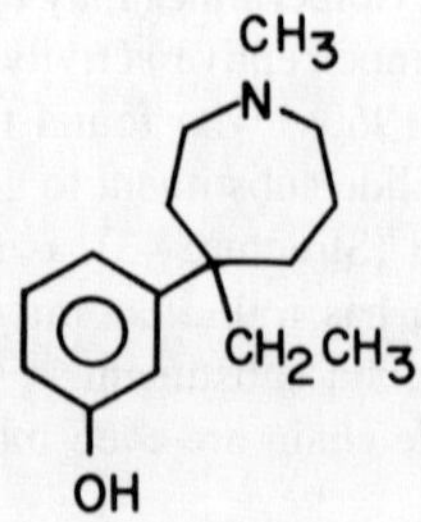

Structure 47

Bridging the piperidine ring can also provide potent antinociceptive agents such as Structure 49.[155]

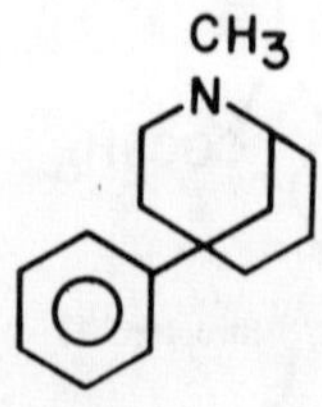

Structure 49

Meperidine Hydrochloride

Pharmacokinetics[28,69,156,157]

Meperidine is half as potent orally as parenterally. Peak plasma levels of 0.52 mg/ℓ at 1 min, 0.20 mg/ℓ at 1 hr, and 0.14 mg/ℓ at 2 hr after 50 mg i.v., i.m., and p.o., respectively were detected; 60% of the drug in the plasma is protein bound and the plasma half-life of the drug is 3 to 4 hr. Onset of analgesia is observed within 15 min of a p.o. dose and 10 min of a s.c. or i.m. dose, and lasts 2 to 4 hr. The maximum analgesic effect is attained 1 hr after s.c. or i.m., and within 2 hr of p.o. administration. About 70% of a therapeutic dose is excreted within 24 hr of administration: 10% unchanged, 10% N-demethylated to normeperidine, 36% hydrolyzed to meperidinic acid and conjugates, 18% demethylated and hydrolyzed to normeperidinic acid and conjugates, and minor amounts of N-hydroxy and hydroxyphenyl metabolites (Figure 9).

Uses and Dosage[67,68]

Meperidine is a narcotic analgesic 1/8 times as potent as morphine and which is slightly faster and shorter acting than the latter drug. It is indicated for the relief of moderate to severe pain, as a preoperative medication, for support of anesthesia, and for obstetrical analgesia. One advantage over morphine is that it retains much of its activity upon oral administration. It causes less smooth muscle spasm, constipation, rise in biliary pressure, depression of cough reflex, stupefaction, and urinary retention than equianalgesic doses of morphine. Patients who experience nausea and vomiting with morphine may not so react to meperidine, and conversely. Tolerance develops more slowly but withdrawal symptoms develop more rapidly, show fewer autonomic effects, and are of shorter duration than those of morphine. Newly born children of mothers who had taken meperidine showed less respiratory depression than those of mothers who had taken equianalgesic doses of morphine or methadone.

The drug is not as effective orally as parenterally. When given as a syrup orally it should be taken with water so as to avoid slight topical anesthesia of the mucous membranes. When administered parenterally, the patient should be lying down so as to minimize side effects. Subcutaneous administration may prove irritating after repeated doses. When the drug is administered i.v. it must be done very slowly and diluted since severe respiratory depression, apnea, hypotension, peripheral circulatory collapse, and cardiac arrest have occurred after rapid i.v. injection. Combinations with A.P.C. or A.P.A.P. are used when enhanced analgesic and antipyretic activity are desired. Combination with promethazine enhances analgesic and sedative activity. Atropine sulfate administered with meperidine imbues the combination with antisecretory activity.

For relief of pain 50 to 150 mg meperidine may be administered every 3 to 4 hr as required and in children 1.00 to 1.75 mg/kg at the same frequency. Preoperatively 50 to 100 mg may be given i.m. or s.c. and 1.00 to 1.75 mg/kg for children. Atropine sulfate may also be used. As a preanesthetic 50 to 100 mg of meperidine combined with 0.4 mg atropine sulfate is given 30 to 90 min prior to initiation of anesthesia. For children the dosage is 50 mg of the drug. For support of anesthesia injections of 10 mg/mℓ i.v. over several minutes or continuous infusion of not more than 1 mg/mℓ/min after an initial loading i.v. dose are used. In obstetrical analgesia the recommended dose is 50 to 100 mg i.m. or s.c. when the pain becomes regular, repeated at 1 to 3 hr intervals as necessary.

Meperidine solutions are incompatible with solutions containing sodium salts of barbiturates, heparin, hydrocortisone succinate, methicillin, methyl prednisolone succinate, nitrofurantoin, bicarbonate, iodine, sulfadiazine, thiamylal, or thiopental. Also with solutions containing aminophylline, ephedrine sulfate, morphine sulfate, oxytetracycline hydrochloride, sulfisoxazole diolamine, tetracycline hydrochloride, potassium iodide, aminosalicyclic acid, or salicylamide.

Availability[158]

Oral (Hydrochloride)

Tablets — Meperidine (various), Pethadol (Halsey), and Demerol (Winthrop) 50 and 100 mg in 100s, 500s, and 1000s.

Syrup — Demerol (Winthrop) 10 mg/mℓ in 1 pt bottles

Parenteral (Hydrochloride)

Available from several manufacturers in strengths ranging from 25 mg/mℓ to 100 mg/mℓ.

Combinations

Demerol APAP Tablets (Breon) 50 mg with 300 mg acetaminophen in 100s. Mepergan Fortis Capsules (Wyeth) 50 mg with 25 mg promethazine hydrochloride in 100s. Mepergan Injection (Wyeth) 25 mg/mℓ with 25 mg/mℓ promethazine in 10-mℓ vials and 2-mℓ syringes.

Toxicity

Meperidine causes a similar degree of sedation and euphoria as morphine. Dizziness may be worse than that of morphine. Other adverse effects include lightheadedness, nausea, vomiting, sweating, dysphoria, weakness, headache, agitation, tremors, uncoordinated muscle movement, psychotomimesis, dry mouth, constipation, biliary tract spasm, facial flushing, tachycardia, bradycardia, palpitations, syncope, respiratory depression, urinary retention, itching, rashes, tissue irritation at the injection site, miosis, corneal anesthesia, and hypotension. When given i.v., heart rate may increase dramatically. Therefore, it is advisable that the drug be given slowly in a diluted solution when administered in this manner. Orthostatic hypotension may occur in ambulatory patients. Meperidine has significant abuse potential, although habituation and addiction may develop more slowly than for morphine. In contrast to other opiates, at toxic doses, symptoms of CNS stimulation may be manifested, probably due to the actions of its major metabolite, normeperidine. Tolerance does not develop to these excitatory effects. Respiratory arrest, shock, and cardiac arrest have also occurred. Overdosage is generally treated with a narcotic antagonist, ventilation, and appropriate support measures, but occasionally an anticonvulsant may also be required.

Warnings — Meperidine is contraindicated in patients hypersensitive to the drug, especially those who have taken MAO inhibitors within 14 days, since a number of fatalities have been thus associated. It should be used very cautiously in patients with respiratory problems or whose ability to maintain blood pressure has been compromised, as well as in those patients with head injuries or a preexisting elevation of intracranial pressure. It should also be used with caution in patients with supraventricular tachycardias because the ventricular response rate may be increased due to the drug's vagolytic effect. The drug should be used carefully and at reduced dosage in the elderly or debilitated as well as in those with impaired hepatic or renal function, hypothyroidism, Addison's disease, prostatic hypertrophy, or urethral stricture. Its safe use in pregnancy prior to labor has not been established. When used as an obstetrical analgesic the neonate should be observed closely for signs of respiratory depression. The drug also appears in the milk of nursing mothers. High doses may induce convulsions in patients prone to them. Patients should be warned against attempting potentially hazardous tasks after taking the drug.

Drug interactions — As is generally observed for narcotic analgesics, meperidine should only be used at reduced dosage with such classes as other narcotic analgesics, general anesthetics, phenothiazines or other tranquilizers, sedative-hypnotics, tricyclic antidepressants, or other CNS depressants including alcohol because the CNS effects are additive. Meperidine has also been reported to aggravate the side effects of isoniazid.

Physical - Chemical and Other Data[159,160]

The structure of meperidine is shown in the pharmacokinetics section (Figure 9, R=Me, R′=Et). Ethyl 1-methyl-4-phenyl-4-piperidine carboxylate hydrochloride; $C_{15}H_{21}NO_2 \cdot HCl$; Demerol (Winthrop); Pethadol (Halsey); odorless, white crystalline powder; m.p. 186 to 189°C; pKa 8.7 (20°C); solubility — 1 g in 20 mℓ ethanol; very soluble in water, soluble in chloroform, slightly soluble in acetone and ether.

Fentanyl Citrate

Pharmacokinetics[161-163]

The serum concentration of fentanyl decreases rapidly after i.v. administration to about 1.4% of the original level within 60 min. However, the terminal elimination phase is slow due to gradual return of the unchanged drug to the central compartment from the peripheral tissues; 70 to 80% of the fentanyl in the blood is bound to serum proteins. The onset of analgesia occurs virtually immediately after i.v. administration and 7 to 15 min after i.m. injection. The peak analgesic effect after i.v. administration occurs within 3 to 5 min and analgesia generally lasts 30 to 60 min. After i.m. injection analgesia lasts 1 to 2 hr. The drug and its metabolites are slowly excreted with 9% being found in the feces and 76% in the urine after 72 hr. Less than 8% of the drug is excreted unchanged, but little is known about the metabolites.

Uses and Dosage[28,67,69,164]

Fentanyl is a narcotic analgesic 100 times as potent as morphine and which is rapid in onset and short acting. It is indicated as an analgesic supplement to the anesthetic technique of neuroleptanalgesia which is the method of choice for certain surgical procedures. Fentanyl is also indicated for analgesic action of short duration during anesthesia (premedication, induction, and maintenance), as well as postoperatively to control pain, tachypnea, and emergence delirium. In neuroleptanalgesia fentanyl is used together with a muscle relaxant to produce general calm, reduced motor activity, and profound analgesia, generally without complete loss of consciousness. Fentanyl causes less nausea and vomiting than morphine or meperidine.

Fentanyl is generally administered i.v. or i.m. Preoperatively 0.05 to 0.1 mg i.m. is administered 30 to 60 min prior to the procedure. In the induction period 0.05 to 0.1 mg is administered i.v. at 2 to 3 min intervals with the dosage lowered to 0.025 to 0.05 mg in poor risk and very old patients. For maintenance 0.025 to 0.05 mg is administered i.v. Postoperatively 0.05 to 0.1 mg is given i.m. every 1 to 2 hr as needed. In children 0.002 to 0.003 mg/kg is used for induction and maintenance. As an adjunct to regional anesthesia 0.05 to 0.1 mg i.m. or by slow i.v. is used. In the technique of neuroleptanalgesia 0.05 to 0.2 mg fentanyl with 2.5 to 10 mg droperidol is administered 30 to 60 min prior to surgery, 0.005 mg/kg with 0.25 mg/kg droperidol i.v. in the induction period, and 0.025 to 0.05 mg i.v. or i.m. for maintenance.

Availability

Injection

Sublimaze (Janssen) — 0.05 mg/mℓ in 2, 5, 10, and 20 mℓ ampules.

Combinations

Innovar (Janssen) — 0.05 mg/mℓ with 2.5 mg/mℓ droperidol in 2- and 5-mℓ ampules.

Toxicity[68]

Miosis, euphoria, bradycardia, hypotension, muscle rigidity, bronchial and laryngeal constriction, and respiratory depression may be observed following fentanyl administration. Nausea and vomiting may also occur. The drug has little hypnotic effect and generally does not cause histamine release. The bradycardia it sometimes causes can be reversed by atropine.

Warnings — Fentanyl when used in combination with a muscle relaxant is contraindicated in patients hypersensitive to either component. It should be avoided when possible in patients prone to respiratory depression, with head injuries, elevated intracranial pressure, or lowered ability to maintain blood pressure. Patients administered the drug should be monitored to avoid apnea or hypotension. The drug should be used cautiously in patients with impairment of pulmonary function. The dosage should be reduced and caution should be used when fentanyl is administered to the elderly, the very young, poor risk patients, or those suffering from Parkinson's disease, liver or kidney dysfunction, or cardiac arrythmias. Fentanyl has not been proven safe in pregnant patients or those under 2 years of age. When the drug is used during labor the neonate should be monitored for signs of respiratory depression. Care should be taken when a patient on fentanyl is moved because of the possiblity of orthostatic hypotension.

Drug interactions — As for all narcotic analgesics, the dosage should be reduced in the presence of other CNS depressants such as barbiturates, tranquilizers, narcotics, and general anesthetics. Fentanyl should not be used in patients who have taken MAO inhibitors within 14 days.

Physical - Chemical and Other Data[165]

The structure of fentanyl (Structure 43) is given in the SAR section. *N*-Phenyl-*N*-[1-(2-phenylethyl)-4-piperidinyl]-propanamide 2-hydroxy-1,2,3-propanetricarboxylate (1:1); $C_{22}H_{28}N_2O \cdot C_6H_8O_7$; Sublimaze (Janssen); air stable white crystalline powder; m.p. between 147 and 152°C; pKa 8.3; solubility — 1 g in 40 mℓ water, 140 mℓ ethanol, 10 mℓ methanol, 350 mℓ chloroform; slightly soluble in ether.

DIARYLPROPYLAMINES

In the 1940s the diarylpropylamines became the second major group of totally synthetic narcotic analgesics after the phenylpiperidines.[145] The first compound of this group reported to have potent analgesic activity was methadone (Structure 50). This compound was shown to be slightly more potent than morphine and slightly less sedating. Despite the many analogues of this compound which have been made over the years, it remains the principal ketone derivative in this class in medical use.

The antinociceptive activity of the diarylpropylamines is quite sensitive to variation in the chain.[52] Moving the methyl group from the α- to the β-position relative to the amine provides isomethadone (Structure 51) and removing it altogether provides normethadone (Structure 52). Both of these derivatives are slightly less potent than methadone (Structure 50). Methyl substitution at both the α- and β-positions in the erythro orientation gives an antinociceptive agent (Structure 53) with five times the potency of methadone. The threo isomers are inactive. Substitution of the α- and β-positions by groups other than methyl generally lowers activity as does lengthening or shortening of the alkyl chain.

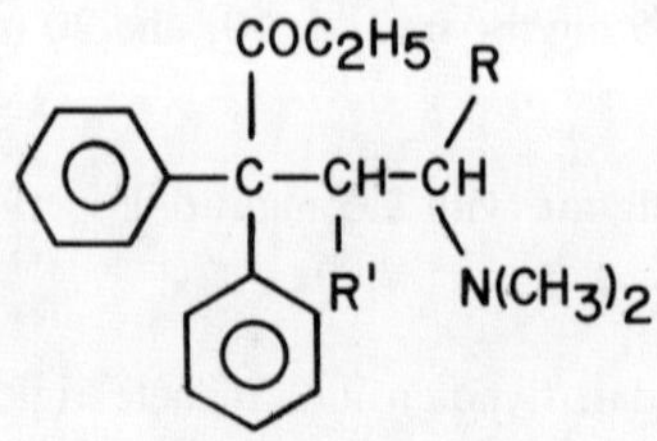

Structure 50; R=Me, R'=H

Structure 51; R=H, R'=Me

Structure 52; R=R'=H

Structure 53; R=R'=Me

An ethyl ketone is not required at the phenyl substituted carbon. Reduction of the ketone to the methadols (Structure 54) lowered activity but acetylation of these alcohols provided the acetylmethadols (Structure 55) which are more potent and longer acting analgesics than methadone. The formyl and propionyl esters are less active as are esters of isomethadol, the reduction product of isomethadone. Methyl and propyl ketone analogues of methadone are less potent as are the alcohols and esters derived from them. Replacement of the ethyl ketone at the quaternary carbon by an alkylsulfone, ethyl ester, propionate, amino alkyl, hydrogen, or hydroxy group decreases or eliminates activity. However, tertiary amides can replace the ketone without loss of antinociceptive activity, particularly in the isomethadone series, as in the clinically active analgesic dextromoramide (Structure 56).

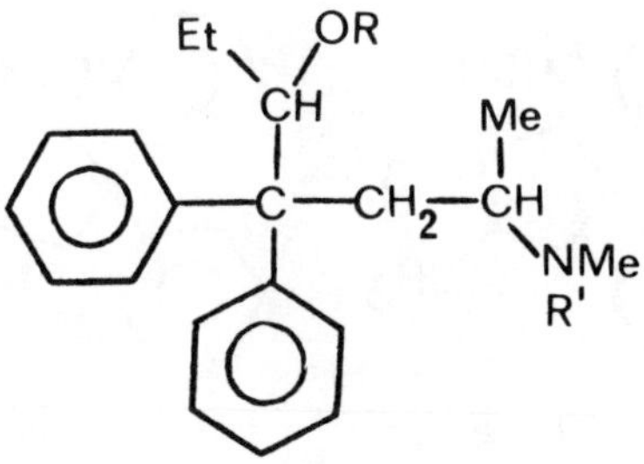

Structure 54; R=H, R′=Me

Structure 55; R=Ac, R′=Me

Structure 57; R=R′=H

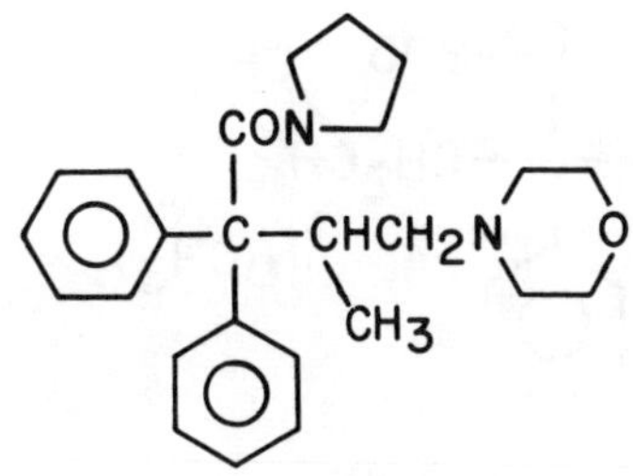

Structure 56

Only a fairly limited group of substituents on the amine retain potency in the methadone series. The other amino groups which retain activity include diethylamino, piperidino, pyrrolidino, and morpholino. When other changes are also made in the methadone structure other substituents may be allowed. For example, normethadol (Structure 57), the N-monodemethylated derivative of methadol, is more potent than the parent alcohol.

Diphenyl substitution at the quaternary carbon is not required for antinociceptive activity since replacing one phenyl group by a methyl substituent causes little decrease in activity.[166] However tetrasubstitution is required since the compound with H replacing phenyl has much less activity. Benzyl substitution at the quaternary carbon and replacement of the ethyl ketone by a propionyloxy group provides propoxyphene (Structure 58), which, although considerably less potent than methadone, is marketed as an orally effective analgesic with low addiction potential. Another group of analgesics which significantly differ from methadone at the aryl-substituted carbon are the thiambutenes (Structure 59), in which the phenyl groups have been replaced by 2-thienyl groups and are attached to a double bond. However, these compounds had no clinical advantage over methadone.

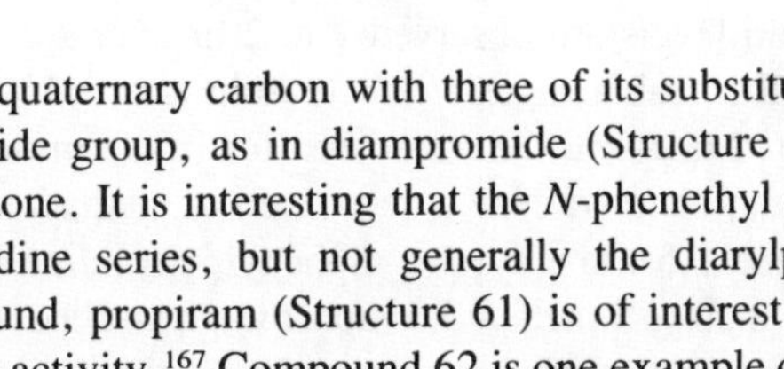

Structure 58

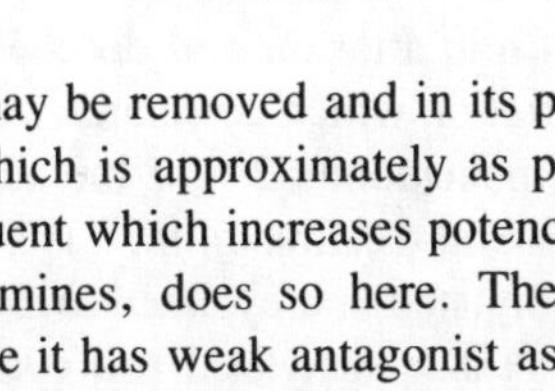

Structure 59

The quaternary carbon with three of its substituents may be removed and in its place put an anilide group, as in diampromide (Structure 60), which is approximately as potent as methadone. It is interesting that the *N*-phenethyl substituent which increases potency in the meperidine series, but not generally the diarylpropylamines, does so here. The related compound, propiram (Structure 61) is of interest because it has weak antagonist as well as agonist activity.[167] Compound 62 is one example of a compound which incorporates portions of the arylpiperidine and diarylpropylamine structures. This compound is about 150 times as potent as morphine.[168]

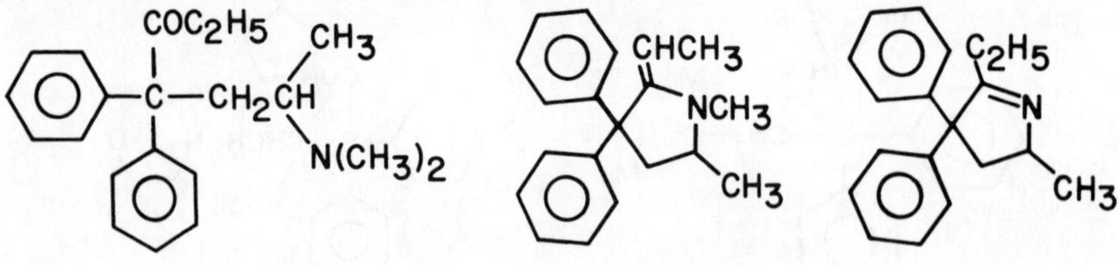

methadone 30%EDDP 1% EMDP

FIGURE 10. Metabolites of methadone.

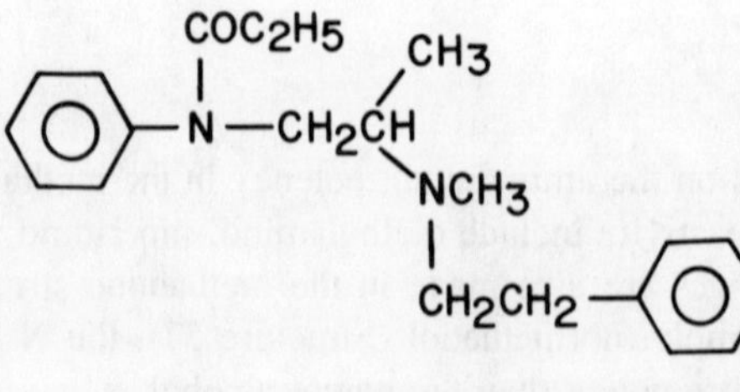

Structure 60

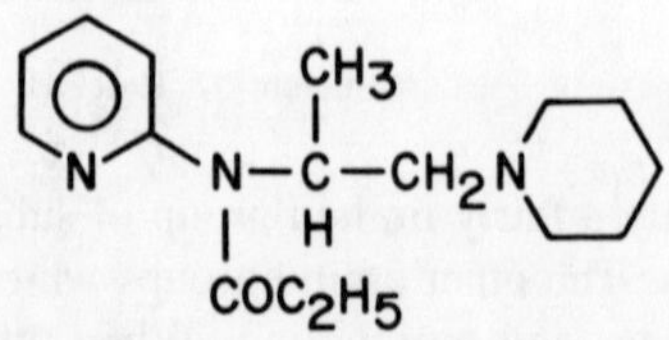

Structure 61

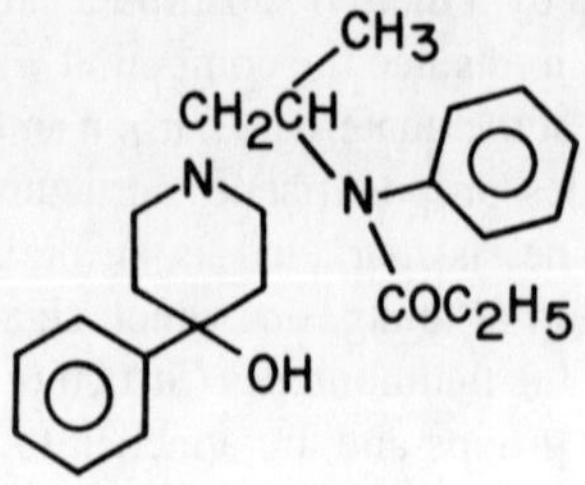

Structure 62

Methadone Hydrochloride

Pharmacokinetics[28,169-172]

Methadone is half as potent orally as parenterally. After a 15-mg oral dose peak plasma levels of 0.075 mg/ℓ are obtained after 4 hr. The plasma half-life of the drug is 15 hr and 85% of the drug in the plasma is protein bound. Methadone gradually accumulates in the tissues and is slowly released back to the plasma, accounting for its long duration of effect: 36 to 48 hr. Onset of analgesia is observed 10 to 20 min following a parenteral dose and 30 to 60 min after an oral dose. Peak brain levels are observed 1 to 2 hr after s.c. or i.m. administration which correlates well with the peak analgesic effect; 96 hr after a 15-mg oral dose of methadone 52% of the administered drug can be accounted for in the urine, about half of this in the first 24 hr; 21% of the administered dose is recovered as methadone, 30% N-demethylated and cyclized to 2-ethylidene-1,5-dimethyl-3,3-diphenylpyrrolidine (EDDP) and 1% N-didemethylated and cyclized to 2-ethyl-5-methyl-3,3-diphenyl-1-pyrroline (EMDP) (Figure 10). Small amounts of aromatic hydroxylated products, ketone reduction to methadol, and reduction and dealkylation to normethadol are also seen. Small amounts of methadone and EDDP have also been detected in the feces.

Uses and Dosage[67,69,173,174]

Methadone is a narcotic analgesic of similar potency parenterally to morphine but more active orally. It is indicated for the relief of moderate to severe pain, especially over a long period of time, and detoxification and temporary maintenance of narcotic addicts. Oral administration of methadone slows its onset and duration of action and lowers its peak effect. Because it is orally active and long acting it is especially suited to long-term pain relief. It is less sedating than morphine and is thus inferior as a preanesthetic. Most of the analgesic activity of methadone resides in the *l*-isomer which is 8 to 50 times as potent as the *d*-isomer. Methadone has persistent effects after repeated administration partially due to its gradual accumulation in the tissues. The drug is also an antitussive.

Methadone is less euphorigenic than morphine and its abstinence syndrome develops more slowly and is less intense and more prolonged than for other narcotics. It is also cross-tolerant with opiates, and can prevent an abstinence syndrome from developing. The combination of these effects causes methadone to be a useful drug in the treatment of narcotic addiction. It is used for detoxification of heroin addicts but only for 21 days, after which time it is categorized as maintenance therapy and can only be undertaken in approved programs. Detoxification can only be repeated after at least 28 days has elapsed since the completion of the initial course.

Local irritation may be observed when methadone is administered s.c. or i.m. As an analgesic, it may be administered i.m., s.c., or p.o. at a dosage of 2.5 to 10 mg 4 to 6 times daily. In children the recommended dose is 0.175 mg/kg 4 times daily. For the detoxification of narcotic addicts the drug is given orally, 15 to 40 mg once daily. The maintenance dosage for addicts is 40 to 120 mg once daily and is gradually decreased when the patient has stabilized.

Availability

Injection

Dolophine (Lilly) — 10 mg/mℓ in 1-mℓ ampules and 20-mℓ vials (with 0.5%) chlorobutanol).

Oral

Dolophine Tablets (Lilly) 5 mg in 100s and 1000s, 10 mg in 100s. Methadone Hydrochloride Diskets (Lilly) 40 mg in 100s. Methadone Hydrochloride Oral Solution (Philips Roxane) 1 mg/mℓ in 500-mℓ bottles.

Toxicity

Lightheadedness, dizziness, sedation, nausea, vomiting, and sweating are commonly observed after methadone administration, especially in ambulatory patients. Other untoward effects include euphoria or dysphoria, weakness, headache, insomnia, agitation, disorientation, visual disturbances, dry mouth, anorexia, constipation, biliary tract spasm, respiratory depression, miosis, facial flushing, bradycardia, palpitation, faintness, urinary retention or hesitancy, reduced libido or potency, pruritis, rashes, and edema. An oral dose of the drug generally does not alter heart rate, blood pressure, or EKG, and causes less sedation than morphine. Although methadone is used to treat narcotic addiction, it itself has significant abuse potential. Severe overdosage may lead to respiratory or cardiac arrest and circulatory collapse. The condition should be treated with a narcotic antagonist, ventilation, and appropriate support measures. Repeated doses of an antagonist may be required to counter its long duration of action.

Warnings — Methadone is contraindicated in patients hypersensitive to it. It should be used cautiously in patients with head injuries, increased intracranial pressure, impaired pulmonary function, or reduced ability to maintain blood pressure. It should be used with caution and at lower dosage in elderly or debilitated patients, and in those with impaired liver or kidney function, hypothyroidism, Addison's disease, prostatic hypertrophy, or urethral stricture. The safe use of methadone in pregnancy has not been established relative to fetal development. It is also not recommended for obstetric analgesia since its long duration of action increases the likelihood of respiratory depression in the neonate. It is not recom-

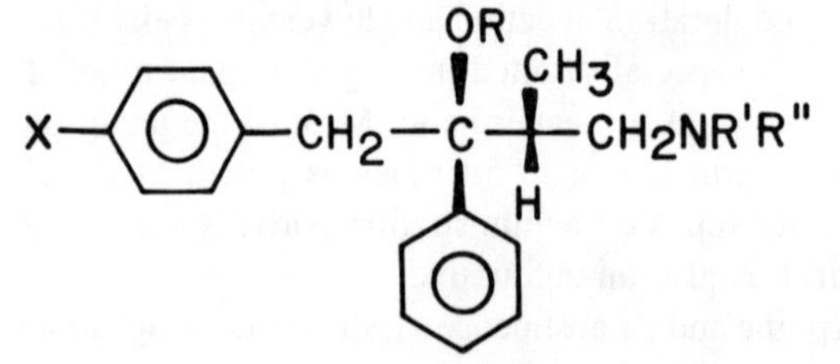

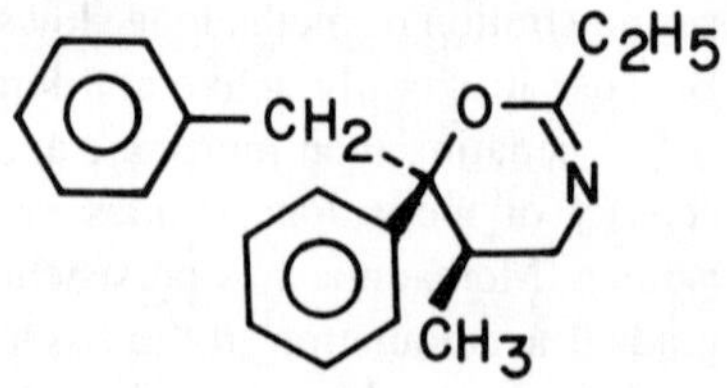

X=H	R=C(O)Et	R′=R″=Me	propoxyphene
X=H	R=C(O)Et	R′=Me R″=H	norpropoxyphene
X=H	R′=C(O)Et	R′=H R″=H	dinorpropoxyphene
X=OH	R=C(O)Et	R′=Me R″=Me	
X=OH	R=C(O)Et	R′=Me R″=H	
X=H	R=H	R′=R″=Me	
X=H	R=H	R′=Me R″=H	
X=H	R=H	R′=R″=H	

cyclic dinorpropoxyphene

FIGURE 11. Metabolites of propoxyphene.

mended for use in children. Patients who are under the influence of the drug should be warned against attempting potentially hazardous tasks.

Drug interactions — Methadone should be used with care in patients who have taken MAO inhibitors within 14 days. Its CNS effects are additive to other narcotic analgesics, general anesthetics, phenothiazines or other tranquilizers, sedative hypnotics, tricyclic antidepressants, or other depressants including alcohol. When used with rifampin, methadone's metabolism is accelerated thus leading to a shorter duration of action.

Physical - Chemical and Other Data[175,176]

The structure of methadone is given in the section on pharmacokinetics (Figure 10). 6-(Dimethylamino)-4,4-diphenyl-3-heptanone hydrochloride; $C_{21}H_{27}NO \cdot HCl$; Dolophine (Lilly); odorless, white crystalline powder; m.p. 233 to 236°C; pKa 8.3; solubility — 1 g in 13 mℓ water, 8 mℓ ethanol, 3 mℓ chloroform, 350 mℓ acetone; insoluble in ether or glycerin.

Propoxyphene

Pharmacokinetics[67,177-179]

Propoxyphene is readily absorbed after oral or parenteral administration. After a 65-mg oral dose of the hydrochloride peak plasma concentrations of 0.02 to 0.15 mg/ℓ are observed in 1 to 3 hr and the plasma half-life is 3.5 hr. The same dose given i.v. or i.m. gives peak levels of 0.17 to 0.49 mg/ℓ in 15 to 30 min and 0.02 to 0.18 mg/ℓ in 2 to 3 hr, respectively. Its major metabolite is norpropoxyphene which is half as active an analgesic as propoxyphene and a weaker CNS depressant. Serum levels of norpropoxyphene are significant and some of the effects of propoxyphene are probably due to this metabolite. Peak analgesia is attained 2 to 2.5 hr after an oral dose; 60 to 70% of an oral dose is excreted in the urine within 5 days. Only about 1% of the propoxyphene is recovered as the free drug and the predominant metabolite is the N-demethylated product norpropoxyphene. The N-didemethylated product dinorpropoxyphene and the cyclic derivative of the latter compound are also observed. *Para*-hydroxylation of the benzyl group occurs as well as demethylation of this material. Hydrolysis of the ester and subsequent N-mono- and -didemethylation also are observed (Figure 11).

Uses and Dosage[28,180,181]

Propoxyphene is a narcotic analgesic having 2/3 the potency of codeine orally and is indicated for the relief of mild to moderate pain. Parenterally it has 1/50 to 1/25 the activity of morphine. Only the *α-dextro* isomer of propoxyphene is active as an analgesic and is the marketed isomer. It has less antitussive and antidiarrheal effect and less addiction liability than codeine, although dependence on the drug has been reported. At therapeutic doses no clinically significant changes in respiration, blood pressure, or pulse rate were observed. The drug has been reported to be possibly less efficacious in cigarette smokers.

Propoxyphene napsylate is less water soluble and provides more stable liquid and tablet formulations. Propoxyphene is often combined with nonnarcotic analgesics to give increased analgesia as well as antiinflammatory and antipyretic effects. The drug is administered orally 3 to 6 times daily as the napsylate (50 to 100 mg) or the hydrochloride (32 to 65 mg).

Availability

Hydrochloride (capsules)

Propoxyphene Hydrochloride (various) — 32 mg in 100s, 500s, and 1000s, 65 mg in 100s, 250s, 500s and 1000s.

Darvon (Lilly) — 32 mg in 100s and 500s, 65 mg in 30s, 50s, and 500s.

Dolene (Lederle) — 65 mg in 25s, 60s, 100s, and 500s.

Pargesic 65 (Parmed) — 65 mg in 500s and 1000s.

Proxagesic (Tutag) — 65 mg in 500s.

Proxene (Bowman) — 65 mg in 500s.

SK-65 (SKF) — 65 mg in 100s and 500s.

Napsylate

Darvon N (Lilly) Tablets — 100 mg in 30s, 50s, 100s and 500s.

Oral Suspension — 10 mg/mℓ in 10-mℓ unit doses and pints.

Combinations

Also available are various combinations of propoxyphene hydrochloride with aspirin, phenacetin, caffeine, and acetaminophen and propoxyphene napsylate with aspirin or acetaminophen.

Toxicity[139]

Less than 1% of the individuals who take the drug at recommended doses experience side effects. Dizziness, sedation, paradoxical excitement and insomnia, light-headedness, headache, weakness, liver dysfunction, nausea, vomiting, constipation, euphoria, dysphoria, respiratory depression, miosis, and skin rash have been observed. Tolerance to, and psychological and physical dependence on, its effects have been observed, although it will not support morphine dependence. Intravenous or subcutaneous abuse of the drug can lead to severe damage to veins and soft tissues. Chronic ingestion of large doses can cause toxic psychosis and convulsions. Acute overdosage may be accompanied by respiratory depression, somnolence, pupillary constriction, circulatory collapse, cardiac arrythmias, pulmonary edema, and convulsions. Overdosage should be treated with a narcotic antagonist, ventilation, and appropriate support measures including possibly an anticonvulsant. Gastric lavage may be helpful and activated charcoal can absorb significant amounts of ingested drug. Fatalities attributed to propoxyphene poisoning have increased in the last few years, in part due to its use in combination with alcohol in suicides.

Warnings — Propoxyphene is contraindicated in persons hypersensitive to it. It should be prescribed cautiously to patients with a history of emotional disturbances, suicidal tendencies, or drug abuse. Its safe use in pregnancy has not been established relative to fetal development and withdrawal symptoms have been observed in neonates whose mothers were taking the drug during pregnancy. The drug has also been detected in the milk of nursing mothers. It is not recommended for use in children. Individuals should be warned against attempting potentially hazardous tasks after taking the drug.

Drug interactions — Its CNS depressant effects are additive to those of alcohol, tranquilizers, sedative-hypnotics, narcotics, and other CNS depressants. Its use with orphenadrine should be avoided since confusion, anxiety, and tremors have been reported.

Physical - Chemical and Other Data

The structure of propoxyphene is given in the section on pharmacokinetics (Figure 11).

Hydrochloride:[182-184] [S-(R*,S*)]-α-[2-(Dimethylamino)-1-methylethyl]-α-phenylbenzene-ethanol propanoate (ester) hydrochloride; $C_{22}H_{29}NO_2 \cdot HCl$; Darvon (Lilly); Dolene (Lederle); Pargesic (Parmed); Proxagesic (Tutag); Proxene (Bowman); white crystalline powder; m.p. in a 3° range between 163.5 and 168.5°C; α(1% in H_2O) +36 to +40°; solubility — 1 g in 0.3 mℓ water, 1.5 mℓ ethanol, 0.6 mℓ chloroform; soluble in acetone and insoluble in ether.

Napsylate[185] $C_{22}H_{29}NO_2 \cdot C_{10}H_8O_3S \cdot H_2O$; odorless, white, crystalline solid; m.p. in a 4° range between 158 and 165°C; α (1% solution in $CHCl_3$) +35 to +43°; solubility — 1 g in 13 mℓ ethanol or methanol, 3 mℓ chloroform; soluble in acetone and slightly soluble in water.

OPIOID PEPTIDES

The existence of opiate receptors had long been a speculation of medicinal chemists and pharmacologists. The idea grew out of, and was supported by, the study of literally thousands of semisynthetic and fully synthetic compounds with opiate-like structures and/or activity. Perhaps the most outstanding characteristic of this activity is the stereochemical requirement: the optical antipodes of active compounds are in general only weakly active or not active at all. Goldstein[186] attempted to use this phenomenon to demonstrate the existence of opiate-binding sites through the displacement of a bound radioactive opiate by its cold counterpart but not by its cold optical antipode. These workers observed that only 2% of total binding was stereospecific. Using the same strategy but changing the tactics to the use of potent radiolabeled opiates of high specific activity Pert and Snyder,[7] Simon et al.,[187] and Terenius[188] were able to demonstrate unequivocally the existence of stereospecific-binding sites. These sites were subsequently observed in all vertebrate species studied. The latter fact seemed to support the speculation that these binding sites play an important role in the physiology of the animal and that there must, therefore, exist a specific endogenous ligand(s) for the sites. It wasn't long before Hughes,[189] and Terenius and Wahlstrom[190] reported the observations of opiate-like activity in brain extracts. Shortly thereafter, Hughes et al.[8] reported the isolation from pig brain, structure determination, synthesis, and opiate properties of two pentapeptides which they named Met-enkephalin (TyrGlyGlyPheMet) and Leu-enkephalin (TyrGlyGlyPheLeu). The authors noted with some interest that the Met-enkephalin sequence was present as residues 61 to 65 of β-lipotropin (β-LPH), a pituitary peptide of unknown function. Given the relatively small size of the morphine molecule and the slightly larger enkephalins the discovery[191] of potent opiate activity in the so called C-fragment of β-LPH (residues 61 to 91, now known as β-endorphin) was of considerable interest. Other opiate peptides have since been isolated from a variety of natural sources. Among the more interesting are dermorphin (Tyr-D-Ala-Phe-Gly-Tyr-Pro-Ser-NH_2) isolated from frog skin,[192] and β-casomorphin (Tyr-Pro-Phe-Pro-Gly-Pro-Ile-OH) isolated from β-casein.[193] Dermorphin is of particular interest because it is the first example of a natural peptide containing D-alanine.

The announcements of the discoveries of the various opiate peptides initiated a considerable effort among peptide chemists to determine structure activity relationships. Hundreds of compounds have been reported and the effort continues unabated. Consequently an in-depth review here is impractical although the major findings will be highlighted. The interested reader is referred to the excellent review by Dewey.[194] In general peptides containing fewer than five amino acids are weakly active or inactive. Exceptions occur among D-Ala^2-peptides containing a bulky substituent on the C-terminal end; for example, Tyr-D-Ala-Gly-N(Me)CH_2CH_2Ph is five times more potent than Met-enkephalin in the guinea pig ileum and ten times more potent in the tail flick.[195] A second and notable exception is the dipeptide Tyr-Arg which is reported to be four times more potent than Met-enkephalin in a tail pinch assay.[196]

With respect to the pentapeptides, little variation of the Tyr^1 residue can be tolerated. Removal of the amino or hydroxyl, methylation or cyclopropylmethylation of the nitrogen or substitution of D-Tyr, Phe, L-Dopa or Trp all drastically decrease or eliminate activity. It is interesting that allylation of nitrogen does not result in antagonist properties. The Gly^2 position may be substituted by a wide range of D-amino acids (but not L-variations) and maintain or enhance potency. Furthermore, the D-amino acids in this position, particularly D-Ala, tend to enhance metabolic stability thereby extending the compounds' durations of action. By contrast no variations of the Gly^3 position have resulted in an increase in potency. Likewise variations of the Phe^4 position have not been very fruitful, although the number of compounds reported is relatively small. On the other hand position 5 has been studied to a large extent. Still, most of these variations tend to decrease activity and the few increases that have been reported are not dramatic. The most extensive group of modifications consist of two or more variations of the enkephalin sequence, usually with D-Ala in position 2. It has been found that multiple changes tend to be additive and some remarkable potencies have been reported. For example, $(MeTyr)^1$-$(D\text{-}Ala)^2$-$(MePhe)^4$-$[Met\text{-}(O)\text{-}ol]^5$ is 68,000 times more potent than Met-enkephalin in the tail flick assay.[197] Structure-activity studies among peptides containing more than five amino acids have been somewhat limited by lack of availability of the compounds, but in general dramatic potency enhancements have not been observed.

Perhaps the most important question following the discoveries of the endogenous opiates was whether among these substances or their analogues there could be found the long sought potent analgesic free of the deleterious side effects of the opiate alkaloids. Shortly after the discovery of the enkephalins Belluzzi and co-workers[198] showed that Met-enkephalin exhibited antinociceptive activity after central administration in the rat. Other more stable analogues have also exhibited activity. β-Endorphin is much more stable and has been evaluated as an analgesic in humans. In a very small open study Catlin and co-workers[199] gave β-endorphin i.v. to patients in pain due to cancer. They used morphine as a positive control and concluded that β-endorphin is analgesic in humans. In another study[200] β-endorphin was shown to elicit long lasting analgesic action in humans. On the other hand Wei[201] has shown that the enkephalins and several of their analogues produce physical dependence and that this liability correlates well with analgesic potency. It also appears that tolerance to the analgesic action of these substances develops on repeated administration.

The first synthetic enkephalin analog to be extensively studied is FK33-824 [Tyr-D-Ala-Gly-MePhe-Met(O)-ol].[202-204] This compound is about 1000 times more potent than morphine in the mouse (i.c.v.) and retains one fifth the activity of morphine when given orally: it also has some addiction potential. In humans the compound had little effect on respiratory rate or blood pressure, but elicited feelings of muscular heaviness, and anaphylactoid and GI symptoms. In addition it had pronounced endocrine effects, raising prolactin and growth hormone levels and lowering LH, FSH, and ACTH levels. A second extensively studied enkephalin analog is metkephamid (Tyr-D-Ala-Gly-Phe-MeMet-NH_2).[205] This compound has a higher preference for the so-called δ-receptor and is currently undergoing clinical evaluation. Finally a third compound, FW34-569 [MeTyr-D-Ala-Gly-MePhe-Met(O)-ol] has been studied in humans[206] The compound increased tolerance to experimentally induced pain but did not alter pain threshold. In conjunction with its side effects it was concluded that this compound had no therapeutic potential. Thus at this point in time no peptide with clear therapeutic advantages over currently used analgesics has emerged and many investigators in the field consider the likelihood of such an emergence rather small. Nevertheless the search and the hope for success continues.

REFERENCES

1. **Tainter, M. L.,** Pain, in *Newer Synthetic Analgesics,* Miner, R. W., Ed., New York Academy of Sciences, New York, 1948, 3.
2. **Gates, M. and Tschudi, G.,** The synthesis of morphine, *J. Am. Chem. Soc.,* 74, 1109, 1952.
3. **Archer, S. and Harris, L. S.,** Narcotic antagonists, in *Progress in Drug Research,* Vol. 8, Jucker, E., Ed., Birkhauser-Verlag, Basel, 1965, 261.
4. **Archer, S. and Michne, W. F.,** Recent progress in research on narcotic antagonists, in *Progress in Drug Research,* Vol. 20, Jucker, E., Ed., Birkhauser-Verlag, Basel, 1976, 45.
5. **Casy, A. F.,** Analgesics and their antagonists: recent developments, in *Progress in Medicinal Chemistry,* Vol. 7, Ellis, G. P. and West, G. B., Eds., Appleton-Century-Crofts, New York, 1970, 229.
6. **Casy, A. F.,** Analgesics and their antagonists, in *Progress in Drug Research,* Vol. 22, Jucker, E., Ed., Birkhauser-Verlag, Basel, 1978, 149.
7. **Pert, C. B. and Snyder, S. H.,** Opiate receptor: demonstration in nervous tissue, *Science,* 179, 1011, 1973.
8. **Hughes, J., Smith, T. W., Kosterlitz, H. W., Fothergill, L. A., Morgan, B. A., and Morris, H. R.,** Identification of two related pentapeptides from the brain with potent opiate agonist activity, *Nature (London),* 258, 577, 1975.
9. **Koster, R., Anderson, M., and deBeer, E. J.,** Acetic acid for analgesic screening, *Fed. Proc.,* 18, 412, 1959.
10. **Collier, H. O. J., Dineen, L. C., Johnson, C. A., and Schneider, C.,** The abdominal constriction response and its suppression by analgesic drugs in the mouse, *Br. J. Pharmacol.,* 32, 295, 1968.
11. **Hendershot, L. C. and Forsaith, J.,** Antagonism of the frequency of phenylquinone-induced writhing in the mouse by weak analgesics and non-analgesics, *J. Pharmacol. Exp. Ther.,* 125, 237, 1959.
12. **Chernov, H. I., Wilson, D. E., Fowler, F., and Plummer, A. J.,** Nonspecificity of the mouse writhing test, *Arch. Int. Pharmacodyn. Ther.,* 167, 171, 1967.
13. **Loux, J. J., Smith, S., and Salem, H.,** Comparative analgesic testing of various compounds in mice using writhing techniques, *Arzneim. Forsch.,* 28, 1644, 1978.
14. **D'Armour, F. E. and Smith, D. L.,** A method for determining loss of pain sensation, *J. Pharmacol. Exp. Ther.,* 72, 74, 1941.
15. **Woolfe, G. and MacDonald, A. D.,** The evaluation of the analgesic action of pethidine hydrochloride (Demerol), *J. Pharmacol. Exp. Ther.,* 80, 300, 1944.
16. **Aceto, M. D., Harris, L. S., Dewey, W. L., and May, E. L.,** Annual report: dependence studies of new compounds in the rhesus monkey (1980), in Problems of Drug Dependence 1980, Harris, L. S., Ed., U.S. Government Printing Office, Washington, D.C., 1981, 297.
17. **Teiger, D. G.,** Induction of physical dependence on morphine, codeine, and meperidine in the rat by continuous infusion, *J. Pharmacol. Exp. Ther.,* 190, 408, 1974.
18. **Gyang, E. A. and Kosterlitz, H. W.,** Agonist and antagonist actions of morphine-like drugs on the guinea pig isolated ileum, *Br. J. Pharmacol.,* 27, 514, 1966.
19. **Henderson, G., Hughes, J., and Kosterlitz, H. W.,** A new example of a morphine sensitive neuro-effector junction: adrenergic transmission in the mouse vas deferens, *Br. J. Pharmacol.,* 46, 764, 1972.
20. **Lord, J. A. H., Waterfield, A. A., Hughes, J., and Kosterlitz, H. W.,** Endogenous opioid peptides: multiple agonists and receptors, *Nature (London),* 267, 495, 1977.
21. **Pert, C. B. and Snyder, S. H.,** Properties of opiate-receptor binding in rat brain, *Proc. Natl. Acad. Sci. U.S.A.,* 70, 2243, 1973.
22. **Pert, C. B. and Snyder, S. H.,** Opiate receptor binding of agonists and antagonists affected differentially by sodium, *Mol. Pharmacol.,* 10, 868, 1974.
23. **deStevens, G.,** *Analgetics,* Academic Press, New York, 1965.
24. **Loh, H. H. and Ross, D. H., Eds.,** *Neurochemical Mechanisms of Opiates and Endorphins. Advances in Biochemical Psychopharmacology,* Vol. 20, Raven Press, New York, 1979.
25. **Way, E. L., Ed.,** *Endogenous and Exogenous Opiate Agonists and Antagonists,* Pergamon Press, New York, 1980.
26. **Beers, R. F. and Bassett, E. G., Eds.,** *Mechanism of Pain and Analgesic Compounds,* Raven Press, New York, 1979.

27. **Johnson, M. R. and Milne, G. M.,** Analgetics, in *Burger's Medicinal Chemistry, Fourth Edition, Part III,* Wolff, M. E., Ed., John Wiley & Sons, New York, 1981, 699.
28. **Jaffe, J. H. and Martin, W. R.,** Opioid analgesics and antagonists, in *The Pharmacological Basis of Therapeutics,* 6th ed., Gilman, A. G., Goodman, L. S., and Goodman, A., **Eds**., Macmillan, New York, 1980, 494.
29. **Fields, H. L. and Basbaum, A. I.,** Brainstem control of spinal pain-transmission neurons, in *Annual Review of Physiology,* Vol. 40, Knobil E., Ed., Annual Reviews Inc., Palo Alto, Calif., 1978, 217.
30. **Kerr, F. W. L. and Wilson, P. R.,** Pain, in *Annual Review of Neuroscience,* Vol. 1, Cowan, W. M., Ed., Annual Reviews Inc., Palo Alto, Calif., 1978, 83.
31. **Bowman, W. C. and Rand, M. J.,** *Textbook of Pharmacology,* Blackwell Scientific Publications, Oxford, 1980.
32. **Way, E. L.,** Review and overview of four decades of opiate research, in *Neurochemical Mechanisms of Opiates and Endorphins. Advances in Biochemical Psychopharmacology,* Loh, H. H. and Ross, D. H., Eds., Raven Press, New York, 1979, 3.
33. **Rosenfeld, G. C., Strada, S. J., and Robinson, G. A.,** Cyclic nucleotides and the central effects of opiates: An overview of current research, in *Neurochemical Mechanisms of Opiates and Endorphins. Advances in Biochemical Psychopharmacology,* Vol. 20, Loh, H. H. and Ross, D. H., Eds., Raven Press, New York, 1979, 261.
34. **Sharma, S. K., Klee, W. A., and Nirenberg, M.,** Dual regulation of adenylate cyclase accounts for narcotic dependence and tolerance, *Proc. Natl. Acad. Sci. U.S.A.,* 72, 3092, 1975.
35. **Jessell, T. M. and Iversen, L. L.,** Opiate analgesics inhibit substance P release from rat trigeminal nucleus, *Nature (London),* 268, 549, 1977.
36. **Domino, E. F.,** Opiate interactions with cholinergic neurons, in *Neurochemical Mechanisms of Opiates and Endorphins. Advances in Biochemical Psychopharmacology,* Vol. 20, Loh, H. H. and Ross, D. H., Eds., Raven Press, New York, 1979, 339.
37. **Iwamoto, E. T. and Way, E. L.,** Opiate actions and catecholamines, in *Neurochemical Mechanisms of Opiates and Endorphins. Advances in Biochemical Psychopharmacology,* Vol. 20, Loh, H. H. and Ross, D. H., Eds., Raven Press, New York, 1979, 357.
38. **Brase, D. A.,** Roles of serotonin and γ-aminobutyric acid in opioid effects, in *Neurochemical Mechanisms of Opiates and Endorphins. Advances in Biochemical Psychopharmacology,* Vol. 20, Loh, H. H. and Ross, D. H., Eds., Raven Press, New York, 1979, 409.
39. **Bentley, K. W.,** The morphine alkaloids, in *The Alkaloids,* Manske, R. H. F., Ed., Academic Press, New York, 1971, 1.
40. **Bentley, K. W. and Lewis, J. W.,** The relationship between structure and activity in the 6,14-*endo*ethenotetrahydrothebaine series of analgesics, in *Agonist and Antagonist Actions of Narcotic Analgesic Drugs,* Kosterlitz, H. W., Collier, H. O. J., and Villarreal, J. E., Eds., University Park Press, Baltimore, 1973, 7.
41. **Lewis, J. W., Bentley, K. W., and Cowan, A.,** Narcotic analgesics and antagonists, in *Annual Review of Pharmacology,* Vol. 11, Elliott, H. W., Ed., Annual Reviews Inc., Palo Alto, Calif., 1971, 241.
42. **Lewis, J. W.,** Ring C-bridge derivatives of thebaine and oripavine, in *Narcotic Antagonists, Advances in Biochemical Psychopharmacology,* Vol. 8, Braude, M. C., Harris, L. S., May, E. L., Smith, J. P., and Villareal, J. E., Eds., Raven Press, New York, 1974, 123.
43. **Hutchins, C. W., Cooper, G. K., Pürro, S., and Rapaport, H.,** 6-Demethoxythebaine and its conversion to analgesics of the 6,14-ethenomorphinan type, *J. Med. Chem.,* 24, 773, 1981.
44. **Bullingham, R. E. S., McQuay, H. J., Dwyer, D., Allen, M. C., and Moore, R. A.,** Sublingual buprenorphine used postoperatively: clinical observations and preliminary pharmocokinetic analysis, *Br. J. Clin. Pharm.,* 12, 117, 1981.
45. **Bullingham, R. E. S., McQuay, H. J., Moore, A., and Bennett, M. R. D.,** Buprenorphine kinetics, *Clin. Pharmacol. Ther.,* 28, 667, 1980.
46. **Prous, J. R. Ed.,** Buprenorphine, *Drugs Today,* 14, 397, 1978.
47. **Heel, R. C., Brogden, R. N., Speight, T. M., and Avery, G. S.,** Buprenorphine: a review of its pharmacological properties and therapeutic efficacy, *Drugs,* 17, 81, 1979.
48. **Jasinski, D. R., Pevnick, J. S., and Griffith, J. D.,** Human pharmacology and abuse potential of the analgesic buprenorphine, *Arch. Gen. Psychiatr.,* 35, 501, 1978.
49. **Coltart, D. J. and Alasdain, D. M.,** Pharmacological and clinical importance of narcotic antagonists and mixed antagonists-use in cardiology, *Br. J. Clin. Pharm.,* 7, 309S, 1979.
50. **Moore, R. A., McQuay, H. J., and Bullingham, R. E. S.,** Reversal of postoperative hyperglycemia by buprenorphine, *Lancet,* 2, 597, 1980.
51. **Gulland, J. M. and Robinson, R.,** Constitution of codeine and thebaine, *Mem. Proc. Manchester Lit. Phil. Soc.,* 69, 79, 1925.
52. **May, E. L. and Sargent, L. J.,** Morphine and its modifications, in *Analgetics,* deStevens, G., Ed., Academic Press, New York, 1965, 123.

53. **Janssen, P. A. J.,** A review of the chemical features associated with strong morphine-like activity, *Brit. J. Anaesth.*, 34, 260, 1962.
54. **Reden, J., Reich, M. F., Rice, K. C., Jacobson, A. E., Brossi, A., Streaty, R. A., and Klee, W. A.,** Deoxymorphines: role of the phenolic hydroxyl in antinociception and opiate receptor interactions, *J. Med. Chem.*, 22, 256, 1979.
55. **Braude, M. C., Harris, L. S., May, E. L., Smith, J. P., and Villareal, J. E., Eds.,** *Narcotic Antagonists, Advances in Biochemical Psychopharmacology,* Vol. 8, Raven Press, New York, 1973.
56. **Kosterlitz, H. W., Collier, H. O. J., and Villareal, J. E., Eds.**, *Agonist and Antagonist Actions of Narcotic Analgesic Drugs,* University Park Press, Baltimore, 1973.
57. **Pohl, J.,** Ueber das N-allylnorcodeine, einen antagonisten des morphins, *Z. Exp. Pathol. Ther.*, 17, 370, 1915.
58. **Kotick, M. P.,** Analgesic narcotic antagonists, 7β, 8β-Methano- and 7β, 8β-epoxydihydrocodeinone, *J. Med. Chem.*, 24, 722, 1981.
59. **Kotick, M. P., Leland, D. L., Polazzi, J. O., and Schut, R. N.,** Analgesic narcotic antagonists. 8β-Alkyl-, 8β-acyl-, and 8β-(tertiary alcohol) dihydrocodeinones and -dihydromorphinones, *J. Med. Chem.*, 23, 166, 1980.
60. **Buckett, W. R.,** Exceptional opiate potency of 14β-alkyl derivatives of morphinone and codeinone in vitro and in vivo, *Br. J. Pharm.*, 76, 269P, 1982.
61. **Osei-Gyimah, P. and Archer, S.,** Synthesis and analgesic activity of some 14β-substituted analogues of morphine, *J. Med. Chem.*, 23, 162, 1980.
62. **Pasternak, G. W. and Hahn, E. F.,** Long-acting opiate agonists and antagonists: 14-hydroxydihydromorphinone hydrazones, *J. Med. Chem.*, 23, 674, 1980.
63. **Portoghese, P. S., Larson, D. L., Sayre, L. S., Fries, D. S., and Takemori, A. E.,** A novel opioid receptor site directed alkylating agent with irreversible narcotic antagonistic and reversible agonistic activities, *J. Med. Chem.*, 23, 233, 1980.
64. **Baselt, R. C.,** *Disposition of Toxic Drugs and Chemicals in Man,* Vol. 1, Biomedical Publications, Canton, Conn., 1978, 25.
65. **Lewis, A. J., Ed.**, *Modern Drug Encyclopedia and Therapeutic Index,* 16th ed., Yorke Medical Books, New York, 1981, 625.
66. **Clarke, E. G. C., Ed.**, *Isolation and Identification of Drugs,* The Pharmaceutical Press, London, 1969, 431.
67. **Halpern, L. M. and Bonica, J. J.,** Analgesics, in *Drugs of Choice 1980—1981,* Modell, W., Ed., C. V. Mosby, St. Louis, 1980, 199.
68. **McEvoy, G. K., Ed.**, *American Hospital Formulary Service,* American Society of Hospital Pharmacists, Bethesda, Md., 1982.
69. **Swinyard, E. A.,** Analgesics and antipyretics, in *Remington's Pharmaceutical Sciences,* 16th ed., Osol, A., Ed., Mack Publishing Co., Easton, Pa., 1980, 1043.
70. **Boyd, J. R., Ed.**, *Facts and Comparisons,* Facts and Comparisons Division, J. B. Lippincott, St. Louis, 1982, 242e.
71. *The Pharmaceutical Codex,* 11th ed., The Pharmaceutical Press, London, 1979, 575.
72. **Wade, A., Ed.**, *Martindale the Extra Pharmacopoeia,* 27th ed., The Pharmaceutical Press, London, 1977, 970.
73. *The United States Pharmacopeia,* 20th Revision, The United States Pharmacopeial Convention, Inc., Rockville, Md., 1979, 536.
74. *European Pharmacopoeia,* Vol. 1, Maisonneuve, S. A., France, 1969, 327.
75. **Baselt, R. C.,** *Disposition of Toxic Drugs and Chemicals in Man,* Vol. 1, Biomedical Publications, Canton, Conn., 1978, 3.
76. **Lewis, A. J., Ed.**, *Modern Drug Encyclopedia and Therapeutic Index,* 16th ed., Yorke Medical Books, New York, 1981, 214.
77. **Bickerman, H. A.,** Antitussive drugs, in *Drugs of Choice 1980—1981,* Modell, W., Ed., C. V. Mosby, St. Louis, 1980, 450.
78. **Boyd, J. R., Ed.**, *Facts and Comparisons,* Facts and Comparisons Division, J. B. Lippincott, St. Louis, 1982, 195b and 245.
79. **Clarke, E. G. C., Ed.**, *Isolation and Identification of Drugs,* The Pharmaceutical Press, London, 1969, 268.
80. **Wade, A., Ed.**, *Martindale The Extra Phamacopoeia,* 27th ed., The Pharmaceutical Press, London, 1977, 953.
81. *European Pharmacopoeia,* Vol. 1, Maisonneuve, S. A., France, 1969, 275.
82. *The United States Pharmacopeia,* 20th Revision, The United States Pharmacopeial Convention, Inc., Rockville, Md., 1979, 165.
83. *The Pharmaceutical Codex,* 11th ed., The Pharmaceutical Press, London, 1979, 215.

84. **Lewis, A. J., Ed.**, *Modern Drug Encyclopedia and Therapeutic Index,* 16th ed., Yorke Medical Books, New York, 1981, 632.
85. **Miller, R. R.,** Evaluation of nalbuphine hydrochloride, *Am. J. Hosp. Pharm.,* 37, 942, 1980.
86. **Prous, J. R., Ed.**, Nalbuphine hydrochloride, *Drugs Today,* 16, 53, 1980.
87. **Boyd, J. R., Ed.**, *Facts and Comparisons,* Facts and Comparisons Division, J. B. Lippincott, St. Louis, 1982, 246f.
88. **Lewis, A. J., Ed.,** *Modern Drug Encyclopedia and the Therapeutic Index,* 16th ed., Yorke Medical Books, New York, 1981, 636.
89. **Boyd, J. R., Ed.,** *Facts and Comparisons,* Facts and Comparisons Division, J. B. Lippincott, St. Louis, 1982, 246.
90. **Ngai, S. H., Berkowitz, B. A., Yang, J. C., Hempstead, J., and Spector, S.,** Pharmacokinetics of naloxone in rats and man, *Anesthesiology,* 44, 398, 1976.
91. **Weinstein, S. H., Pfeffer, M., and Schor, J. M.,** Metabolism and pharmacokinetics of naloxone, in *Narcotic Antagonists, Advances in Biochemical Psychopharmacology,* Vol. 8, Braude, M. C., Harris, L. S., May, E. L., Smith, J. P., and Villarreal, J. E., Eds., Raven Press, New York, 1973, 525.
92. **Wade, A., Ed.**, *Martindale The Extra Pharmacopoeia,* 27th ed., The Pharmaceutical Press, London, 1977, 987.
93. **Ostrowski, N. L., Rowland, N., Foley, T. L., Nelson, J. L., and Reid, L. D.,** Morphine antagonists and consummatory behaviors, *Pharmacol. Biochem. Behav.,* 14, 549, 1981.
94. **Verebey, K., Volavka, J., and Clouet, D.,** Endorphins in psychiatry: an overview and a hypothesis, *Arch. Gen. Psychiatr.,* 35, 877, 1978.
95. **Lightman, S. L. and Jacobs, H. S.,** Naloxone: non-steroidal treatment for postmenopausal flushing?, *Lancet,* 2, 1071, 1979.
96. **Jeffcoate, W. J., Herbert, M., Cullen, M. H., Hastings, A. G., and Walder, C. P.,** Prevention of effects of alcohol intoxication by naloxone, *Lancet,* 2, 1157, 1979.
97. **Martin, W. R., Gorodetzky, C. W., and McClane, T. K.,** An experimental study in the treatment of narcotic addicts with cyclazocine, *Clin. Pharmacol. Ther.,* 7, 455, 1966.
98. **Holaday, J. W. and Faden, A. I.,** Naloxone reversal of endotoxin hypotension suggests role of endorphins in shock, *Nature, (London),* 275, 450, 1978.
99. **Faden, A. I. and Holaday, J. W.,** Opiate antagonists: a role in the treatment of hypovolemic shock, *Science,* 205, 317, 1979.
100. **Clarke, E. G. C., Ed.**, *Isolation and Identification of Drugs,* Vol. 2, The Pharmaceutical Press, London, 1975, 1066.
101. **Grewe, R.,** Sythetische Arzneimittel mit Morphin-Wirkung, *Angew. Chem.,* 59, 194, 1947.
102. **Jacobson, A. E., Hsu, F., Rozwadowska, M. D., Schmidhammer, H., Atwell, L., Brossi, A., and Medzihradsky, F.,** Structure-activity relationships of oxygenated morphinans. 4-Mono- and 3,4-dimethoxy-N-methylmorphinans and -N-methylmorphinan-6-ones with unusually high antinociceptive potency, *Helv. Chim. Acta,* 64, 1298, 1981.
103. **Schmidhammer, H., Jacobson, A. E., Atwell, L., and Brossi, A.,** Structure-activity relationship of oxygenated morphinans. Narcotic agonist and antagonist activity in the 14-hydroxymorphinan series, *Helv. Chim. Acta,* 64, 2540, 1981.
104. **Prous, J. R., Ed.**, *Drugs Future,* 4, 438, 1979.
105. **Pachter, I. J.,** Synthetic 14-hydroxymorphinan narcotic antagonists, in *Narcotic Antagonists, Advances in Biochemical Psychopharmacology,* Vol. 8, Braude, M. C., Harris, L. S., May, E. L., Smith, J. P., and Villareal, J. E., Eds., Raven Press, New York, 1974, 57.
106. **Prous, J. R., Ed.**, Proxorphan, *Drugs Future,* 6, 633, 1981.
107. **Kametani, T., Kigasawa, K., Hiiraga, M., Wagatsuma, N., Kusama, O., and Uryu, T.,** Studies on the synthesis of analgesics. Optical resolution of (+)-N-cyclopropylmethyl-3-hydroxy-9-azamorphinan, *Chem. Pharm. Bull.,* 24, 2563, 1976.
108. **Shiotani, S. and Kametani, T.,** B.C.-*cis*- and -*trans*-1,3,4,9,10,10a-Hexahydro-2H-10,14a-methanoiminoethanophenanthrene (homo- and homo-isomorphinan derivatives as analgesics, *J. Med. Chem.,* 20, 976, 1977.
109. **Belleau, B., Conway, T., Ahmed., F. R., and Hardy, A. D.,** Importance of the nitrogen lone pair orientation in stereospecific opiates, *J. Med. Chem.,* 17, 907, 1974.
110. **Lewis, A. J., Ed.**, *Modern Drug Encyclopedia and Therapeutic Index,* 16th ed., Yorke Medical Books, New York, 1981, 124.
111. **Vandam, L. D.,** Butorphanol, *N. Engl. J. Med.,* 302, 381, 1980.
112. **Gaver, R. C., Vasiljev, M., Wong, H., Monkovic, I., Swigor, J. E., VanHarken, D. R., and Smyth, R. D.,** Disposition of parenteral butorphanol in man, *Drug Metab. Disp.,* 8, 230, 1980.
113. **Boyd, J. R., Ed.**, *Facts and Comparisons,* Facts and Comparisons Division, J. B. Lippincott, St. Louis, 1982, 246d.

114. **Prous, J. R., Ed.**, Butorphanol tartate, *Drugs Today,* 14, 197, 1978.
115. **Heel, R. C., Brogden, R. N., Speight, T. M., and Avery, G. S.,** Butorphanol: a review of its pharmacological properties and therapeutic efficacy, *Drugs,* 16, 473, 1978.
116. **Barnhart, J. W.,** The urinary excretion of dextromethorphan and three metabolites in dogs and humans, *Toxicol. Appl. Pharmacol.,* 55, 43, 1980.
117. **Barnhart, J. W. and Massad, E. N.,** Determination of dextromethorphan in serum by gas chromatography, *J. Chromatog.,* 163, 390, 1979.
118. **Swinyard, E. A.,** Respiratory drugs, in *Remington's Pharmaceutical Sciences,* 16th ed., Osol, A., Ed., Mack Publishing Co., Easton, Pa., 1980, 804.
119. **Boyd, J. R., Ed.**, *Facts and Comparisons,* Facts and Comparisons Division, J. B. Lippincott, St. Louis, 1982, 196.
120. **Nelemans, F. A.,** Drugs used in bronchial asthma and cough, in *Side Effects of Drugs,* Annual 5, Dukes, M. N. G., Ed., Excerpta Medica, Amsterdam, 1981, 169.
121. **Clarke, E. G. C., Ed.**, *Isolation and Identification of Drugs,* The Pharmaceutical Press, London, 1969, 289.
122. **Wade, A., Ed.**, *Martindale The Extra Pharmacopoeia,* 27th ed., The Pharmaceutical Press, London, 1977, 1248.
123. *The Pharmaceutical Codex,* 11th ed., The Pharmaceutical Press, London, 1979, 257.
124. *British Pharmacopoeia, 1980,* Vol. 1, Her Majesty's Stationery Office, London, 1980, 143.
125. **Jacobson, A. E. and May, E. L.,** Structures related to morphine. 2′-Substituted benzomorphans, *J. Med. Chem.,* 8, 563, 1965.
126. **Villareal, J. E.,** The effects of morphine agonists and antagonists on morphine-dependent rhesus monkeys, in *Agonist and Antagonist Actions of Narcotic Analgesic Drugs,* Kosterlitz, H. W., Collier, H. O. J., and Villareal, J. E., Eds., University Park Press, Baltimore, 1973, 73.
127. **Albertson, N. F.,** Cyclazocine and congeners, in *Narcotic Antagonists, Advances in Biochemical Psychopharmacology,* Vol. 8, Braude, M. C., Harris, L. S., May, E. L., Smith, J. P., and Villareal, J. E., Eds., Raven Press, New York, 1974, 63.
128. **Albertson, N. F.,** Effect of 9-hydroxylation on benzomorphan, *J. Med. Chem.,* 18, 619, 1975.
129. **Janssen, P. A. J.,** Potent long-acting antagonists of the 9,9-dialkylbenzomorphan series, in *Narcotic Antagonists, Advances in Biochemical Psychopharmacology,* Vol. 8, Braude, M. C., Harris, L. S., May, E. L., Smith, J. P., and Villareal, J. E., Eds., Raven Press, New York, 1974, 109.
130. **Michne, W. F., Lewis, T. R., Michalec, S. J., Pierson, A. K., and Rosenberg, F. J.,** (2,6-Methano-3-benzazocin-11β-yl) alkanones. Arylalkanones: a new series of N-methyl derivatives with novel opiate activity profiles, *J. Med. Chem.,* 22, 1158, 1979.
131. **Katsuura, K., Ohta, M., and Mitsuhashi, K.,** Heteroaromatic analogs of benzomorphan. Synthesis of novel thiazolo[4,5-f]morphans, *Chem. Pharm. Bull.,* 29, 1780, 1981.
132. **Takeda, M. and Kugita, H.,** Homologs of benzomorphan derivatives, *J. Med. Chem.,* 13, 630, 1970.
133. **Shiotani, S., Kometani, T., Mitsuhashi, K., Nozawa, T., Kurobe, A., and Futsukaichi, O.,** 10-Hydroxy-4-methyl-2,3,4,5,6,7-hexahydro-1,6-methano-1H-4-benzazonine derivatives (homobenzomorphans) as analgesics, *J. Med. Chem.,* 19, 803, 1976.
134. **Martin, W. R., Eades, C. G., Thompson, J. A., Huppler, R. E., and Gilbert, P. E.,** The effects of morphine- and nalorphine-like drugs in the nondependent and morphine-dependent chronic spinal dog, *J. Pharmacol. Exp. Ther.,* 197, 519, 1976.
135. **Römer, D., Büscher, H., Hill, R. C., Maurer, R., Petcher, T. J., Welle, H. B. A., Bakel, H. C. C. K., and Akkerman, A. M.,** Bremazocine: a potent, long-acting opiate kappa-agonist, *Life Sci.,* 27, 971, 1980.
136. **Lewis, A. J., Ed.**, *Modern Drug Encyclopedia and Therapeutic Index,* 16th ed., Yorke Medical Books, New York, 1981, 725.
137. **Baselt, R. C.,** *Disposition of Toxic Drugs and Chemicals in Man,* Vol. 1, Biomedical Publications, Canton, Conn., 1978, 30.
138. **Boyd, J. R., Ed.**, *Facts and Comparisons,* Facts and Comparisons Division, J. B. Lippincott, St. Louis, 1982, 247.
139. **Bewley, T. H. and Ghodse, A. H.,** Opioid analgesics and narcotic antagonists, in *Side Effects of Drugs,* Annual 5, Dukes, M. N. G., Ed., Excerpta Medica, Amsterdam, 1981, 64.
140. *The National Formulary-Fourteenth Edition,* American Pharmaceutical Association, Washington, D. C., 1975, 540.
141. *British Pharmacopoeia, 1980,* Vol. 1, Her Majesty's Stationery Office, London, 1980, 330.
142. **Wade, A., Ed.**, *Martindale The Extra Pharmacopoeia,* 27th ed., The Pharmaceutical Press, London, 1977, 977.
143. *The Pharmaceutical Codex,* 11th ed., The Pharmaceutical Press, London, 1979, 653.
144. **Eisleb, O. and Schaumann, O.,** Dolantin, a new antispasmodic and analgesic, *Deut. Med. Wochshr.,* 65, 967, 1939.

145. **Hardy, R. A. and Howell, M. G.,** Synthetic analgetics with morphine-like actions, in *Analgetics,* de Stevens, G., Ed., Academic Press, New York, 1965, 179.
146. **Janssen, P. A. J.,** A review of the chemical features associated with strong morphine-like activity, *Br. J. Anaesth.,* 34, 260, 1962.
147. **Pert, C. B., Snyder, S. H., and Portoghese, P. S.,** Correlation of opiate receptor affinity with analgetic effects of meperidine homologues, *J. Med. Chem.,* 19, 1249, 1976.
148. **Jensen, K. A., Lindquist, F., Rekling, E., and Wolffbrandt, C. G.,** A new type of piperidine derivative with analgesic and spasmolytic activity, *Dansk Tids. Farm.,* 17, 173, 1943.
149. **Ziering, A. and Lee, J.,** Piperidine derivatives. 1,3-dialkyl-4-aryl-4-acyloxypiperidines, *J. Org. Chem.,* 12, 911, 1947.
150. **VanBever, W. F. M., Niemegeers, C. J. E., and Janssen, P. A. J.,** N-4-Substituted 1-(2-arylethyl)-4-piperidinyl-N-phenylpropanamides, a novel series of extremely potent analgesics with unusually high safety margin, *Arzneim. Forsch.,* 26, 1548, 1976.
151. **Yang, T. T., Srulevitch, D. B., and Lien, E. J.,** QSAR of synthetic narcotic analgetic agents related to fentanyl, *Acta Pharm. Jugoslav.,* 31, 77, 1981.
152. **Van Bever, W. F. M., Niemeggeers, C. J. E., and Janssen, P. A. J.,** Sythetic analgesics. Synthesis and pharmacology of the diastereomers of N-[3-methyl-1-(2-phenylethyl)-4-piperidyl]-N-phenylpropanamide and N-[3-methyl-1-(1-methyl-2-phenylethyl)-4-piperidyl]-N-phenylpropanamide, *J. Med. Chem.,* 17, 1047, 1974.
153. **McCarthy, D. A.,** Arylpyrrolidines as narcotic antagonists, in *Narcotic Antagonists, Advances in Biochemical Psychopharmacology,* Vol. 8, Braude, M. C., Harris, L. S., May, E. L., Smith, J. P., and Villareal, J. E., Eds., Raven Press, New York, 1973, 145.
154. **Bradley, G., Cavalla, J. F., Edington, T., Sheperd, R. G., White, A. C., Bushell, B. J., Johnson, J. R., and Weston, G. O.,** Synthetic studies on meptazinol. Anion chemistry in the synthesis of α-aryl lactams, *Eur. J. Med. Chem.,* 15, 375, 1980.
155. **Rogers, M. E. and May, E. L.,** Improved synthesis and additional pharmacology of the potent analgetic (-)-5-*m*-hydroxyphenyl-2-methylmorphan, *J. Med. Chem.,* 17, 1328, 1974.
156. **Baselt, R. C.,** *Disposition of Toxic Drugs and Chemicals in Man,* Vol 1, Biomedical Publications, Canton, Conn., 1978, 18.
157. **Lewis, A. J., Ed.,** *Modern Drug Encyclopedia and Therapeutic Index,* 16th ed., Yorke Medical Books, New York, 1981, 557.
158. **Boyd, J. R., Ed.,** *Facts and Comparisons,* Facts and Comparisons Division, J. B. Lippincott, St. Louis, 1982, 243c.
159. *The Pharmaceutical Codex,* 11th ed., The Pharmaceutical Press, London, 1979, 673.
160. **Wade, A., Ed.,** *Martindale The Extra Pharmacopoeia,* 27th ed., The Pharmaceutical Press, London, 1977, 980.
161. **Lewis, A. J., Ed.,** *Modern Drug Encyclopedia and Therapeutic Index,* 16th ed., Yorke Medical Books, New York, 1981, 386.
162. **McClain, D. A. and Hug, C. C.,** Intravenous fentanyl kinetics, *Clin. Pharmacol. and Ther.,* 28, 106, 1980.
163. **Fung, D. L. and Eisele, J. H.,** Fentanyl pharmacokinetics in awake volunteers, *J. Clin. Pharmacol.,* 20, 652, 1980.
164. **Boyd, J. R., Ed.,** *Facts and Comparisons,* Facts and Comparisons Division, J. B. Lippincott, St. Louis, 1982, 244a.
165. *British Pharmacopoeia, 1980,* Vol. 1, Her Majesty's Stationery Office, London, 1980, 191.
166. **Hassan, M. M. A.,** Phenylmethyl propylamines: new potent analgetics, *Eur. J. Med. Chem.,* 16, 95, 1981.
167. **Hiltman, V. R., Hoffmeister, F., Niemers, E., Schlichting, U., and Wollweber, H.,** 2-Acylaminopyridine-Derivative mit Morphin-Agonistischer und -antagonistischer Wirksamkeit, *Arzneim. Forsch.,* 24, 584, 1974.
168. **Casy, A. F.,** Analgesic receptors, in *A Guide to Molecular Pharmacology-Toxicology, Part 1,* Featherstone, R. M., Ed., Marcel Dekker, New York, 1973, 221.
169. **Clarke, E. G. C., Ed.,** *Isolation and Identification of Drugs,* The Pharmaceutical Press, London, 1969, 408.
170. **Baselt, R. C.,** *Disposition of Toxic Drugs and Chemicals in Man, Vol. 1,* Biomedical Publications, Canton, Conn., 1978, 21.
171. **Inturrisi, C. E. and Verebely, K.,** Disposition of methadone in man after a single oral dose, *Clin. Pharmacol. Ther.,* 13, 923, 1972.
172. **Verebely, K., Volavka, J., Mule, S., and Resnick, R.,** Methadone in man: pharmacokinetic and excretion studies in acute and chronic treatment, *Clin. Pharmacol. Ther.,* 18, 180, 1975.
173. **Boyd, J. R., Ed.,** *Facts and Comparisons,* Facts and Comparisons Division, J. B. Lippincott, St. Louis, 1982, 243a.

174. **Lewis, A. J., Ed.**, *Modern Drug Encyclopedia and Therapeutic Index,* 16th ed., Yorke Medical Books, New York, 1981, 576.
175. *The Pharmaceutical Codex,* 11th ed., The Pharmaceutical Press, London, 1979, 545.
176. Specifications for the Quality Control of Pharmaceutical Preparations, World Health Organization, Geneva, 1967, 338.
177. *The Pharmaceutical Codex,* 11th ed., The Pharmaceutical Press, London, 1979, 258.
178. **McMahon, R. E., Sullivan, H. R., Due, S. L., and Marshall, F. J.,** The metabolite pattern of *d*-propoxyphene in man. The use of heavy isotopes in drug disposition studies, *Life Sci.*, 12, 463, 1973.
179. **Baselt, R. C.,** *Disposition of Toxic Drugs and Chemicals in Man,* Vol. 1, Biomedical Publications, Canton, Conn., 1978, 34.
180. **Lewis, A. J., Ed.**, *Modern Drug Encyclopedia and Therapeutic Index,* 16th ed., Yorke Medical Books, New York, 1981, 820.
181. **Boyd, J. R., Ed.**, *Facts and Comparisons,* Facts and Comparisons Division, J. B. Lippincott, St. Louis, 1982, 247c.
182. **Clarke, E. G. C., Ed.**, *Isolation and Identification of Drugs,* The Pharmaceutical Press, London, 1969, 290.
183. *The United States Pharmacopeia,* 20th Revision, The United States Pharmacopeial Convention, Inc., Rockville, Md., 1979, 674.
184. **Wade, A., Ed.**, *Martindale The Extra Pharmacopoeia,* 27th ed., The Pharmaceutical Press, London, 1977, 956.
185. *The National Formulary-Fourteenth Edition,* American Pharmaceutical Association, Washington, D.C., 1975, 617.
186. **Goldstein, A., Lowney, L. I., and Pal, B. K.,** Stereospecific and nonspecific interactions of the morphine congener levorphanol in subcellular fractions of mouse brain, *Proc. Natl. Acad. Sci. U.S.A.*, 68, 1742, 1971.
187. **Simon, E. J. and Hiller, J. M.,** The opiate receptors, *Ann. Rev. Pharmacol. Toxicol.*, 18, 371, 1978.
188. **Terenius, L.,** Stereospecific interaction between narcotic analgesics and a synaptic plasma membrane fraction of rat cerebral cortex, *Acta Pharmacol. Toxicol.*, 32, 317, 1973.
189. **Hughes, J.,** Isolation of an endogenous compound from the brain with pharmacological properties similar to morphine, *Brain Res.*, 88, 295, 1975.
190. **Terenius, L. and Wahlstrom, A.,** Inhibitor(s) of narcotic receptor binding in brain extracts and cerebrospinal fluid. *Acta Pharmacol. Toxicol.*, 35 (Suppl. 1), 55, 1974.
191. **Cox, B. M., Goldstein, A., and Li, C. H.,** Opioid activity of a peptide, β-lipotropin(61-91), derived from β-lipotropin, *Proc. Natl. Acad. Sci. U.S.A.*, 73, 1821, 1976.
192. **Montecucchi, P. C., DeCastiglione, R., Piani, S., Gozzini, L., and Erspamer, V.,** Amino acid composition and sequence of dermorphin, a novel opiate-like peptide from the skin of Phyllomedusa Sauvagei, *Int. J. Peptide Protein Res.*, 17, 275, 1981.
193. **Henschen, A., Brantl, B., Teschemacher, H., and Lottspeich, F.,** β-Casomorphins — novel opioid peptides derived from bovine casein — isolation and structure, in *Endogenous and Exogenous Opiate Agonists and Antagonists,* Way, E. L., Ed., Permagon Press, Elmsford, 1980, 233.
194. **Dewey, W. L.,** Structure-activity relationships, in *Endorphins, Chemistry, Physiology, Pharmacology, and Clinical Relevance,* Malick, J. B. and Bell, R. M. S., Eds., Marcel Dekker, New York, 1982, 23.
195. **Morgan, B. A., Bower, J. D., Guest, K. P., Handa, B. K., Metcalf, G., and Smith, C. F. C.,** Structure-activity relationships of enkephalin analogs, in *Peptides: Proc. 5th Am. Peptide Symp.*, Goodman, M. and Meienhofer, J., Eds., John Wiley & Sons, New York, 1977, 111.
196. **Takagi, H., Satoh, M., Shiomi, H., Akaike, A., Ueda, H., Kawajiri, S., Yamamoto, M., and Amano, H.,** Identification of a new opioid peptide from the bovine brain and its mechanism of action, in *Endogenous and Exogenous Opiate Agonists and Antagonists,* Way, E. L., Ed., Permagon Press, Elmsford, 1980, 201.
197. **Pless, J., Bauer, W., Cardinaux, F., Classe, A., Hauser, D., and Huguenin, R.,** Structure-activity relationship of enkephalin analogs, in *Molecular Endocrinology,* MacIntyre, I. and Szeke, M., Eds., Elsevier/North-Holland, Amsterdam, 1977, 279.
198. **Belluzzi, J. D., Grant, N., Garsky, V., Sarantakis, D., Wise, C. D., and Stein, L.,** Analgesia induced *in vivo* by central administration of enkephalin in rat, *Nature (London),* 260, 625, 1976.
199. **Catlin, D. H., Hui, K. K., Loh, H. H., and Li, C. H.,** Pharmacologic activity of β-endorphin in man, *Commun. Psychopharmacol.*, 1, 493, 1977.
200. **Hosobuchi, Y. and Li, C. H.,** The analgesic activity of human β-endorphin in man, *Commun. Psychopharmacol.*, 2, 33, 1978.
201. **Wei, E.,** Enkephalin analogs: correlation of potencies for analgesia and physical dependence, in *Characteristics and Function of Opioids,* van Ree, J. M., and Terenius, L., Eds., Elsevier/North Holland, Amsterdam, 1978, 445.

202. **Roemer, D., Buescher, H. H., Hill, R. C., Pless, J., Bauer, W., Cardinaux, F., Closse, A., Hauser, D., and Huguenin, R.,** A synthetic enkephalin analogue with prolonged parenteral and oral analgesic activity, *Nature (London)*, 268, 547, 1977.

203. **von Graffenried, B., DelPozo, E., Roubicek, J., Krebs, E., Poldinger, W., Burmeister, P., and Kerp, L.,** Effects of the synthetic enkephalin analogue FK 33-824 in man, *Nature (London)*, 272, 729, 1978.

204. **Stubbs, W. A., Jones, A., Edwards, C. R. W., Delitala, G., Jeffcoate, W. J., Ratter, S. J., Besser, G. M., Bloom, R. S., and Alberti, K. G. M. M.,** Hormonal and metabolic responses to an enkephalin analogue in normal man, *Lancet*, p. 1225, 1978.

205. **Frederickson, R. C. A., Smithwick, E. L., Shuman, R., and Bemis, K. G.,** Metkephamid, a systemically active analog of methionine enkephalin with potent opioid δ-receptor activity, *Science*, 211, 603, 1981.

206. **Lindeburg, T., Larsen, V., Kehlet, H., and Jacobsen, E.,** Respiratory, analgesic and endocrine responses to an enkephalin analogue in normal man, *Acta Anaesthesiol. Scand.*, 25, 254, 1981.

ANTIDEPRESSANTS

Dennis J. Hlasta, Dean R. Haubrich, and Daniel Luttinger

INTRODUCTION

Historical

For many years barbiturates and other sedatives were used to treat some of the symptoms of depression, such as insomnia and agitation. In the late 1930s CNS stimulants, such as amphetamines, were used to relieve the symptoms associated with milder forms of depression.[1] During this period, electroconvulsive therapy was introduced as a means of treating severely depressed patients, and this is still the most effective treatment for the management of severe depression. Currently electroconvulsive shock is used only in severely suicidal patients who do not respond to other types of treatment.

In the early 1950s, phenothiazines and reserpine were used to treat depression,[2] but these compounds did not relieve the principal features of the disorder. The antidepressant effects of imipramine and iproniazid were discovered almost simultaneously in the mid-1950s. The CNS stimulant properties of iproniazid were originally discovered in 1951 during a clinical trial to determine its efficacy against tuberculosis. In 1952, iproniazid was found to inhibit monoamine oxidase, an enzyme whose importance in the metabolism of norepinephrine was becoming apparent at this time. A successful clinical trial with iproniazid was conducted in 1958, and the results of this study led to its introduction to general psychiatric use. At the present time, monoamine oxidase inhibitors are used only in patients who are refractory to the tricyclic compounds or who exhibit severe endogenous atypical depression. Monoamine oxidase inhibitors are hepatotoxic and cause hypertensive crises when ingested with foods that contain pressor amines.

The antidepressant action of imipramine, which is chemically related to chlorpromazine, was discovered almost by accident when Kuhn,[3] in 1958, found that the compound improved some of the depressive symptoms in a number of schizophrenic patients being treated for their psychosis. A subsequent trial confirmed the antidepressant action of this compound, and it was then officially introduced for this purpose.

Subsequent laboratory studies have demonstrated that the depressive syndrome induced in rodents by treatment with reserpine was abolished by the tricyclic agents. With the availability of the reserpine model, a number of new antidepressant compounds have been discovered and introduced into therapy, and these compounds, discussed below, are now the treatment of choice for most depressions. In addition, several newer compounds that do not share the classical profile of activity in laboratory animal tests as the tricyclic compounds, and are not monoamine oxidase inhibitors, have been found to relieve the symptoms of depression in the clinic. A few of these compounds, referred to as "second generation" antidepressants, are also discussed below.

Classification and Symptoms of Depression

Depression is one of the affective disorders. It is estimated that 8 to 20% of the population suffers from depression,[4,5] Only about 70% of patients treated with antidepressant drugs will improve, and this often includes a 50% placebo effect. Furthermore, improvement requires at least 4 weeks of treatment with antidepressant drugs. Affective disorders do not include states of sadness that occur commonly in every human being. In addition, they are generally differentiated from reactive depression, which is usually precipitated by a significant life stress, such as loss of a loved one. Affective disorders are characterized by a dysphoric mood that is pervasive and for which there is no obvious reason or precipitating factor.

Affective disorders characterized only by depression are called unipolar depression. In some types of depression, the victim oscillates between states of depression and mania. This is called manic-depressive psychosis, or bipolar depression.

The symptoms of depression include the following: a loss of interest and pleasure in surroundings, a feeling of hopelessness and helplessness, lack of concern for self and inability to respond to the environment in a meaningful way. Social relationships of depressed patients usually suffer, particularly family ties and friendships, and victims are unable to work effectively and creatively. Vegetative symptoms also occur, including severe disturbance of appetite which usually leads to a marked gain or loss of weight. There is a disturbance of sleep, a lack of energy, disinterest in ones surroundings and impotence. There is a feeling of self-reproach, diminution of intellectual function, and lapses of memory. People with severe affective disorders usually have recurrent thoughts of death and may consider the possibility of suicide.

Victims of manic disorders are usually irritable, with expansive moods, excessive talkativeness, flight of ideas, delusions of grandeur, and a decreased need for sleep. In addition, they have difficulty concentrating, tend to lack judgment, and are irresponsible.

Most affective disorders are cyclic, and in that sense are self-limiting. As much as half the patient population improves spontaneously without the need for treatment. Thus, the goal of therapy is to shorten and abort the depressive episode. About 10 to 15% of patients are chronically depressed, so that their level of function is never satisfactory. Depression recurs in about 40 to 50% of patients. Interruption of severe depression is essential, however, in order to minimize the possibility of suicide, which may be as high as 25% in chronic, repeated severe depressive patients. Suicide risk is especially high in adolescents and in patients over 45 years of age, who have suffered from significant traumatic events.

Theory of Antidepressant Drug Mechanism

The most widely held theories of depression suggest an aberation in one of the biogenic amines. Norepinephrine (NE) appears to be the most likely candidate. This is due to the apparent mechanism of action of clinically effective antidepressants. Drugs that are effective in the treatment of depression produce their therapeutic action by increasing the amount of norepinephrine at selected receptor sites within the brain. This effect results from enhancement of transmitter availability from noradrenergic nerves. The most important group of antidepressants are the tricyclic compounds. Tricyclic antidepressants (TCA) increase the amount of norepinephrine within the synapse by blocking the transport or reuptake of catecholamines into the presynaptic nerve terminal from which they were released. This reuptake process is one of the major steps involved in the control of noradrenergic transmission, since reuptake of the neurotransmitter leads to its physiologic inactivation. Another group of antidepressant compounds, the monoamine oxidase inhibitors, also cause an increase in the concentration of norepinephrine at its receptors but, in contrast to the tricyclic compounds, they do so by inhibition of the enzyme that controls the rate at which released norephinephrine is metabolized.

Serotonin (5-HT) has also been implicated in the mechanism of action of antidepressant drugs. Monoamine oxidase inhibitors decrease serotonin metabolism as well as norepinephrine. In addition, the tricyclic antidepressants generally inhibit the reuptake of serotonin. The various tricyclic antidepressants show a spectrum of relative potencies for inhibition of serotonin and norepinephrine uptake; thus, some possess specificity for inhibiting serotonin uptake and others, a specificity for inhibiting norepinephrine uptake.

Some generalities have been drawn between the chemical structure of the tricyclic antidepressants and their mechanism of action in alleviating depression, i.e., their specificity for the inhibition of NE or 5-HT reuptake. The secondary amine tricyclic antidepressants (desipramine, nortriptyline, protriptyline, and amoxapine) are more selective for NE uptake inhibition, while the tertiary amines (imipramine, amitriptyline, and doxepin) are more selective for inhibition of 5-HT uptake. Out of this group, amitriptyline is the most selective 5-HT uptake inhibitor. Maprotiline, a tetracyclic antidepressant and a secondary amine, is a specific NE uptake inhibitor. Some of the newer ''second generation'' antidepressants agree with the above structure-selectivity correlation; however, their chemical structures are too divergent to make any comparison meaningful.

The secondary amine tricyclic antidepressants, in addition to being more selective for NE uptake inhibition, are also less sedative and cause fewer anticholinergic side effects than the tertiary amines, with the exception of maprotiline, a secondary amine that is both sedative and produces significant anticholinergic side effects. A similar lack of correlation between secondary and tertiary amines and side effects exists for the ''second generation'' antidepressants. Nonetheless, there is a general agreement among the secondary amines with respect to selectivity for NE uptake inhibition and their reduced propensity for causing sedation and anticholinergic effects. This correlation can be drawn only for the tricyclic antidepressants and not for the ''second generation'' antidepressants (e.g., trazodone, zimelidine, nomifensine, etc.).

Since there are antidepressants that are selective uptake inhibitors for NE (maprotiline, desipramine, and nomifensine) or for 5-HT (trazodone, amitriptyline, and zimelidine), attempts have been made to identify subtypes of depression by biochemical markers and therapeutic response. These subtypes are presumed to be NE or 5-HT deficient depressed patients. Measurements of urinary or CSF levels of the major metabolites of NE and 5-HT tend to support this contention.

These studies suggest that NE deficient patients can be characterized by their low pretreatment levels of 3-methoxy-4-hydroxyphenylglycol (MHPG), the major central NE metabolite. These patients should respond well to the more selective NE uptake inhibitors (i.e., the secondary amine TCA's, maprotiline, nomifensine, and imipramine which is metabolized to desipramine). Patients with 5-HT deficiencies should respond well to the more selective 5-HT uptake inhibitors (e.g., amitriptyline, trazodone, and zimelidine).

Side Effects

Significant toxic effects of the tricyclic antidepressants are relatively common, occurring in up to 5% of patients. These effects have recently been reviewed by Blackwell.[6]

The cardiovascular side effects of the tricyclic antidepressants has been an area of major concern. Carney[7] has described four separate effects on cardiac function that appear to be mediated by three different mechanisms, anticholinergic-vagal inhibition, quinidine-like effect on His bundle, and α-adrenergic receptor blockade. These effects are impaired contractility, increased heart rate, delayed conduction, and postural hypotension.

Aside from the therapeutic effect of the tricyclic antidepressants by an action on the CNS, adverse effects on the nervous system have also been noted. This includes lowering of seizure threshold, disorientation, hyperpyrexia, and agitation.

Agranulocytosis and skin rashes have also been reported to occur with the tricyclic antidepressants.

The anticholinergic (muscarinic) effects of the tricyclic compounds account for a great number of the side effects. Increased heart rate and disorientation are due to the anticholinergic effects. In addition, the GI system is affected with resultant dry mouth, constipation, paralytic ileus, and hiatus hernia. The genitourinary system is affected by the anticholinergic effects resulting in hesitancy, retention, and impaired sexual function. The anticholinergic effect of tricyclic antidepressants can result in blurred vision and precipitation of glaucoma.

Because a major part of the population being treated with tricyclic antidepressants is suicidal most of the reports on acute intoxication are from attempted suicides.[8] Many of the effects of overdose are an exacerbation of the side effects, principally the anticholinergic and cardiovascular. In adults treated with imipramine there is approximately a tenfold difference between therapeutic and lethal doses.

Drug Interactions

In addition to the adverse effects of tricyclics when given alone, these compounds are known to interact with a variety of drugs.[6] These include monoamine oxidase inhibitors, sympathomimetic agents, alcohol, phenothiazines, barbiturates, oral contraceptives, cigarette smoking, and some antihypertensive agents. Barbiturates, oral contraceptives, and cigarette smoking decrease and phenothazines increase the plasma concentration of tricyclic antidepressants. The hypertensive effects of phenylephrine, norepinephrine, and methylphenidate are potentiated and the hypotensive effects of clonidine, reserpine, alpha-methyl dopa, and guanethidine are blocked. The sedative effects of alcohol are increased. Thus, care must be taken when prescribing tricyclic antidepressants.

Testing Methods

Methods used for detecting ''antidepressant'' activity in animals have developed with the discovery of various classes of therapeutically efficacious antidepressants. Very few animal models for depression exist, and those that do are generally either a form of reactive depression or drug-induced depression. Other preclinical testing methods rely on describing the effects in animals of known antidepressants and searching for new compounds with similar effects. As new classes of antidepressants are developed the testing methods are refined. Thus, several methods that were previously used for preclinical screening for antidepressants do not detect the newer (second generation) antidepressants, such as mianserin. The usefulness of a method is greatly reduced if it does not detect second generation antidepressants.

In Vitro Methods

The in vitro testing methods for antidepressants rely heavily on the biogenic amine hypothesis of affective disorders. Inhibition of reuptake of norepinephrine (NE) and/or serotonin (5-HT) is a common action shared by the tricyclic antidepressants. Thus, in vitro testing can determine if a potential antidepressant may function by this mechanism. However, monoamine oxidase inhibitors (MAOI), such as phenelzine, tranylcypromine, and iproniazid, are also antidepressants, but do not act by inhibition of reuptake. Rather, they act by inhibiting MAO. Because of the different mechanisms of action of antidepressants, in vitro screening has limited utility in the discovery of new classes of antidepressants but is extremely useful in determining the mechanism of action and in the complete characterization of the drug effects.

Ex Vivo Methods

In ex vivo studies, the drug is injected into the intact animal and biochemical determinations are made after sacrifice of the animal. Since the biogenic amine hypothesis of depression suggests that increases in synaptic NE would alleviate depression, the activity of noradrenergic neurons is measured by a variety of techniques. Norepinephrine ''turnover'' can be determined by measuring the rate of decline of NE after synthesis inhibition with alpha-methyl-tyrosine (AMT), or measurement of the concentrations of NE and metabolites, such as 3-methoxy-4-hydroxyphenylglycol (MHPG). No single approach is sufficient for measuring ''turnover'' changes induced by antidepressants. For instance, MHPG concentrations would not increase if MAO is inhibited because MAO acts in one of the steps in the metabolism of NE to MHPG. Nevertheless, MAOIs are antidepressants, and they do increase NE concentrations.

Increased noradrenergic neuronal activity can also be measured by observing postsynaptic effects. One such approach is to inject the test drug and at some time following injection the animal is sacrificed and the sensitivty of NE stimulated adenylate cyclase is determined.[9] Acute administration of most antidepressants increase the sensitivity of NE stimulated adenylate cyclase. However, many antidepressants, including desipramine do not produce this effect.[10]

Clinically efficacious antidepressants exhibit a delayed onset of therapeutic action. Recent studies have observed effects which occur after chronic but not acute administration of antidepressants. Thus, a wide variety of antidepressants when administered chronically decrease the activity of NE coupled adenylate cyclase. These compounds include uptake inhibitors (chlorimipramine, imipramine, amitriptyline, desipramine, nisoxetine), MAO inhibitors (pargyline, nialamid, tranylcypromine) and atypical antidepressants (iprindole, mianserin). In addition electroconvulsive therapy (ECT) which is also used to treat depression reduced the sensitivity of NE sensitive adenylate cyclase.[11] Chronic treatment with a wide variety, but not all (e.g., mianserin, nisoxetine) antidepressants, has been found to reduce beta-receptor density in the cerebral cortex; alpha-1 receptors are seldom changed.[12-14] The density of alpha-2 receptors, believed to be localized on the presynaptic nerve terminal, have been found to increase after chronic antidepressant treatment.[14] This was also observed with the second generation antidepressant, mianserin. Serotonin ($5\text{-}HT_2$), but not $5\text{-}HT_1$, receptors have been noted to decrease after chronic antidepressant treatment.[12] In addition, mianserin has been reported to inhibit the $5\text{-}HT_2$ receptors.[12]

In Vivo Methods

The in vivo methods for testing antidepressant activity can be broken down into three subclasses. These are (1) using drugs which are known to induce depression in humans and determining if potential antidepressants antagonize the effects in animals, (2) attempts to induce depression in animals by manipulation of their environment, and (3) observing the effects of antidepressants and testing new compounds for similar effects, although these effects may not have an obvious correlation with depression.

Clinically, reserpine induces depression in humans. This has served as a rationale for the testing of antagonism of reserpine-induced effects. In addition tetrabenazine, a synthetic benzoquinolizine derivative, with a similar mechanism of action as reserpine has been used. Reserpine-induced hypothermia, ptosis, and sedation as well as tetrabenazine-induced ptosis and sedation are antagonized by most antidepressants.[15] However, mianserin is a notable exception.

Several techniques have been used to induce depression in animals by manipulating their environment. The separation of two animals that have formed an affectional bond has been utilized as a model of depression.[16] Suomi et al.[17] observed that chronic imipramine abolished the symptoms of separation. Kraemer et al.[18] observed similar effects with the MAOI, isoniazid. This procedure is not routinely used since the effects of various psychotropic compounds have not been fully characterized and it is expensive to do in both time and animals.

Two different procedures have been used for inducing "depression" in rodents. These are learned helplessness[19] and behavioral despair.[20] Learned helplessness entails presentation of an unavoidable aversive event (e.g., shock). The animals are then tested in a situation where they can escape from shock; "helpless" animals do not escape the shock as frequently. Some pharmacological testing has suggested that chronic but not acute imipramine ingestion reduced the degree of learned helplessness.[21] The behavioral despair procedure is similar to learned helplessness and has been thoroughly characterized pharmacologically.[20] Briefly, this procedure, which has been successfully used with rats and mice, entails placing the animal in a container of water from which it cannot escape. After actively swimming for a few minutes the animals exhibit a clearly defined immobile posture. The duration of immobility is quantified. Immobility in this case is thought to represent a depressed condition. Tricyclic antidepressants, MAOI, and atypical antidepressants, such as iprindole and mianserin, reduce the amount of immobility. Anticholinergics, antihistaminergics, and stimulants produce a positive result in this test. Many of these false positives can be eliminated by assessing the effects of the compound on general activity. Results have indicated that antidepressants mostly decrease motor activity at doses that decrease immobility whereas most of the false positives increase general activity.

Several models which are hard to directly relate to human depression have been used for the testing of antidepressants. One of these is muricide (i.e., mouse killing by rats). Tricyclic antidepressants, MAOI, and mianserin as well as electroconvulsive shock therapy inhibit muricide.[22-24] As with the behavior despair test psychomotor stimulants, anticholinergics, and antihistaminergics produce false positives.

A variety of electrophysiological techniques have been used to screen antidepressants. Monosynaptic spinal reflexes have been studied and several antidepressants potentiate 5-HT induced facilitation of the monosynaptic reflex. However, the atypical antidepressants iprindole and mianserin do not work in this test. Another neurophysiological technique is kindling. The effects of antidepressants in this model however vary with class. Tricyclics suppress amygdaloid seizures, MAOI prolong seizures, and atypicals (iprinidole and mianserin) are without effect.[25] Thus, these tests yield little hope as a preclinical screen, but may help in teasing apart various classes of antidepressants.

Finally, O'Donnell and Seiden[26] have observed improved performance on a specific operant schedule (DRL-72 sec schedule). Antidepressants, but not major tranquilizers or antihistaminergics, improved performance. A wide variety of antidepressants were active, including tricyclics, MAOI, and atypicals (iprindole and mianserin). This paradigm at present seems to be one of the most specific for detecting antidepressant activity. Its utility in discovering new, novel antidepressants remains to be determined.

Clinical Evaluation

The Early Clinical Drug Evaluation program (ECDEU) assessment manual lists the following instruments as those most frequently used by drug investigators to evaluate antidepressants:

1. Hamilton Depression Scale (HAMD)
2. Clinical Global Impressions (CGI)
3. Depression Status Inventory (DSI)
4. Self Rating Depression Scale (SDS)

The ECDEU assessment manual[27] should be consulted for details on how the subjects are rated. As an example of one diagnostic system, the Hamilton Depression Scale, which is one of the most widely used methods for assessment of depressive states, is used to assess the severity of a patient's condition quantitatively and to assess changes in that condition.[28] It is a 23 item scale with a format that can be used with the General Scoring Sheet. The scale has been used in a number of different versions and therefore, comparisons between laboratories are sometimes difficult. In the original method, the subject is scored on such items as depressed mood, guilt, suicide, insomnia, work, retardation, agitation, anxiety, somatic symptoms, and weight changes. The subject is scored by a rater; therefore, the value of the system depends upon the skill and experience of the rater and the adequacy of the information available to him. Each symptom is rated as to degree of severity.

In the clinical assessment of antidepressant drugs, the subject may be evaluated before and after treatment or several treatments may be compared with each other. One of the problems in assessing antidepressant efficacy is the high rate of spontaneous recovery that

Imipramine

Phenothiazines

R'=Cl, R''=CH_3 Chlorpromazine

FIGURE 1. Structural similarities of imipramine with the phenothiazines.

Amitriptyline

Thioxanthenes

R'=Cl, R''=CH_3 Chlorprothixene

FIGURE 2. Structural similarities of amitriptyline with the thioxanthenes.

occurs during the 3- to 6-week period normally required for the pharmacologic effect to occur. The difference between a 50% recovery in placebo treated vs. 70% in drug-treated individuals is often difficult to assess.

DRUG MONOGRAPHS

The clinically useful antidepressants that are marketed in the U.S. as of 1983 will be discussed and their structure-activity relationships (SARs), pharmacokinetics, uses and dosage, toxicology, and physical-chemical data will be presented in monograph form. The antidepressants are subdivided by chemical structure into various classes. The MAO inhibitors are discussed in the accompanying chapter of this volume. This chapter describes the tricyclic antidepressants and the second generation antidepressants, which are newer agents with nontricyclic structures that act by mechanisms other than inhibition of monoamine reuptake. This latter class was led by the introduction of trazodone in the U.S. market in 1982.

The discovery of the antidepressant activity of the monoamine uptake inhibitors or tricyclic antidepressants occurred serendipitously when Kuhn[29,30] in 1958 reported that during the clinical evaluation of imipramine as a potential antipsychotic he noted an antidepressant therapeutic response in depressed patients. Imipramine and its derivatives bear structural similarities to the phenothiazine class of tranquilizers and were originally prepared as analogues of the phenothiazines by Hafliger and Schindler.[31] Their originally claimed utility was for use as antihistaminics, sedatives, analgesics, and antiparkinsonian drugs. Imipramine and the phenothiazines differ in structure only in the central ring (Figure 1) where the thiazine ring of the phenothiazines is replaced with an azepine ring, yet imipramine possesses entirely different activity in man than the phenothiazines. Imipramine and its derivatives are ineffective in agitated psychotic patients, but imipramine is useful in the treatment of endogenous depression.[3,30]

Similarly, structural analogues of the thioxanthene antipsychotics were shown to possess antidepressant activity. This group was led by amitriptyline which differs in structure from the thioxanthenes only in the central ring (Figure 2) where the cyclohexane ring is replaced with a cycloheptane ring. As was the case with imipramine, amitriptyline possesses an entirely different spectrum of activity in man than the antipsychotic thioxanthenes.

Although the structures of the tricyclic antidepressants (e.g., imipramine) and the phenothiazines (e.g., chlorpromazine) appear to be similar, these compounds possess entirely different clinical activity. Wilhelm[32] has attributed these differences to the ability of imipramine to assume a bent conformation where the flexure angle of the two aromatic rings is 55°. The structure of chlorpromazine is more planar than imipramine and the flexure angle is 25°. Therefore, chlorpromazine is unable to assume the postulated proper conformation to cause an antidepressant response.

Since the discovery of the antidepressant activity of imipramine many experimental and clinically useful antidepressants have been found which have a diversity of chemical structure. Research had been directed toward finding drugs with the same mechanism of action as imipramine, that is, monoamine reuptake inhibition. More recently the product of research has revealed a second generation of antidepressants with various mechanism of actions.

The SAR of antidepressants have been described in great detail.[33-35] Pharmacokinetics,[36,37] metabolism,[38] and toxicology[8,39-41] of antidepressants have been discussed in several reviews. The most recently marketed antidepressants, doxepin,[42] amoxapine,[43,44] maprotiline,[45,46] and trazodone,[47,48] have been reviewed as well.

In the following discussions of the SARs of the antidepressants, the various agents are distinguished by a "*" for drugs marketed in the U.S., by a "+" for agents that are clinically effective but not marketed, and by a "**" for drugs that are marketed in foreign countries but not the U.S. In the following sections on pharmacokinetics, the metabolites which occur in man are distinguished by "solid" arrows in the figures and the metabolites which occur in animals are distinguished by "dashed" arrows.

Dibenzazepines (SARs)

The prototype drug of this class is imipramine.[49] The antidepressant effects that it manifests in man have been attributed to its ability to inhibit norepinephrine and serotonin reuptake by nerve terminals,[50] thus increasing synaptic concentrations of these neurotransmitters in the CNS. Imipramine is rapidly metabolized to desipramine (see Figure 3) which may contribute to the antidepressant action of imipramine, since desipramine has been shown to be a clinical effective antidepressant.[51] Desipramine is claimed to have a faster onset of action and to be pharmacologically more active than imipramine. The introduction of a beta-methyl group on the imipramine side chain gave trimipramine, another clinically active antidepressant.[52] Trimipramine exhibits sedative and anxiolytic effects in addition to its antidepressant action. Two antidepressant compounds were found when the tertiary amine of imipramine was varied. The first clinically effective compound, lofepramine, is rapidly metabolized to desipramine[53] and exhibits similar effects on norepinephrine and serotonin reuptake in rat nerve terminals.[54] In the clinic lofepramine was comparable with imipramine and is claimed to produce fewer side effects.[55,56] Opipramol, which contains a 10,11-double bond, retains the antidepressant action of imipramine and is also anxiolytic.[57,58]

Substitution of the aromatic nucleus gives variable results on antidepressant activity[59] in the reversal of reserpine-like sedation caused by RO-4-1284 in rats. The 3-cloro derivative, chlorimipramine, was less active than imipramine in this test (see Figure 4). In human platelets norepinephrine reuptake inhibition was the same for chlorimipramine and imipramine; however serotonin reuptake inhibition was increased for chlorimipramine.[60] Chlorimipramine is as effective as imipramine in the clinic.[61] Ketipramine, the 10-keto derivative, is clinically useful in depression.[62] Blocking the metabolic deactivating 2-hydroxylation of imipramine with 2,8-dimethyl substituents gave a compound less active than imipramine in the prevention of reserpine-like sedation in rats.[59] Replacement of the aromatic ring of imipramine with a pyridine ring gave an analogue less active in the tetrabenazine sedation model in rats and increased toxicity.[63]

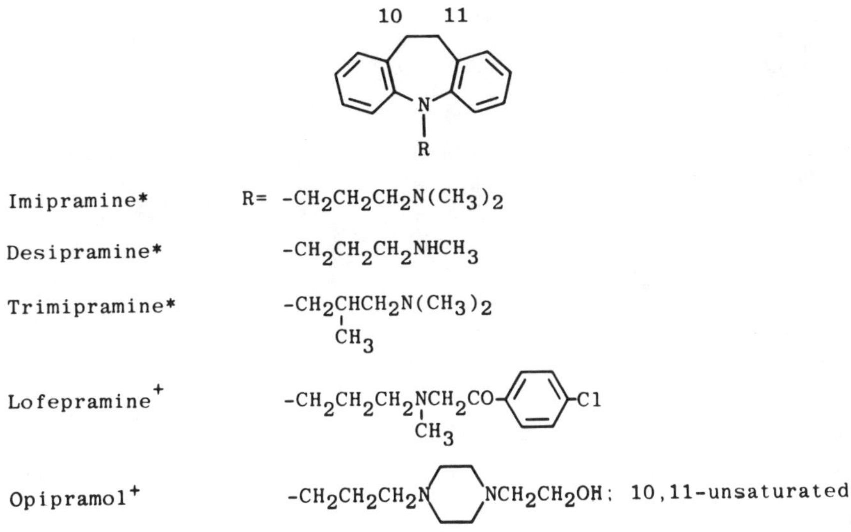

FIGURE 3. Basic side-chain modifications of the dibenzazepines.

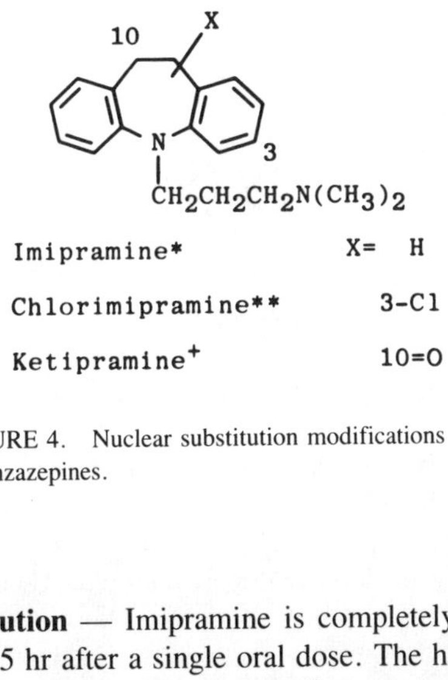

FIGURE 4. Nuclear substitution modifications of the dibenzazepines.

Imipramine

Pharmacokinetics

Absorption and distribution — Imipramine is completely absorbed and peak plasma levels are achieved in 2 to 5 hr after a single oral dose. The half-life of imipramine is 8 to 16 hr[64] and is 89 to 94% bound to serum protein.[65] Extensive first-pass metabolism takes place in the liver to desipramine, the primary clinically active metabolite.[64] Imipramine and its metabolites are widely distributed throughout the body. The therapeutically effective plasma concentration of imipramine plus desipramine is in the range of 180 to 240 ng/mℓ.[36,66,67]

Metabolism — Imipramine is rapidly metabolized to the primary active metabolite, desipramine. Further metabolism of imipramine and desipramine takes place to give the oxidized, hydroxylated, and dealkylated metabolites shown in Figure 5. Following a single dose of imipramine about 70% of the dose is excreted within 3 days in the urine as free or conjugated metabolites. Only trace amounts of imipramine are excreted unchanged.[64,68-71]

Uses and Dosage

Indications — For the relief of the symptoms of depression. Endogenous depression is more likely to be alleviated than other forms of depression. Also used as temporary adjuctive

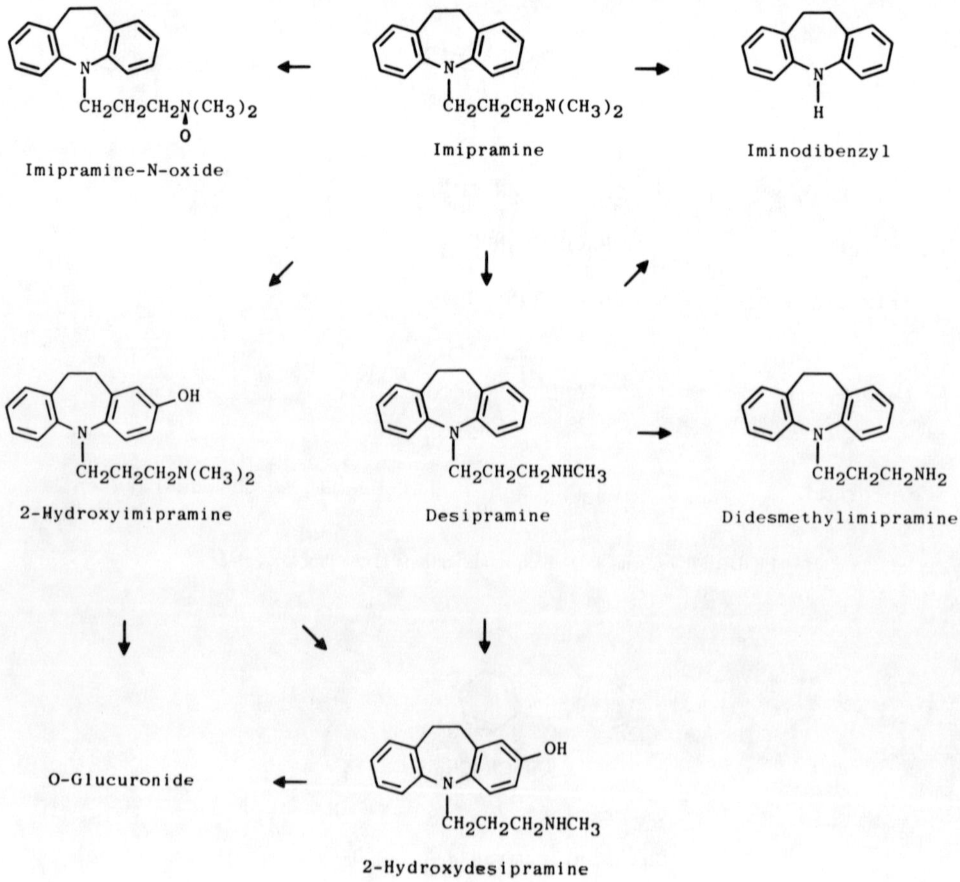

FIGURE 5. Metabolism of imipramine.

therapy in reducing enuresis (bed wetting) in children aged 6 years and older after possible organic causes have been excluded.[72]

Dosage — In hospitalized patients initially 100 mg/day in divided doses is administered orally and gradually increased to 200 mg/day as required. The dose is increased to 250 to 300 mg/day if there is no response after 2 weeks. In patients unable or unwilling to use oral medication, therapy is started with parenteral administration up to 100 mg/day i.m. in divided doses. The oral form should replace the parenteral form as soon as possible. In outpatients initially 75 mg/day is given orally and is increased to 150 mg/day. After remission the maintenance dose is 50 to 150 mg/day. Adolescent and elderly patients are given 30 to 40 mg/day initially and it is generally not necessary to exceed 100 mg/day.[72]

Dosage forms — Imipramine hydrochloride is available in tablet form [10, 25, and 50 mg as Imavate® (Robins), Janimime® (Abbott), SK-Pramine® (SKF), Tofranil® (Geigy), or by nonproprietary name] and in injectable form [25 mg/2 mℓ as Tofranil® (Geigy)]. Another salt form of imipramine, imipramine pamoate, is available in capsule form [75, 100, 125, and 150 mg of imipramine hydrochloride as Tofranil-PM® (Geigy)].[72] The advantage of imipramine pamoate over imipramine hydrochloride is yet to be established.

Toxicity

The oral LD_{50} of imipramine is 666 mg/kg in mice[73] and 305 to 682 mg/kg in rats.[74] The minimum reported lethal dose of imipramine by an adult is 625 mg and the maximum

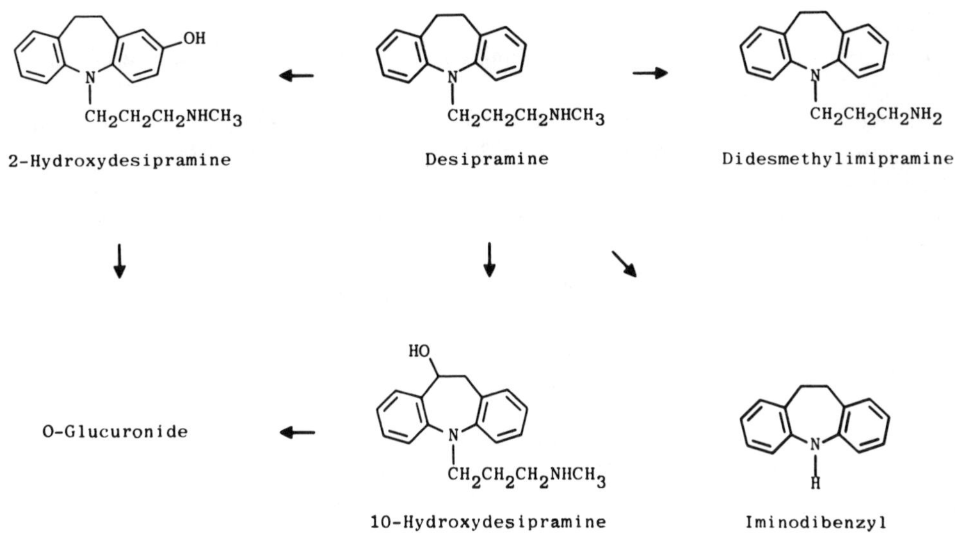

FIGURE 6. Metabolism of desipramine.

survived dose is 5375 mg.[39] When plasma levels of imipramine and desipramine exceed 1 μg/mℓ severe overdose symptoms occur.[40]

The symptoms of acute poisoning with imipramine and the other tricyclic antidepressants are generally the same and for the most part are an extension of the side effects of these compounds. The CNS or anticholinergic effects include initial motor disturbance, dry mouth, mydriasis, agitation with visual and auditory hallucinations followed by coma, convulsions, and respiratory depression. Other effects are hyperreflexis, hyperpyrexia, urinary retention, paralytic ileus, ataxia, tremors, delirum, muscle rigidity, acidosis, and hypokalemia. The cardiotoxic effects are tachycardia, arrthymias, atrial fibrillation, depressed myocardial contractility, bundle-branch block, profound hypotension, hypoxemia, and cardiac arrest.[8,40,41]

Physical-Chemical Data

Chemical name and formula — 10,11-Dihydro-*N*,*N*-dimethyl-5*H*-dibenz[b,f]azepine-5-propanamine monohydrochloride; $C_{19}H_{24}N_2 \cdot HCl$ (316.9).[75]

Physical properties — Imipramine hydrochloride is an odorless white to off-white crystalline powder with a melting point of 170 to 174°C and a pKa of 9.4. It is soluble in water and alcohol and is very slightly soluble in ether. Imipramine hydrochloride turns yellow to red on exposure to light; however, if moisture and light are excluded, imipramine hydrochloride is stable for several years.[72,75,76]

Desipramine

Pharmacokinetics

Absorption and distribution — Desipramine is well absorbed when administered orally. The half-life of desipramine is 30 hr[77] and is about 92% bound to serum protein.[78] Distribution in man is available from a fatal overdose case where desipramine concentrations in the brain, liver, and kidney were ten times the concentrations found in plasma.[79] There appears to be a correlation between desipramine plasma levels and therapeutic effect with effective plasma concentrations in the range of 37 to 163 ng/mℓ.[80,81]

Metabolism — Desipramine is the primary metabolite of imipramine. In the absence of a quantitative study in man, desipramine is believed to be similarily metabolized (see Figure 6). The hydroxylated metabolites are eliminated in the urine primarily as glucuronides.[69] Only trace amounts of desipramine are excreted in the urine unchanged.[82]

Uses and Dosage

Indications — For the relief of the symptoms of depression, particularly useful in endogenous depression.[72]

Dosage — The adult oral dose of 75 to 200 mg/day can be administered in divided or once daily doses and may be increased to 300 mg/day as required. When a satisfactory response is achieved, a lower maintenance once daily dosage is continued for at least 2 months. Adolescent and geriatric patients are given 25 to 50 mg/day initially and may be increased to 100 mg/day as required.[72]

Dosage forms — Desipramine hydrochloride is available in tablet form [25, 50, 75, 100, and 150 mg as Norpramin® (Merrell Dow)] and in capsule form [25 and 50 mg as Pertofane® (USV)].[72]

Toxicity

The oral LD_{50} of desipramine is 315 to 935 mg/kg in mice and 320 to 625 mg/kg in rats.[73,74] Reports of acute poisoning with desipramine are limited.[40] The symptoms are generally the same as with the other tricyclic antidepressants (see toxicity section for imipramine).

Physical-Chemical Data

Chemical name and formula — 10,11-Dihydro-*N*-methyl-5*H*-dibenz[b,f]azepine-5-propanamine monohydrochloride; $C_{18}H_{22}N_2 \cdot HCl$ (302.8).

Physical properties — Desipramine hydrochloride is an odorless white crystalline powder with a melting point of 208 to 218° C and pKa's of 1.5 and 10.2. It is soluble in water, alcohol and chloroform and is insoluble in ether.[75,76]

Trimipramine

Pharmacokinetics

Absorption and distribution — Trimipramine is rapidly absorbed following oral administration. Peak plasma levels are reached in 2 hr after a single oral dose. The half-life of trimipramine is about 7.4 hr.[83,84]

Metabolism — Trimipramine is a beta-methyl analogue of imipramine and shares common metabolic processes. Metabolism takes place by oxidative demethylation, hydroxylation, and dealkylation as shown in Figure 7. The free or conjugated hydroxylated metabolites are the major components of the urinary metabolites.[83,85]

Uses and Dosage

Indications — For the relief of the symptoms of depression. Endogenous depression is more likely to be alleviated than other forms of depression.[72]

Dosage — The adult outpatient dose is 75 mg/day in divided doses and may be increased to 150 mg/day. In hospitalized patients initially 100 mg/day in divided doses is administered orally and gradually increased to 200 mg/day as required. The dose is increased to 250 to 300 mg/day if there is no improvement in 2 to 3 weeks. Adolescent and geriatric patients are given orally 50 mg/day initially and may be gradually increased up to 100 mg/day as required. Following remission the lowest dose that maintains remission is administered once a day at bedtime. To minimize relapse maintenance therapy is continued for about 3 months.[72]

Dosage forms — Trimipramine maleate is available in capsule form [25 and 50 mg as Surmontil® (Ives)].[72]

Toxicity

The oral LD_{50} of trimipramine is 250 to 475 mg/kg in mice and 800 mg/kg in rats.[73] The symptoms of acute poisoning with trimipramine are similar to the other tricyclic antidepressants (see toxicity section for imipramine).

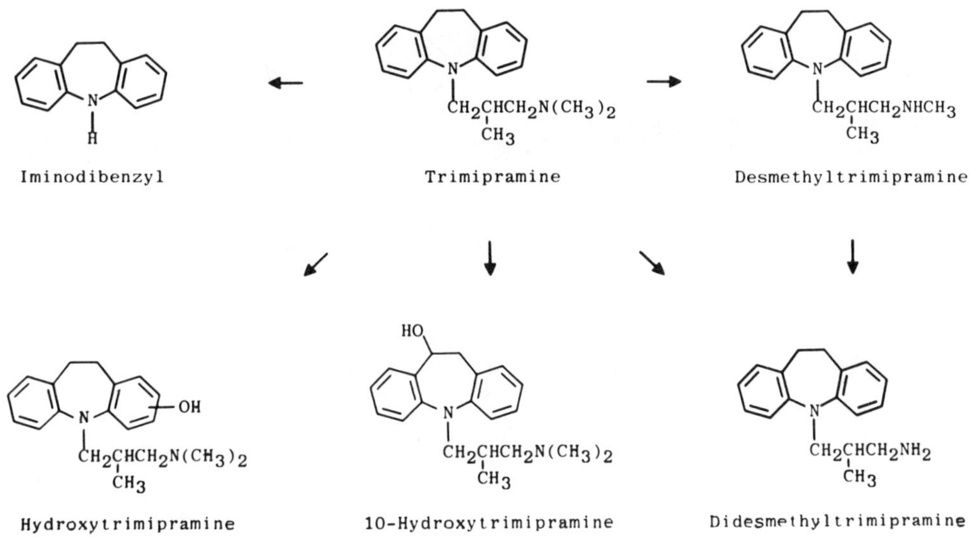

FIGURE 7. Metabolism of trimipramine.

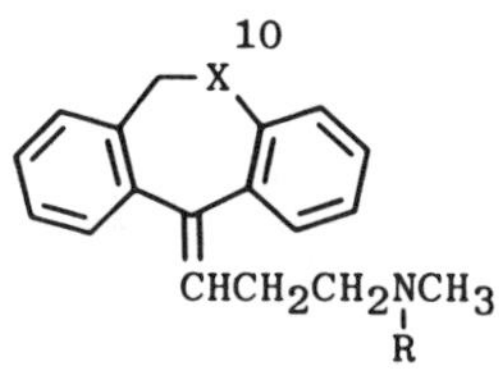

Amitriptyline*	X=CH_2	R=CH_3
Nortriptyline*	CH_2	H
Doxepin*	O	CH_3
Dothiepin**	S	CH_3

FIGURE 8. Dibenzocycloheptane antidepressants.

Physical-Chemical Data

Chemical name and formula — 10,11-Dihydro-β,*N*,*N*-trimethyl-5*H*-dibenz[b,f]azepine-5-propanamine (Z)-2-butenedioate; $C_{20}H_{26}N_2{\cdot}C_4H_4O_4$(410.5).

Physical properties — Trimipramine maleate is an odorless white crystalline powder with a melting point of 140 to 144°C. It is soluble in methanol and chloroform, is slightly soluble in alcohol and acetone, and is very slightly soluble in ether and water.[72,75]

Dibenzocycloheptanes (SARs)

The mechanism of action of this class is the same as that of the dibenzazepines. These drugs inhibit the reuptake of norepinephrine and serotonin into nerve terminals, thereby increasing synaptic concentrations of these neurotransmitters in the CNS. The prototype drug of this class is amitriptyline,[49] which is extensively metabolized and is deactivated principally by C-10 hydroxylation. The major metabolite, nortriptyline (see Figure 8) apparently contributes to the antidepressant effect of amitriptyline, since nortriptyline is a clinical useful antidepressant.[86] Modification of the cycloheptane ring of amitriptyline by replacement of the C-10 methylene with oxygen and sulfur lead to the clinically effective antidepressants, doxepin[87-89] and dothiepin.[90] Doxepin is a mixture of *cis*- and *trans*-isomers and not only is an antidepressant but also has anxiolytic activity.

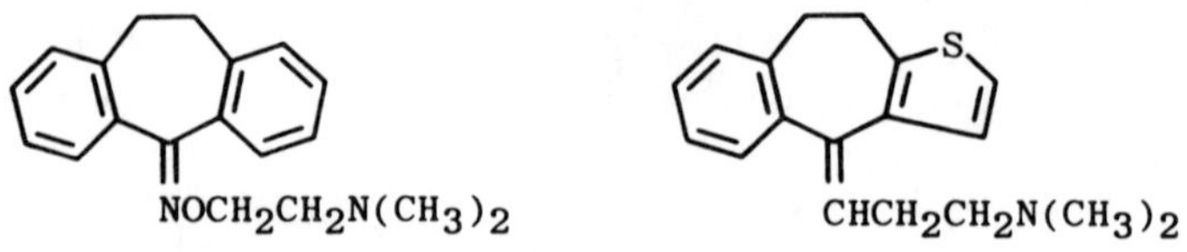

Noxiptilin+ IB-503+

FIGURE 9. Heteroatom analogues of amitriptyline.

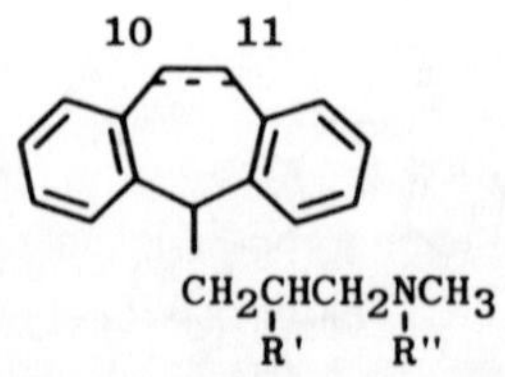

Protriptyline*	R'=H	R"=H	10,11-unsaturated
Butriptyline**	CH3	CH3	10,11-saturated

FIGURE 10. Modifications in site and degree of unsaturation of the dibenzocycloheptanes.

Numerous modifications have been made in the structure of amitriptyline to give compounds that are pharmacologically active,[33] but as yet no clear advantage over amitriptyline has been shown. The oximino analogue, noxiptilin,[91] is clinically useful in depression and may have a more favorable effect on mood than amitriptyline (see Figure 9). Antidepressant activity is retained in derivatives where a benzene ring is replaced by aromatic heterocycles.[35] One example is IB-503 which is a mixture of *cis-trans* isomers.[92]

Changes in the site and degree of unsaturation present in the amitriptyline nucleus lead to two clinically useful antidepressants shown in Figure 10. Butriptyline has a beta-methyl group on the amitriptyline side chain, and has a similar profile of biological activity to imipramine. It is more sedating than amitriptyline and may also have anxiolytic activity.[93,94] Protriptyline is one of the most potent antidepressants and is less sedating than amitriptyline.[95] This potency is attributed to the slow metabolism of protriptyline, since the primary site of metabolic deactivation of amitriptyline, i.e., hydroxylation at C-10, is blocked by the double bond. Substituents on the aromatic rings of protriptyline reduced potency as did the introduction of an exocyclic double bond.[96]

Amitriptyline

Pharmacokinetics

Absorption and distribution — Amitriptyline is rapidly absorbed from the GI tract and is extensively metabolized. Peak plasma levels are reached 2 to 4 hr after administration of a single 50-mg oral dose. The primary psychoactive metabolite is nortriptyline.[97] The half-life of amitriptyline is in the range of 9 to 25 hr.[98,99] Approximately 96% of amitriptyline is bound to plasma protein.[78] Information on distribution in man is available from four fatal overdose cases.[100] The highest levels were found in the liver and lower levels in the brain, kidney, urine, and plasma. In a study of 18 outpatients a linear correlation was shown between amitriptyline plus nortriptyline levels and therapeutic response. The therapeutically effective plasma concentration of amitriptyline plus nortriptyline is in the range of 160 to 240 ng/mℓ.[101]

Metabolism — Amitriptyline is extensively metabolized, and within 7 days after a single dose of amitriptyline 76 to 91% of the dose is eliminated in the urine as metabolites. Only trace amounts of unchanged drug are found.[102-104] The 10-hydroxylated metabolites are the major components found in human urine.[105] The major plasma metabolite is the psychoactive demethylated compound, nortriptyline. The major routes of metabolism are shown in Figure 11, which include oxidation, aliphatic hydroxylation, and demethylation.[104,106-108]

Uses and Dosage

Indications — Amitriptyline is prescribed for the relief of the symptoms of depression. Endogenous depression is more likely to be alleviated than other forms of depression. Amitriptyline in combination with the tranquilizer perphenazine or with chlordiazepoxide is recommended for the treatment of patients that exhibit both depression and anxiety.[72]

Dosage — In hospitalized patients, the initial dose is 100 mg/day administered orally, and this is gradually increased to 200 mg/day as required. In patients unwilling or unable to use oral medication, 20 to 30 mg is administered i.m. four times a day. The oral form should replace the parenteral form as soon as possible. In outpatients initially 75 mg/day in divided doses is given orally and is increased to 150 mg/day as required. The maintenance dosage is 50 to 100 mg/day and may be given in one dose at bedtime. Adolescent and elderly patients are given 10 mg three times a day with 20 mg at bedtime.[72]

Dosage forms — Amitriptyline hydrochloride is available in tablet form [10, 25, 50, 75, 100, and 150 mg as Amitid® (Squibb), Amitril® (Parke-Davis), Elavil® (MSD), Endep® (Roche), SK-Amitriptyline® (SKF), or by nonproprietary name] and in injectable form[10 mg/mℓ as Elavil® (MSD) or by nonproprietary name].[72] Amitriptyline hydrochloride is available in tablet form in combination with the tranquilizer perphenazine as Etrafon® (Schering) and Triavil® (MSD) and with chlordiazepoxide as Limbitrol® (Roche).

Toxicity

The oral LD_{50} of amitriptyline is 27 to 289 mg/kg in mice and 530 mg/kg in rats.[73] The minimum reported lethal dose of amitriptyline by an adult is 950 mg.[39] Overdose patients have survived ingestion of 2 g but suffered severe overdose symptoms including coma, convulsions, and arrhythmias. When plasma levels of amitriptyline and nortriptyline exceed 1 μg/mℓ severe overdose symptoms occur.[40] The symptoms of amitriptyline overdose are similar to the other tricyclic antidepressants (see toxicity section for imipramine).

Physical-Chemical Data

Chemical name and formula — 3-(10,11-Dihydro-5*H*-dibenzo[a,d]cyclohepten-5-ylidene)-*N*,*N*-dimethyl-1-propanamine monohydrochloride; $C_{20}H_{23}N \cdot HCl$ (313.9).

Physical properties — Amitriptyline hydrochloride is an odorless white or practically white crystalline powder with a melting point of 195 to 199°C and a pKa of 9.4. It is very soluble in water, alcohol, chloroform, and methanol and is insoluble in ether. Amitriptyline should be protected from light and air.[72,75,76]

Nortriptyline

Pharmacokinetics

Absorption and distribution — Nortriptyline is well absorbed from the GI tract. Peak plasma levels are reached in 5 hr following a single oral dose. The half-life of nortriptyline is 27 hr.[109] Approximately 95% is bound to plasma protein.[78] A curvilinear relationship exists between nortriptyline plasma levels and therapeutic response.[36] Patients with plasma levels in the range of 50 to 175 ng/mℓ show significant therapeutic response; however, patients above or below this range do not respond as well.[86,110,111]

Metabolism — Nortriptyline is extensively metabolized by hydroxylation and oxidative demethylation as is shown in Figure 11.[106,107] After a single dose of nortriptyline within 6 days nearly the entire dose is accounted for in the urine[112] with about 80% of the metabolites as free or conjugated 10-hydroxynortriptyline. Smaller amounts (17%) of 10-hydroxydes-

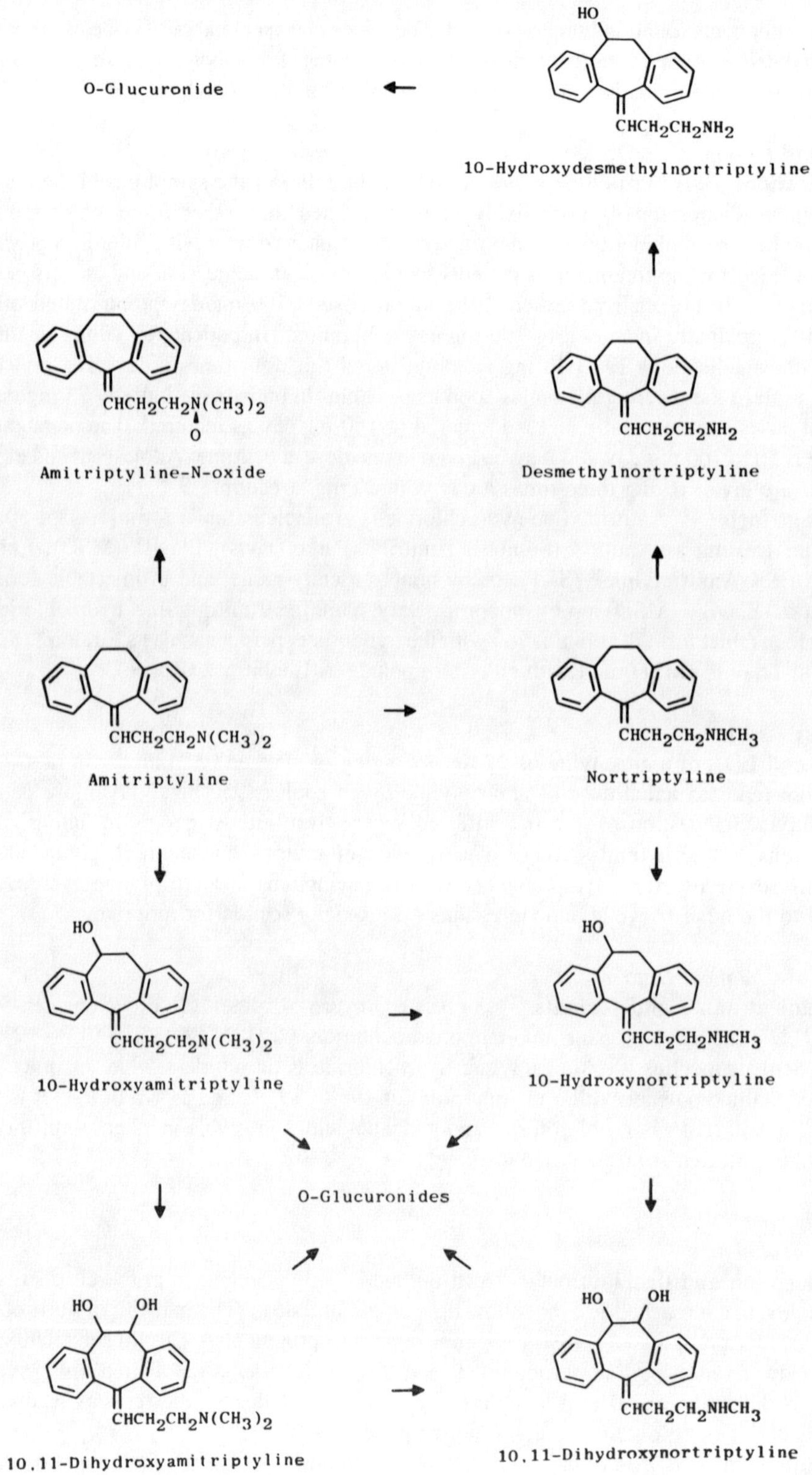

FIGURE 11. Metabolism of amitriptyline and nortriptyline.

methylnortriptyline and 1 to 2% of unchanged nortriptyline and desmethylnortriptyline are present in the urinary metabolites.[113]

Uses and Dosage

Indications — For the relief of the symptoms of depression. Endogenous depression is more likely to be alleviated than other forms of depression.[72]

Dosage — In adults 25 mg three or four times a day is administered orally in capsule or liquid form. The maximum recommended dose is 100 mg/day. Adolescent and elderly patients are given 30 to 50 mg/day in divided doses.[72]

Dosage form — Nortriptyline hydrochloride is available in liquid form [10 mg/5 mℓ as Aventyl® (Lilly) and Pamelor® (Sandoz)] and in capsule form [10, 25, and 75 mg as Aventyl® (Lilly) and Pamelor® (Sandoz)].[72]

Toxicity

The oral LD_{50} of nortriptyline is 275 to 329 mg/kg in mice and 503 mg/kg in rats.[73] Reports of acute poisoning with nortriptyline are limited and the symptoms of overdose are similar to the other tricyclic antidepressants (see toxicity section for imipramine). In one unsuccessful suicide attempt 2.5 g of nortriptyline was ingested with plasma levels of 0.9 μg/mℓ being obtained.[82]

Physical-Chemical Data

Chemical name and formula — 3-(10,11-Dihydro-5*H*-dibenzo[a,d]cyclohepten-5-ylidene)-*N*-methyl-1-propanamine monohydrochloride; $C_{19}H_{21}N{\cdot}HCl$ (299.8).

Physical properties — Nortriptyline hydrochloride is a white to off-white powder with a slight characteristic odor with a melting point of 213 to 215°C[114] and a pKa of 9.73. It is soluble in water, ethanol, chloroform, and methanol and is practically insoluble in ether. Nortriptyline should be protected from light.[75,76]

Protriptyline

Pharmacokinetics

Absorption and distribution — Protriptyline is well absorbed from the GI tract and peak plasma levels are reached in 8 to 12 hr.[72] The half-life of protriptyline is 78 ± 11 hr. It is 92% bound to plasma protein.[78] In rats and mice protriptyline rapidly penetrates the blood-brain barrier and is present almost entirely unchanged in the brain.[72] Preliminary studies show a correlation between therapeutic response and protriptyline plasma levels in the range of 70 to 270 ng/mℓ.[115,116]

Metabolism — Protriptyline is extensively metabolized but slowly excreted. Two hours after an oral dose of radioactive protriptyline, significant amounts of radioactivity appear in the urine. However, within 16 days only 50% of the dose is accounted for in the urine.[72] The slow excretion of protriptyline is attributed to slow metabolism to easily eliminated metabolites.[117] Metabolism has not been studied extensively in man. In animals hydroxylation, demethylation and 10,11-epoxidation are the major metabolic routes (see Figure 12).[118,119] Desmethylprotriptyline is the major metabolite in dogs. In rats the 10,11-epoxides of protriptyline and desmethylprotriptyline account for 40% of the urinary metabolites.[118]

Uses and Dosage

Indications — For the treatment of the symptoms of depression. The activating properties of protriptyline make it particularly useful for withdrawn and anergic patients.[72]

Dosage — In adults, initially 15 to 40 mg/day in divided doses is administered orally and increased to 60 mg/day as required. Adolescent and elderly patients are given 5 mg three times a day and gradually increased as required.[72]

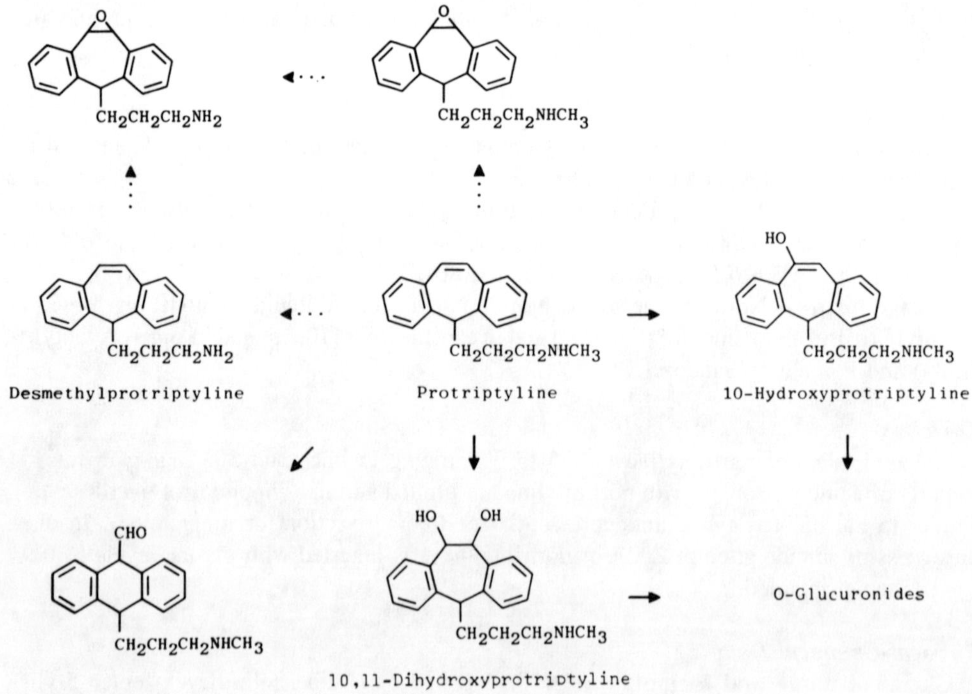

FIGURE 12. Metabolism of protriptyline.

Dosage forms — Protriptyline hydrochloride is available in tablet form [5 and 10 mg as Vivactil® (MSD)].[72]

Toxicity

The oral LD_{50} of protriptyline is 211 to 269 mg/kg in mice and 299 to 720 mg/kg in rats.[73,74] The symptoms of acute poisoning with protriptyline are similar to the other tricyclic antidepressants (see toxicity section for imipramine).

Physical-Chemical Data

Chemical name and formula — *N*-Methyl-5*H*-dibenzo[a,d]cycloheptene-5-propanamine monohydrochloride; $C_{19}H_{21}N{\cdot}HCl$ (299.8).

Physical properties — Protriptyline hydrochloride is an odorless white to yellowish-white powder with a melting point of 169 to 171°C. It is soluble in water, alcohol, chloroform, and methanol.[75,76,114]

Doxepin

Pharmacokinetics

Absorption and distribution — Doxepin is well absorbed with peak plasma levels being reached within 4 hr after oral administration. The half-life is approximately 17 hr. The major pharmacologically active metabolite, desmethyldoxepin, after oral administration of doxepin reaches peak plasma levels in 2 to 10 hr with a half-life of 51 hr.[120] Therapeutically effective plasma concentrations of doxepin plus desmethyldoxepin in the range of 19 to 154 ng/mℓ have been reported.[121] In ten patients with mild to moderate depression efficacy of doxepin correlated with desmethyldoxepin levels but not with doxepin levels.[122] In another study doxepin plus desmethyldoxepin levels of 110 ng/mℓ were needed for therapeutic response.[123]

Metabolism — Desmethyldoxepin is the primary metabolite in man and is pharmacologically active. Doxepin-*N*-oxide, didesmethyldoxepin and hydroxydoxepin and its glu-

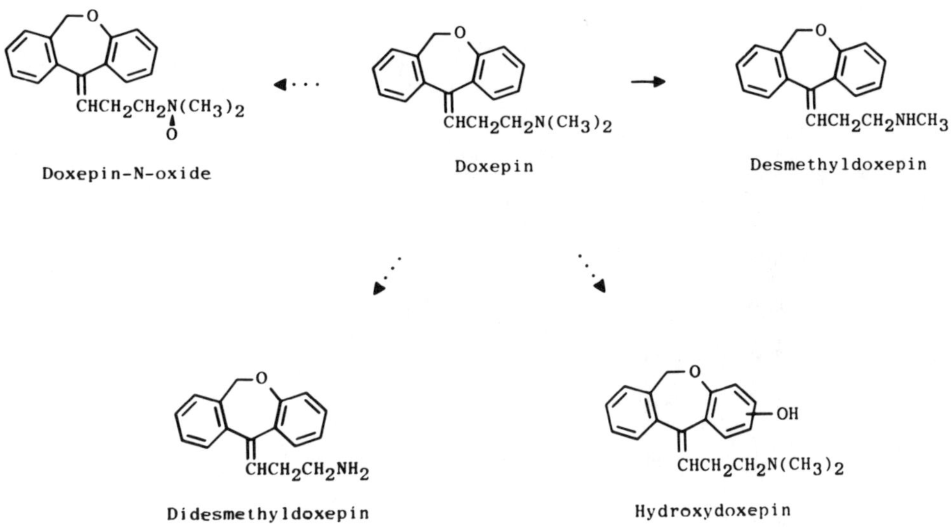

FIGURE 13. Metabolism of doxepin.

curonide (see Figure 13) are postulated metabolites based on animal studies.[124,125] Only 1% of the administered dose of doxepin was excreted in the urine of volunteers as unchanged doxepin.[125]

Uses and Dosage

Indications — For the treatment of patients with psychoneurotic anxiety and/or depressive reactions; with depression and/or anxiety associated with alcoholism or with organic disease; and with psychotic depressive disorders including involutional depression and manic-depressive reactions.[72]

Dosage — In patients with mild to moderately severe conditions initially 25 mg three times a day is administered orally and increased to the optimum dosage of 75 to 150 mg/day. In more severely ill patients higher initial dosages may be needed and gradually increased to 300 mg/day. In patients with mild symptoms or emotional symptoms associated with organic disease, 25 to 50 mg/day may be sufficient.[72]

Dosage forms — Doxepin hydrochloride is available in capsule form [10, 25, 50, 75, 100, and 150 mg as Adapin® (Pennwalt) and Sinequan® (Roerig)] and in oral concentrate form [10 mg/mℓ as Sinequan® (Roerig)].[72]

Toxicity

The oral LD_{50} of doxepin is 117 to 178 mg/kg in mice and 114 to 460 mg/kg in rats.[42,73] The minimum reported lethal dose of doxepin by an adult is about 1.3 g[42] and the maximum survived dose is 3.75 g.[126] The symptoms of acute poisoning with doxepin are similar to the other tricyclic antidepressants (see toxicity section for imipramine).

Physical-Chemcial Data

Chemical name and formula — 3-(Dibenz[b,e]oxepin-11(6*H*)-ylidene)-*N*,*N*-dimethyl-1-propanamine monohydrochloride; $C_{19}H_{21}NO \cdot HCl$ (315.8).

Physical properties — Doxepin hydrochloride is a white crystalline powder with a melting point of 185 to 191°C. Due to the unsymmetrical nature of this compound it exists as a mixture *cis*- and *trans*-geometric isomers. Doxepin is freely soluble in water and alcohol and should be protected from sunlight.[75,76]

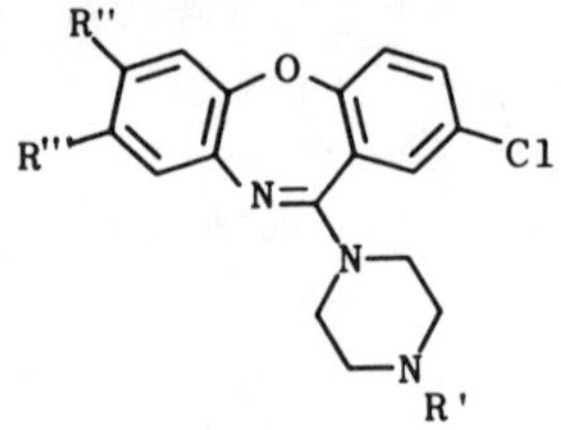

Amoxapine*	R'=H	R''=H	R'''=H
Loxapine	CH_3	H	H
7-Hydroxyamoxapine	H	OH	H
8-Hydroxyamoxapine	H	H	OH

FIGURE 14. Amoxapine and loxapine.

Dibenzoxazepines (SARs)

Amoxapine is the sole member of this class that is clinically effective and marketed for use in depression.[127-129]

Amoxapine is the desmethyl derivative (see Figure 14) of the neuroleptic, loxapine, and is claimed to have anxiolytic properties. The mechanism of antidepressant action appears to be the same as imipramine. Amoxapine selectively inhibits norepinephrine uptake in rat brain,[130,131] inhibits tetrabenazine-induced depression,[132] and is not an MAO inhibitor.[131] The primary metabolites, 7- and 8-hydroxyamoxapine, are inhibitors of norepinephrine uptake in rat brain.[130]

Amoxapine

Pharmacokinetics

Absorption and distribution — Amoxapine is rapidly absorbed and peak plasma levels are achieved 1 to 2 hr after a single oral dose.[133] The half-life is approximately 8 hr and the major pharmacologically active metabolite, 8-hydroxyamoxapine, has a half-life of 30 hr. In rats the highest concentrations are found in the lungs, spleen, kidney, heart, and brain with tissue concentrations ten times those found in serum.[131] The correlation between serum levels and clinical efficacy has not been established.

Metabolism — Amoxapine is metabolized (shown in Figure 15) to 8-hydroxyamoxapine and 7-hydroxyamoxapine, both are pharmacologically active. In a 6-day period after a 150-mg oral dose, amoxapine was excreted unchanged in the urine (2%) and feces (1%). The metabolites, 8-hydroxyamoxapine and 7-hydroxyamoxapine, were excreted in the urine (33 and 25%, respectively) and feces (4 and 2%, respectively).[44,131]

Uses and Dosage

Indications — For the treatment of the symptoms of depression in patients with neurotic or reactive depression as well as endogenous and psychotic depressions.[72]

Dosage — In adults initially 50 mg three times a day is administered orally and depending on tolerance the dosage may be increased to 100 mg three times a day. Increasing the dosage above 300 mg/day should be done only if the desired effect is not observed after 2 weeks of therapy. Elderly patients are given initially 25 mg three times a day and, depending on tolerance, the dosage may be increased to 50 mg three times a day. Some elderly patients may require careful increases of the dosage up to 300 mg/day. For maintenance therapy at doses of 300 mg/day or less, a single dose at bedtime is recommended.[72]

Dosage forms — Amoxapine is available in tablet form [50, 100, and 150 mg as Asendin® (Lederle)].[72]

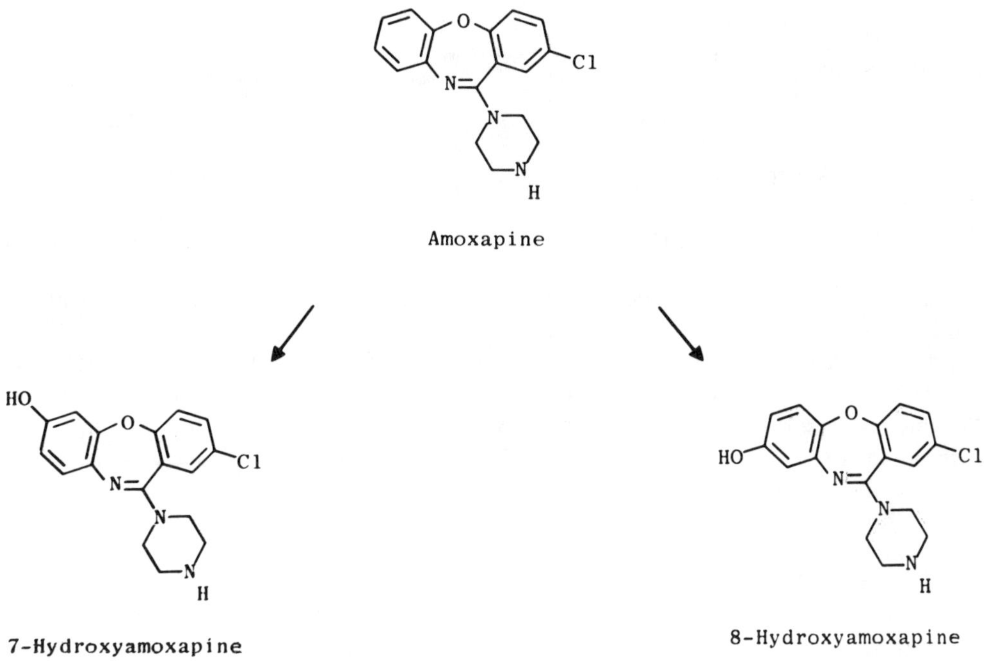

FIGURE 15. Metabolism of amoxapine.

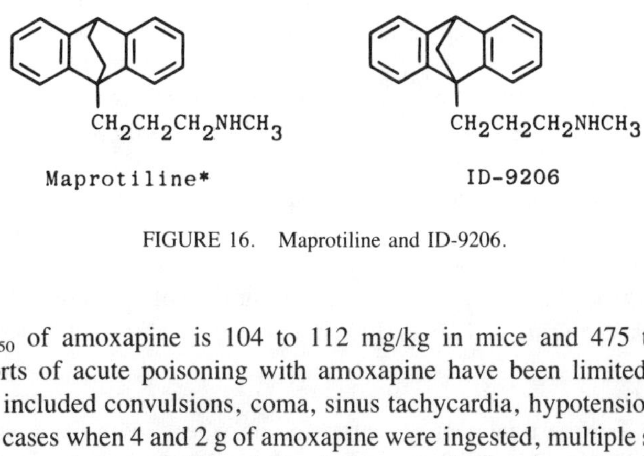

FIGURE 16. Maprotiline and ID-9206.

Toxicity

The oral LD_{50} of amoxapine is 104 to 112 mg/kg in mice and 475 to 730 mg/kg in rats.[131,132] Reports of acute poisoning with amoxapine have been limited.[134,135] Overdose symptoms have included convulsions, coma, sinus tachycardia, hypotension, and metabolic acidosis. In two cases when 4 and 2 g of amoxapine were ingested, multiple seizures occurred and the patients were comatose. Neurologic damage resulted and was present 1 to 3 months later.[135]

Physical-Chemical Data

Chemical name and formula — 2-Chloro-11-(1-piperazinyl)dibenz[b,f][1,4]oxazepine; $C_{17}H_{16}ClN_3O$ (313.8).

Physical properties — Amoxapine is a crystalline powder with a melting point of 175 to 176°C.[136]

Dibenzobicyclooctadienes (SARs)

Maprotiline, the only clinically effect derivative of this class[46,137-139] possesses a pharmacological profile similar to imipramine. It inhibits norepinephrine but not serotonin uptake in rat nerve terminals.[140,141]

Variation of the side chain length of maprotiline (see Figure 16) leads to loss of activity.[142]

FIGURE 17. Metabolism of maprotiline.

However, the bicycloheptane derivative, ID-9206, retains the pharmacological activity of maprotiline.[143]

Although the structure of maprotiline is considerably different from imipramine the flexure angle of the two aromatic rings coincides with the angle postulated by Wilhelm[32] that is required for antidepressant activity. This flexure angle in maprotiline is rigidly fixed by the ethano-bridge, whereas imipramine is flexible and is postulated to assume a bent conformation when exerting its pharmacological effect.

Maprotiline

Pharmacokinetics

Absorption and distribution — Maprotiline hydrochloride is slowly absorbed after oral administration. Peak serum levels are achieved 9 to 16 hr after a single oral dose of 50 mg.[46] Maprotiline had a half-life in 6 volunteers ranging from 27 to 58 hr, and was about 88% bound to serum protein.[144] Distribution in man is available from suicide overdose cases with the highest concentrations of maprotiline and its metabolites being found in the liver, lungs, brain, kidney, adrenal gland, heart, and muscle.[46] There appears to be a correlation between serum levels and clinical efficacy. In one study of 15 patients, the seven that responded totally at day 21 had higher maprotiline serum levels (453.6 ± 163.1 ng/mℓ than other patients (285.4 ± 108.4 ng/mℓ) who showed at least one indicator of depression at day 21.[45,145]

Metabolism — The primary metabolite is the pharmacologically active desmethylmaprotiline shown in Figure 17. Both maprotiline and desmethylmaprotiline appear to be slowly metabolized to various trace metabolites by deamination, aromatic hydroxylation, and oxidation of the aliphatic side chain. After i.v. administration of maprotiline about 57% of the dose is excreted in the urine and about 30% in the feces over a 21-day period.[144]

Uses and Dosage

Indications — For the treatment of depression in patients with depressive neurosis and manic-depressive illness (depressed type). It is indicated for the relief of anxiety associated with depression.[72]

Dosage — In adults initially 75 mg/day is suggested for mild-to-moderate depression and, depending upon tolerance, the dosage may be increased to 225 mg/day as required. More severely depressed hospitalized patients are given initially 100 to 150 mg/day. If required and tolerated, the dosage can be increased to 300 mg/day. In elderly patients, who do not tolerate higher doses, 50 to 75 mg/day is sufficient.[72]

Dosage forms — Maprotiline hydrochloride is available in tablet form [25 and 50 mg as Ludiomil® (Ciba)].[72]

Toxicity

The oral LD_{50} of maprotiline is 660 to 900 mg/kg in mice and 760 to 1050 mg/kg in rats.[146] Of the reported cases of maprotiline overdose, the maximum dose ingested with recovery was 5 g. In five recorded suicides one victim ingested 3 g and the other four ingested 2 to 10 g of maprotiline combined with tricyclic antidepressants, sedatives, or

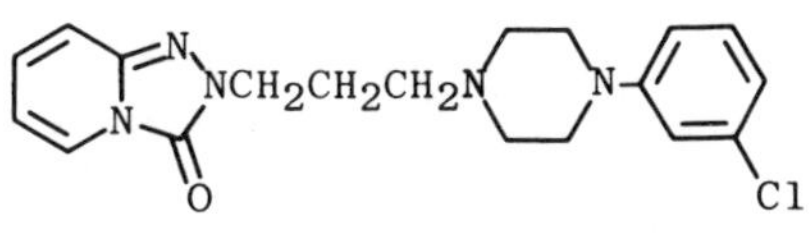

Trazodone*

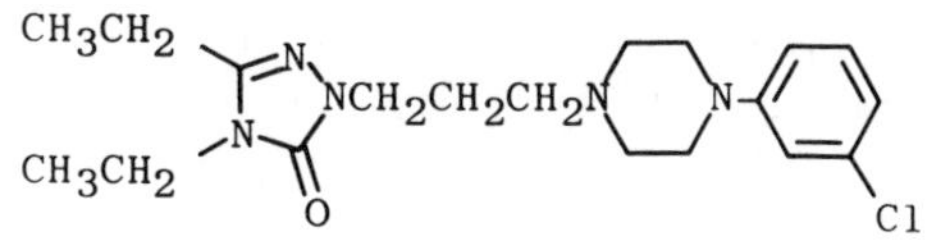

Etoperidone

FIGURE 18. Trazodone and etoperidone.

hypnotics.[147,148] The symptoms of acute poisoning are drowsiness, coma, convulsions, and urinary retention followed by confusion, agitation, disorientation, and visual hallucinations during recovery.[149] Comparison of maprotiline and tricyclic antidepressant overdose cases suggests that delirium and convulsions are more prevalent with maprotiline and cardiac arrhythmias are more prevalent with tricyclics.[46,149-151]

Physical-Chemical Data

Chemical name and formula — *N*-Methyl-9,10-ethanoanthracene-9(10*H*)-propranamine monohydrochloride; $C_{20}H_{23}N{\cdot}HCl$ (313.9).

Physical properties — Maprotiline hydrochloride is an odorless white crystalline powder with a melting point of 236 to 246°C. It is slightly soluble in water.[72]

Triazoles (SARs)

Trazodone is the first of a group of second generation antidepressants to be introduced in the U.S. Trazodone is structurally and pharmacologically different from the currently used antidepressants, and has a lower incidence of adverse cardiovascular effects[152] and of anticholinergic effects.[153,154] The antidepressant activity of trazodone is attributed to selective inhibition of serotonin uptake into nerve terminals[155] thereby increasing synaptic concentrations of serotonin and availability of serotonin at serotoninergic receptors. An early study of trazodone revealed a pharmacological profile in animals suggestive of antidepressant activity.[156]

Little is known of the SARs of trazodone. Fission of the pyridinyl ring of trazodone gave the clinically effective triazole analogue etoperidone[157,158] shown below in Figure 18. It exhibits similar pharmacological effects as trazodone on serotonin reuptake.

Trazodone

Pharmacokinetics

Absorption and distribution — Trazodone is rapidly and completely absorbed after oral administration and peak plasma levels are achieved in 1 to 2 hr.[159,160] When taken after a meal, absorption is delayed and peak plasma levels are decreased. Elimination is reported to be biphasic with an earlier phase in 3 to 10 hr (half-life of 4.4 hr) and a later phase in 10 to 34 hr (half-life of 7.5 hr) after dose administration.[47,48] At concentrations attained with therapeutic doses, trazodone was 89 to 95% protein bound in vitro.[161] Distribution data in man is unavailable; however, in animals trazodone has been shown to cross the blood-brain barrier[47,162] with tissue levels significantly higher than plasma levels.[163] A correlation between serum levels and clinical effects has not been established.

Metabolism — Trazodone is metabolized extensively in the liver by hydroxylation and oxidation as shown in Figure 19.[164] Following the oral administration of 25 mg, unchanged trazodone is the major component found in plasma[164] while only trace amounts are excreted unchanged in the urine.[159] The primary urinary metabolites, oxotriazolopyridinepropionic acid, dihydrodioltrazodone, and hydroxytrazodone account for 35, 16, and 19%, respectively, of a one-dose administration. Trazodone and metabolites are excreted mainly in the urine (70 to 75%) after a single dose within 72 hr. The remainder is excreted in the feces.[164]

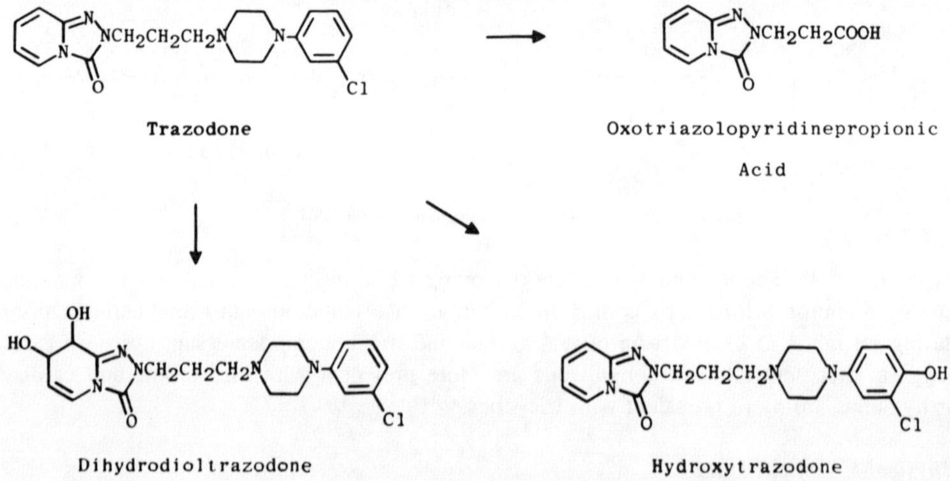

FIGURE 19. Metabolism of trazodone.

Uses and Dosage

Indications — For the treatment of the symptoms of depression in patients with and without prominent anxiety.[160]

Dosage — In adults initially 150 mg/day is administered orally and may be increased by 50 mg/day every 3 to 4 days up to the usual maximum outpatient dose of 400 mg/day in divided doses. In-patients or more severely depressed patients may be given up to 600 mg/day in divided doses.[160]

Dosage forms — Trazodone hydrochloride is available in tablet form [50 and 100 mg as Desyrel® (Mead Johnson)].[160]

Toxicity

The oral LD_{50} of trazodone is 610 mg/kg in mice, 486 mg/kg in rats, and 560 mg/kg in rabbits. Trazodone appears to pose less of a threat in overdose than the other antidepressants. Death by overdose has not occurred with trazodone. In one suicide attempt 7.5 g was ingested. The patient was unconscious but arousable and following emesis recovery was uneventful. Overdose with trazodone may cause an increase in the severity of the adverse effects.[160]

Physical-Chemical Data

Chemical name and formula — 2-[3-[4-(3-Chlorophenyl)-1-piperazinyl]propyl]-1,2,4-triazolo[4,3-a]pyridin-3(2*H*)-one monohydrochloride; $C_{19}H_{22}ClN_5O{\cdot}HCl$ (408.3).

Physical properties — Trazodone hydrochloride is an odorless white crystalline solid with a melting point of 223°C and a pKa of 6.14. It is soluble in chloroform and is slightly soluble in alcohol, methanol, and water.[114,165]

INVESTIGATIONAL AGENTS

The tricyclic antidepressants were developed over 20 years ago and, since the initial discovery of the antidepressant effects of imipramine, the drugs developed in this class have not led to a major reduction of side effects or reduced potential for suicidal overdose. In recent years many new agents have been discovered that differ from the tricyclic antidepressants and MAO inhibitors both in chemical structure and in mechanism of action. These new agents exert a wide variety of effects on neurotransmitter systems that differ from the

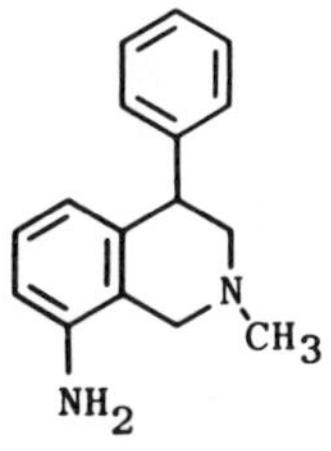

Nomifensine

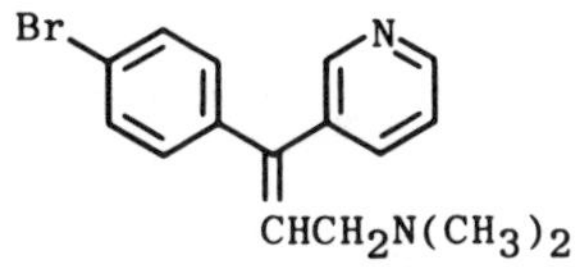

Zimelidine

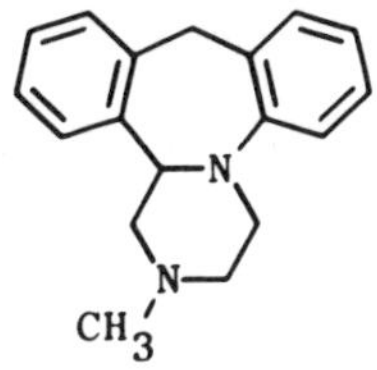

Mianserin

FIGURE 20. Second generation antidepressants.

classical tricyclic antidepressants. These so-called "second generation" antidepressants[166] appear to be as effective as the tricyclics with fewer adverse effects. The major advantage over current therapy lies in their reduced propensity for causing anticholinergic side effects, cardiotoxicity, convulsions, and overdose. Some claims have also included faster onset of antidepressant effect. The common advantage of these agents clearly resides in their reduced toxicity; however, these claims have yet to be substantiated in clinical practice in large patient populations. Some of the most promising new agents as of 1983 are discussed below and are shown in Figure 20.

Nomifensine

The pharmacological profile of nomifensine is similar to imipramine. It antagonizes reserpine-induced syndrome[167] and is a potent norepinephrine reuptake inhibitor. However, nomifensine is unique in its ability to inhibit the reuptake of dopamine.[168,169] Clinically, nomifensine is at least as effective as imipramine[170] and amitriptyline.[171] It has a rapid onset of action relative to the tricyclics,[171,172] lower cardiotoxicity, and fewer anticholinergic side effects. Nomifensine is nonsedative, but appears to have stimulant effects. For this reason the use of nomifensine in antidepressant therapy may lie in retarded apathetic depression and may be less desirable for use in agitated depressions.

Zimelidine

The selective inhibition of serotonin reuptake both in vitro[173] and in vivo[174] and its novel structure distinguishes zimelidine from the tricyclic antidepressants. The mechanism of action of zimelidine is thought to be due to the ability to increase serotonergic transmission in the CNS by inhibition of serotonin reuptake into presynaptic nerve terminals. In animal models zimelidine increases serotonin synthesis and turnover, while exerting little effect on norepinephrine and dopamine uptake.[173,175,176] Zimelidine is rapidly metabolized by monodemethylation to the principal metabolite norzimelidine, which may contribute to the antidepressant effect of zimelidine. Norzimelidine is a potent inhibitor of serotonin uptake.[173,174] The SARs of zimelidine analogs have been discussed.[177]

Zimelidine in the clinic is as effective as amitriptyline, desipramine, and maprotiline.[178-180] It exhibited fewer and milder anticholinergic side effects and was nonsedative.

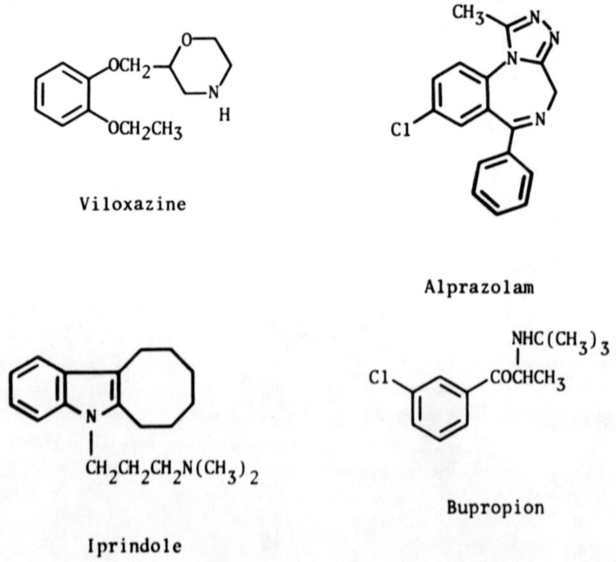

FIGURE 21. Clinically active second generation antidepressants.

At therapeutic doses and overdose zimelidine appears to be much less cardiotoxic than the tricyclics.[178,181]

Mianserin

This tetracyclic compound was originally synthesized for its antihistamine and antiserotonin activity[182] and was later examined in the clinic for its antidepressant properties. The SAR of mianserin has been discussed.[183] Mianserin unlike the tricyclics is a weak inhibitor of monoamine reuptake.[184-186] The mechanism of action of mianserin appears to involve blockade of the presynaptic alpha-receptor, thus increasing norepinephrine release and adrenergic transmission in the brain.[187-189] In single unit cell studies in rats, mianserin increased the firing rate of norepinephrine neurons in the locus and antagonized the inhibitory effect of the alpha-agonist clonidine.[190]

Mianserin is as effective as the tricyclic antidepressants in the clinic with few anticholinergic side effects and minimal cardiotoxicity.[191-193] The sedative and anxiolytic properties of mianserin may find use in the treatment of agitated or anxious depressives. Mianserin appears to be safe in overdose.

There are many other second generation antidepressants that are being evaluated in the clinic.[35,166] Some of the agents that have shown clinical activity have been reviewed separately. These include viloxazine,[194] alprazolam,[195] iprindole,[196] lithium,[197] and bupropion (Figure 21).[198]

REFERENCES

1. **Alexander, L.,** *Treatment of Mental Disorders,* W. B. Saunders, Philadelphia, 1953.
2. **Denber, H. C. B. and Bird, E. G.,** Chlorpromazine in the treatment of mental illness. III. The problem of depression, *Am. J. Psychiatr.,* 112, 1021, 1956.
3. **Kuhn, R.,** Treatment of depressive states with G-22355 (imipramine hydrochloride), *Am. J. Psychiat.,* 115, 459, 1958.

4. **Schildkraut, J. J.,** *Neuropsychopharmacology and the Affective Disorders,* Little, Brown, Boston, 1970.
5. **Ayd, F. J., Ed.**, *Clinical Depressions: Diagnostic and Therapeutic Challenges,* Ayd Medical Communications, Baltimore, 1980.
6. **Blackwell, B.,** Adverse effects of antidepressant drugs. 1. Monoamine oxidase inhibitors and tricyclics, *Drugs,* 21, 201, 1981.
7. **Carney, M. W. P.,** Tricyclics and the heart, *Br. J. Psychiatr.,* 134, 637, 1979.
8. **Thomann, P. and Hess, R.,** Toxicology of antidepressant drugs, in *Handbook of Experimental Pharmacology, Psychotropic Agents,* Vol. 55/I, Hoffmeister, F. and Stille, G., Eds., Springer-Verlag, New York, 1980, chap. 22.
9. **Sulser, F. and Vetulani, J.,** The functional post-synaptic norepinephrine receptor system and its modification by drugs which either alleviate or precipitate depression, in *Models in Psychiatry and Neurology,* Usdin, E. and Hanin, I., Eds., Pergamon Press, New York, 1977, 189.
10. **Vetulani, J., Stawarz, R. J., Dingell, J. V., and Sulser, F.,** A possible common mechanism of action of antidepressant treatments, *Naunyn-Schmiedeberg's Arch. Pharmacol.,* 293, 109, 1976.
11. **Sulser, F. and Mobley, P. L.,** Regulation of central noradrenergic receptor function: new vistas on the mode of action of antidepressant treatments, in *Neuroreceptors: Basic and Clinical Aspects,* Usdin, E., Davis, J. M., and Bunney, W. E., Eds., John Wiley & Sons, London, 1981.
12. **Enna, S. J., Mann, E., Kendall, D., and Stancel, G. M.,** Effect of chronic antidepressant administration on brain neurotransmitter receptor binding, in *Antidepressants: Neurochemical, Behavioral and Clinical Perspectives,* Enna, S. J., Malick, J. B., and Richelson, E., Eds., Raven Press, New York, 1981, 91.
13. **Mobley, P. L. and Sulser, F.,** Down-regulation of the central noradrenergic receptor system by antidepressant therapies: biochemical and clinical aspects, in *Antidepressants: Neurochemical, Behavioral and Clinical Perspectives,* Enna, S. J., Malick, J. B., and Richelson, E., Eds., Raven Press, New York, 1981, 31.
14. **Asakura, M., Tsukamoto, T., and Hasegawa, K.,** Modulation of rat brain alpha-2 and beta adrenergic receptor sensitivity following long term treatments with antidepressants, *Brain Res.,* 235, 192, 1982.
15. **Howard, J. L., Soroko, F. E., and Cooper, B. R.,** Empirical behavioral models of depression, with emphasis on tetrabenazine antagonism in *Antidepressants: Neurochemical, Behavioral and Clinical Perspectives,* Enna, S. J., Malick, J. B., and Richelson, E., Eds., Raven Press, New York, 1981, 107.
16. **McKinney, W. T.,** Biobehavioral models of depression in monkeys, in *Animal Models in Psychiatry and Neurology,* Hanin, I. and Usdin, E., Eds., Pergamon Press, New York, 1977, 117.
17. **Suomi, S. J., Seaman, S. F., Lewis, J. K., DeLizio, R. D., and McKinney, W. T.,** Effects of impramine treatment of separation-induced social disorders in rhesus monkeys, *Arch. Gen. Psychiatr.,* 35, 321, 1978.
18. **Kraemer, G. W., McKinney, W. T., Prange, A. J., Breese, G. R., McMurray, T. M., and Kemnitz, J.,** Isoniazid: behavioral and biochemical effects in rhesus monkeys, *Life Sci.,* 19, 49, 1976.
19. **Seligman, M. E. P.,** *Helplessness: On Depression, Development and Death,* Freeman, San Francisco, 1975.
20. **Porsolt, R. D.,** Behavioral despair, in *Antidepressants: Neurochemical, Behavioral and Clinical Perspectives,* Enna, S. J., Malick, J. B., and Richelson, E., Eds., Raven Press, New York, 1981, 121.
21. **Sherman, A. D., Allers, G. L., Petty, F., and Henn, F. A.,** A neuropharmacologically-relevant animal model of depression, *Neuropharmacology,* 18, 891, 1979.
22. **Horovitz, Z. P., Piala, J. J., High, J. P., Burke, J. C., and Leaf, R. C.,** Effects of drugs on the mouse-killing (muricide) test and its relationship to amygdaloid function, *Int. J. Neuropharmacol.,* 5, 405, 1966.
23. **Brogden, R. N., Heel, R. C., Speight, T. M., and Avery, G. S.,** Mianserin: a review of its pharmacologic properties and therapeutic efficacy in depressive illness, *Drugs,* 16, 273, 1978.
24. **Vogel, J. R. and Haubrich, D. R.,** Chronic administration of electroconvulsive shock: effects on mouse-killing activity and brain monoamines in rats, *Physiol. Behav.,* 11, 725, 1973.
25. **Babington, R. G.,** Neurophysiologic techniques and antidepressive activity, in *Antidepressants: Neurochemical, Behavioral and Clinical Perspectives,* Enna, S. J., Malick, J. B., and Richelson, E., Eds., Raven Press, New York, 1981, 157.
26. **O'Donnell, J. M. and Seiden, L. S.,** Differential-reinforcement-of-low-rate 71-second schedules: selective effects of antidepressant drugs, *J. Pharmacol. Exp. Ther.,* 224, 80, 1983.
27. **Guy, W.,** ECDEU assessment manual for psychopharmacology, in U.S. Dept. Health Education and Welfare Publication No. (ADM) 76-338, 1976.
28. **Hamilton, M.,** Development of a rating scale for primary depressive illness, *Br. J. Soc. Clin. Psychol.,* 6, 278, 1967.
29. **Kuhn, R.,** Uber die behandlung depressiver zustande mit einem iminodibenzylderivat, *Schweiz. Med. Wochenschr.,* 87, 1135, 1957.
30. **Kuhn, R.,** The imipramine story, in *Discoveries in Biological Psychiatry,* Ayd, F. J., Jr. and Blackwell, B., Eds., Lippincott, Philadelphia, 1970, chap. 16.
31. **Haeflinger, F. and Schindler, W.,** Tertiary Aminoalkyl-Iminobenzyls, U.S. Patent No. 2,554,736, 1951.

32. **Wilhelm, M.,** The chemistry of polycyclic psycho-active drugs-serendipity or systematic investigation?, in *Depressive Illness,* Kielholz, P., Ed., Hans Huber Publishing, Bern, 1972, 129.
33. **Kaiser, C. and Setler, P. E.,** Antidepressant agents, in *Burger's Medicinal Chemistry,* Part 3, 4th ed., Wolff, M. E., Ed., Interscience, New York, 1981, chap. 58.
34. **Finch, N.,** Chemical approaches to the development of antidepressants, in *Industrial Pharmacology. Antidepressants,* Vol. 2, Fielding, S. and Lal, H., Eds., Futura Publishing, Mount Kisco, 1975, chap. 1.
35. **Zeelen, F. J.,** Chemistry (structure and activity), in *Handbook of Experimental Pharmacology, Psychotropic Agents,* Vol. 55/I, Hoffmeister, F. and Stille, G., Eds., Springer-Verlag, New York, 1980, chap. 16.
36. **Risch, S. C., Janowsky, D. S., and Huey, L. Y.,** Plasma levels in tricyclic antidepressants and clinical efficacy, in *Antidepressants: Neurochemical, Behavioral and Clinical Perspectives,* Enna, S. J., Malick, J. B., and Richelson, E., Eds., Raven Press, New York, 1981, 183.
37. **Baselt, R. C.,** *Disposition of Toxic Drugs and Chemicals in Man,* Vol. 1, Biomedical Publications, Canton, 1978, section VI.
38. **Bickel, M. H.,** Metabolism of antidepressants, in *Handbook of Experimental Pharmacology, Psychotropic Agents,* Vol. 55/I, Hoffmeister, F. and Stille, G., Eds., Springer-Verlag, New York, 1980, chap. 23.
39. **Bickel, M. H.,** Poisoning by tricyclic antidepressant drugs, *Int. J. Clin. Pharmacol. Ther. Toxicol.,* 11, 145, 1975.
40. **Biggs, J. T., Spiker, D. G., Petit, J. M., and Ziegler, V. E.,** Tricyclic antidepressant overdose, *JAMA,* 238, 135, 1977.
41. **Manoguerra, A. S. and Weaver, L. C.,** Poisoning with tricyclic antidepressant drugs, *Clin. Toxicol.,* 10, 149, 1977.
42. **Pinder, R. M., Brogden, R. N., Speight, T. M., and Avery, G. S.,** Doxepin up-to-date: a review of its pharmacological properties and therapeutic efficacy with particular reference to depression, *Drugs,* 13, 161, 1977.
43. **Jue, S. G., Dawson, G. W., and Brogden, R. N.,** Amoxapine: a review of its pharmacology and efficacy in depressed states, *Drugs,* 24, 1, 1982.
44. **Dugas, J. E. and Weber, S. S.,** Amoxapine, *Drug Intell. Clin. Pharm.,* 16, 199, 1982.
45. **Stimmel, G. L.,** Maprotiline, *Drug Intell. Clin. Pharm.,* 14, 585, 1980.
46. **Pinder, R. M., Brogden, R. N., Speight, T. M., and Avery, G. S.,** Maprotiline: a review of its pharmacological properties and therapeutic efficacy in mental depressive states, *Drugs,* 13, 321, 1977.
47. **Brogden, R. N., Heel, R. C., Speight, T. M., and Avery, G. S.,** Trazodone: a review of its pharmacological properties and therapeutic use in depression and anxiety, *Drugs,* 21, 401, 1981.
48. **Rawls, W. N.,** Trazodone, *Drug Intell. Clin. Pharm.,* 16, 7, 1982.
49. **Klerman, G. L. and Cole, J. O.,** Clinical pharmacology of imipramine and related antidepressant compounds, *Pharmacol. Rev.,* 17, 101, 1965.
50. **Koe, B. K.,** Antidepressant drugs on brain catecholamines and serotonin, in *Industrial Pharmacology. Antidepressants,* Vol. 2, Fielding, S. and Lal, H., Eds., Futura Publishing, Mount Kisco, 1975, chap. 7.
51. **Lapolla, A. and Jones, H.,** Placebo-control evaluation of desipramine in depression, *Am. J. Psychiatr.,* 127, 335, 1970.
52. **Lambert, P. A. and Guyotat, J.,** A new sedative antidepressive derivative of iminodibenzyl: RP-7162. Therapeutic trials. *Presse Med.,* 69, 1425, 1961.
53. **Forshell, G. P., Siwers, B., and Tuck, J. R.,** Pharmacokinetics of lofepramine in man: relationship to inhibition of noradrenaline uptake, *Eur. J. Clin. Pharmacol.,* 9, 291, 1976.
54. **Segawa, T., Nomura, Y., Tanaka, A., and Murakami, H.,** Effect of lofepramine and other depressants on the uptake of 5-hydroxytryptamine and noradrenaline into rat brain monoaminergic neurons, *J. Pharm. Pharmacol.,* 29, 139, 1977.
55. **Eriksoo, E. and Rohie, O.,** Chemistry and pharmacology of a new potential antidepressant, *Arzneim. Forsch.,* 20, 1561, 1970.
56. **Tuck, J. R., Kahan, E., and Siwers, B.,** Biological and pharmacokinetic evidence for generic equivalence of three imipramine preparations: comparison with a new imipramine analogue, *Acta Pharmacol. Toxicol.,* 32, 304, 1973.
57. **Murphy, J. E., Donald, J. F., and Beaumont, G.,** A comparative clinical trial of opipramol and chlordiazepoxide employing a method of patient self-assessment, *Clin. Trials J.,* 8, 28, 1971.
58. **Roger, S. C., Davies, F. J., and Galbraith, A. W.,** A study of depression in two general practices, including a double-blind comparison of desimipramine and opipramol, *Clin. Trials J.,* 6, 5, 1969.
59. **Bickel, M. H. and Brodie, B. B.,** Structure and antidepressant activity of imipramine analogues, *Int. J. Neuropharmacol.,* 3, 611, 1964.
60. **Todrick, A. and Tait, A. C.,** The inhibition of human platelet 5-hydroxytryptamine uptake by tricyclic antidepressive drugs. The relation between structure and potency, *J. Pharm. Pharmacol.,* 21, 751, 1969.
61. **Symes, M. H.,** Monochlorimipramine: a controlled trial of a new antidepressant, *Br. J. Psychiatr.,* 113, 671, 1967.

62. **Poldinger, W., Gehring, A., Schaublin, A., and Thilges, R.,** Vergleichende klinische erfahrungen mit chlorimpramin und ketimipramin, *Arzneim. Forsch.,* 19, 492, 1969.
63. **Villani, F. J. and Mann, T. A.,** 5-(Dimethylaminopropyl)-10,11-dihydro-5*H*-benzo[2,3]pyrido[6,7-b]azepine, *J. Med. Chem.,* 11, 894, 1968.
64. **Gram, L. F. and Christiansen, J.,** First-pass metabolism of imipramine in man, *Clin. Pharmacol. Ther.,* 17, 555, 1975.
65. **Piafsky, K. M. and Borga, O.,** Plasma protein binding of basic drugs. II. Importance of alpha-1-acid glycoprotein for interindividual variation, *Clin. Pharmacol. Ther.,* 22, 545, 1977.
66. **Glassman, A. H., Perel, J. M., Shostak, M., Kantor, S. J., and Fleiss, J. L.,** Clinical implications of imipramine plasma levels for depressive illness, *Arch. Gen. Psychiatr.,* 34, 197, 1977.
67. **Reisby, N., Gram, L. F., Bech, P., Nagy, A., Peterson, G. O., Ortmann, J., Ibsen, I., Dencker, S. J., Jacobson, O., Krautwald, O., Sondergaard, I., and Christiansen, J.,** Imipramine: clinical effects and pharmacokinetic variability, *Psychopharmacology,* 54, 263, 1977.
68. **Christiansen, J., Gram, L. F., Kofod, B., and Rafaelsen, O. J.,** Imipramine metabolism in man, *Psychopharmacologia,* 111, 255, 1967.
69. **Crammer, J. L. and Scott, B.,** New metabolites of imipramine, *Psychopharmacologia,* 8, 461, 1966.
70. **Gram, L. F., Kofod, B., Christiansen, J., and Rafaelsen, O. J.,** Imipramine metabolism: pH-dependent distribution and urinary excretion, *Clin. Pharmacol. Ther.,* 12, 239, 1971.
71. **Crammer, J. L., Scott, B., and Rolfe, B.,** Metabolism of 14C-imipramine. II. Urinary metabolites in man, *Psychopharmacologia,* 15, 207, 1969.
72. *Physicians' Desk Reference,* 36th ed., Medical Economics, Oradell, 1982, section 6.
73. **Usdin, E. and Efron, D. H.,** Psychotropic Drugs and Related Compounds, 2nd ed., Publ. No. (HSM) 72-9074, Department of Health, Education and Welfare, Washington, D.C., 1972.
74. **Goldenthal, E. I.,** A compilation of LD_{50} values in newborn and adult animals, *Toxicol. Appl. Pharmacol.,* 18, 185, 1971.
75. *The Pharmaceutical Codex,* 11th ed., The Pharmaceutical Press, London, England, 1979.
76. **Osol, A., Ed.,** *Remington's Pharmaceutical Sciences,* 16th ed., Mack Publishing, Easton, 1980, chap. 58.
77. **Nagy, A. and Johansson, R.,** Plasma levels of imipramine and desipramine in man after different routes of administration, *Naunyn-Schmiedeberg's Arch. Pharmacol.,* 290, 145, 1975.
78. **Borga, O., Azarnoff, D. L., Plym Forshell, G., and Sjoqvist, F.,** Plasma protein binding of tricyclic antidepressants in man, *Biochem. Pharmacol.,* 18, 2135, 1969.
79. **Bickel, M. H., Brochon, R., Friolet, B., Herrmann, B., and Stofer, A. R.,** Clinical and biochemical results of a fatal case of desipramine intoxication, *Psychopharmacologia,* 10, 431, 1967.
80. **Friedel, R. O., Veith, R. C., Bloom, V., and Bielski, R. J.,** Desipramine plasma levels and clinical response in depressed outpatients, *Commun. Psychopharmacol.,* 3, 81, 1979.
81. **Khalid, R., Amin, M. M., and Ban, T. A.,** Desipramine plasma levels and therapeutic response, *Psychopharmacol. Bull.,* 14, 43, 1978.
82. **Sjoqvist, F., Berglund, F., Borga, O., Hammer, W., Andersson, S., and Thorstrand, C.,** The pH-dependent excretion of monomethylated tricyclic antidepressants, *Clin. Pharmacol. Ther.,* 10, 826, 1969.
83. **Demiaux, J. P., Motin, J., Roche, L., and Badinand, A.,** Etude de l'emination de derives de la dibenzoazepine au cours d'intoxications aigues, *C. R. Soc. Biol.,* 160, 1470, 1966.
84. **Caille, G., Besner, J. -G., Lacasse, Y., and Vezina, M.,** Pharmacokinetic characteristics of two different formulations of trimipramine determined with a new GLC method, *Biopharm. Drug Dispos.,* 1, 187, 1980.
85. **Populaire, P., Terlain, B., Pascal, S., Lebreton, G., and Decouvelaere, B.,** Resorption, excretion et biotransformation de la trimeprimine (7162 R.P.) chez le chien et le lapin. Contribution a l'identification de plusieurs metabolites. Biotransformation chez l'homme, *Prod. Probl. Pharm.,* 25, 632, 1970.
86. **Asberg, M., Cronholm, B., Sjoqvist, F., and Tuck, D.,** Relationship between plasma levels and therapeutic effect of nortriptyline, *Br. Med. J.,* 3, 331, 1971.
87. **Gomez-Martinez, I.,** Preliminary double-blind clinical trial with a new antidepressive doxepin, *Curr. Ther. Res. Clin. Exp.,* 10, 116, 1968.
88. **Goldstein, B. J. and Pinosky, D. G.,** Clinical evaluation of doxepin in anxious depressed outpatients, *Curr. Ther. Res. Clin. Exp.,* 11, 169, 1968.
89. **Bauer, G. and Nowak, H.,** Doxepin, ein neves Antidepressivum: Wirkungsvergleich mit Amitriptylin, *Arzneim. Forsch.,* 19, 1642, 1969.
90. **Lipsedge, M. S. and Rees, W. L.,** A double-blind comparison of dothiepin and amitriptyline for the treatment of depression with anxiety, *Psychopharmacologia,* 19, 153, 1971.
91. Bayer Symposium: Bericht uber das Symposion analaesslich der einfuhrung des Antidepressivums Agedal, *Arzneim. Forsch.,* 19, 835, 1969.
92. **Bastian, J. M., Ebnother, A., Jucker, E., Rissi, E., and Stoll, A. P.,** Beitrage zur chemie des 4,5-dihydro-10*H*-benzo[5,6]cyclohepta[1,2-b]thiophens, *Helv. Chim. Acta,* 54, 277, 1971.
93. **Lippmann, W.,** Ayerst symposium: butriptyline, *J. Med. Exp. Clin.,* 2, 250, 1971.

94. **Voith, K. and Herr, F.,** Psychopharmacological evaluation of a new antidepressant: butriptyline, *Arch. Int. Pharmacodyn. Ther.*, 182, 318, 1969.
95. **Daneman, E. A.,** Clinical experience and a double-blind study of a new anti-depressant, Vivactil hydrochloride, *Psychosomatics*, 6, 342, 1965.
96. **Engelhardt, E. L., Christy, M. E., Colton, C. D., Freedman, M. B., Boland, C. C., Halpern, L. M., Vernier, V. G., and Stone, C. A.,** Antidepressants. Tetrabenazine-antagonizing activity in a series of 5*H*-dibenzo[a,d]cycloheptene-5-propylamines, *J. Med. Chem.*, 11, 325, 1968.
97. **Garland, W. A.,** Quantitative determination of amitriptyline and its principal metabolite, nortriptyline, by GLC - chemical ionization mass spectrometry, *J. Pharm. Sci.*, 66, 77, 1977.
98. **Jorgensen, A. and Hansen, V.,** Pharmacokinetics of amitriptyline infused intravenously in man, *Eur. J. Clin. Pharmacol.*, 10, 337, 1976.
99. **Jorgensen, A. and Staer, P.,** On the biological half-life of amitriptyline, *J. Pharm. Pharmacol.*, 28, 62, 1976.
100. **Baselt, R. C.,** unpublished data, 1977, in *Disposition of Toxic Drugs and Chemicals in Man,* Vol. 1, Biomedical Publications, Canton, 1978, 172.
101. **Ziegler, V. E., Co, B. T., Taylor, J. R., Clayton, P. J., and Biggs, J. T.,** Amitriptyline plasma levels and therapeutic response, *Clin. Pharmacol. Ther.*, 19, 795, 1976.
102. **Diamond, S.,** Human metabolization of amitriptyline tagged with carbon 14, *Curr. Ther. Res. Clin. Exp.*, 7, 170, 1965.
103. **Braithwaite, R. A. and Whatley, J. A.,** Specific gas chromatographic determination of amitriptyline in human urine following therapeutic doses, *J. Chromatogr.*, 49, 303, 1970.
104. **Eschenhof, E. and Rieder, J.,** Untersuchungen uber das Schicksal des Antidepressivums Amitriptylin im Organismas des ratte und des Menschen, *Arzneim. Forsch.*, 19, 957, 1969.
105. **Santagostino, G., Facino, R. M., and Pirillo, D.,** Urinary excretion of amitriptyline N-oxide in humans, *J. Pharm. Sci.*, 63, 1690, 1974.
106. **Hammar, C. G., Alexanderson, B., Holmstedt, B., and Sjoqvist, F.,** Gas chromatography - mass spectrometry of nortriptyline in body fluids in man, *Clin. Pharmacol. Ther.*, 12, 496, 1971.
107. **De Leenheer, A. and Heyndrickx, A.,** Identification of a major metabolite of nortriptyline in human urine, *J. Pharm. Sci.*, 60, 1403, 1971.
108. **Hucker, H. B.,** Metabolism of amitriptyline, *Pharmacologist*, 4, 171, 1962.
109. **Alexanderson, B.,** Pharmacokinetics of desmethylimipramine and nortriptyline in man after single and multiple oral doses — a cross-over study, *Eur. J. Clin. Pharmacol.*, 5, 1, 1972.
110. **Kragh-Sorenson, P., Asberg, M., and Eggert-Hansen, C.,** Plasma-nortriptyline levels in endogenous depression, *Lancet*, 1, 113, 1973.
111. **Kragh-Sorenson, P., Eggert-Hansen, C., and Asberg, M.,** Plasma levels of nortriptyline in the treatment of endogenous depression, *Acta Psychiatr. Scand.*, 49, 444, 1973.
112. **Gram, L. F. and Overo, K. F.,** First-pass metabolism of nortriptyline in man, *Clin. Pharmacol. Ther.*, 18, 305, 1975.
113. **Alexanderson, B. and Borga, O.,** Urinary excretion of nortriptyline and five of its metabolites in man after single and multiple oral dose, *Eur. J. Clin. Pharmacol.*, 5, 174, 1973.
114. *The Merck Index*, 9th ed., Merck, Rahway, 1976.
115. **Biggs, J. T. and Ziegler, V. E.,** Protriptyline plasma levels and antidepressant response, *Clin. Pharmacol. Ther.*, 22, 269, 1977.
116. **Whyte, S. F., MacDonald, A. J., Naylor, G. J., and Moody, J. P.,** Plasma concentrations of protriptyline and clinical effects in depressed women, *Br. J. Psychiatr.*, 128, 384, 1976.
117. **Charalampous, K. D. and Johnson, P. C.,** Studies of 14C-protriptyline in man: plasma levels and excretion, *J. Clin. Pharmacol.*, 7, 93, 1967.
118. **Hucker, H. B., Balletto, A. J., Demetriades, J., Arison, B. H., and Zacchei, A. G.,** Epoxide metabolites of protriptyline in rat urine, *Drug Metab. Disp.*, 3, 80, 1975.
119. **Sisenwine, S. F., Tio, C. O., Shrader, S. R., and Ruelius, H. W.,** The biotransformation of protriptyline in man, pig, and dog, *J. Pharmacol. Exp. Ther.*, 175, 51, 1970.
120. **Ziegler, V. E., Biggs, J. T., Wylie, L. T., Rosen, S. H., Hawf, D. J., and Coryell, W. H.,** Doxepin kinetics, *Clin. Pharmacol. Ther.*, 23, 573, 1978.
121. **Biggs, J. T., Preskorn, S. H., Ziegler, V. E., Rosen, S. H., and Meyer, D. A.,** Dosage schedule and plasma levels of doxepin and desmethyldoxepin, *J. Clin. Psychiatr.*, 39, 740, 1978.
122. **Kline, N. S., Cooper, T., and Johnston, B.,** Doxepin and desmethyldoxepin serum levels and clinical response, in *Pharmacokinetics of Psychoactive Drugs: Blood Levels and Clinical Response,* Gottschalk, L. A. and Merlis, S., Eds., Spectrum Publications, Holliswood, 1976, 221.
123. **Friedel, R. O. and Raskind, M. A.,** Relationship of blood levels of Sinequan to clinical effects in the treatment of depression in aged patients, in *Sinequan® (Doxepin HC1),* Mendels, J., Ed., Excerpta Medica, Amsterdam, 1975, 51.
124. **Hobbs, D. C.,** Distribution and metabolism of doxepin, *Biochem. Pharmacol.*, 18, 1941, 1969.

125. **Kimura, Y., Kume, M., and Kageyama, K.,** Absorption, distribution, excretion and metabolism of doxepin hydrochloride, *Oyo Yakuri,* 6, 955, 1972.
126. **Pebay-Peyroula, F., Bismuth, C., and Autret, A.,** Premier cas francais d'intoxication aigue par la doxepine, *J. Eur. Toxicol.,* 2, 77, 1969.
127. **Gallant, D. M., Bishop, M. P., and Guerrero-Figueroa, R.,** CL-67,772: a preliminary evaluation of a potential antidepressant compound: animal and human correlations, *Curr. Ther. Res. Clin. Exp.,* 13, 364, 1971.
128. **Aberg, A. and Holmberg, G.,** Controlled trial of a new antidepressive, amoxapine, in comparison with amitriptyline, *Curr. Ther. Res. Clin. Exp.,* 22, 304, 1977.
129. **Gallant, D. M., Swanson, W. C., Mielke, D. H., Guerrero-Figueroa, R., and Collins, H.,** Amoxapine: a double-blind evaluation of antidepressant activity, *Curr. Ther. Res. Clin. Exp.,* 15, 56, 1973.
130. **Coupet, J., Rauh, C. E., Szues-Myers, V. A., and Yunger, L. M.,** 2-Chloro-11-(1-piperazinyl)dibenz[b,f][1,4]oxazepine (Amoxapine), an antidepressant with antipsychotic properties — a possible role for 7-hydroxyamoxapine, *Biochem. Pharmacol.,* 28, 2514, 1979.
131. **Greenblatt, E. N., Hardy, R. A., and Kelly, R. G.,** Amoxapine, in *Pharmacological and Biochemical Properties of Drug Substances,* Vol. 2, Golberg, M. E., Ed., American Pharmaceutical Association, Washington, D.C., 1979, 1.
132. **Greenblatt, E. N., Lippa, A. S., and Osterberg, A. C.,** The neuropharmacological actions of amoxapine, *Arch. Int. Pharmacodyn. Ther.,* 233, 107, 1978.
133. **Cooper, T. B. and Kelly, R. G.,** GLC analysis of loxapine, amoxapine and their metabolites in serum and urine, *J. Pharm. Sci.,* 68, 216, 1979.
134. **Ross, S. R. and Rogers, S. R.,** Physostigmine in amoxapine overdose, *Am. J. Hosp. Pharm.,* 38, 1121, 1981.
135. **Goldberg, M. J. and Spector, R.,** Amoxapine overdosage: report of two patients with severe neurologic damage, *Ann. Intern. Med.,* 96, 463, 1982.
136. **Prous, J. R., Ed.,** Amoxapine, *Drugs Future,* I, 511, 1976.
137. **Murphy, J. E., Ed.,** Ludiomil® in general practice, a symposium, *J. Int. Med. Res.,* 5(Suppl. 4), 1977.
138. **Lehmann, H. E., Pecknold, J. C., Ban, T. A., and Orbach, L.,** A double-blind comparative clinical trial with maprotiline (Ludiomil®) and imipramine in newly-admitted depressed patients, *Curr. Ther. Res. Clin. Exp.,* 19, 463, 1976.
139. **Murphy, J. E., Ed.,** Ludiomil® symposium, *J. Int. Med., Res.,* 3(Suppl. 2), 1975.
140. **Baumann, P. A. and Maitre, L.,** Neurobiochemical aspects of maprotiline (Ludiomil®) action, *J. Int. Med. Res.,* 7, 391, 1979.
141. **Greengrass, P. M., Waldmeier, P. C., Imhof, P. R., and Maitre, L.,** Comparison of the effects of maprotiline (Ludiomil®) and clomipramine (Anafranil®) on serotonin uptake and tryptophan-binding in plasma, *Biol. Psychiatr.,* 11, 91, 1976.
142. **Bein, H. J.,** Biological research in pharmaceutical industry with reserpine, in *Discoveries in Biological Psychiatry,* Ayd, F. J., Jr. and Blackwell, B., Eds., Lippincott, Philadelphia, 1970, chap. 11.
143. **Fukushima, H., Nakamura, M., and Yamamoto, H.,** Pharmacology of 9-γ-methylaminopropyl-9,10-dihydro-9,10-methanoanthracene·HCl (ID-9206·HCl), a new potent antidepressant, *Arch. Int. Pharmacodyn. Ther.,* 229, 163, 1977.
144. **Riess, W., Dubey, L., Funfgeld, E. W., Imhof, P., Hurzeler, H., Matussek, N., Rajagopalan, T. G., Raschdorf, F., and Schmid, K.,** The pharmacokinetic properties of maprotiline (Ludiomil®) in man, *J. Int. Med. Res.,* 3, Suppl. 2, 16, 1975.
145. **Fishbach, R.,** Maprotiline in adult depressed patients, *Arzneim. Forsch.,* 29, 352, 1979.
146. **Hess, R., Diener, R. M., Fritz, H., Pericin, C., Heywood, R., Newman, A. J., and Rivett, K. T.,** Tossicita, ed effetti sulla riproduzione nell'animale del nuovo farmaco antidepressivo maprotilina (Ludiomil®), *Boll. Chim. Farm.,* 112, 782, 1973.
147. **Jukes, A. M.,** Maprotiline (Ludiomil®): side-effects and overdose, *J. Int. Med. Res.,* 3(Suppl. 2), 126, 1975.
148. **Szeless, S., Kleinberger, G., Kotzaurek, R., Pall, H., Pichler, M., and Raberger, G.,** Lethal intoxication with the tetracyclic antidepressant Ludiomil®, *Wien. Klin. Wochenschr.,* 87, 254, 1975.
149. **Park, J. and Proudfoot, A. T.,** Acute poisoning with maprotiline and mianserin, *Br. Med. J.,* 1, 1573, 1977.
150. **Meek, D., Bott, M. D., Mamtora, H., and Gabriel, R.,** Convulsion following maprotiline overdose, *Br. Med. J.,* 2, 275, 1975.
151. **Crome, P. and Newman, B.,** Poisoning with maprotiline and mianserin, *Br. Med. J.,* 2, 260, 1977.
152. **Larochelle, P., Hamet, P., and Enjalbert, M.,** Responses to tyramine and norepinephrine after imipramine and trazodone, *Clin. Pharmacol. Ther.,* 26, 24, 1979.
153. **Taylor, D. P., Hyslop, D. K., and Riblet, L. A.,** Trazodone, a new nontricyclic antidepressant without anticholinergic activity, *Biochem. Pharmacol.,* 29, 2149, 1980.

154. **Trapp, G. A., Handorf, C. R., and Larach, V.,** Trazodone in the treatment of depressed inpatients, *Psychopharmacol. Bull.*, 15, 25, 1979.
155. **Stefanini, E., Fadda, F., Medda, L., and Gessa, G. L.,** Selective inhibition of serotonin uptake by trazodone, a new antidepressant agent, *Life Sci.*, 18, 1459, 1976.
156. **Silvestrini, B., Cioli, V., Burberi, S., and Catanese, B.,** Pharmacological properties of AF 1161 a new psychotropic drug, *Int. J. Neuropharmacol.*, 7, 587, 1968.
157. **Carruba, M. O., Parenti, M., Ricciardi, S., Piccoti, G. B., and Mantegazza, P.,** Effect of etoperidone on the depletion of brain 5-hydroxytryptamine and catecholamines induced respectively by *p*-chloromethamphetamine and 6-hydroxydopamine in rats, *Pharmacol. Res. Commun.*, 11, 169, 1979.
158. **Ramacci, M. T., Ghirardi, O., Maccari, F., Pacifici, L., and Sale, P.,** Pharmacological evaluation of Δ_2,1,2,4-triazolin-5-one[1,3(4-m-chlorophenyl-1-piperazinyl)-propyl]-3,4-diethyl-hydrochloride (Etoperidone), a new potential antidepressant drug, *Arzneim. Forsch.*, 29, 294, 1979.
159. **Catanese, B. and Lisciani, R.,** Investigations on the absorption and distribution of trazodone on AF 1161 in rats, dogs and humans, *Boll. Chim. Farm.*, 109, 369, 1970.
160. *Physicians Desk Reference*, 36th ed., Supplement A, Medical Economics, Oradell, 1982, A64.
161. **Koss, F. W. and Busch, U.,** Pharmacokinetics and metabolism of trazodone in different species, in *Depression and the Role of Trazodone in Antidepressant Therapy*, Morozov, G., Saarma, J., and Silvestrini, B., Eds., Edizioni Luigi Pozzi S.p.A., Rome, Italy, 1978, 11.
162. **Koss, F. W. and Busch, U.,** Trazodone as an example — drug levels in the blood and brain, in *Trazodone — A New Broadspectrum Antidepressant. Proc. 11th Congr. Collegium International Neuro-Psychopharmacologicum*, Gershon, Rickels, and Silvestrini, B., Eds., Excerpta Medica, Amsterdam, 1980, 27.
163. **Fujita, T., Yamato, C., Takahashi, T., and Ohtake, S.,** Absorption, tissue distribution, excretion and metabolism of 14C-labelled trazodon hydrochloride in rats and rabbits, in *Modern Problems of Pharmacopsychiatry, Trazodone*, Vol. 9, Ban, T. A. and Silvestrini, B., Eds., Karger, Basel, 1974, 56.
164. **Jauch, V. R., Kopitar, Z., Prox, A., and Zimmer, A.,** Pharmacokinetics and metabolism of trazodone in man, *Arzneim. Forsch.*, 26, 2084, 1976.
165. **Prous, J. R., Ed.,** *Annual Drug Data Report*, Vol. 2, J. R. Prous Publishers, Barcelona, Spain, 1979/80, 233.
166. **Shopsin, B., Cassano, G. B., and Conti, L.,** An overview of new "second generation" antidepressant compounds: research and treatment implications, in *Antidepressants: Neurochemical Behavioral and Clinical Perspectives*, Enna, S. J., Malick, J. B., and Richelson, E., Ed., Raven Press, New York, 1981, 219.
167. **Hoffmann, I., Ehrhart, G., and Schmitt, K.,** 8-Amino-4-phenyl-1,2,3,4,-tetrahydroisochinoline, eine neue gruppe antidepressiver psychopharmaka, *Arzneim. Forsch.*, 21, 1045, 1971.
168. **Hunt, P., Kannergiesser, M. H., and Raynaud, J. P.,** Nomifensine: a new potent inhibitor of dopamine uptake into symaptosomes from rat brain corpus striatum, *J. Pharm. Pharmacol.*, 26, 370, 1974.
169. **Gerhards, H. J., Carenzi, A., and Costa, E.,** Effect of nomifensine on motor activity, dopamine turnover rate, and cyclic 3′,5′-adenosine monophosphate concentrations of rat striatum, *Naunyn-Schmiedeberg's Arch. Pharmacol.*, 286, 49, 1974.
170. **Forrest, A., Hewett, A., and Nicholson, P.,** Controlled randomized group comparison of nomifensine and imipramine in depressive illness, *Br. J. Clin. Pharmacol.*, 4, 215S, 1977.
171. **Grof, P., Saxena, B., Daigle, L., and Mahutte, G.,** Dopaminergic agonist nomifensine compared with amitriptyline: a double-blind clinical trial in acute primary depressions, *Br. J. Clin. Pharmacol.*, 4, 221S, 1977.
172. **Hanks, G. W.,** A profile of nomifensine, *Br. J. Clin. Pharmacol.*, 41, 243S, 1977.
173. **Ogren, S. -O., Ross, S. B., Hall, H., Holm, A. -C., and Renyi, A. L.,** The pharmacology of zimelidine: a 5-HT selective reuptake inhibitor, *Acta Psychiatr. Scand. Suppl.*, 63(290), 127, 1981.
174. **Ross, S. B. and Renyi, A. L.,** Inhibition of the neuronal uptake of 5-hydroxytryptamine and noradrenaline in rat brain by (Z)- and (E)-3-(4-bromophenyl)-*N,N*-dimethyl-3-(3-pyridyl)-allylamines and their secondary analogues, *Neuropharmacology*, 16, 57, 1977.
175. **Carlsson, A. and Lindqvist, M.,** Effects of antidepressant agents on the synthesis of brain monoamines, *J. Neural Transm.*, 43, 73, 1978.
176. **Ross, S. B., Hall, H., Renyi, A. L., and Westerlund, D.,** Effects of zimelidine on serotoninergic and noradrenergic neurons after repeated administration in the rat, *Psychopharmacology*, 72, 219, 1981.
177. **Hogberg, T., Ulff, B., Renyi, A. L., and Ross, S. B.,** Synthesis of pyridylallylamines related to zimelidine and their inhibition of neuronal monoamine uptake, *J. Med. Chem.*, 24, 1499, 1981.
178. **London, J. B., Tiplady, B., Ashcroft, G. W., and Waddell, J. L.,** Zimelidine and amitriptyline in the treatment of depressive illness in general practice, *Acta Psychiatr. Scand. Suppl.*, 63(290), 454, 1981.
179. **Potter, W. Z., Calil, H. M., Extein, I., Gold, P. W., Wehr, T. A., and Goodwin, F. K.,** Specific norepinephrine and serotonin uptake inhibitors in man: a crossover study with pharmacokinetic biochemical, neuroendocrine and behavioural parameters, *Acta Psychiatr. Scand. Suppl.*, 63(290), 152, 1981.
180. **d'Elia, G., Hallstrom, T., Nystrom, C., and Ottosson, J. -O.,** Zimelidine vs. maprotiline in depressed outpatients. A preliminary report, *Acta Psychiatr. Scand. Suppl.*, 63(290), 225, 1981.

181. **Georgotas, A., Mann, J., Bush, D., and Gershon, S.,** Zimelidine vs. amitriptyline vs. placebo in a double-blind study: safety data on zimelidine hydrochloride following overdose, *Acta Psychiatr. Scand. Suppl.,* 63(290), 256, 1981.
182. **van der Burg, W. J., Bonta, I. L., Delobelle, J., Ramon, C., and Vergaftig, B.,** A novel type of substituted piperazine with high antiserotonin potency, *J. Med. Chem.,* 13, 35, 1970.
183. **Nickolson, V. J. and Wieringa, J. P.,** Presynaptic alpha-block and inhibition of noradrenaline and 5-hydroxytryptamine reuptake by a series of compounds related to mianserin, *J. Pharm. Pharmacol.,* 33, 760, 1981.
184. **Doxey, J. C., Everitt, J., and Metcalf, G.,** Mianserin — an analysis of its peripheral autonomic actions, *Eur. J. Pharmacol.,* 51, 1, 1978.
185. **Peet, M. and Behagel, H.,** Mianserin: a decade of scientific development, *Br. J. Clin. Pharmacol.,* 5, 5S, 1978.
186. **Raiteri, M., Angelini, F., and Bertollini, A.,** Comparative study of the effects of mianserin, a tetracyclic antidepressant and of imipramine on uptake and release of neurotransmitters in synaptosomes, *J. Pharm. Pharmacol.,* 28, 483, 1976.
187. **Baumann, P. A. and Maitre, L.,** Blockade of presynaptic alpha-receptors and of amine uptake in the rat brain by the antidepressant mianserin, *Naunyn-Schmiedeberg's Arch. Pharmacol.,* 300, 31, 1977.
188. **Robson, R. D., Antonaccio, M. J., Saelens, J. K., and Liebman, J.,** Antagonism by mianserin and classical alpha-adrenoceptor blocking drugs of some cardiovascular and behavioral effects of clonidine, *Eur. J. Pharmacol.,* 47, 431, 1978.
189. **Harper, B. and Hughes, I. E.,** Presynaptic alpha-adrenoceptor blocking properties among tri- and tetracyclic antidepressant drugs, *Br. J. Pharmacol.,* 67, 511, 1979.
190. **Engberg, G. and Svensson, T. H.,** Mianserin: direct activation of brain norepinephrine neurons by blocking alpha-2 receptors, *Commun. Psychopharmacol.,* 4, 233, 1980.
191. **Jaskari, M. O., Ahlfors, U. G., Ginman, L., Lydeken, K., and Tienari, P.,** Three double-blind comparative trials of mianserin (Org GB.-94) and amitriptyline in the treatment of depressive illness, *Pharmakopsychiatr. Neuropsychopharmakol.,* 10, 101, 1977.
192. **Kopera, H.,** Anticholinergic and blood pressure effects of mianserin, amitriptyline and placebo, *Br. J. Clin. Pharmacol.,* 5, 29S, 1978.
193. Proceeding of a symposium on mianserin, *Br. J. Clin. Pharmacol.,* 5, Suppl. 1, 1978.
194. **Pinder, R. M., Brogen, R. N., Speight, T. M., and Avery, G. S.,** Viloxazine: a review of its pharmacological properties and therapeutic efficacy in depressive illness, *Drugs,* 13, 401, 1977.
195. **Evans, R. L.,** Alprazolam, *Drug Intell. Clin. Pharm.,* 15, 633, 1981.
196. **Zis, A. P. and Goodwin, F. K.,** Novel antidepressants and the biogenic amine hypothesis of depression, *Arch. Gen. Psychiatr.,* 36, 1097, 1979.
197. **Goodwin, F. K., Ed.,** The lithium ion: impact on treatment and research, *Arch. Gen. Psychiatr.,* 36, 833, 1979.
198. **Ferris, R. M., White, H. L., Cooper, B. R., Maxwell, R. A., Tang, F. L. M., Beaman, O. J., and Russell, A.,** Some neurochemical properties of a new antidepressant, bupropion hydrochloride (Wellbutrin®), *Drug Dev. Res.,* 1, 21, 1981.
199. **Fabre, L. F. and Crismon, L.,** Efficacy of fluoxetine in outpatients with major depression, *Curr. Ther. Res. Clin. Exp.,* 37, 115, 1985.

MONOAMINE OXIDASE INHIBITORS

Beng T. Ho and Patricia M. Kralik

INTRODUCTION

Monoamine oxidase (MAO) [monoamine :oxygen oxidoreductase (deaminating EC 1.4.3.4] is an enzyme capable of oxidizing primary and secondary aralkylamines to their corresponding aldehydes, which, due to their relative instability, are either further oxidized by a second enzyme, aldehyde dehydrogenase, to their acids, or reduced by other enzymes, alcohol dehydrogenase or aldehyde reductase, to their alcohols. The absence of a nitrogen atom in the aldehyde products of MAO leads to the proposed imino (Schiff's base) intermediates[1] (Figure 1), and the overall oxidation process is designated oxidative deamination.

MAO exists in all vertebrates, and can be found in varying peripheral and central tissues (for review, see pp. 31 to 32 of Reference 2). Specifically, its distribution in the human brain has been reported.[3,4] The enzyme is located in the outer membrane of mitochondria, although some activity has also been found in microsomal preparations.[5] Mitochondria-containing MAO is present both intraneuronally (i.e., within the presynaptic nerve terminals) and extraneuronally (possibly located postjunctionally).[6] The function of MAO is therefore to regulate the levels of endogenous monoamines, such as epinephrine, norepinephrine, dopamine, tyramine, tryptamine, and serotonin. It is this regulation of monoamine concentration within the presynaptic nerve terminals that implicates the importance of inhibitors of MAO (MAOIs) in animal and human behavior, as well as in their clinical use as antidepressants.

MAO DIFFERENTIATION BY SELECTIVE INHIBITORS

In 1968, Johnston[7] reported that when tyramine or dopamine (DA) was used as substrate for MAO, varying concentrations of clorgyline [*N*-(2,4-dichlorophenoxypropyl)-*N*-methyl-2-propynylamine] showed a double-sigmoidal inhibition curve, suggesting the existence of two types MAO. He then designated these activities a type-A and type-B MAO. Since then, numbers of other inhibitors have been found to exert specificity for the deamination of certain substrates (see also listing in Table 3 and text under classification of MAOIs), and selection of the propert inhibitor has become a useful means in differentiating the two types of MAO. Based on the selective action of clorgyline in inhibiting type-A enzyme, serotonin (5-HT), norepinephrine (NE), and normetanephrine are generally classified as preferred substrates of type-A MAO (Table 1). On the other hand phenethylamine (PEA) and benzylamine are preferred substrates of type-B MAO because they are predominantly oxidized by the (−)-deprenyl [*N*-(1-phenylisopropyl)-*N*-methyl-2-propynylamine]-sensitive type-B enzyme. Other monoamines, such as tyramine and tryptamine, have been shown to be oxidized by both types. Although 5-HT and PEA are preferred substrates of type-A and type-B MAO, respectively, recent in vitro and in vivo findings have indicated that at high concentrations the above substrates can be deaminated by both types of MAO.[9-14] Interspecies variation exists, as DA is selective for type-B MAO in human platelets and certain brain areas, but in rat brain it is oxidized predominantly by type A. There are also exceptions to this substrate specificity due to tissue variations in activity. For example, it has been reported that in rat heart[15] and in rat kidney[16] PEA is oxidized by type-A MAO, and that benzylamine, another type-B substrate, is metabolized by both types in rat heart.[15]

The existence of two types of MAO was further demonstrated in the intact organism. In the perfused rabbit lung, using radioactively labeled substrates 5-HT and PEA, harmaline (7-methoxy-1-methyl-3,4-dihydro-β-carboline) (Figure 5), a type-A inhibitor, decreased the

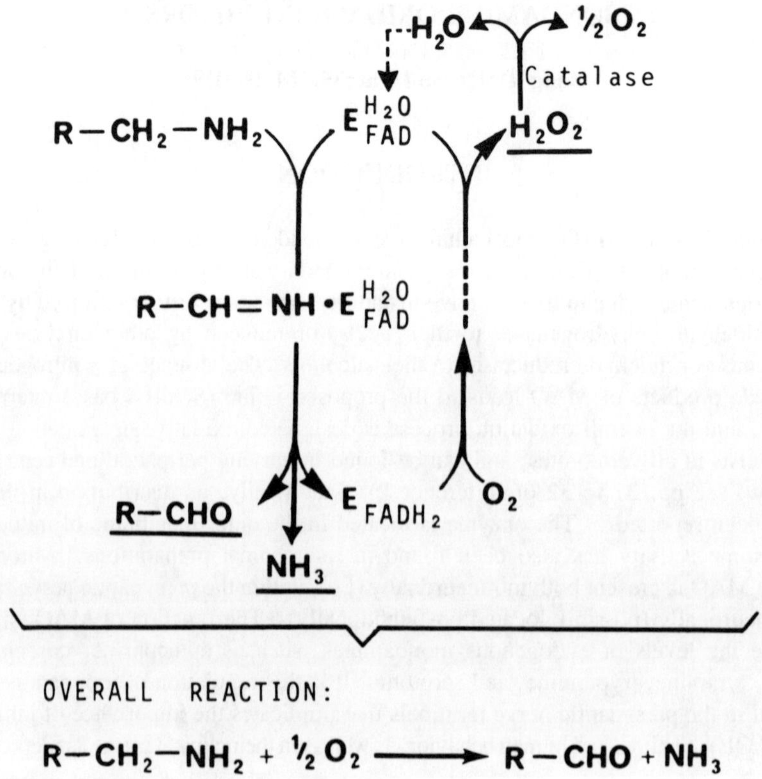

FIGURE 1. MAO catalyzed oxidative deamination. (From Gey, K. F. and Pletscher, A., *Pharmacol. Ther. B.*, 2, 137, 1976. With permission.)

Table 1
SUBSTRATES OF TYPE-A AND TYPE-B MONOAMINE OXIDASE

	MAO	
	Type A	Type B
Common substrates	Dopamine Tyramine Tryptamine	
Preferred substrates	Serotonin Norepinephrine Normethanephrine	β-Phenylethylamine Benzylamine

From Neff, N. H. and Yang, H.-Y. T., *Life Sci.*, 14, 2061, 1974. With permission.

deaminated product only from 5-HT but not from PEA.[17] Bakhle and Youdim[18] have demonstrated the presence of type-B MAO in perfused rat lung, as deprenyl was shown to be more effective in preventing the deamination of PEA than inhibitors of type-A enzyme. Further support of the two types of MAO was found from the work of Yang and Neff,[19] who have reported that 5-hydroxyindoleacetic acid (a metabolite of 5-HT) and 3,4-dihydroxyphenylacetic acid (a metabolite of DA) decreased following injection of clorgyline, whereas only a decrease in 3,4-dihydroxyphenylacetic acid was observed with deprenyl treatment.

Table 2
PREDOMINANT TYPE OF MAO IN TISSUES OF VARIOUS SPECIES

	MAO type
Rabbit lung	A
Rat lung	B
Rat heart	A
Rat kidney	A
Rat pineal gland	B
Mouse liver	B
Bovine liver	B
Human small intestine	A
Human platelets	B

Both types of the enzyme MAO are not present in all tissues. While most tissues of different species contain both types of MAO, Table 2 lists those sources containing only one type. The type-B enzyme was found exclusively in mouse liver,[20] bovine liver,[21] and human blood platelets.[22] In human brain, 80% of MAO is type B, compared to the small intestine which contains predominantly type-A enzyme.[20,23] Type-A MAO is associated with synaptic nerve endings.[24] In rat pineal gland, which contains high levels of 5-HT, the predominant activity is type B.[8]

Gel electrophoresis separates solubilized rat liver mitochondrial MAO into multiple bands associated with varying levels of phospholipids. The bands which are more resistant to inhibition by clorgyline contain the least phospholipid material and are suggested to be type-B MAO.[25,26] After treatment of rat liver mitochondria with chaotropic agents for the removal of phospholipids, sensitivity to clorgyline is diminished, indicating the type-A enzyme activity is apparently lost.[27,28] These results have demonstrated the influence of phospholipids in the regulation of enzyme activities and have led to speculation that the surrounding phsopholipid levels actually influence or determine a favorable enzyme conformation, which accounts for the specificity of one type MAO to its substrate. Data accumulated to date has not yet resolved whether the two types of MAO are separate proteins or only one. While some investigators have provided evidence supporting the existence of two individual proteins,[29-31] there are others who suggest that MAO is a single enzyme protein with two adjacent sites whereby the phospholipid environment determines the substrate and inhibition specificity, depending on their concentrations.[32,33]

CLASSIFICATION OF MAOIS

Structure Variation

Monoamine oxidase inhibitors (MAOIs) can be grouped according to their chemical structure as follows:[34-39]

1. Hydrazine derivatives, e.g., phenelzine (2-phenethylhydrazine), isocarboxazid (5-methyl-3-isoxazolecarboxylic acid 2-benzylhydrazide), and iproniazid (4-pyridinecarboxylic acid 2-isopropylhydrazide) (Figure 2)
2. Acetylenic amines (propargylamines), e.g., clorgyline [*N*-(2,4-dichlorophenoxypropyl)-*N*-methyl-2-propynylamine], deprenyl [*N*-(1-phenylisopropyl)-*N*-methyl-2-propynylamine], and pargyline (*N*-benzyl-*N*-methyl-2-propynylamine) (Figure 3)
3. Arylcyclopropylamines, e.g., tranylcypromine (*trans*-2-phenylcyclopropylamine) (Figure 4)

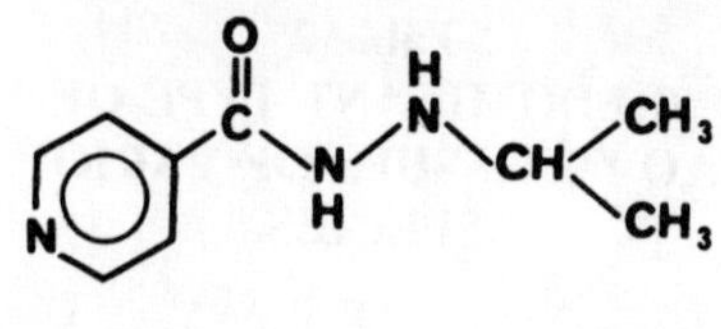

Iproniazid

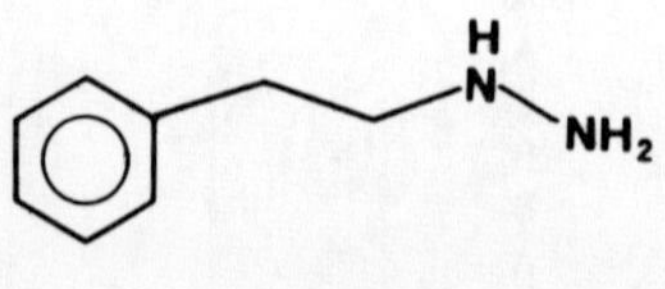

Phenelzine

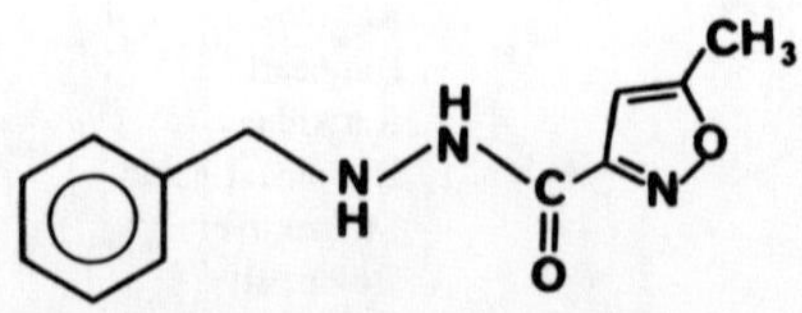

Isocarboxazid

FIGURE 2. Hydrazine derivatives as MAOI.

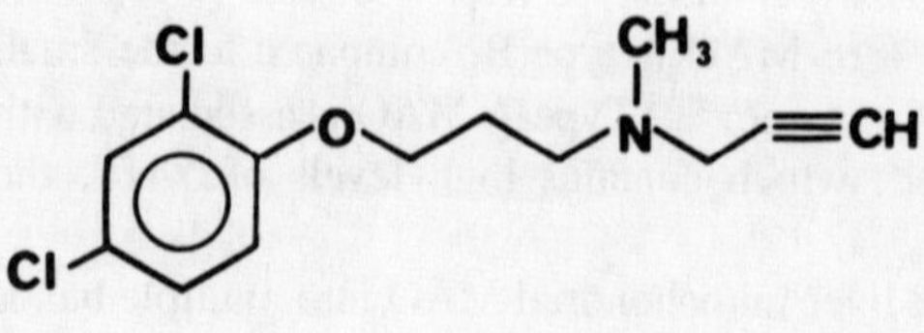

Clorgyline

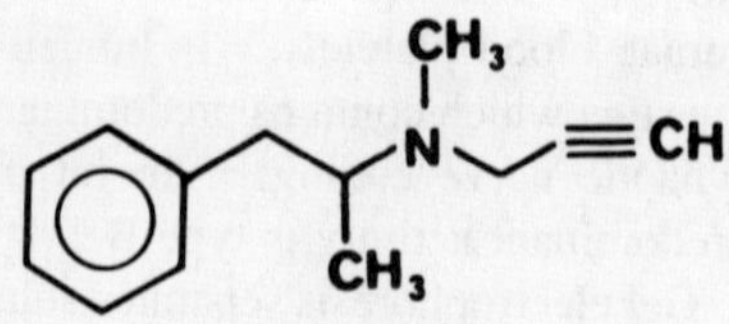

Deprenyl

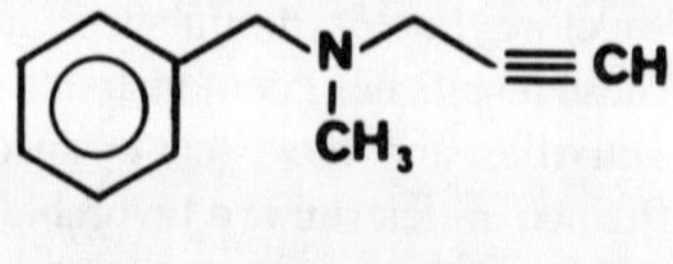

Pargyline

FIGURE 3. Acetylenic amines as MAOI.

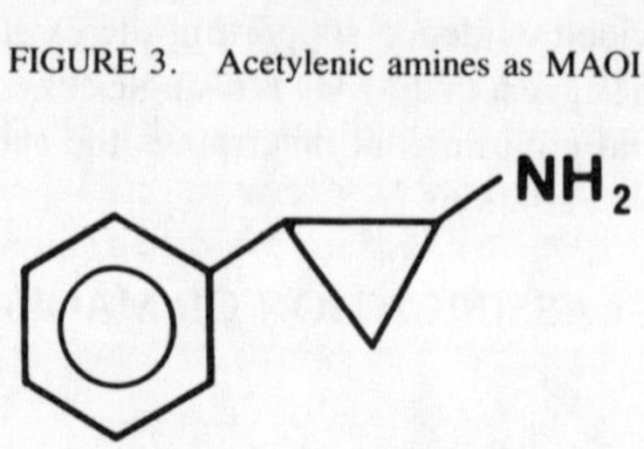

Tranylcypromine

FIGURE 4. Arylcyclopropylamines as MAOI.

4. Aryloxycyclopropylamines, e.g., 2-chlorophenoxyethylcyclopropylamine (Lilly 51641, Figure 6)
5. α-Alkylated indolealkylamines or α-alkylated aralkylamines, e.g., α-ethyltryptamine (Figure 5)
6. N-Heterocyclic amines, e.g., harmaline (7-methoxy-1-methyl-3,4-dihydro-β-carboline) (Figure 5), 2-methyl-3-piperidinopyrazine, furoxanobenzofuroxan (Figure 6) and

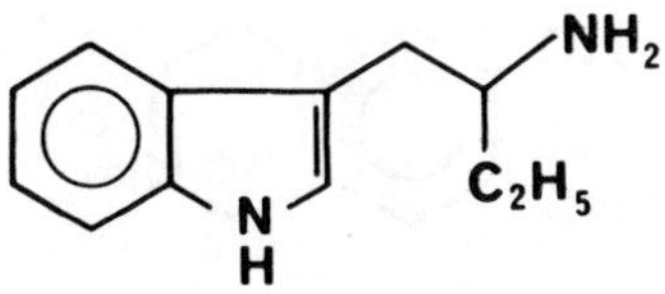

α-Ethyltryptamine

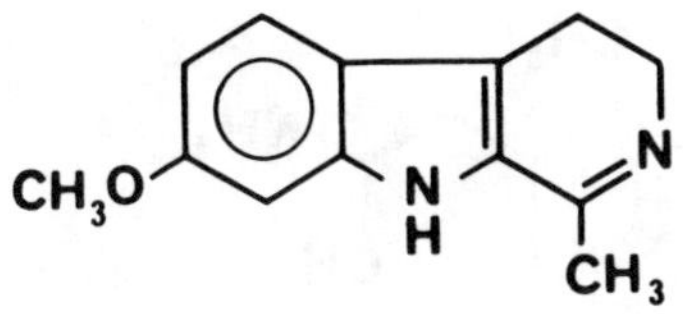

Harmaline

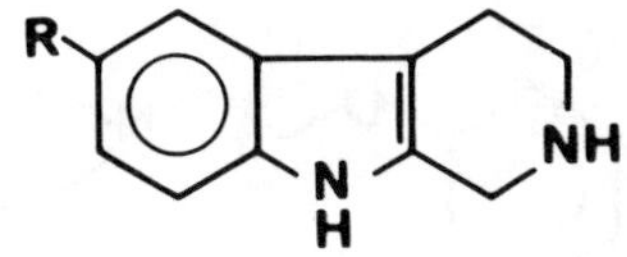

R=H: Tetrahydro-β-carboline

$R=OCH_3$: 6-Methoxytetrahydro-β-carboline

FIGURE 5. Indolealkylamines and β-carbolines as MAOI.

other oxazoles, oxadiazoles, and benzoxadiazoles, 6,7-dihydroxytetrahydroisoquinoline (Figure 6) and other tetrahydroisoquinolines, imipramine and other tricyclic psychotherapeutic agents

7. Amidines and isothioureas, e.g., propamidine and other amidines, and S-*n*-alkylisothioureas[36]

Reversibility and Irreversibility

When considering the kinetics of inhibition, MAOIs can be classified into reversible and irreversible inhibitors. Most of the irreversible inhibitors of MAO belong to one of the three classes of compounds, namely acetylenic amines, substituted hydrazines, and arylcyclopropylamines. These compounds, once attached to the substrate site, produce irreversible inhibition of the enzyme by forming covalently bonded adducts.[40] Because of their specificity in inducing the inhibition at the active site of the enzyme, they are called active site-directed inhibitors or ''suicide inhibitors''. A prosthetic group flavin is covalently bound to MAO through its 8-α-position to a cysteine group on the apoenzyme forming a thioether[41] (Figure 8). Acetylenic amines (e.g., deprenyl) act by forming a flavocyanine adduct with N(5) of the covalently bound flavin component[42] (Figure 9). The reaction is characterized by bleaching of the absorption band between 450 to 455 nm due to oxidized flavin, and the resulting adduct shows a large band at 410 nm.[40,43] The suicide reaction between substituted hydrazines and MAO involves the irreversible interaction of diazenes, which are oxidized products of hydrazines formed by MAO, to the C(4a) of the flavin[40,44] (Figure 10). This reaction results in a loss of N_2 from the diazene, a bleaching of absorption between 450 to 455 nm due to the oxidized form of flavin, and an appearance of a 410 nm band of the adduct. There is also evidence that part of the diazene residue may be attached to the thiol group at the substrate site of MAO.[40]

The inactivation of MAO by arylcyclopropylamines such as tranylcypromine occurs in a fashion different from that by either acetylenic amines or substituted hydrazines. It has been proposed that tranylcypromine acts on the thiol group at the substrate site of the enzyme to form a thioaminoketal or thiohemiketal[40,45] (Figure 11). Arguments against the attachment of the inhibitor to either N(5) or C(4a) of the flavin is supported by findings of a reappearance of the 450 nm abosrbance, due to the oxidized form of flavin, when the enzyme-inhibitor

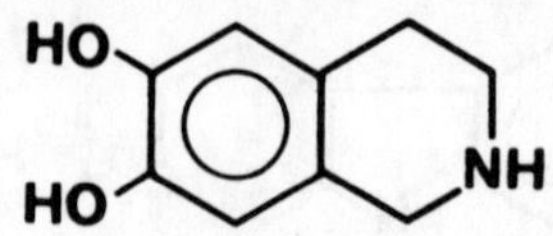

6,7-Dihydroxytetrahydro-
isoquinoline

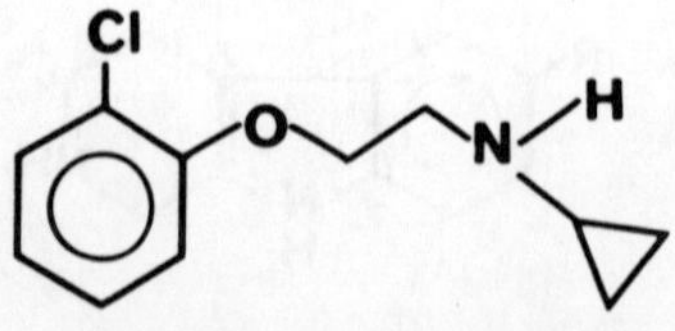

Lilly 51641

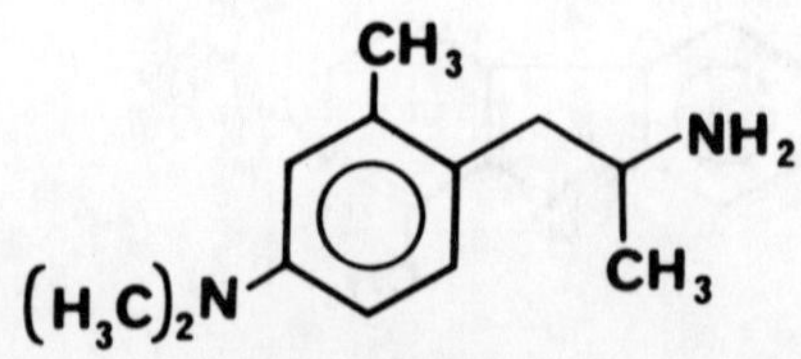

FLA 336

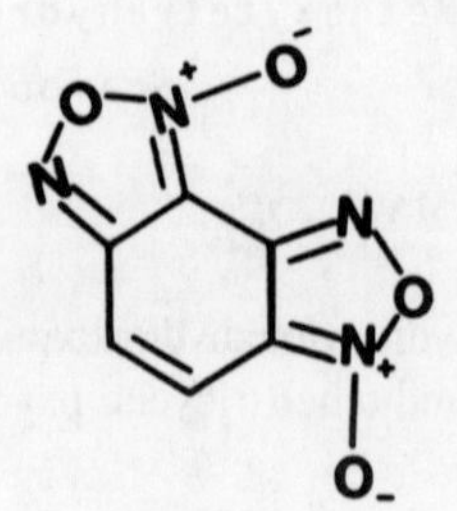

Furoxanobenzofuroxan

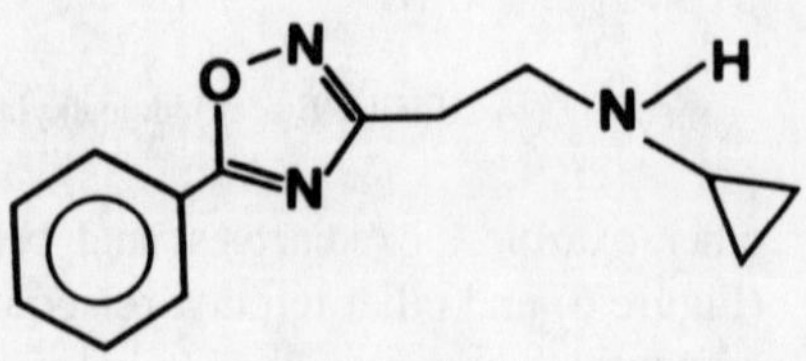

PCO

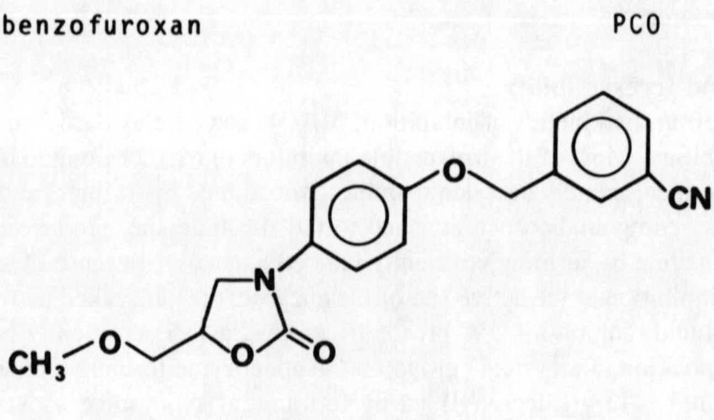

MD 780515

FIGURE 6. Some selective inhibitors as type-A MAO.

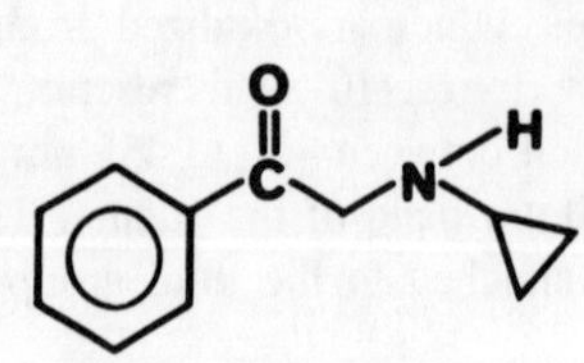

L-54761

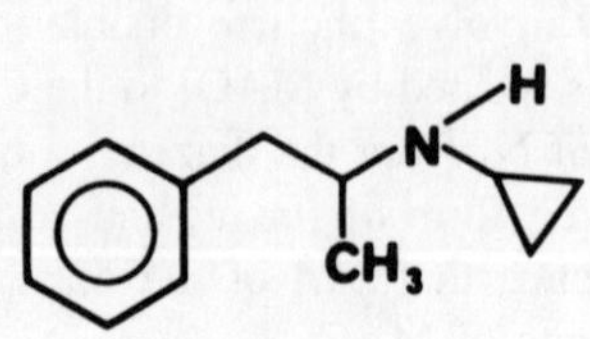

L-54748

FIGURE 7. Some selective inhibitors of type-B MAO.

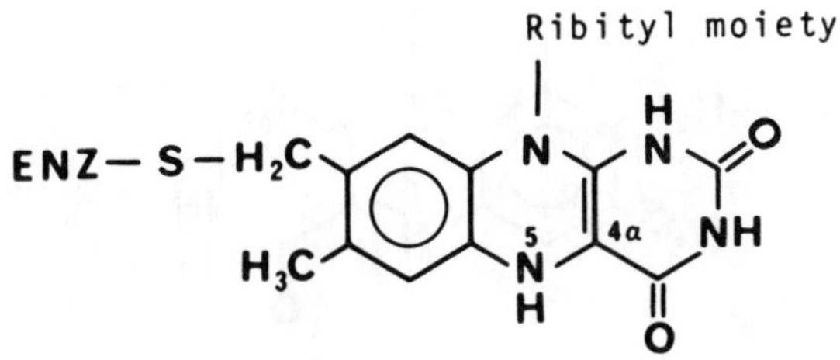

FIGURE 8. Attachment of flavin to the apoprotein on MAO.

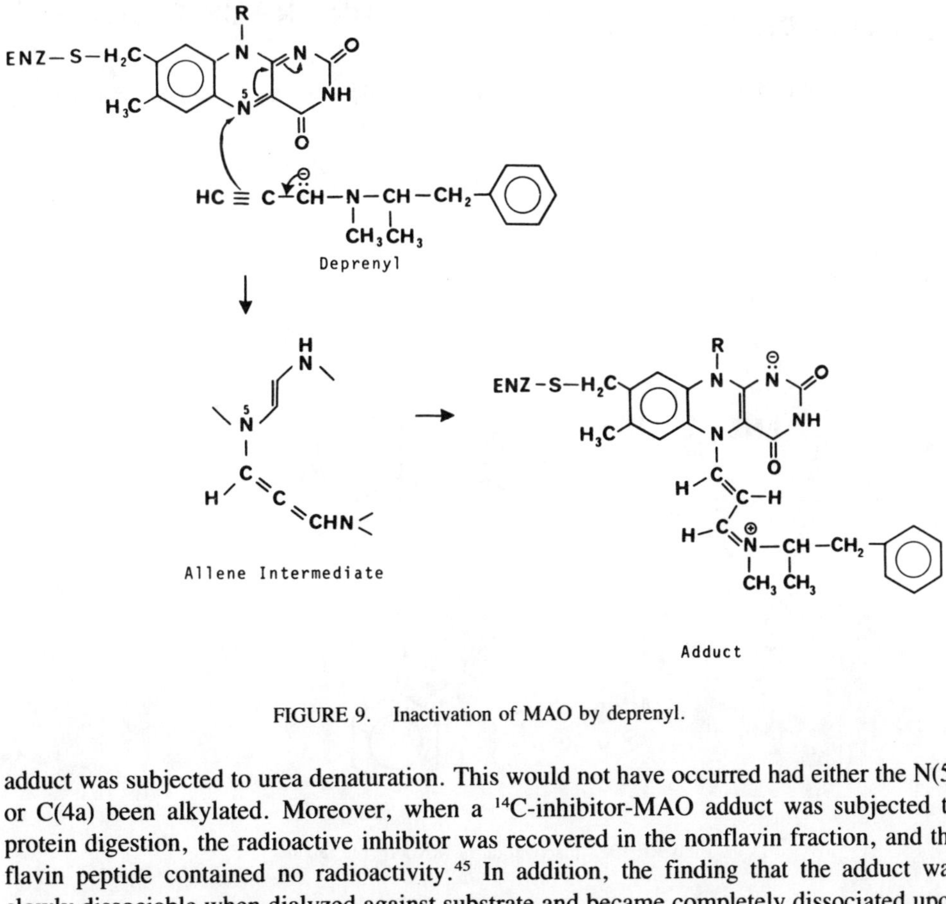

FIGURE 9. Inactivation of MAO by deprenyl.

adduct was subjected to urea denaturation. This would not have occurred had either the N(5) or C(4a) been alkylated. Moreover, when a ^{14}C-inhibitor-MAO adduct was subjected to protein digestion, the radioactive inhibitor was recovered in the nonflavin fraction, and the flavin peptide contained no radioactivity.[45] In addition, the finding that the adduct was slowly dissociable when dialyzed against substrate and became completely dissociated upon denaturation, supports the formation of the proposed thioaminoketal.

The inhibition of MAO by an irreversible inhibitor proceeds with a reversible phase of inhibition prior to irreversible binding. In the initial phase, a reversible enzyme-inhibitor complex, EI, is formed which, with time, becomes nondissociable when the inhibitor becomes irreversibly bound to the enzyme with the formation of a covalent linkage.

Reversible inhibitors of MAO are numerous and cover a wide spectrum of structures. These include harmaline and other β-carbolines, α-alkylated tryptamine, amphetamine, imipramine, and other tricyclic compounds.

Selectivity

MAOIs can also be separated into three groups based on their selective action on the type-

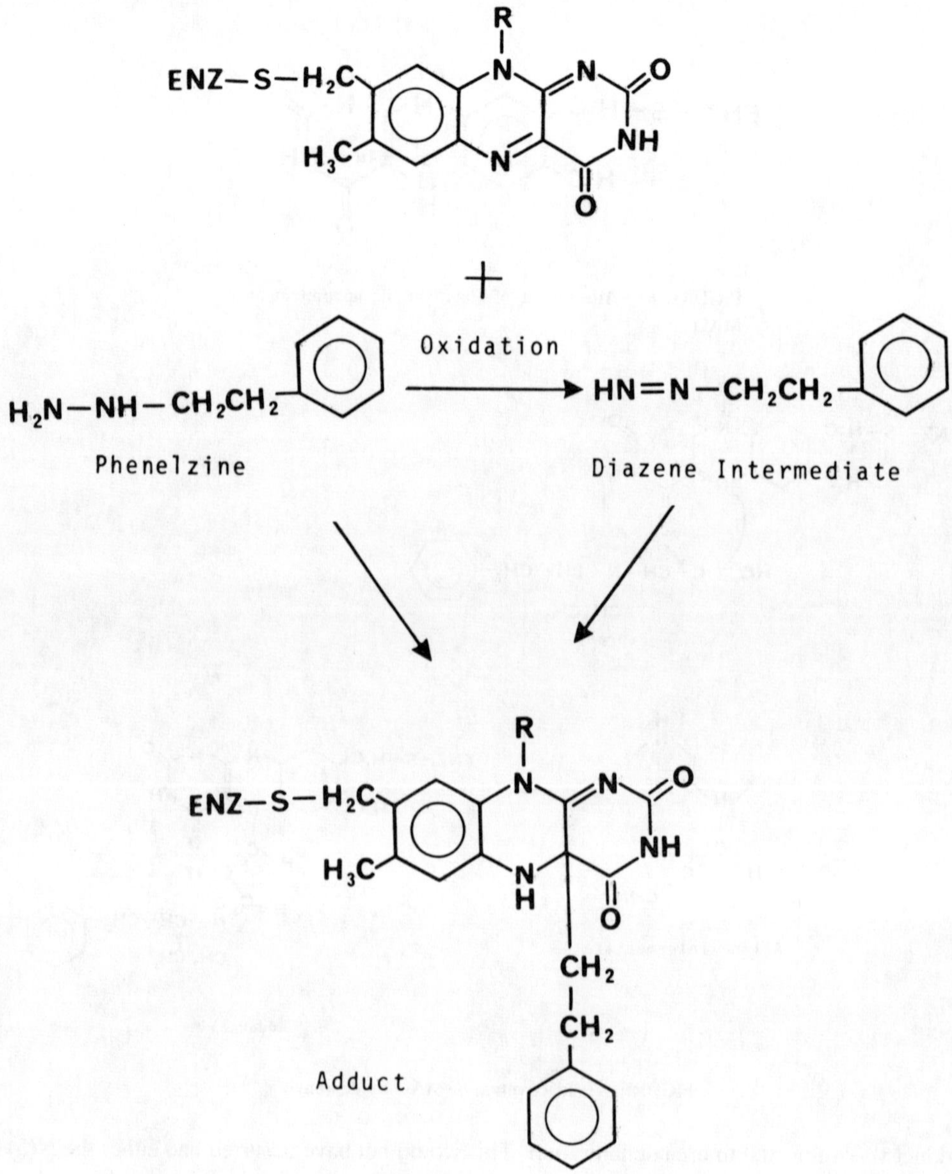

FIGURE 10. Inactivation of MAO by phenelzine.

A or type-B enzyme. Compounds which show preferred inhibitor action on type-A or type-B MAO are listed in Table 3. There is a third group of compounds lacking selectivity with either type of the enzyme. Caution must be taken in choosing concentrations (in vitro) or doses (in vivo) of an inhibitor in attempting to achieve selective inhibition of either type MAO with some irreversible inhibitors, since at high concentrations both types may become inhibited. In vitro experiments have demonstrated that the selective inhibitors such as clorgyline and deprenyl tend to lose their selectivity for one type of MAO when the incubation time is prolonged or the incubation temperature is elevated.[62] During the course of chronic treatment with an MAOI, selectivity of inhibition depends on the doses and duration of treatment.[63] Until the nature of the two types MAO is better understood, and their selective inhibition more easily achieved, it will remain necessary to perform experiments with two

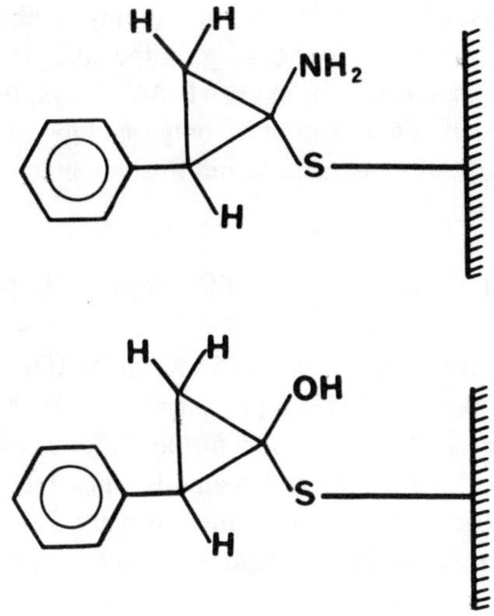

FIGURE 11. Proposed structures of adduct of tranylcypromine. (From Singer, T. P. and Salach, J. I., in *Monoamine Oxidase Inhibitors: The State of the Art,* Youdim, M. B. H. and Paykel, E. S., Eds., John Wiley & Sons, Chichester, 1981, 17. With permission.)

Table 3
SOME SELECTIVE INHIBITORS OF TYPE-A AND TYPE-B MAO

Type-A preferred (5-HT oxidation)		Type-B preferred (PEA oxidation)	
Reversible		Reversible	
d-Amphetamine	Rat brain[46]	Imipramine	Rabbit brain, lung[53]
Harmaline	Rat brain[46]		Human platelets[54]
Tetrahydro-β-carboline	Rat hypothalamus[47]	Amitriptyline	Mouse brain[55]
Tetrahydroisoquinoline	Rat hypothalamus[47,48]		Human platelets[54]
FLA 336[a]	Human brain[49]		
FBF[b]	Guinea pig brain[50]		
MD 780515[c]	Rat brain, liver, heart[51]		
Irreversible		Irreversible	
Clorgyline	Rat brain[7]	Deprenyl	Rat brain, liver[56]
			Human platelets[57]
Lilly 51641[d]	Rat liver[46]	Pargyline	Rat brain[58]
			Human platelets[59]
PCO[e]	Rat brain, liver, heart, kidney[52]	L-54761[f]	Human platelets, rat brain[60]
		L-54748[g]	Human platelets, rat brain[60]
		Phenylhydrazine	Human brain[61]

[a] *FLA 336:* 4-dimethylamino-α,2-dimethylphenethylamine (Figure 6).
[b] *FBF:* Furoxanobenzofuroxan (Figure 6).
[c] *MD 780515:* 3-[4-(3-cyanobenzoxyl)phenyl]-5-(methoxymethyl)-2-oxazolidinone (Figure 6).
[d] Lilly 51641: 2-chlorophenoxyethylcyclopropylamine (Figure 6).
[e] PCO: 5-phenyl-3-[2-(*N*-cyclopropyl)aminoethyl]-1,2,4-oxadiazole (Figure 6).
[f] *L-54761: N*-(benzoylmethyl)cyclopropylamine (Figure 7).
[g] L-54748: *N*-(1-phenylisopropyl)cyclopropylamine (Figure 7).

preferred sustrates, such as 5-HT and PEA, with varying inhibitor concentrations. Also, in some instances, both types of MAO (depending on the source) can oxidize any substrate at high substrate concentrations. Hence, for in vitro MAO assays, the use of the proper substrate concentrations will then assure deamination by only one type of the enzyme, and in addition will eliminate the possibility of substrate inhibition such as that observed with PEA at high concentrations.

STRUCTURE-ACTIVITY RELATIONSHIP

Data on the structure-activity relationship (SAR) of MAOIs were available prior to the introduction of the two types of MAO, type A and type B. Moreover, in some of these early works (see References 34 and 35 for further references) the in vitro assays were performed with a variety of substrates and methods, thus making comparison of potency across chemical classes difficult. The relative in vivo potencies of MAO inhibitors may also vary with the test procedures, species, tissues, and route of administration.[35] Despite these variables, some β-carbolines, tranylcypromine, pheniprazine, and pargyline are generally more potent MAOIs than iproniazid and the α-alkyltryptamines.[35] Pharmacological studies of MAOIs based on their potentiation of tryptamine-induced convulsions in rats, indicated the relative potencies in the following decreasing order: tranylcypromine, pheniprazine, phenelzine, isocarboxazid, nialamide, and iproniazid.[34]

Compounds that are capable of forming a covalent bond with MAO, such as acetylenic amines, substituted hydrazines, and arylcyclopropylamines, are irreversible inhibitors of the enzyme. The reversibility and irreversibility among MAOIs has been described previously under Classification of MAOIs.

Besides their MAO inhibitory activity, a number of hydrazino and nonhydrazino MAOIs possess other pharmacological properties such as alteration of release, uptake, synthesis and receptor activity of NE, DA, and 5-HT. For details, see under Pharmacology of MAOIs.

Methyl substitution on the α-position of either phenethylamine or phenethylhydrazine generally retards the metabolism of the corresponding compounds that exert some MAO inhibitory activity. Because of the structural resemblance to amphetamine, the ring-substituted α-methylphenethylhydrazines and α-methylphenethylamines produce pressor effects and exhibit analeptic activity in experimental animals.[34] Although the α-methylphenethylhydrazines are better MAOIs than α-methylphenethylamines, there exists no correlation between the MAO inhibitory activity and the analeptic effect.

TESTING METHODS

Biochemical Evaluation

The in vitro methods for biochemical assay of MAO activity includes measurement of oxygen uptake by polarographic methods, Warburg technique for measurement of oxygen uptake and ammonia evolution, spectrophotometric estimation of substrate disappearance, and fluorometric or radiometric technique for determining the formation of products, aldehyde and/or acids.[64] For the in vivo determination, either the increase in concentrations of endogenous monoamines or the decrease in the respective metabolites serves as an indication of the enzyme activity after administration of an inhibitor. However, in assessing MAO oxidation of catecholamines, NE or DA, one must account for the *O*-methylation of the deaminated products. A method involving in vitro assay of MAO in tissues obtained after administration of a MAOI in vivo is also widely used.

Pharmacological Evaluation

Reserpine is known to induce a state of mental depression in man,[65] and to produce depressed motor activity in animals. Thus, prevention of reserpine-induced depression in rats is often used in the laboratory to test potential antidepressants, including MAOIs and tricyclic drugs. The test is based on the release by reserpine of endogenous monoamines from neuronal storage.[66,67] The released monoamines are rapidly metabolized by MAO, and the absence of these monoamines then results in motor depression. Pretreatment with a MAOI in animals given reserpine prevents the metabolism of the monoamines which in turn counteracts the reserpine-induced depression. It was found that only the type-A inhibitors such as clorgyline, Lilly 51641, or harmaline suppresses the reserpine syndrome (i.e., depressed motor activity, ptosis, catatonia, and hyperthermia), whereas type-B inhibitors such as pargyline and deprenyl do not.[68,69] The use of tetrabenazine is more advantageous than reserpine in this test,[70] since it has a preferential effect on central stores of monoamines and a more rapid onset of action.[71] A sequential screening of antidepressant drugs reported by Giurgea and Dauby,[72] which includes the potentiation of tryptamine-induced motor seizures and the prolonged protection of rats against catatonia induced by reserpine, tetrabenazine, or phenothiazines, was shown to be specific for MAOIs.

PHARMACOLOGY OF MAOIs

Besides altering monoamine metabolism, MAOIs also possess other pharmacological actions. The ability of MAOIs to alter the release and uptake of ^{3}H-NE and ^{3}H-DA was studied in vitro in rat cerebral cortex[73] and rat brain striatum,[74] respectively. Tranylcypromine and pheniprazine produce a release of ^{3}H-NE from tissues at a potency comparable to that seen with *d*-amphetamine, whereas pargyline and iproniazid are ineffective.[73] As regards ^{3}H-DA release, α-ethyltryptamine and pheniprazine are equipotent to *d*-amphetamine. While tranylcypromine and pheniprazine are weaker than α-ethyltryptamine in releasing 5-HT, pargyline, deprenyl, isocarboxazid, nialamide, and iproniazid are ineffective releasers of either DA or 5-HT.[74]

Tranylcypromine and pheniprazine were reported to inhibit DA uptake.[75-78] These two compounds, as well as phenelzine, inhibit ^{3}H-NE uptake at concentrations approximately ten times that of *d*-amphetamine.[73] Pargyline and iproniazid only slightly inhibited ^{3}H-NE uptake, and, similarly, deprenyl was found to be a weak uptake inhibitor in vas deferens[79] and heart.[80]

Repeated administrations of either MAOIs or tricyclic antidepressants diminish beta-adrenergic receptor binding in rat cerebral cortex.[81-84] However, only repeated treatment with MAOIs, such as clorgyline and nialamide, but not the tricyclic antidepressants (amitriptyline, chlorimipramine, or fluoxetine) lower ^{3}H-5-HT binding in rat cerebral cortex and hippocampus.[85,86] The diminished ^{3}H-5-HT binding after MAOIs is associated with a reduction in the B_{max}, the maximum number of specific binding sites for the ligand. It was further determined that this effect on the 5-HT receptor binding was specific with inhibitors of type-A MAO, since two inhibitors of selective type-B MAO, deprenyl and pargyline, failed to exert a similar effect.[85,86] Whether the differential effect of these two classes of antidepressants (tricyclic drugs and MAOIs) on 5-HT binding is related to their respective clinical actions in endogenous or atypical depression, requires further elucidation.

Type-A MAO within DA-containing neurons appears to play a role in regulating the synthesis of DA. Schoepp and Azzaro[87] reported a decrease of DA synthesis from ^{14}C-tyrosine in rat striatal slices, when type-A MAO was inhibited more than 90% by clorgyline. No effect on DA synthesis was observed at 70% inhibition, nor did the type-B inhibitor, deprenyl, affect DA synthesis, even at concentrations producing 90% inhibition of the type-B activity.

Chronic administration of phenelzine to rats results in an adaptive increase in mesencephalic tegmentum tryptophan hydroxylase activity between 14 to 21 days, but induces no change in striatal or hypothalamic tyrosine hydroxylase.[88] This delayed response of increased tryptophan hydroxylase activity occurs when the MAO activity is 95% inhibited, but the monoamine (5-HT, DA, and NE) concentrations in whole brain have returned to near normal levels. In an earlier study by Mandell,[89] chronic administration of pargyline was shown to decrease mid-brain tyrosine hydroxylase activity. Tranylcypromine was found to have no effect on either tryptophan hydroxylase or tyrosine hydroxylase activity, but it induced an increase of aromatic decarboxylase activity after 14 and 21 days of treatment.[88]

A behavioral syndrome characterized by hyperactivity and tremor is generally observed when laboratory animals are pretreated with a nonselective MAOI, such as tranylcypromine, before receiving large doses (e.g., 400 mg/kg) of L-tryptophan[90] (see the journal article for references). These behavioral effects can also be seen when L-tryptophan-treated rats were pretreated with small doses (1 to 2 mg/kg) of the selective inhibitors clorgyline and deprenyl, in combination but not individually.[90] It would appear that a total inhibition of the two types of MAO is required before these behavioral changes are manifested. Similar observations were also reported by Green et al.,[91] who found that administration of clorgyline with deprenyl (5 mg/kg each), or tranylcypromine (20 mg/kg) alone, followed by either L-dopa or L-tryptophan (50 mg/kg each) in rats resulted in increased locomotor activity. It is apparent from these findings that a correlation exists between behavioral effects and brain monoamine (5-HT, DA, and NE) concentrations.

One common property shared by most MAOIs is the potentiation of the pressor effect of indirectly acting sympathomimetic amines such as tyramine. The hypertensive crisis, observed in MAOI-treated depressive patients after ingestion of foods (e.g., cheese) containing tyramine, has long been considered a hazard in psychiatry. The potentiation of tyramine (the so-called "cheese effect") results from the inhibition of type-A MAO in the GI tract and liver, thus allowing tyramine to release NE for pressor action. However, this potentiation of orally administered tyramine is not observed with (−)-deprenyl.[92,93] This is attributed to the fact that deprenyl is a poor (and initially reversible) inhibitor of the type-A enzyme which predominates in the GI tract, and, as a result, the pressor effects of tyramine are minimized. Moreover, no adverse pressor reaction was observed either in parkinsonian patients on L-dopa-deprenyl combination therapy or in normal or parkinsonian volunteers on deprenyl, when challenged with up to 400 mg of oral tyramine.[94] Very recently, a new type-A MAOI, K-Y 1349, 5-[3-(*N*-methyl-*N*-2-propynylamino)propyl]-10-11-dihydro-5-H-dibenz[b,f]azepine (desmethylimipramine with an *N*-propynyl group) was found to inhibit tyramine responses in the isolated rat vas deferens, suggesting its potential therapeutic efficacy without presenting the cheese effect.[95] One study reported no "cheese effect" when an MAOI (phenelzine) and a tricyclic antidepressant are given together.[96]

PHARMACOKINETICS

The pharmacokinetic data on MAOIs appearing in the literature are far from complete. Most of the earlier work in animals was reviewed by Biel et al.[34] and Zirkle and Kaiser.[35] A few reports in recent years concern clinical studies using phenelzine, today's most popularly prescribed MAOI, and deprenyl, a new inhibitor currently under clinical evaluation in Europe. In this chapter, only those MAOIs presently available in the U.S. (see under Clinical Indications) plus deprenyl, will be summarized.

Robinson et al.[97] have studied plasma levels of phenelzine (60 mg/day) in drug-free depressed patients and in patients chronically treated with phenelzine. Peak plasma levels in drug-free subjects were detected between 2 to 4 hr after ingestion. In the chronic patients, mean trough levels, when measured at 2, 4, and 6 weeks, were shown to progressively increase, and appeared to be higher in the responders than the nonresponders during the 6-week treatment period.

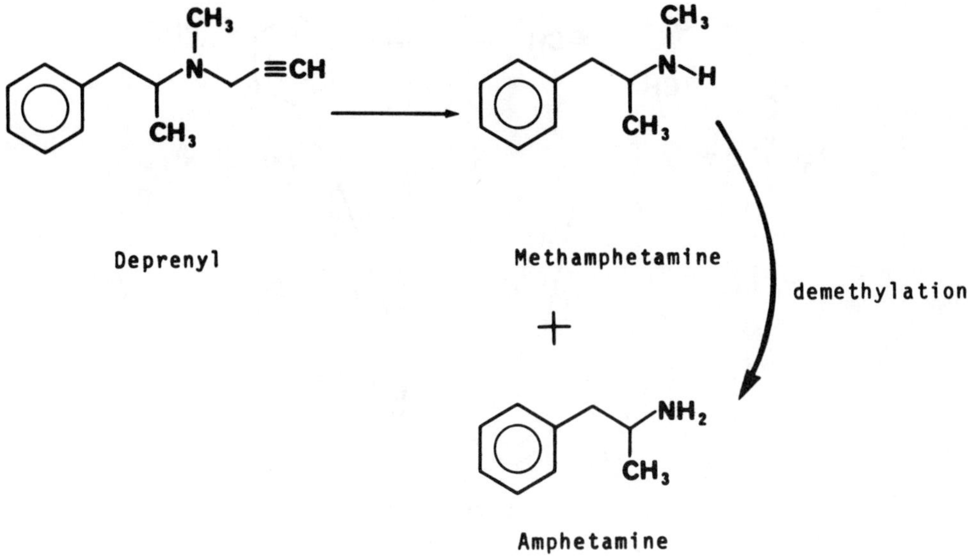

FIGURE 12. Metabolism of (−)-deprenyl in man.

In normal human volunteers receiving 5 or 10 mg of (−)-deprenyl daily for 3 days, an analysis of a 24-hr urine collected on the last day indicated the presence of its metabolites, methamphetamine and amphetamine, in a ratio of 4 to 1 (Figure 12).[98] No unchanged deprenyl was detected. Whether the amphetamine was formed directly from deprenyl first by demethylation, then a loss of the propynyl group, or from methamphetamine via demethylation is not known. An earlier study in rats indicated a fast uptake of deprenyl in brain follwing i.v. adminstration;[99] almost all of the drug was eliminated from the tissues after 20 to 30 min.

When carbon-labeled parglyine (100 mg/kg) was given orally to rats, most tissues attained maximum radioactivity between 1 to 2 hr.[100] A metabolite, which formed slowly, accounted for 3% of the total dose. Although not characterized, it is probably the same metabolite, benzylamine, as that derived from parglyine in vitro and in vivo.[101] Benzylamine can be formed either via a hydroxylamine intermediate (Figure 13) or directly from pargyline after loss of the methyl and propynyl groups, similar to the conversion of deprenyl to amphetamine as described above.

Metabolism of carbon-labeled isocarboxazid in rats is rapid; a large portion of the drug was found to be converted to hippuric acid (glycine conjugate of benzoic acid) and excreted in urine.[102,103] Benzylhydrazine appears to be the intermediate for this metabolic process (Figure 14).

Carbon-labeled tranylcypromine administered to rats showed rapid absorption in the gut, and 90% of the dose was excreted after 3 days.[104] Of the 70 to 75% excreted in the urine and feces within the first 24 hr, the unchanged drug and hippuric acid (from benzoic acid) accounted for 4 and 13%, respectively. Benzoic acid may be formed either by oxidation at the α-carbon and opening of the cyclopropane ring of tranylcypromine, or by hydrolytic cleavage of the cyclopropane ring to hydrocinnamaldehyde followed by oxidation via hydrocinnamic acid (Figure 15).

In summary, the absorption of most MAOIs in animals and man is generally rapid, and maximum tissue concentrations are attained within 1 to 3 hr. Excretion is normally complete within 2 to 3 days, while the inhibition of MAO activity persists after the disappearance of most of the drug from the body.

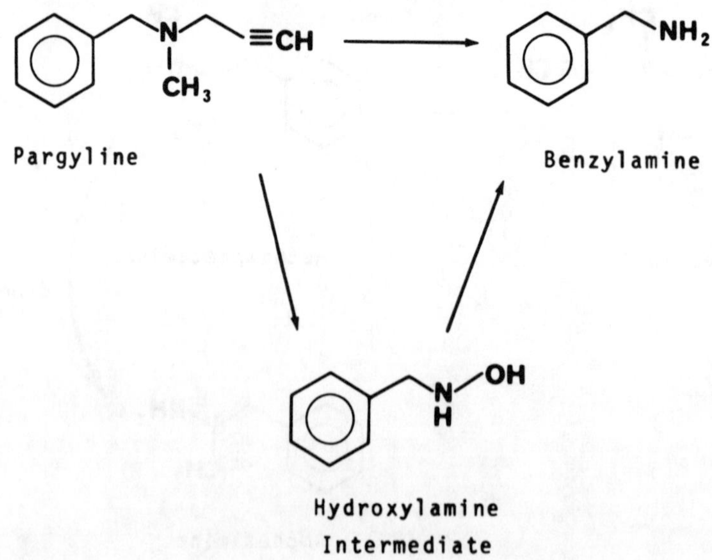

FIGURE 13. Metabolism of pargyline.

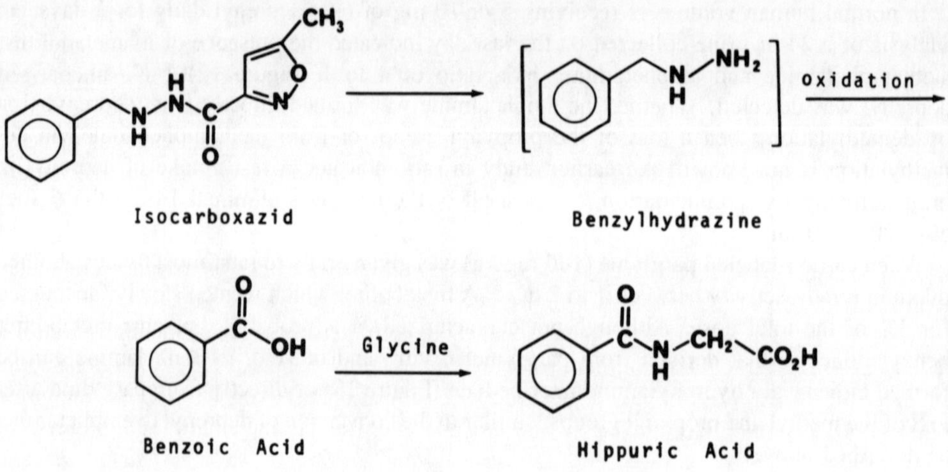

FIGURE 14. Metabolism of isocarboxazid.

CLINICAL INDICATIONS

MAOIs for Depression

The antidepressant effect of MAOIs was discovered by accident. Concurrent with the discovery in 1952 by Zeller et al.[105] of the inhibitory property of iproniazid (Marsilid), it was reported that the drug, then used to treat tuberculosis patients, produced euphoria, increased psychomotor activity, and increased social behavior.[106,107] This mood-elevating effect of iproniazid led to two independent studies in depressed patients,[108,109] with auspicious results. The hepatic toxicity of iproniazid has since warranted its withdrawal from the market, and in the interim many other hydrazine and nonhydrazine MAOIs have been developed. Table 4 lists the currently available MAOIs and their effective daily dose ranges. Although isocarboxazid (Marplan) is currently available, like nialamide (Niamid, a hydrazine drug not available in the U.S.), it has not proved consistently more effective than placebo in

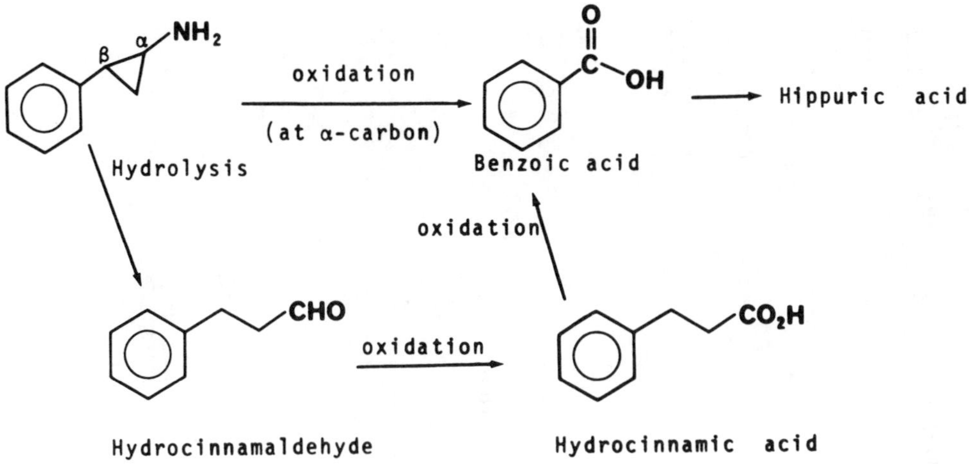

FIGURE 15. Metabolism of tranylcypromine.

controlled studies. This left pehnelzine (Nardil) and tranylcypromine (Parnate), the two more commonly used MAOIs, which were found to have application in the treatment of depressive illness.

While some patients treated with MAOIs develop hypertensive crises (''cheese effect'', see also under Pharmacology of MAOIs) which occasionally prove fatal, phenelzine and tranylcypromine offer less liability than other MAOIs. Even so, tranylcypromine was once briefly withdrawn from the U.S. market in 1964, and its present use is subjected to restrictive guidelines. Deprenyl (presently in use in Europe) does not produce the ''cheese effect''; however, its effectiveness in depressive illness is related to its potentiation of the antidepressant action of 5-HTP.[117]

In contrast to the tricyclic antidepressants, phenelzine and tranylcypromine have minimal, if any, sedative and anticholinergic effects. The cardiotoxic effects sometimes associated with the tricyclic antidepressants have not been seen with the use of MAOIs.

As with the tricyclic antidepressants, phenelzine's effectiveness is not manifested until 2 to 3 weeks of treatment. In contrast, tranylcypromine provides rapid onset of action (about 10 days).[118,119] This may be due to its wide spectrum of pharmacological effects, including blockade of monoamine reuptake by neurons and exerting an amphetamine-like release of monoamines from the storage.[120,121] Because of the prolonged action of phenelzine and tranylcypromine, they can be administered on a one-a-day basis.

Because of occasional hypertensive crises reported with the use of MAOIs, tricyclic antidepressants became the drugs of choice when antidepressant medication was indicated. Nevertheless, only 70% of all depressed patients have shown a favorable response to the tricyclic drugs, and some depressed patients unresponsive to tricyclic therapy may respond to a MAOI. Specifically, patients with ''atypical depression'', characterized by fatigue, somatic complaints, phobic anxiety, and with the absence of classical endogenous depression such as guilt, insomnia, and weight loss, may improve dramatically with MAOIs.[122,123] The suggested guidelines delineating the use of MAOIs are therefore: (1) first drug of choice in ''atypical depression'', and (2) in depressed patients with typical symptoms who are refractory to tricyclic antidepressants, and in whom electroconvulsive treatment is not considered a priority.[123]

MAOIs for Parkinson's Disease

Although the type-B MAOI, deprenyl, does not appear to be effective in the treatment of depression,[124] it has shown promise as an adjunct to L-dopa therapy of Parkinson's disease.

Table 4
DOSAGE AND PHYSICAL-CHEMICAL DATA OF MAO INHIBITORS

Chemical class	Generic name	Trade name	Manufacturer	Effective daily dose range[110] (mg)			m.p.(°C)	pKa	Solubility[111]
				Low	Moderate	High			
Hydrazide	Isocarboxazid	Marplan	Roche	10—20	20—40	40—60	105—106 (111)	10.4 (112)	Very sl.sol.(0.05%) in hot H_2O Sl.Sol.(1—2%) in 95% alc., glycerol, and propylene glycol
Hydrazine	Phenelzine (sulfate)	Nardil	Parke-Davis	15—30	30—45	45—75	164—168 (113)	NA	Sol. in H_2O
Aralkylamine	Tranylcypromine (sulfate)	Parnate	Smith, Kline & French	10—20	20—30	30—40	218 (114)	8.2 (115, 116)	Sol. in H_2O
Acetylenic amine	Pargyline[a] (hydrochloride)	Eutonyl	Abbott	25—50	50—75	75—100	154—155 (111)	6.6 (116)	Sol. in H_2O

[a] Primarily used in the treatment of hypertension.

When given in combination with Madopar (L-dopa in combination with the peripheral DOPA decarboxylase inhibitor, benserazide, Ro 4-4602), a definite potentiation of the antiakinetic effects of Madopar has been indicated.[125-128] Deprenyl was reported to be especially beneficial in reducing the frequency and severity of "on-off" effects seen in parkinsonism. In addition, the combination of deprenyl and Madopar allows a reduction of the daily dose requirements of Madopar when given alone. Whether deprenyl's action is associated with inhibition of type-B MAO or with facilitating the utilization of synthesized DA from L-dopa, remains to be determined.

MAOIs for Other Disorders

Pargyline (Eutonyl) has been used in the treatment of hypertension and angina pectoris. Phenelzine is an effective MAOI for the treatment of hysteroid-dysphoric states, and is particularly useful as a prophylactic in this disorder[129] (see pp. 288 to 289 of Reference 129). MAOIs, either alone or in combination with chlordiazepoxide or a tricyclic antidepressant, have been found beneficial in the treatment of phobias and phobic-anxiety.[130] Migraine headaches have reportedly been relieved by MAOI treatment.[131]

MAO Inhibition and Therapeutic Effect

Studies concerning the extent of inhibition by a given MAOI and its time-course of action, have provided valuable information as to its effectiveness as an antidepressant, as well as serving as an index of a patients's clinical response.

MacLean et al.[132] and Pare[133] have reported correlations between the onset of antidepressant effects and brain MAO inhibition. In patients given MAOIs for the first time, it is generally 2 to 3 weeks before MAO inhibition can be detected; the same interval usually corresponds to the onset of therapeutic effect.

Inhibition of platelet MAO has been suggested as a peripheral index of brain MAO inhibition. Robinson et al.[134,135] studied the relationship between the degree of platelet MAO inhibition and clinical improvement. Using phenelzine, it was found that patients show a better response to MAOI treatment when their platelet MAO is inhibited between 80 to 100%. Inhibition after 4 weeks with a 60 mg/day dose of phenelzine was reported to be predictive of a high probability of favorable clinical response in depressed outpatients with significant nonendogenous symptomatology. Giller and Lieb,[136] designating inhibition of 80% or greater as an estimate of therapeutic efficacy, found the effective daily dose for patients with phenelzine to be 45 mg, isocarboxazid 20 mg, and tranylcypromine 10 mg or less.

Combined MAOI-Tricyclic Antidepressant Treatment

Reports of hyperpyrexia, hypertensive crises, convulsions, and even death in patients who were on combinations of tricyclics and MAOIs most likely added to the reluctancy of clinicians to utilize combination therapy. During the past 2 years, results of controlled studies in the U.S. and Europe have indicated that the combination treatment is both safe and effective and that it produces no more adverse effects than either drug individually. It is now recommended that combined treatment be used in patients who are refractory to usual doses of either drug alone.[137] However, certain precautions should be taken to insure the safety of such combinations. White and Simpson[137] suggested to avoid the introduction of an MAOI during an ongoing course of a tricyclic drug. Instead, the patient should be removed from either drug for at least 1 week and then initiate treatment with both types together. Spiker and Pugh[138] further recommended adding the MAOI in minimal dosage 1 or 2 days after starting a low dose of tricyclic drug. Over the course of some weeks, dosages of both can be gradually increased to a level which is approximately half the dosage of the individual drug used alone. Thus, the combination of 150 mg amitriptyline and 30 mg phenelzine seems to be a safe and effective regimen.

SIDE EFFECTS AND TOXICITY[139]

"Cardiovascular" side effects of MAOIs, such as orthostatic (postural) hypotension, tachycardia, and palpitation, are commonly observed and can be fatal.

"Autonomic" side effects of MAOIs are quite similar to those occurring during treatment with tricyclic antidepressants. These include dry mouth, blurred vision, decreased GI activity and bladder tone, impotency, dry skin, etc.

"GI" side effects are manifested by constipation, anorexia, diarrhea, and nausea.

"Central effects" may result from overstimulation. They are expressed clinically as agitation, restlessness, dizziness, insomnia, anxiety, and irritability, and, like tricyclic drugs, may precipitate manic or psychotic episodes in some individuals.

"Hepatotoxicity" is more likely to result from hydrazine MAOIs than nonhydrazine MAOIs. Severe jaundice and liver failure can occur.

"Overdose" of MAOIs is initially apparent by a gradual onset of headache, often accompanied by nausea, vomiting, tachycardia, mydriasis, and photophobia. Severe overdose can cause CNS stimulation and paradoxical hypertension, which can also precipitate pulmonary edema, circulatory collapse, or intracranial hemorrhage. Phentolamine (a short-acting α-adrenergic blocking agent), or alternatively, chlorpromazine (a neuroleptic) is generally used to manage serious overdose of MAOIs. Like the tricyclics, MAOIs have a narrow margin of safety and fatalities can result from overdosage. Approximately ten times the therapeutic dose of phenelzine can be lethal, and with tranylcypromine the margin of safety is even less, with six times the therapeutic dose being potentially fatal.

INTERACTIONS

Interactions with MAOIs can be grouped into two categories. The first involves the interaction of MAOIs with either monoamine-containing foodstuffs or medications containing sympathomimetic amines, and generally results in a hypertensive crisis. The second category consists of MAOIs' ability to potentiate the action of certain drugs, possibly via the hepatic drug-metabolizing enzymes.[140]

Hypertensive crises are occasionally observed in patients on MAOIs (pargyline, tranylcypromine, phenelzine, etc.) who ingest foods containing tyramine (see also under Pharmacology of MAOIs) or medications containing sympathomimetic agents. Certain items of the major food groups to be avoided include aged cheeses (e.g. boursault, camembert, cheddar, emmenthaler, gruyere, stilton), meat and fish (chicken livers, caviar, fermented meats, pickled herring and lox), alcoholic beverages (beer and red wine), vegetables (fava beans, which also contain dopamine), fruits (bananas, canned figs) and yeast extracts (e.g., bovril or marmite). Caffeine-containing beverages and chocolate (containing phenethylamine) should not be taken in large amounts because of their sympathomimetic effects. Although the overall incidence of hypertensive crisis is low, it is potentially fatal, and the dietary restrictions must be reinforced. Medications which interact with MAOIs are amphetamine or congeners (a component of diet pills), ephedrine, mephentermine, metaraminol, phenylephedrine, phenylpropanolamine, etc. (used in cold or sinus remedies). These sympathomimetic agents induce hypertension either by the release of norepinephrine or by acting on adrenergic receptors. Hypertensive crises can also result from the combination of MAOIs with L-dopa (for parkinsonism) or reserpine (for hypertension).

By their action on the hepatic drug-metabolizing enzymes, MAOIs potentiate the CNS depressant effects of alcohol, sedative-hypnotics (barbiturates, chloral hydrate), general anesthetics, benzodiazepines, phenothiazines, and antihistamines. MAOIs may act by retarding the metabolism of these agents, making more drugs available for pharmacological actions. The narcotic effects of analgesics (morphine or meperidine) may also be potentiated by MAOIs, and death from cardiovascular collapse has been reported. Interactions between MAOIs and tricyclic antidepressants can result in hypertensive crises, hyperpyrexia, tremor, convulsions, and even death. Other drugs potentiated by MAOIs are thiazides, insulin, and hypoglycemics.

CONCLUSION

While significant advances have been made in the area of MAO research over the past several years, some of these findings have indicated the necessity of reexamining previous data or revision in designing new studies. For example, the question of specificity of selective inhibitors or substrates for the two types of the enzyme has been confounded by recent findings that these specificities are concentration dependent. Accordingly, crucial experiments presently underway in several laboratories will be essential in defining the molecular basis of the two types of MAO. This information will be instrumental in designing new inhibitors for biochemical studies and as potential therapeutic agents. Moreover, it should help to explain why and how factors such as sex, age, pituitary hormones, and iron affect the enzyme's activity.

Since the development of tricyclic antidepressants in the late 1950s, they have become the drug of choice for the treatment of depression, replacing the early MAOIs which were responsible for the hypertensive crises observed in some patients. Although the recent development of safer inhibitors has renewed an interest in their clinical use, the mechanism of the antidepressant action of MAOIs has yet to be fully elucidated. In the 1960s, the "catecholamine and indoleamine hypotheses" proposed the accumulation of brain synaptic monoamine concentrations via either inhibition of MAO (i.e., MAOIs) or blockade of amine reuptake (tricyclic antidepressants) in alleviating depressive symptoms. These proposals were based predominantly on acute studies in which MAO inhibition or uptake blockade was observed within relatively short time intervals; however, at the clinical level, improvement in depressed patients is observed only after several weeks of drug treatment. In the latter part of the 1970s, data began to emerge demonstrating the effects of chronic administration of MAOIs and other antidepressants on adrenergic and serotonergic receptors, which advanced the understanding of the clinical actions of MAOIs.

Alterations in platelet MAO activity (B-type) have been reported in a number of mental disorders including bipolar and unipolar depression and chronic schizophrenia. Due to difficulties related to accessibility of the A-type activity, its status in these disorders is not yet known. In any case, a causal relationship between levels of MAO and symptomatology of these disorders has not been established. However, as stated earlier, the significant clinical response seen in atypical depression after MAOI treatment warrants further investigation into possible alterations of MAO and the result of its inhibition in these patients.

In an attempt to avert the "hypertensive crises" observed with the use of most existing MAOIs, new compounds lacking the tyramine potentiating effect (the "cheese effect") appear to be in demand. One such compound, deprenyl, has been shown to lack the "cheese effect", and was found to be beneficial as an adjunct to the L-dopa therapy for Parkinson's disease. However, its effect as an antidepressant has not been established. Another new type-A MAOI, K-Y 1349, which apparently inhibits tyramine responses in in vitro preparations, may have clinical potential. In view of our present knowledge, perhaps the more ideal MAOIs as antidepressants should be selective, "reversible" inhibitors of the enzyme, lacking the cheese effect. With a shorter duration of action, some of the unwanted side effects common to those MAOIs currently available may be avoided. With the development of new inhibitors which are pharmacologically specific and clinically beneficial, their use in the treatment of depression may prove to be as good as, if not better than, the tricyclic antidepressants.

REFERENCES

1. **Gey, K. F. and Pletscher, A.,** Monoamine oxidase inhibitors and the extrapyramidal system, *Pharmacol. Ther. B,* 2, 137, 1976.
2. **Kapeller-Adler, R.,** *Amine Oxidases and Methods for their Study,* John Wiley & Sons, New York, 1970, 31.
3. **Youdim, M. B. H., Collins, G. G. S., Sandler, M., Bevan Jones, A. B., Pare, C. M. B., and Nicholson, W. J.,** Human brain monoamine oxidase: multiple forms and selective inhibitors, *Nature (London),* 236, 225, 1972.
4. **Owen, F., Cross, A. J., Lofthouse, R., and Glover, V.,** Distribution and inhibition characteristics of human brain monoamine oxidase, *Briochem. Pharmacol.,* 28, 1077, 1979.
5. **Greenawalt, J. W.,** Localization of monoamine oxidase in rat liver mitochondria, in *Monoamine Oxidases — New Vistas, Advances in Biochemical Psychopharmacology,* Vol. 5, Costa, E. and Sandler, M., Eds., Raven Press, New York, 1972, 207.
6. **Horita, A. and Lowe, M. C.,** On the extraneuronal nature of cardiac monoamine oxidase in the rat, in *Monoamine Oxidases — New Vistas, Advances in Biochemical Psychopharmacology,* Vol. 5, Costa, E. and Sandler, M., Eds., Raven Press, New York, 1972, 227.
7. **Johnston, J. P.,** Some observations upon a new inhibitor of monoamine oxidase in brain tissue, *Biochem. Pharmacol.,* 17, 1285, 1968.
8. **Neff, N. H. and Yang, H. -Y. T.,** Another look at the monoamine oxidases and the monoamine oxidase inhibitor drugs, *Life Sci.,* 14, 2061, 1974.
9. **Ekstedt, G.,** Substrate specificity of the different forms of monoamine oxidase in rat liver mitochondria, *Biochem. Pharmacol.,* 25, 1133, 1976.
10. **Achec, F. M. and Gabay, S.,** Studies on monoamine oxidase: inhibition of bovine brain MAO in intact mitochondria by selective inhibitors, *Biochem. Pharmacol.,* 26, 1637, 1977.
11. **Kinemuchi, H., Wakui, Y., Toyoshima, Y., Hayashi, N., and Kamijo, K.,** β-Phenylethylamine (PEA), a concentration-dependent preferential substrate for multiple forms of MAO, in *Monoamine Oxidase: Structure, Function and Altered Functions,* Singer, T. P., Von Korff, R. W., and Murphy, D. L., Eds., Academic Press, New York, 1979, 205.
12. **Suzuki, O., Katsumate, Y., and Oya, M.,** Characterization of some biogenic monoamines as substrates for type A and type B monoamine oxidases, in *Monoamine Oxidase: Structure, Function and Altered Functions,* Singer, R. P., Von Korff, R. W., and Murphy, D. L., Eds., Academic Press, New York, 1979, 197.
13. **Kawatra, M. M. and Sourkes, T. L.,** Muscle and liver monoamine oxidases: effects of iron and riboflavin deficiencies in the rat, in *Monoamine Oxidase: Structure, Function and Altered Functions,* Singer, T. P., Von Korff, R. W., and Murphy, D. L., Eds., Academic Press, New York, 1979, 279.
14. **Tipton, K. F. and Mantle, T. J.,** Inhibition of rat liver monoamine oxidase by clorgyline and deprenyl, in *Monoamine Oxidase Inhibitors: The State of the Art,* Youdim, M. B. H. and Paykel, E. S., Eds., John Wiley & Sons, Chichester, 1981, 3.
15. **Parkinson, D. and Callingham, B. A.,** Substrate and inhibitor selectivity of human heart monoamine oxidase, *Biochem. Pharmacol.,* 28, 1639, 1979.
16. **Lyles, G. A. and Shaffer, C. J.,** Substrate specificity and inhibitor sensitivity of monoamine oxidase in rat kidney mitochondria, *Biochem. Pharmacol.,* 28, 1099, 1979.
17. **Roth, J. A.,** Multiple forms of monoamine oxidase and their interaction with tricyclic psychomimetic drugs, *Gen. Pharmacol.,* 7, 381, 1976.
18. **Bakhle, Y. S. and Youdim, M. B. H.,** Metabolism of phenylethylamine in rat isolated perfused lung: evidence for monoamine oxidase type-B in lung, *Br. J. Pharmacol.,* 56, 125, 1976.
19. **Yang, H. -Y. T. and Neff, N. H.,** The monoamine oxidases of brain: selective inhibition with drugs and the consequences for the metabolism of the biogenic amines, *J. Pharmacol. Exp. Ther.,* 189, 733, 1974.
20. **Squires, R. G.,** Multiple forms of monoamine oxidase in intact mitochondria as characterized by selective inhibitors and thermal stability: a comparison of eight mammalian species, in *Monoamine Oxidases — New Vistas, Advances in Biochemical Psychopharmacology,* Vol. 5, Costa, E. and Sandler, M., Eds., Raven Press, New York, 1972, 355.
21. **Salach, G. I., Detmer, D., and Youdim, M. B. H.,** The reaction of bovine and rat liver monoamine oxidase with ^{14}C-clorgyline and ^{14}C-deprenyl, *Mol. Pharmacol.,* 16, 234, 1979.
22. **Robinson, D. S., Lovenberg, W., Keiser, H., and Sjoerdsma, A.,** Effects of drugs on human blood platelet and plasma amine oxidase activity *in vitro* and *in vivo, Biochem. Pharmacol.,* 17, 109, 1968.
23. **Youdim, M. B. H.,** Tyramine and psychiatric disorders, in *Neuroregulators and Psychiatric Disorders,* Youdim, M. B. H., Ed., Oxford University Press, New York, 1977, 57.
24. **Neff, N. H. and Fuentes, A.,** The use of selective monoamine oxidase inhibitor drugs for evaluating pharmacological and physiological mechanisms, in *Monoamine Oxidase and its Inhibition,* Ciba Foundation Symp., No. 39, Wolstenholme, G. E. W. and Knight, J., Eds., Elsevier, Amsterdam, 1976, 163.

25. **Youdim, M. B. H. and Collins, G. G. S.,** Properties and physiological significance of multiple forms of mitochondrial monoamine oxidase (MAO), in *Marker Isozymes,* Academic Press, New York, 1975.
26. **Youdim, M. B. H., Holzbauer, M., and Woods, F. H.,** *Neuropsychopharmacology of Mooamines and their Regulatory Enzymes,* Raven Press, New York, 1974, 11.
27. **Houslay, M. D. and Tipton, K. F.,** The nature of the electrophoretically separable multiple forms of rat liver monoamine oxidase, *Biochem. J.,* 135, 173, 1973.
28. **Tipton, K. F., Houslay, M. D. and Garrett, N. J.,** Allotropic properties of human brain monoamine oxidase, *Nature New Biol.,* 24, 213, 1973.
29. **Cawthorn, R. M. and Breakfield, X. O.,** Differences in A and B forms of monoamine oxidase revealed by limited proteolysis and peptide mapping, *Nature (London),* 281, 692, 1979.
30. **Pintar, J. E., Barbosa, J., Francke, U., Castiglione, C. M., Hawkins, M., Jr., and Breakfield, X. O.,** Gene for monoamine oxidase type A assigned to the human X chromosome, *J. Neurosci.,* 1, 166, 1981.
31. **Denney, R. M., Fritz, R. R., Patel, N. T., and Abell, C. W.,** Human liver MAO-A and MAO-B separated by immunoaffinity chromatography with MAO-B-specific monoclonde antibody, *Science,* 215, 1400, 1982.
32. **Russell, S. M., Davey, J., and Mayer, R. J.,** The vectorial orientation of human monoamine oxidase in the mitochondrial outer membrane, *Biochem. J.,* 181, 7, 1979.
33. **Schurr, A., Ho, B. T., and Schoolar, J. C.,** Human brain monoamine oxidase: one molecular entity — multiple binding sites, *J. Pharm. Pharmacol.,* 33, 165, 1981.
34. **Biel, J. H., Horita, A., and Drukker, A. E.,** Monoamine oxidase inhibitors (hydrazines), in *Psychopharmacological Agents,* Vol. 1, Gordon, M., Ed., Academic Press, New York, 1964, 359.
35. **Zirkle, C. L. and Kaiser, C.,** Monoamine oxidase inhibitors (nonhydrazines), in *Psychopharmacological Agents,* Vol. 1, Gordon, M., Ed., Academic Press, New York, 1964, 445.
36. **Kapeller-Adler, R.,** *Amine Oxidases and Methods for their Study,* John Wiley & Sons, New York, 1970, 50.
37. **Zeller, E. A.,** Structure-activity relationship of monoamine oxidase inhibitors, in *Monoamine Oxidase Inhibitors: Relationship Between Pharmacological and Clinical Effect,* Cheymol, J. and Boissier, J. R., Eds., Pergammon Press, New York, 1968, 3.
38. **Ho, B. T.,** Monoamine oxidase inhibitors, *J. Pharm. Sci.,* 61, 821, 1972.
39. **Berger, P. A. and Barchas, J. D.,** Monoamine oxidase inhibitors, in *Psychotherapeutic Drugs, Part II,* Vol. 2, Usdin, E. and Forrest, I. S., Eds., Marcel Dekker, New York, 1977, Chapter VI-2.
40. **Singer, T. P. and Salach, J. I.,** Interaction of suicide inhibitors with the active site of monoamine oxidase, in *Monoamine Oxidase Inhibitors: The State of the Art,* Youdim, M. B. H. and Paykel, E. S., Eds., John Wiley & Sons, Chichester, 1981, 17.
41. **Walker, W. H., Kearney, E. B., Seng, R., and Singer, T. P.,** The covalently-bound flavin of hepatic monoamine oxidase. II. Identification and properties of cysteinyl riboflavin, *Eur. J. Biochem.,* 24, 328, 1971.
42. **Krantz, A. and Lipkowitz, G. S.,** Studies of mitochondrial monoamine oxidase inactivation of the enzyme by isomeric acetylenic and ellenic amines yielding mutally exclusive products, *J. Am. Chem. Soc.* 99, 4156, 1977.
43. **Chuang, H. Y. K., Patek, D. R., and Hellerman, L.,** Mitochondrial monoamine oxidase: inactivation by pargyline. Adduct formation, *J. Biol. Chem.,* 249, 2381, 1974.
44. **Patek, D. R. and Hellerman, L.,** Mitochondrial monoanine oxidase: mechanism of inhibition by phenylhydrazine and by aralkylhydrazines. Role of enzymatic oxidation, *J. Biol. Chem.,* 249, 2373, 1974.
45. **Paech, C., Salach, J. I., and Singer, T. P.,** Suicide inactivation of monoamine oxidase by *trans*-phenylcyclopropylamine, *J. Biol. Chem.,* 255, 2700, 1980.
46. **Fuller, R. W.,** Selective inhibition of monoamine oxidase, in *Monoamine Oxidase — New Vistas, Advances in Biochemical Psychopharmacology,* Vol. 5, Costa, E. and Sandler, M., Eds., Raven Press, New York, 1972, 339.
47. **Meller, E., Friedman, E., Schweitzer, J. W., and Friedhoff, A.,** Tetrahydro-β-carbolines: specific inhibitors of type A monoamine oxidase in rat brain, *J. Neurochem.,* 28, 995, 1977.
48. **Katz, S. and Cohen, G.,** A comparison of 6,7-dihydroxytetrahydroisoquinoline, salsolinol and tetrahydropapaveroline as inhibitors of monoamine oxidase within the adrenergic nerve plexus of the isolated mouse atrium, *Res. Commun. Chem. Pathol. Pharmacol.,* 13, 217, 1976.
49. **Fowler, C. J. and Oreland, L.,** Substrate- and stereoselective inhibition of human brain monoamine oxidase by 4-dimethylamino-α, 2-dimethylphenethylamine, *J. Pharm. Pharmacol.,* 33, 403, 1981.
50. **Bolt, A. G. and Sleigh, M. H.,** Furoxanobenzofuroxan, a selective monoamine oxidase inhibitor, *Biochem. Pharmacol.,* 23, 1969, 1977.
51. **Kan, J. P. and Benedetti, M. S.,** Antagonism between long acting monoamine oxidase inhibitors (MAOI) and MD 780515, a new specific and reversible MAOI, *Life Sci.,* 26, 2165, 1980.

52. **Long, R. F., Mantle, T. J., and Wilson, K.,** Substrate-selective inhibition of monoamine oxidase by some cyclopropylamino substituted oxadiazoles, *Biochem. Pharmacol.*, 25, 247, 1976.
53. **Roth, J. A. and Gillis, C. N.,** Deamination of β-phenethylamine by monoamine oxidase — inhibition by imipramine, *Biochem. Pharmacol.*, 23, 2537, 1974.
54. **Edwards, D. J. and Burns, M.,** Effects of tricyclic antidepressants upon human platelet monoamine oxidase, *Life Sci.*, 15, 2045, 1974.
55. **Edwards, D. J.,** Inhibition of human platelet monoamine oxidase (MAO) by tricyclic antidepressants, *Pharmacologist*, 16, 265, 1974.
56. **Knoll, J. and Magyer, K.,** Some puzzling pharmacological effects of monoamine oxidase inhibitors, in *Monoamine Oxidase — New Vistas, Advances in Biochemical Psychopharmacology*, Vol. 5, Costa, E. and Sandler, M., Eds., Raven Press, New York, 1972, 393.
57. **Donnelly, C. and Murphy, D.,** Substrate- and inhibitor-related characteristics of human platelet monoamine oxidase, *Biochem. Pharmacol.*, 26, 853, 1977.
58. **Fuller, R. W., Warren, B. J., and Molloy, B. B.,** Selective inhibition of monoamine oxidase in rat brain mitochondria, *Biochem. Pharmacol.*, 19, 2934, 1970.
59. **Robinson, D. S., Lovenberg, W., Keiser, H., and Sjoerdsma, A.,** Effects of drugs on human blood platelet and plasma amine oxidase activity *in vitro* and *in vivo*, *Biochem. Pharmacol.*, 17, 109, 1968.
60. **Murphy, D. L., Donnelly, C. H., Richelson, E., and Fuller, R. W.,** N-substituted cyclopropylamines as inhibitors of MAO-A and -B forms, *Biochem. Pharmacol.*, 27, 1767, 1978.
61. **Roth, J. A., Eddy, B. J., Pearce, B., and Mulder, K. M.,** Phenylhydrazine: selective inhibition of human brain type B monoamine oxidase, *Biochem. Pharmacol.*, 30, 945, 1981.
62. **Egashira, T., Ekstedt, B., and Oreland, L.,** Inhibition by clorgyline and deprenyl of the different forms of monoamine oxidase in rat liver mitochondria, *Biochem. Pharmacol.*, 25, 2583, 1976.
63. **Ekstedt, B., Magyar, K., and Knoll, J.,** Does the B form selective monoamine oxidase inhibitor lose selectivity by long term treatment?, *Biochem. Pharmacol.*, 28, 919, 1979.
64. **Kapeller-Adler, R.,** *Amine Oxidases and Methods for their Study*, John Wiley & Sons, New York, 1970, 165.
65. **Achor, R. W., Hanson, N. O., and Gifford, R. W.,** Hypertension treated with *Rauwolfia serpentina* (whole root) and with reserpine: controlled study disclosing occasional severe depression, *JAMA*, 159, 841, 1955.
66. **Brodie, B. B., Pletscher, A., and Shore, P. A.,** Evidence that serotonin has a role in brain function, *Science*, 122, 968, 1955.
67. **Holzbauer, M. and Vogt, M.,** Depression by reserpine of the noradrenaline concentration in the hypothalamus of the cat, *J. Neurochem.*, 1, 8, 1956.
68. **Neff, N. F., Yang, H. -Y. T., and Fuentes, J. A.,** The use of selective monoamine oxidase inhibitor drugs to modify amine metabolism in brain, in *Neuropsychopharmacology of Monoamines and Their Regulatory Enzymes*, Usdin, E., Ed., Raven Press, New York, 1974, 49.
69. **Fuentes, J. A. and Neff, N. H.,** Selective monoamine oxidase inhibitor drugs as aids in evaluating the role of type A and B enzymes, *Neuropharmacology*, 14, 819, 1975.
70. **Christman, A. J., Coulson, C. J., Maxwell, D. R., and Riddell, D.,** A comparison of the pharmacological and biochemical properties of substrate-selective monoamine oxidase inhibitors, *Br. J. Pharmacol.*, 45, 490, 1972.
71. **Quinn, G. P., Shore, P. A., and Brodie, B. B.,** Biochemical and pharmacological studies of Ro 1-9569 (tetrabenazine), a non-indole tranquillising agent with reserpine-like effects, *J. Pharmacol. Exp. Ther.*, 127, 103, 1959.
72. **Giurgea, C. and Dauby, J.,** Sequential screening of antidepressant drugs (a critical survey), in *Methods in Drug Evaluation*, Mantegazza, P. and Piccinini, F., Eds., North-Holland, Amsterdam, 1966, 251.
73. **Ziance, R. J., Moxley, K., Mullis, M., and Gray, W.,** Influence of MAO inhibitors on uptake and release of norepinephrine in rat brain *in vitro*, *Arch. Int. Pharmacodyn.*, 288, 30, 1977.
74. **Baker, G. B., Hiob, L. E., and Dewhurst, W. G.,** Effects of monoamine oxidase inhibitors on release of dopamine and 5-hydroxytryptamine from rat striatum *in vitro*, *Cell. Mol. Biol.*, 26, 182, 1980.
75. **Horn, A. S. and Snyder, S. H.,** Steric requirements for catecholamine uptake by rat brain synaptosomes: studies with rigid analogues of amphetamine, *J. Pharmacol. Exp. Ther.*, 180, 523, 1972.
76. **Ferris, R. M., Howard, J. L., and White, H. L.,** A relationship between clinical efficacy and various biochemical parameters of monoamine oxidase inhibitors, *Pharmacologist*, 17, 451, 1975.
77. **Baker, G. B., McKim, H. R., Calverley, D. G., and Dewhurst, W. G.,** Effects of the monoamine oxidase inhibitors tranylcypromine, phenelzine and pheniprazine on the uptake of catecholamines in slices from rat brain regions, *Proc. Eur. Soc. Neurochem.*, 1, 536, 1978.
78. **Dyck, L. E. and Boulton, A. A.,** Effects of some monoamine oxidase inhibitors (MAOIs) on the deamination and uptake of *meta*-tyramine (*m*-TA), *para*-tyramine (*p*-TA) and dopamine (DA) by rat striatum *in vitro*, *Proc. Am. Soc. Neurosci.*, 4, 314, 1978.

79. **Finberg, J. P. M., Tenne, M., and Youdim, M. B. H.,** Tyamine antagonistic properties of AGN 1135 — an irreversible inhibitor of monoamine oxidase type B, *Br. J. Pharmacol.*, 73, 65, 1981.
80. **Simpson, L. L.,** Evidence that deprenyl, a type B monoamine oxidase inhibitor, is an indirectly acting sympathomimetic amine, *Biochem. Pharmacol.*, 27, 1591, 1978.
81. **Banerjee, S. P., Kung, L. S., Riggi, S. J., and Chanda, S. K.,** Development of β-adrenergic receptor subsensitivity by antidepressants, *Nature (London)*, 268, 455, 1977.
82. **Sarai, K., Frazer, A., Brunswick, D., and Mendels, J.,** Desmethylimipramine-induced decrease in β-adrenergic receptor binding in rat cerebral cortex, *Biochem. Pharmacol.*, 27, 2179, 1978.
83. **Wolfe, B. B., Harden, K., Sporn, J. R., and Molinoff, P. B.,** Presynaptic modulation of *beta*-adrenergic receptors in rat cerebral cortex after treatment with antidepressants, *J. Pharmacol. Exp. Ther.*, 207, 446, 1978.
84. **Sellinger, M., Frazer, A., and Mendels, J.,** Central *beta*-adrenergic receptor subsensitivity develops after repeated administration of antidepressant drugs, in *Catecholamines: Basic and Clinical Frontiers*, Vol. I, Usdin, E., Kopin, I. J., and Barchas, J., Eds., Pergamon Press, New York, 1979, 743.
85. **Savage, D. D., Frazer, A., and Mendels, J.,** Differential effects of monoamine oxidase inhibitors and serotonin reuptake inhibitors on ^{3}H-serotonin receptor binding in rat brain, *Eur. J. Pharmacol.*, 58, 87, 1979.
86. **Savage, D. D., Mendels, J., and Frazer, A.,** Monoamine oxidase inhibitors and serotonin uptake inhibitors: differential effects on [^{3}H]serotonin binding sites in rat brain, *J. Pharmacol. Exp. Ther.*, 212, 259, 1980.
87. **Schoepp, D. D. and Azzaro, A. J.,** Alteration of dopamine synthesis in rat striatum subsequent to selective type A monoamine oxidase inhibition, *J. Neurochem.*, 37, 527, 1981.
88. **Robinson, D. S., Campbell, I. C., Walker, M., Statham, N. J., Lovenberg, W., and Murphy, D. L.,** Effects of chronic monoamine oxidase inhibitor treatment on biogenic amine metabolism in rat brain, *Neuropharmacology*, 18, 771, 1979.
89. **Mandell, A. J.,** Neurobiological mechanisms of presynaptic metabolic adaptation and their organization: implications for a pathophysiology of the affective disorders, in *Neurobiological Mechanisms of Adaptation and Behavior*, 1st ed., Mandell, A. J., Ed., Raven Press, New York, 1975, 1.
90. **Squires, R. F. and Lassen, J. B.,** The inhibition of A and B forms of MAO in the production of a characteristic behavioral syndrome in rats after 1-tryptophan loading, *Psychopharmacologia*, 41, 145, 1975.
91. **Green, A. R., Mitchell, B. D., Tordoff, A. F. C., and Youdim, M. B. H.,** Evidence for dopamine deamination by both type A and type B monoamine oxidase in rat brain *in vivo* and for the degree of inhibition of enzyme necessary for increased functional activity of dopamine and 5-hydroxytryptamine, *Br. J. Pharmacol.*, 60, 343, 1977.
92. **Knoll, J.,** Analysis of the pharmacological effects of selective monoamine oxidase inhibitors, in *Monoamine Oxidase and its Inhibition*, Ciba Foundation Symposium 39, Wolstenholme, G. E. W. and Knight, J., Eds., Elsevier/Excerpta Medica/North Holland, Amsterdam, 1976, 135.
93. **Finberg, J. P. M., Tenne, M., and Youdim, M. B. H.,** Selective irreversible propargyl derivative inhibitors of monoamine oxidase (MAO) without the cheese effect, in *Monoamine Oxidase Inhibitors: the State of the Art*, Youdim, M. B. H., and Paykel, E. S., Eds., John Wiley & Sons, Chichester, 1981, 31.
94. **Elsworth, J. D., Glover, V., Reynolds, G. P., Sandler, M., Lees, A. J., Phuapradit, P., Shaw, K. M., Stern, G. M., and Kumar, P.,** Deprenyl administration in man: a selective monoamine oxidase B inhibitor without the "cheese effect", *Psychopharmacology*, 57, 33, 1978.
95. **Youdim, M. B. H. and Finberg, J. P. M.,** Use of selective acetylenic MAO inhibitors in depression, *Adv. Biol. Psychiatr.*, 7, 126, 1981.
96. **Pare, C. M. B.,** Monoamine oxidase inhibitors in resistant depression, *Int. Pharmacopsychiatr.*, 14, 101, 1979.
97. **Robinson, D. S., Nies, A., Corcella, J., and Cooper, T. B.,** MAO inhibitors, new biochemical and clinical findings, *Psychopharmacol. Bull.*, 17, 154, 1981.
98. **Reynolds, G. P., Elsworth, J. D., Blau, K., Sandler, M., Lees, A. J., and Stern, G. M.,** Deprenyl is metabolized to methamphetamine and amphetamine in man, *Br. J. Clin. Pharmacol.*, 6, 542, 1978.
99. **Magyar, K., Skolnik, J., and Knoll, J. V.,** *Conferentia Hungarica Pro Therapia et Investigatione in Pharmacologica*, Leszkovszky, E. P., Ed., Akademiai Kiado, Budapest, 1971, 103.
100. **Taylor, J. D. and Krause, R. A.,** Distribution and metabolism of the monoamine oxidase inhibitor (N-methyl-N-benzyl-2-propynylamine hydrochloride) in the rat and dog, *Fed. Proc.*, 21, 181, 1962.
101. **Durden, D. A., Philips, S. R., and Boulton, A. A.,** Identification and distribution of benzylamine in tissue extracts isolated from rats pretreated with pargyline, *Biochem. Pharmacol.*, 25, 858, 1976.
102. **Schwartz, M. A.,** The metabolism of isocarboxazid (Marplan) in the rat, *J. Pharmacol. Exp. Ther.*, 130, 157, 1960.
103. **Schwartz, M. A.,** Isocarboxazid metabolism in rat liver homogenate, *Proc. Soc. Exp. Biol. Med.*, 107, 613, 1961.
104. **Alleva, J. J.,** Metabolism of tranylcypromine-C^{14} and *dl*-amphetamine-C^{14} in the rat, *J. Med. Chem.*, 6, 621, 1963.

105. **Zeller, E. A., Barsky, J., Fouts, J. R., Kirchheimer, W. F., and Van Orden, L. S.,** Influence of isonicotinic acid hydrazide (INH) and 1-isonicotinyl-2-isopropyl hydrazide (IIH) on bacterial and mammalian enzymes, *Experientia,* 8, 349, 1952.
106. **Selikoff, I. J., Robitzek, E. H., and Ornstein, G. G.,** Toxicity of hydrazine derivatives of isonicotinic acid in the chemotherapy of human tuberculosis, *Q. Bull. Sea View Hosp.,* 13, 17, 1952.
107. **Bloch, R. G., Doonief, A. S., Buchberg, A. S., and Spellman, S.,** The clinical effect of isoniazid and iproniazid in the treatment of pulmonary tuberculosis, *Ann. Intern. Med.,* 40, 881, 1954.
108. **Crane, G. E.,** Iproniazid (Marsilid) phosphate, a therapeutic agent for mental disorders and debilitating disease, *Psychiatr. Res. Rep.,* 8, 142, 1957.
109. **Kline, N. S.,** Clinical experience with iproniazid (Marsilid), *J. Clin. Exp. Psychopathol.,* 19 (Suppl. 1), 72, 1958.
110. **Honigfeld, G. and Howard, A.,** *Psychiatric Drugs,* Academic Press, New York, 1973, 6.
111. *The Merck Index,* 9th ed., Windholz, M., Ed., Merck, New Jersey, 1976, 675.
112. **Rudy, B. C. and Senkowski, B. A.,** Isocarboxazid, in *Analytical Profiles of Drug Substances,* Vol. 2, Florey, K., Ed., Academic Press, New York, 1972, 307.
113. **Daly, R. E.,** Phenelzine sulfate, in *Analytical Profiles of Drug Substances* Vol. 2, Florey, K., Ed., Academic Press, New York, 1972, 393.
114. **Seinyard, E. A.,** Psychopharmacologic agents, in *Remington's Pharmaceutical Sciences,* 15th ed., Osol, A. and Hoover, J. E., Eds., Mack Publishing, Easton, Pa., 1975, 1033.
115. **Newton, D. W. and Kluza, R. B.,** PK_a values of medicinal compounds in pharmacy practice, *Drug Intell. Clin. Pharm.,* 12, 546, 1978.
116. **Mc Ewen, C. M., Sasaki, G., and Jones, D. C.,** Human liver mitochondrial monoamine oxidase. III. Kinetic studies concerning time-dependent inhibition, *Biochemistry,* 8(10), 3963, 1969.
117. **Mendlewicz, J. and Youdim, M. B. H.,** Antidepressant potentiation of 5-hydroxytryptophan by L-deprenyl, an MAO type B inhibitor, *J. Neural Transm.,* 43, 279, 1978.
118. **Cole, J. O. and Davis, J. M.,** Antidepressant drugs, in *Comprehensive Textbook of Psychiatry/II,* Freedman, A. M., Kaplan, H. I., and Sadock, B. J., Eds., Williams & Wilkins, Baltimore, 1975, 1941.
119. **Lesse, S.,** Tranylcypromine (Parnate) — a study of 1000 patients with severe agitated depressions, *Am. J. Psychother.,* 32, 220, 1978.
120. **Hendley, E. D. and Snyder, S. H.,** Relationship between the action of monoamine oxidase inhibitors on the noradrenaline uptake system and their antidepressant efficacy, *Nature (London),* 220, 1330, 1968.
121. **Goldberg, L. I.,** Monoamine oxidase inhibitors, *JAMA,* 190, 456, 1964.
122. **West, E. D. and Dally, P. J.,** Effects of iproniazid in depressive syndromes, *Br. Med. J.,* 1, 1491, 1959.
123. **Quitkin, F., Rifkin, A., and Klein, D. F.,** Monoamine oxidase inhibitors, *Arch. Gen. Psychiatr.,* 36, 749, 1979.
124. **Mendis, N., Pare, C. M. B., Sandler, M., Glover, V., and Stern, G.,** Is the failure of (−)-deprenyl, a selective monoamine oxidase B inhibitor, to alleviate depression related to freedom from the cheese effect?, *Psychopharmacology (Berlin),* 73, 87, 1981.
125. **Birkmayer, W., Riederer, P., Youdim, M. B. H., and Linauer, W.,** Potentiation of the antiakinetic effect of L-DOPA treatment by an inhibitor of MAO-B, deprenil, *J. Neural Transm.,* 36, 303, 1975.
126. **Ambrozi, L., Birkmayer, W., Riederer, P., and Youdim, M. B. H.,** L-DOPA and (−)-deprenil in the treatment of Parkinson's disease: a long-term study, *Br. J. Pharmacol.,* 58, 423, 1976.
127. **Birkmayer, W., Riederer, P., Ambrozi, L., and Youdim, M. B. H.,** Implications of combined treatment with "Madopar" and L-deprenyl in Parkinson's disease: a long-term study, *Lancet,* 1, 439, 1977.
128. **Lees, A. J., Shaw, K. M., Kohout, L. J., Stern, G. M., Elsworth, J. D., Sandler, M., and Youdim, M. B. H.,** Deprenyl in Parkinson's disease, *Lancet,* 2, 791, 1977.
129. **Klein, D. F. and Shader, R. I.,** The borderline state: psychopharmacologic treatment approaches to the undiagnosed case, *Manual of Psychiatric Therapeutics,* Shader, R. I., Ed., Little, Brown, Boston, 1975, 281.
130. **Kelly, D., Guirguis, W., Frommer, E., Mitchell-Heggs, N., and Sargent, W.,** Treatment of phobic states with antidepressants, *Br. J. Psychiatr.,* 116, 387, 1970.
131. **Anthony, M. and Lance, J. W.,** Monoamine oxidase inhibition in the treatment of migraine, *Arch. Neurol.,* 21, 263, 1969.
132. **MacLean, R., Nicholson, W. J., Pare, C. M. B., et al.,** Effect of monoamine oxidase inhibitors on the concentration of 5-hydroxytryptamine in the human brain, *Lancet,* 2, 205, 1965.
133. **Pare, C. M. B.,** Some clinical aspects of antidepressant drugs, in *The Scientific Basis of Drug Therapy in Psychiatry,* Marks, J. and Pare, C. M. B., Eds., Pergamon Press, New York, 1965, 103.
134. **Robinson, D. S., Nies, A., and Ravaris, C. L.,** Clinical pharmacology of phenelzine: MAO activity and clinical response, in *Psychoharmacology: A Generation of Progress,* Lipton, M., DiMascio, A., and Killam, K. F., Eds., Raven Press, New York, 1978, 961.
135. **Robinson, D., Nies, A., Ravaris, C. L., Ives, J. O., and Bartlett, D.,** Clinical pharmacology of phenelzine, *Arch. Gen. Psychiatr.,* 35, 629, 1978.

136. **Giller, E., Jr. and Lieb, J.,** MAO inhibitors and platelet MAO inhibition, *Commun. Psychopharmacol.*, 4, 79, 1980.
137. **White, K. and Simpson, G.,** Combined MAOI-tricyclic antidepressant treatment: a reevaluation, *J. Clin. Psychopharmacol.*, 1, 264, 1981.
138. **Spiker, D. G. and Pugh, D. D.,** Combining tricyclic and monoamine oxidase inhibitor antidepressants, *Arch. Gen. Psychiatr.*, 33, 828, 1976.
139. **Bassuk, E. L. and Schoonover, S. C.,** Depression, in *The Practitioner's Guide to Psychoactive Drugs*, Plenum Medical Book Company, New York, 1977, 40.
140. **Blackwell, B.,** Adverse effects of antidepressant drugs. I. Monoamine oxidase inhibitors and tricyclics, *Drugs*, 21, 201, 1981.

HALLUCINOGENS

David E. Nichols

INTRODUCTION

This chapter on hallucinogens may be unique to this volume. For, while hallucinogens have no presently accepted medical value, they probably are the oldest drugs known to man. Various hallucinogens such as *Cannabis sativa, Amanita muscaria,* the ''magic drugs'' of Mexico, and others have clearly played a significant role in many ancient societies, particularly in Asia and South America. These natural materials provided alternate states of reality around which various social and religious rites were structured.

Without delving into philosophical and anthropological issues, it is safe to say that hallucinogens exert a powerful and most unique effect upon the human psyche. Reading, for example, the accounts of R. G. Wasson, A. Hofmann, or A. Huxley as they described their various encounters with teonanacatl, LSD, or mescaline, one gains considerable appreciation for Lewin's proposal to classify these drugs in a category to be known as ''Phantastica''.[1]

Definitions

The first task is to define the term hallucinogen. Based on Hollister's definitions,[2] these are compounds which produce changes in thought, perception, and mood, without causing major disturbances in the autonomic nervous system. Impairment of memory and intellectual function is minimal. Stupor, narcosis, or excessive stimulation are not an integral part of the action. There is usually good recall of the experience afterward. Hallucinogens are not addictive. Although the term hallucinogen implies production of hallucinations, this is not true for all compounds. For some this is dose dependent, with changes in thought and perception being most evident at low doses. Furthermore, the ability to provoke hallucinations does not make a compound an hallucinogen. Deliriants, scopolamine for example, can cause true hallucinations, but these are considered to be manifestations of a toxic response to the drug.

Other terms which have been proposed or used to describe this class of drugs include; delusionogens, dysleptics, eidetics, misperceptinogens, mysticomimetics, pharnerothymes, phantasticants, psychedelics, psychodysleptics, psychogens, psychosomimetics, psychoticants, psychotogens, psychotomimetics, and schizogens. Of these, only psychedelic and psychotomimetic have been used with any regularity. The designations psychotomimetic and hallucinogen are used nearly interchangeably in the present literature.

The basic clinical syndrome observed with hallucinogens would typically include the following features:[3]

1. Somatic symptoms — dizziness, weakness, tremors, nausea, drowsiness, paresthesias, and blurred vision
2. Perceptual symptoms — altered shapes and colors, difficulty in focusing on objects, a sharpened sense of hearing, and sometimes synesthesias
3. Psychic symptoms — alterations in mood (happy, sad, or irritable at varying times), tension, distorted time sense, difficulty in expressing thoughts, depersonalization, dreamlike feelings, and visual hallucinations (depending on drug and dose)

The most common physical effects are dilated pupils (mydriasis), hyperreflexia, increased muscle tension, incoordination, and ataxia.

So much has been published on hallucinogens that the present treatment must be limited in scope. Therefore, the monographs which follow cover only those compounds which presently seem most important, either from a medicinal or a drug abuse standpoint. thus, no information is provided for SOMA (*Amanita muscaria*), ibogaine, or the various beta-carbolines such as harmaline. There is also no treatment of psychotomimetic anticholinergics (e.g., piperidyl benzilates) or psychotomimetic opiate derivatives. For a more complete overview interested readers should consult the excellent book by Schultes and Hofmann.[4]

For the purposes of a general discussion, tetrahydrocannabinol (''THC'', from marijuana) and phencyclidine (PCP) clearly have different mechanisms of action from LSD, mescaline, psilocybin, and their related analogues. Therefore, specific comments on mechanism of action for THC and PCP are reserved for the monographs on these drugs. The following discussion is meant to apply ''only'' to LSD, mescaline, psilocybin, and the derivatives described in their monographs.

Value

At the present time the value of hallucinogens lies mainly in their use as pharmacological tools. The discovery of LSD helped to catalyze a virtual revolution in the neurosciences, which continues to the present. The efforts to elucidate the mechanism of action for hallucinogens have stimulated intense investigation by numerous groups over the past 25 years.

Hallucinogens seem to have tremendous potential as tools to assist in understanding a number of cognitive processes. The effects of hallucinogens have sometimes been likened to a sort of conscious dreaming.[5] Thus, the mechanism and significance of the dreaming process may someday yield to studies implemented with hallucinogens.

Although hallucinogens are no longer believed to produce a model psychosis (''psychotomimetics''), the possibility they may be somehow utilized as tools in the practice of psychiatry remains a tantalizing and worthwhile goal. It is a genuine tragedy that the drug abuse problem has led to a climate which continues effectively to stifle the use of hallucinogens in experimental psychotherapy.

Models

For most investigators development of human test situations is simply out of the question. In order to circumvent this barrier, a number of model systems have been studied. As with any pharmacological model, these represent assays where an unknown drug is examined for its effects in a system where a known hallucinogen, such as LSD, elicits a particular or unique constellation of effects. Thus, similarity of effects in the model system may be used to infer that the unknown has hallucinogen-like properties. One is on the safest ground when this similarity is evident in a number of different model assays. However, the true assessment of a compound as ''hallucinogenic'' can, by definition, only be determined in man.

Although many model systems have been reported, only a few have found utility. They mainly possess the advantages of economy, ease of set-up or data analysis, or relatively straightforward training procedure when whole animals are used. The more common models include the following; mouse head twitch,[6] chronic spinal dog (stepping and flexor reflex),[7] rabbit hyperthermia,[8] discriminated Sidman avoidance,[9] cat limb flick,[10] mouse ear scratch,[11] and drug discrimination in rats.[12]

All of the above suffer from various disadvantages, but the cat limb flick and drug discrimination models may be the best. The limb flick is the simplest, since the cats do not require training. Its drawbacks are that a significant percentage of cats do not respond and false positives can be obtained. The drug discrimination assay in rats requires extensive training and a modest investment in instrumentation. It has also yielded instances of false positives. However, this latter model has proven to be a very powerful tool in the study of many behaviorally active drugs. The rank order of potencies in these two models generally correlates well with their activity in man.

Mechanisms of Action

The sequence of pharmacological events which leads to the unique effects of these drugs is still somewhat a mystery. Certainly a number of processes have been studied and many biochemical changes which occur have been mapped. It appears that some key target sites for these drugs have been identified. Yet, it should come as no surprise that drugs which produce such complex behavioral action will require a good deal of additional study before anyone has definitive answers.

It is appropriate to reiterate at this point that the present discussion does not apply to the action of tetrahydrocannabinol (THC) or phencyclidine (PCP). These agents have a different action and the following comments apply only to LSD, mescaline, psilocybin, and related compounds. The monographs on PCP and THC should be consulted for details of their action.

LSD, mescaline, and psilocybin are all considered to have the same or a very similar mechanism of action. In man, the three compounds have very similar subjective effects.[13,14] Rapid tolerance and cross-tolerance develops between LSD, mescaline, and psilocybin, in man and in animal models.[15-18] The ability of LSD and mescaline to serve as discriminative stimuli in rats was reported by Hirschhorn and Winter.[19] The two possess stimulus properties that are similar, if not identical. Subsequently, Schechter and Rosecrans[20] showed that when LSD, mescaline, and psilocybin are administered to rats in functionally equivalent doses, they produce qualitatively similar interoceptive cues. These facts, along with results from many other studies have consistently shown that these three hallucinogens, and certain closely related structures, share a high degree of pharmacological similarity.

The earliest ideas about mechanism of action related to the possibility that LSD blocked serotonin (5-HT) receptors in the brain.[21,22] The finding that the nonhallucinogenic 2-bromo-LSD (BOL) was a more potent 5-HT antagonist than LSD was the major evidence which led to abandonment of this theory.[23] However, hallucinogens are potent "agonists" at 5-HT receptors. In vivo there is observed increased brain levels of 5-HT,[24] with a concomitant decrease in 5-HT turnover, as measured by production of 5-hydroxyindole acetic acid.[25] Numerous studies, employing a variety of approaches, consistently have implicated interaction with serotonin systems as a major effect of hallucinogens.[26]

The densest regions of 5-HT containing cell bodies in the brain are the raphe nuclei in the midline of the brainstem. Cell bodies of the dorsal raphe nucleus (DRN) send ascending projections to various parts of the forebrain, including dense projections to the visual and limbic regions. In these areas, the release of 5-HT from serotonergic terminals inhibits activity of the postsynaptic target neuron.[27] The indole hallucinogens, such as LSD, depress the firing rate of raphe neurons,[28] through a serotonin-like action at 5-HT autoreceptors on the neuron.[29]

Thus, a simplistic view which may be taken is that suppression of these 5-HT cell bodies, and consequent cessation of 5-HT release onto target cells in the forebrain, removes a tonic inhibitory influence and results in disinhibition of the target cells.[30,31] Why and how this leads to hallucinogenesis is not clear and will require much study. However, the identification of target areas for these 5-HT neurons in the visual and limbic systems of the brain is consistent with this type of action.

This simplistic explanation is complicated by the fact that mescaline and phenethylamine hallucinogens do not reliably depress raphe activity.[28] Their mechanism may therefore differ somewhat from the indole hallucinogens. The neurochemical evidence does, however, clearly indicate an impact on brain serotonin systems in the actions of the phenethylamine type hallucinogens.

Although a serotonergic mechanism is believed to be central to the actions of hallucinogens, observed dopamine-like actions for LSD and certain other hallucinogens may modulate the potency or amplitude of the effect.[32,33]

The reader should consult References 4 and 34 to 37 for reviews which summarize additional findings about hallucinogens.

LSD-25

Mechanism of Action

The story of Albert Hofmann's first accidental discovery of the potent central effects of LSD is by now classic.[4] Although the literature prior to that time contained numerous works related to hallucinogens, the year in which Hofmann's discovery occurred, 1943, must mark the beginning of the present era of interest in hallucinogenic drugs.

Lysergic acid diethylamide (Structure 1) (LSD), remains the most potent drug within this class. It was the 25th in a series of semisynthetic lysergic acid derivatives prepared by Hofmann and is thus often referred to as "LSD-25". Its mechanism of action is generally believed to be similar to that for the phenethylamine and indolealkylamine hallucinogens. A major component of that mechanism appears to be an agonist action at presynaptic serotonin autoreceptors located on neurons in the DRN.[29] LSD, as well as other ergolines, also possess potent dopamine-like effects.[38] The relative importance of this action has not beenassessed. The general aspects of the mechanism of action are further discussed in the introduction to this chapter.

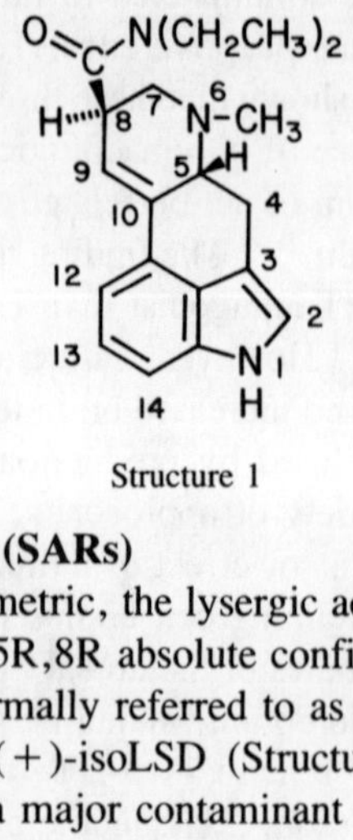

Structure 1

Structure-Activity Relationships (SARs)

Since both C5 and C8 are asymmetric, the lysergic acid structure possesses four possible isomers. Only the amide with the 5R,8R absolute configuration is biologically active. This isomer is dextrorotatory and is normally referred to as *d*-LSD or (+)-LSD. Epimerization at C8 yields the 5R,8S isomer, (+)-isoLSD (Structure 2). This is a chemically facile transformation and appears to be a major contaminant of illicit LSD. The (−)-isomers of LSD and iso-LSD have been clinically evaluated and are devoid of activity. Comparison of *d* and *l*-LSD in receptor binding assays using ^{3}H-serotonin or ^{3}H-dopamine as radioligands has shown *d*-LSD to have much higher affinity than its enantiomer.[39,40] It is of some interest that ololiuqui, a preparation from the seeds of *Rivea corymbosa,* was used by the ancient Aztecs as a potion and ingredient of various "magical" ointments. In 1960 Hofmann and Tscherter reported that the psychoactive components in these seeds are ergot alkaloids, principally lysergic acid amide (ergine).[41]

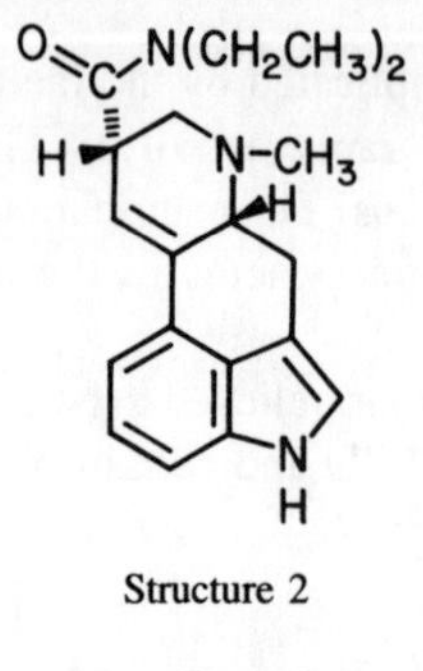

Structure 2

The structural complexity of lysergic acid has precluded extensive structure-activity studies. The major transformations which have been studied include variation of the amide substituent at C8, and substitution at N1 and C2.

Alteration of the amide substituent invariably decreases potency. The dimethylamide, pyrrolidide, and morpholide analogues possess only about 10% of the activity of LSD. However, as reported by Bach et al.[42] complete removal of the 8-carboxy group to yield 8-descarboxy LSD still allows the retention of an LSD-like pharmacological profile in mice.

The only modifications which retain good potency are the *N*(1)-acetyl and the *N*(1)-methyl diethylamides. These have approximately 100 and 33% of the potency of LSD, respectively. It is generally believed that the *N*(1)-acetyl is hydrolyzed in vivo to yield the nonacetylated LSD.

Reduction of the 2,3-bond to give 2,3-dihydro-LSD reduces activity to about one eighth that of LSD, with a delayed onset and increased duration of action. It has been speculated that this compound may be rearomatized to LSD in vivo. Reduction of the 9,10-double bond abolishes activity. Bromination of LSD at C(2) yields 2-bromo-LSD or BOL, a compound which is inactive as an hallucinogen. The SARs of the lysergamides have been summarized by Brimblecombe and Pinder.[37]

Little subsequent structure-activity work has been done on lysergamides. A noteworthy exception concerns the alteration of the alkyl substituent at N(6). The *N*-ethyl, *N*-allyl, and *N*-*n*-propyl-*N*(6)-norLSD homologues all possess LSD-like action in the rabbit and also contract the rat uterus.[43,44] In these assays they are more potent than LSD itself. However, no studies have been reported which would indicate whether they have a comparable clinical potency.

Pharmacokinetics

Effects following oral LSD administration usually develop from the 40-min point and continue to intensify to about 1.5 hr after ingestion. Peak effects then last to about the 4-hr point with a gradual recovery over the next 4 hr. The total duration of action is about 8 hr or slightly longer.

LSD is rapidly absorbed from the gut following oral administration. It is almost completely metabolized, mainly by the liver. Metabolites are secreted in the bile and excreted via the gut, with negligible amounts of unchanged drug found in urine or feces. Although the half-life of LSD in rodents is less than 10 min, the plasma half-life in cat, monkey, and man is 130, 100, and 109 min, respectively.[45] Wagner et al.[45] followed plasma levels of LSD following i.v. administration of 2 μg/kg, in man. Within 30 min plasma levels reached 7 ng/mℓ, gradually declining over 8 hr to 2 ng/mℓ. This parallels the course of psychic events and suggests that LSD is active per se, and not simply an initiator of a sequence of events which leads to intoxication, as has sometimes been suggested. Only a small amount of the drug actually reaches the brain, on the order of 1 to 1.5%. Indeed, Axelrod et al.[46] have calculated that LSD produces behavioral effects in man at an average brain concentration of about 0.5 ng/g of brain tissue.

The metabolism of LSD is highly species dependent. Thus, there is questionable relevance to a discussion of studies carried out in lower mammals. However, metabolism seems generally to follow several pathways. Aromatic hydroxylation leads to 13- and 14-hydroxy LSD, which are primarily excreted as the glucuronides.[47] Although it was initially believed that aromatic hydroxylation led to 12-hydroxy LSD, this was found to be erroneous.[47,48] There is no evidence that 12-hydroxy LSD is a metabolite. Deethylation at the diethylamide to yield lysergic acid monoethylamide also occurs. In some species, for example the guinea pig, LSD is extensively oxidized at C2 to give 2-oxo-2,3-dihydro-LSD. The latter is biologically inert.

Uses and Dosage

Following his accidental ingestion of an unknown quantity of LSD, Hofmann subsequently performed a self-experiment with 0.25 mg of LSD tartrate. This would now be considered a relatively large dose. Doses which were routinely employed in clinical studies usually involved about 0.5 to 1 μg/kg of body weight, with total oral doses as low as 20 to 25 μg detectable in some individuals. Oral doses of 100 to 150 μg of the tartrate salt were common in many clinical studies. Rapid tolerance develops to the effects of LSD.

LSD was studied quite intensively in search of medical applications. Between 1950 and the mid 1960s over 1000 clinical papers, discussing 40,000 patients, appeared. Numerous books and six internationsl symposia were devoted to the use of hallucinogens in therapy. Several recent books have provided overviews of this period.[49-51] In therapy, two basic approaches were utilized. One was termed psycholytic therapy and essentially used low to moderate doses of LSD as an adjuvant to psychotherapy. The second procedure, known as psychedelic therapy, employed relatively large (200 to 450 μg) doses of LSD to bring about a single overwhelming experience.[52]

The literature on the effectiveness of LSD, or other hallucinogens, for treatment of mental illness, drug, or alcohol addiction is contradictory. Some investigators claim unique effectiveness for LSD, while others dismiss it as worthless. As noted by McCabe et al.[52] nearly all the clinical literature is subject to some criticism. Typical problems were lack of appropriate control groups, inadequate group size and description of patient populations, or improper statistical and assessment procedures. However, it is quite clear that there are cases where LSD has been used to bring about remarkable improvement in patients resistant to conventional approaches.[51] Although LSD was normally employed in conjunction with conventional therapeutic approaches, a report by Brandrup and Vanggaard documents a case where LSD alone, with minimal psychotherapy, brought about a dramatic recovery in a patient with severe compulsive neurosis.[53] These authors point out that there are no means known other than LSD for the successful treatment of a genuine compulsive neurosis.

A second area of clear utility for LSD, and related to the discussion above, is in the treatment of terminal cancer patients.[54,55] In general, such patients, following LSD sessions, require less narcotic medication for pain and experience improved mood and relief from anxiety.

Although LSD psychotherapy is a specialized technique which requires intensive training of the therapist for proper use, the sensationalism of LSD by the media, as well as a number of legal and procedural barriers have gradually brought to a halt any significant research in this area.

Toxicity

Used in a medical setting with appropriate safeguards, LSD is a safe drug. A dose of 10 mg of LSD has been ingested without fatal outcome and the LD_{50} in man has been estimated to be 0.2 mg/kg, based on extrapolation of data from other species.[56] Reports of unsupervised LSD users ''jumping out of windows'', etc. were greatly exaggerated. However, the nonmedical use of LSD can have severe, very damaging psychological consequences. LSD can catalyze the onset of severe emotional problems, including psychosis, which may require hospitalization. Suicide following LSD-precipitated reactions is not unknown.

Also of great concern was the possibility that LSD produced chromosomal damage in users. Few serious researchers still believe that the medical use of ''pure'' LSD represents any genetic hazard.[52,57]

Physical-Chemical and Other Data

Chemical name — 9,10-didehydro-*N*,*N*-diethyl-6-methylergoline-8-β-carboxamide; (*N*,*N*,-diethyl lysergamide; *d*-lysergic acid diethylamide; Lysergide; LSD-25; Delysid); $C_{20}H_{25}N_{30}$

(323.42). The free base has m.p. 80 to 85°C after recrystallization from benzene or diisopropyl ether. (+)-LSD has $[\alpha]_D = +17°$ (c = 0.5, pyridine).[58] The salt with D-tartaric acid has m.p. 198 to 200°C, $[\alpha]_D = +30°$ (c = 1, H_2O). The mass spectrum of LSD is published.[59] The pKa of LSD is reported as 7.8[60]

MESCALINE AND PHENETHYLAMINES

Mechanism of Action

Mescaline is generally accepted to be the hallucinogenic component contained in peyote cactus, *Lophophora williamsii* (Lemaire) Coulter. This cactus, native to the American Southwest, contains about 3.5 to 4% total alkaloids by weight in the dried, sliced tops. Numerous alkaloids are present consisting mainly of structurally similar phenethylamines and tetrahydroisoquinolines.[61] Mescaline represents approximately one fourth of this total, or about 1% of the dry weight.

Although there continues to be speculation that some of the minor alkaloids in peyote probably contribute to, or modify the overall intoxication,[62] it is firmly established that mescaline is the component present in adequate concentration to bring about the profound mental effects which are characteristic of peyote ingestion.

The mechanism of action for mescaline is believed to be similar to that for LSD and psilocybin. This is based on their similar subjective effects in man, the development of cross-tolerance, and other similarities between them in various in vitro and animal behavioral models. Detailed comments on mechanism of action are included in the introduction to this chapter.

SARs

Mescaline (Structure 3) is quite important from a historical point of view. It was the first of the hallucinogens to receive serious scientific investigation. Mescaline was isolated and identified as the active component of peyote by Heffter, in 1896.[63] Its structure was elucidated and mescaline was subsequently synthesized by Ernst Späth in 1919.[64] Thus, it was the first hallucinogen to become available in a chemically pure form. It was the subject of a good deal of scientific study, especially during the early 1900s.

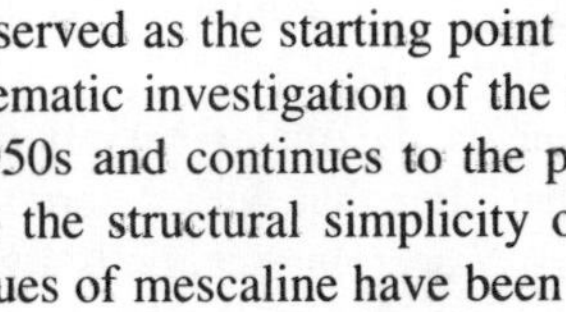

Structure 3

Most importantly, mescaline served as the starting point for extensive, detailed programs of structural modification. Systematic investigation of the SARs for mescaline-like phenethylamines began in the late 1950s and continues to the present. This series has been the most amenable to study due to the structural simplicity of the phenethylamines. Several hundred phenethylamine analogues of mescaline have been studied, with perhaps about 100 having been evaluated in man. Reviews of the SARs in general, and as they apply to clinical studies have recently appeared.[36,65] These should be consulted for detailed information. However, an optimally active hallucinogenic phenethylamine derivative would possess the following structural features:

1. A two-carbon separation between the aromatic ring and the amino function
2. A primary amino group; N-alkylation usually destroys activity

3. A methyl group attached to the alpha side chain carbon; an alkyl larger than methyl abolishes activity
4. 2,5-Dimethoxy substitution on the aromatic ring
5. A hydrophobic substituent located at the 4-position of the ring, i.e., para to the side chain; typically this is an *n*-alkyl, halogen, or alkylthio

Simply replacing the 4-methoxy group of mescaline with a more hydrophobic group, as in modification "5" above, is sufficient to impart greatly increased potency to the resulting compounds.

The alpha-methyl substituted phenethylamines are often referred to as hallucinogenic "amphetamines", although this nomenclature is technically incorrect. In these, the alpha side chain carbon is asymmetric and it is the enantiomer with the R absolute configuration which possesses the highest biological activity.[66] This isomer is levorotatory, to be contrasted with amphetamine itself, where it is the S-(+) isomer that is most active.

The compound receiving perhaps the most notoriety, as the result of widespread abuse, was 1-(2,5-dimethoxy-4-methylphenyl)-2-propanamine, known as DOM or STP. It has about 80 to 100 times the potency of mescaline and is active at an oral dosage of 2 to 5 mg. The most potent derivative yet reported is the 4-bromo homologue of DOM, known as DOB, which has about 400 times the potency of mescaline.[65]

Pharmacokinetics

The effects following an oral dose of mescaline begin to appear within about 30 to 40 min. There is often a period of nausea or vomiting, usually during the first hour after administration. This normally disappears with the clear onset of mental effects. Psychological effects continue to develop in intensity to a peak at about 4 hr, with a gradual tapering off for the next 4 to 6 hr. The total duration of action is in the range of 8 to 12 hr, considered to be slightly longer acting than LSD.

Mescaline is readily absorbed from the intestinal tract. Distribution and metabolism studies have been carried out in several animal species. Work through 1965 has been reviewed by Patel.[67] The most complete study of human metabolism seems to be that of Charalampous et al.;[68] 500 mg of mescaline hydrochloride containing 250 μCi of ^{14}C-mescaline was administered to human volunteers. An average of 87% of the radioactivity was excreted into the urine in the first 24 hr. Within 48 hr an average of 92% of the administered dose of radioactivity had been eliminated. The biological half-life of the radioactivity was approximately 6 hr. In this study approximately 60% of the radioactivity was unchanged mescaline. About 30% of the excreted activity consisted of 3,4,5-trimethoxybenzoic acid, with a minor amount of an O-demethylated hydroxydimethoxybenzoic acid. This O-demethylated acid, together with its conjugates, amounted to about 1% of the excreted activity. *N*-acetylmescaline represented about 0.1% of the radioactivity. The remaining mescaline metabolites account for only 10% of the radioactivity. About half of this is a conjugate of *N*-acetyl-5-hydroxy-3,4-dimethoxy-β-phenethylamine.

Although unchanged mescaline is the major excretion product in the first hour after administration, it drops somewhat after that with a corresponding increase in the level of trimethoxybenzoic acid excretion. *N*-acetylmescaline is a very minor metabolite which only appears in the urine much later.

Charalampous et al.[68] also studied plasma and cerebrospinal fluid (CSF). In the CSF, unmetabolized mescaline accounts for about 50% of the radioactivity. In contrast to urinary excretion, the major metabolites in the CSF are *N*-acetyl and *N*-acetyl-5-*O*-demethylated mescaline. The radioactivity level in the CSF reaches a peak at a time when the behavioral effects are maximal. Levels remain high through at least the 9.5 hr point after drug administration. The mental effects are therefore probably due to the action of mescaline per se, rather than to initiation of a sequence of events which continues after the drug is gone. As with LSD, this latter suggestion has been made based either on plasma drug levels or on data from rodent studies.

A comparison of the metabolism of mescaline with the nonhallucinogenic 3,4-dimethoxy-β-phenethylamine showed that the latter is much more rapidly and extensively deaminated.[69] Semicarbazide-sensitive diamine oxidases appear to be largely responsible for the oxidative deamination of mescaline.[70]

Uses and Dosage

Mescaline presently has no accepted medical use. Although it was investigated for possible benefit in treatment of mental illness, most studies of hallucinogens in psychotherapy employed LSD. This discussion is presented in the monograph on LSD.

On the other hand, mescaline, as utilized in the form of the dried tops of the peyote cactus ("peyote buttons") is used as a sacrament in religious services of the Native American Church. This church was chartered in Oklahoma in 1918 and has about 250,000 members.[71] The dose of mescaline ingested in one of these ceremonies varies; from three to twelve of the dried peyote buttons may be consumed over the course of the all-night proceedings. This would be expected to provide a dose of perhaps 200 mg of mescaline, in addition to the other alkaloids present in the cactus. For pure mescaline, as the sulfate salt, the oral dose is about 5 mg/kg, or a total dose of about 350 mg.[62]

Within the context of use by the Native American Church, peyote may serve a useful role. Alcohol and heroin addiction have been successfully treated, once the individual became a participating member of the church.[72] One cannot conclude whether this result is due to the drug, the social structure and support of the church, or a combination of these. Based on the treatment approaches utilizing LSD psychotherapy, it seems likely that the drug does play a significant role in the process.

The most cited description of the effects of mescaline is probably provided by Aldous Huxley's book *The Doors of Perception*.[73] This brief work provides description of a mescaline self-experiment, untainted by the social and legal prejudices which later resulted from widespread hallucinogen abuse.

Toxicity

The toxicity of mescaline is low. Underhill reports a case where 8 g of mescaline was taken without apparent ill effect.[74] As with LSD, psilocin, and related compounds, tolerance develops rapidly to the effects of mescaline. Cross-tolerance is produced between mescaline and LSD or psilocybin.[75] Any relevant example of chronic toxicity would likely be found among members of the Native American Church. However, no reports have appeared which would suggest that peyote-using Indians suffer from any unusual pathology. Dorrance et al.[76] did study this group for chromosomal abnormalities. The subjects of this study were a group of Huichol Indians in Northern Mexico who used peyote ceremonially up to a maximum of 35 times yearly. Peyote use begins as early as 6 years of age and any members of society, including pregnant women, may participate. A group of nonpeyote-using Huichols was used as a control. No serious chromosome damage was detected and the authors concluded that multigenerational ingestion of peyote is not associated with abnormalities in lymphocyte chromosomes.

Physical-Chemical and Other Data

Chemical name — 2-(3,4,5-trimethoxyphenyl)ethanamine; 3,4,5-trimethoxy-β-phenethylamine; $C_{11}H_{17}NO_3$ (211.25). The free base has m.p. 35 to 36°C and b.p. 180°C/12 mm. The HCl salt has m.p. 181°C, the picrate m.p. is 219 to 220°C, and hemisulfate m.p. is 158°C.[67] The pKa of mescaline is reported as 9.60[77] and the octanol-water partition coefficient for mescaline is 0.78.[78] The mass spectrum has been published by Bellman.[79]

PSILOCYBIN, PSILOCIN, AND RELATED TRYPTAMINES

Mechanism of Action

Spanish chroniclers who came to Mexico soon after Cortez' conquest mentioned numerous plants used by the Indians which had intoxicating properties. Among these was Teonanacatl, or "flesh of the gods". Along with peyotl and ololiuqui, it held special importance for the Aztecs and neighboring Indians in religious and medical practices. Teonanacatl was identified in the 1950s as mushrooms of the genus *Psilocybe, Stropharia,* and *Conocybe,*[80] The use and worship of these mushrooms probably dates back at least several thousand years.[81]

The active component of teonanacatl was subsequently isolated from *psilocybe mexicana*, identified, and synthesized by Hofmann et al.[82,83] The major psychoactive compound is psilocybin (Structure 4), accompanied by trace amounts of its dephosphorylated analogue psilocin (Structure 5).

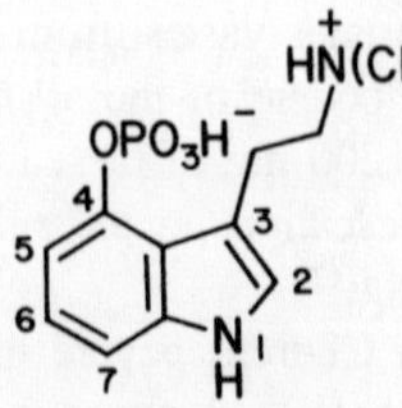

Structure 4

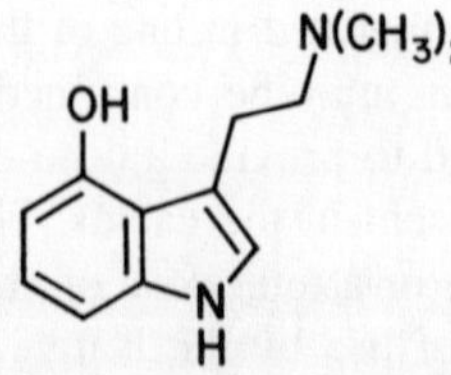

Structure 5

The mechanism of action for psilocybin is regarded to be the same or similar to that of both LSD and mescaline. This assessment is based on similar clinical effects in man, development of cross-tolerance to LSD,[75] and similarities in a number of in vitro and animal behavioral models. Further details of the mechanism of action are provided in the introduction.

SARs

Little is available for SARs of psilocybin analogues in humans. Although extensive series of N-alkylated psilocin homologues have been prepared,[84,85] no biological data have been reported. There are other series, also closely related in structure to psilocin and psilocybin, for which animal data are reported but which have not been evaluated clinically.[86,87] The most closely related congeners which have been tested in humans are the *N,N*-diethyl homologues of psilocin and psilocybin, CZ 74 and CEY 19, respectively.[88] These two compounds have potencies and qualitative effects in man which are virtually identical to those of psilocin and psilocybin. The major difference is the shorter duration of action for the *N,N*-diethyl homologues. Most subjects could resume normal mental activity after 4 hr.[88]

Most other tryptamines have major differences from psilocybin either in the type of effect produced or the route of administration required. The tryptamine derivatives which have been evaluated in man are briefly summarized by Kantor et al.[89] and this work can be consulted for further references. The SARs can be summarized as follows. First of all, moving the oxygen function to positions other than the 4- or 5-position of the indole ring leads to inactive compounds. If the oxygen is at the 5-position it cannot be as the free hydroxy. That is, 5-hydroxy-*N,N*-dimethyltryptamine, known as bufotenin, is inactive as an hallucinogen. This loss of activity has been explained as a result of failure of bufotenin to penetrate the blood-brain barrier.[90] On the other hand, 5-methoxy-*N,N*-dimethyltryptamine is active, but only by parenteral administration.

N,N-dialkyltryptamines in general are not active other than by parenteral administration. This includes *N,N*-dimethyl, *N,N*-diethyl, and *N,N*-dipropyl tryptamine. These three are all psychoactive, with the duration of action increasing as the length of the *N*-alkyl increases.

However, both *N,N*-diisopropyl and *N,N*-diisopropyl-5-methoxytryptamine are orally active.[91] Tryptamine itself, and 5-methoxytryptamine are inactive as hallucinogens, even by parenteral administration. This has been partially attributed to rapid degradation by monoamine oxidase.[92] Evidently the *N,N*-dialkylated compounds have increased resistance to this transformation and/or sufficiently increased lipid solubility to give active brain levels of compound. Addition of the 4-hydroxy somehow further increases metabolic stability so that oral activity is observed. No reports on clinical evaluation of the *N*-monomethyl or primary amino psilocin analogues have appeared, although these are known components of several mushroom species and are referred to as baeocystin and norbaeocystin, respectively.[93]

Addition of an alpha methyl to tryptamine and 5-methoxytryptamine leads to compounds which are very potent. Indeed, alpha-methyl-5-methoxytryptamine is the most potent tryptamine hallucinogen known, other than the lysergamide derivatives.[92]

Pharmacokinetics

The major difference between psilocybin (or psilocin) and LSD or mescaline is the short duration of action for psilocybin. The central effects appear about 20 min following oral ingestion and develop very rapidly over the next 20 to 30 min. A plateau of peak effects rarely lasts more than an hour, with a gradual recovery over the next 2 to 3 hr. The total duration of action is 4 to 6 hr, about half that of mescaline.[75]

Psilocybin and psilocin are indistinguishable when the doses are corrected for the molecular weight difference ratio of about 1.4.[75] Horita and Weber[94] found that psilocybin is rapidly dephosphorylated to psilocin by mammalian alkaline phosphatase. The further report that mouse brain accumulates psilocin following psilocybin administration supports the concept that it is psilocin which is the active species following psilocybin administration.[95]

Cross-tolerance is produced between psilocybin and LSD.[15] However, psilocybin does not produce the degree of tolerance that is observed with LSD. One communication has even reported a lack of cross-tolerance between psilocybin and LSD, but only two psychoneurotic patients were used as subjects.[96]

Studies on the metabolic fate of psilocybin and psilocin in man seem to be lacking. In the rat, oral dosing with ^{14}C-psilocin leads to about 50% absorption from the GI tract and relatively uniform distribution throughout the body.[97] Approximately 65% of the dose is excreted in the urine in 24 hr; 15 to 20% appears in the bile and feces. Although most of the excretion occurs in the first 8 hr, metabolites were still detected in the urine 1 week after administration of drug. About 25% of the psilocin was excreted unchanged in the urine. Only 4% undergoes side chain oxidation to 4-hydroxyindole acetic acid. The remainder of the metabolites were unidentified highly water-soluble compounds.[97]

However, this may not accurately reflect human metabolism. Bufotenin, the 5-hydroxy isomer of psilocin, is metabolized to 5-hydroxyindole acetic acid (5-HIAA) only to the extent of about 15% in rats.[98] By contrast, in humans approximately 70% of bufotenin is converted into 5-HIAA.[99] This is an area where much work remains to be done.

Usage and Dosage

Dried specimens of *Psilocybe mexicana* contain 0.2 to 0.4% psilocybin by weight. The median oral dose of psilocybin, as the pure material, is 4 to 8 mg,[81] consistent with the effective dose of about 2 g of the dried mushrooms.

Psilocybin received some attention for possible use in psychiatry. However, most investigators using hallucinogens in psychotherapy employed LSD. It is interesting to note that attention which focused on psilocybin experiments, carried out by Leary during the early 1960s, was an important factor in initiating the era of hallucinogen abuse which occurred during the 1960s and 1970s. It is somewhat of a paradox, therefore, that these compounds seem to be the least studied of all the hallucinogens. This is unfortunate since they are probably the most likely candidates for clinical use, due to their considerably shortened duration of action when compared with LSD or mescaline.

Toxicity

Psilocybin is regarded as having low toxicity, although very little about this is actually known. In mice, psilocybin is 2.5 times less toxic than mescaline, as measured by the LD_{50}.[81] Since psilocybin is about 50 times more potent than mescaline in man, it is implied that the therapeutic index is high. Although psilocybin and psilocin may have low systemic toxicity, the use of these materials can lead to the same types of emotional and psychological disturbances associated with LSD and mescaline.

Physical-Chemical and Other Data

Chemical name — 3-[2-(dimethylamino)ethyl]indol-4-ol; 4-hydroxy-*N*,*N*-dimethyltryptamine (psilocin) and 3-[2-(dimethylamino)ethyl]indol-4-ol dihydrogen phosphate ester; O-phosphoryl-4-hydroxy-*N*,*N*-dimethyltryptamine (psilocybin).

Psilocin — $C_{12}H_{16}N_{20}$ (204.27) m.p. 173 to 176°C (from methanol). Unstable in solution, especialy at basic pH. The crystal structure has been published.[100] The pKa is reported as 8.47 (amino) and 11.33 (OH).[90]

Psilocybin — $C_{12}H_{17}N_2O_4P$ (284.27) m.p. 220 to 228°C (from water) and m.p. 185 to 195°C (from methanol). The crystal structure has been published.[101] Psilocin forms three polymorphs which have different melting points and infrared spectral properties. Psilocybin forms solvates.[102]

PHENCYCLIDINE

Mechanism of Action

Phencyclidine (PCP) (Structure 6) was first marketed in the early 1960s as a surgical anesthetic for human use under the name Sernyl.[103,104] However, patients showed marked anxiety, agitation, and hallucinations and responses to the drug were unpredictable. Human use was discontinued and, in 1967, phencyclidine was marketed for veterinary use in non-human primates as Sernylan. This same year PCP appeared on the illicit market under the name ''Peace Pill'' and ''Hog''. It has since been available under various names, including angel dust, dust, or crystal.

Human subjects report the phencyclidine ''high'' to be very intense, much stronger than marijuana, and more comparable to LSD, although the action is much shorter in duration.[105]

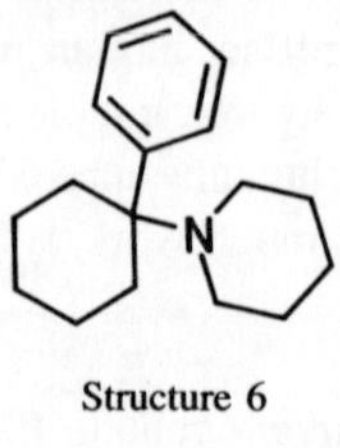

Structure 6

Although subjects compare PCP to LSD, it is described as different and in a class by itself. This categorization is supported by animal studies, where PCP produces discriminative stimuli which differ from those produced by a wide variety of other classes of psychoactive drugs, including LSD and tetrahydrocannabinol.[106] Furthermore, PCP perhaps comes closest to being a true ''psychotomimetic'' since PCP overdose is often misdiagnosed as an acute schizophrenic episode.[107] Based on these and related findings, phencyclidine cannot be placed into the same pharmacological classification with the hallucinogens which have been discussed earlier. It seems rather to be the prototype of a novel class of drugs which differ substantially in the nature of the effects produced.

A great deal of literature on the chemistry and pharmacology of PCP has appeared. A recent National Institute on Drug Abuse monograph provides a good general overview of the problem up to about 1978.[108]

It seems unlikely that the action of phencyclidine can be attributed to an alteration in function for a single neurotransmitter system. Various studies have implicated serotonin, dopamine, and norepinephrine as well as acetyl choline and GABA. Interaction with opiate receptors has also been suggested as a component of action. However, most of these studies have been done with whole brain where little about functional aspects of a particular transmitter system will be revealed. Very little is really known in an unambiguous way about effects on brain monoamines.

Symptoms resulting from a moderate dosage (5 to 10 mg) of PCP might typically include: coma or stupor with eyes remaining open, pupils in midposition and reactive nystagmus, vomiting, hypersalivation, repetitive motor movements, myoclonus (shivering), muscle rigidity on stimulation, flushing, diaphoresis, fever, and decreased peripheral sensation (pain, touch, and position).[109]

SARs

Although PCP has been available on the street since the late 1960s, it has only gained real popularity in recent years. It is still not clear why a drug with such a poor street reputation should enjoy such popularity. It may be a result of change in route of administration from oral to smoking, where effects are easier to "titrate" and the possibility of overdose is reduced. An important factor clearly is availability, as a result of the ease with which PCP or its many active congeners can be manufactured in illicit laboratories. The essential chemical intermediates are relatively accessible and the synthetic procedures can be carried out with little sophisticated chemical knowledge or technique.

One of the aspects of PCP abuse which has caused much alarm is the fact that PCP seems to be merely the prototype of a whole series of chemically related structures which possess similar pharmacological activity. Many of these have already appeared on the illicit market. Although potency comparisons in humans are not available, many appear nearly as active as PCP, with some perhaps even more potent. Shannon[106] has recently reported comparative data in the rat for several analogues and metabolites of PCP. Using a two-choice, discrete trial avoidance procedure, rats were trained to discriminate saline from a dose of 3 mg/kg PCP. Ability of the drugs to produce PCP-like discriminable stimuli was then measured. Replacement of the piperidine ring of PCP by methylamino or *n*-propylamino gave compounds with potency comparable to PCP. Replacing the piperidine with an ethylamino (PCE) gave a compound with nearly six times the potency of PCP. Replacement of the benzene ring with a 2-thienyl gives TCP (Structure 7), with a potency in rats about 1.3 times that of PCP. Both of the latter compounds have been available on the illicit market and user's subjective responses suggest that both are more potent than PCP in man.

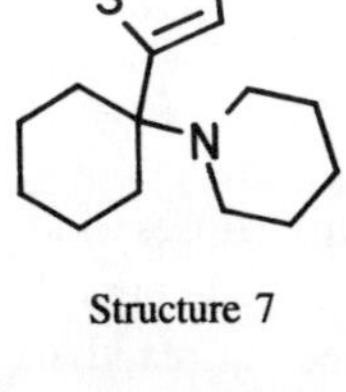

Structure 7

Replacement of the piperidine ring of either PCP or TCP with a pyrrolidine reduced potency only slightly from the parent. Contraction of the cycohexane ring to a cyclopentane reduces potency.[110] Substitution of the aromatic benzene ring with electron donating substitutents either in the 3- or 4-position increases activity, while an electron withdrawing group abolishes activity.[111] Kalir et al.[111] have attributed this to change in pK_a. Vincent et al.[112,113] and Geneste et al.[114,115] have examined stereoisomeric PCP derivatives containing alkyl or alkoxy group substitution in the cyclohexane ring. Molecules better able to adopt a conformation with the phenyl group axial and the piperidine equatorial proved to be more active.

A related compound, ketamine (Ketalar) (Structure 8) was developed as a structural analogue of phencyclidine. It was introduced in 1970, as the hydrochloride, as an anesthetic agent with rapid onset and brief duration following parenteral administration. Although only about one tenth the potency of PCP, it has also recently been abused. It produces a dissociative state and mental effects similar to PCP. Siegel has presented a brief review on ketamine.[116]

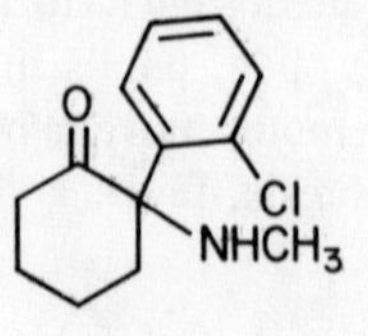

Structure 8

Pharmacokinetics

Using radioactive PCP Ober et al.[117] found that, in monkeys, about 60% of the radioactivity appeared in the urine within 12 hr. Nearly all the activity was in the form of metabolites, with the main one being a conjugated nonphenolic dihydroxyphencyclidine. However, in humans no dihydroxymetabolite was detected. Following enzymatic hydrolysis, human urine yielded two monohydroxy-metabolites from subjects acutely intoxicated with PCP.[118] Oxidative hydroxylation is the major mode of metabolism, similar to that observed in the rat.[119] However, PCP is less extensively metabolized in man than in the rat. Most metabolites are excreted as conjugates.

The two monohydroxy metabolites of PCP have been compared with PCP using rats in a two-choice drug discrimination assay.[106] Hydroxylation at the 4-position of the piperidine ring (PCHP) reduced potency to about one third that of PCP, but this metabolite retained full efficacy in producing PCP-like stimuli. By contrast, hydroxylation at the 4-position of the cycohexane ring (PPC) reduced both potency and efficacy.

Tissue distribution studies in animals have demonstrated relatively high and persistent concentrations of PCP in adipose and brain tissue. Blood concentrations were lower and fell rapidly.[118,120] Thin-layer, GLC, and GC-MS methods have been developed for the quantitation of PCP in body fluids and tissues.[121-123]

Uses and Dosage

As indicated earlier, PCP has no recognized therapeutic value in clinical medicine. Low to moderate dosages are generally considered to be in the range 1 to 5 mg, with moderate doses from 5 to 10 mg, and high doses in excess of 10 mg.

Toxicity

Acute

According to Aronow and Done,[109] high doses of PCP (in excess of 10 mg) would generally produce the following symptoms: prolonged coma (from 12 hr to many days, depending on dose), eyes closed, variable pupil size, but reactive, hypertension, opisthotonic posturing, decerebrate positioning, repetitive motor movements, muscular rigidity, convulsions (doses of 0.5 to 1 mg/kg), absent peripheral sensation, decreased or absent gag and corneal reflexes, diaphoresis, hypersalivation, flushing, and fever.

Treatment of overdosage is initially symptomatic, often with intensive medical management; there is no specific PCP antagonist. Acidification of the urine will hasten elimination. If the drug was recently ingested gastric drainage is indicated. Most clinicians advocate placing the patient in an isolated environment to reduce sensory stimulation so as to minimize excitability, irritability, anxiety, paranoia, and violence which often follow the comatose phase. Restraints and the assistance of several strong orderlies may be required.

Chronic

Lerner and Burns[105] report that chronic phencyclidine users experienced persistent cognitive and memory problems. Speech difficulties included stuttering, poor speech articulation, and difficulties in expression, with effects lasting 6 months to a year after prolonged daily PCP use. Mood disorders also occurred such as depression, anxiety, and violent behavior.

Although there is no strong evidence that PCP has a teratogenic or mutagenic effect, Golden et al.[124] described the case of a mother who used PCP throughout pregnancy. Subsequently she gave birth to a male child suffering from dysmorphology and cerebral palsy. The authors suggest that the observed deficits may have resulted from prenatal exposure to PCP.

Physical-Chemical and Other Data

Chemical name — 1-(1-phenylcyclohexyl)piperidine; $C_{17}H_{25}N$ (243.38). The free base has a m.p. of 46 to 46.5°C with a b.p. of 135 to 137°C at 1 mmHg. It was originally marketed as the hydrochloride salt under the names Sernyl and Sernylan. The hydrochloride salt has a m.p. of 243 to 244°C; the hydrobromide m.p. is 214 to 218°C. Phencyclidine has a reported pKa of 8.5[125] or 8.29.[114]

TETRAHYDROCANNABINOLS

Mechanism of Action

Among the compounds described in this chapter, the cannabinoids are chemically unique in that they do not contain a basic nitrogen atom. Indeed, it is unusual to observe potent central effects in any nonnitrogenous drug. Such drugs are usually assumed to interact with receptors for the natural monoamine transmitters such as dopamine or serotonin, which have a basic amino functionality.

Cannabinoids, particularly the tetrahydrocannabinols (THCs), are the psychoactive principles contained in various marijuana (*Cannabis*) preparations. In the normal context of marijuana use, the drug might more properly be considered to be a euphoriant. It is more often used in a manner similar to alcohol. As such, its effects are not considered to be debilitating, at least not to the extent that LSD or related hallucinogens impair functioning. However, it is also clear that very large doses of tetrahydrocannabinol can produce hallucinations and disorientation.

At present there is no accepted mechanism of action for the cannabinoids. Their relatively high potency would suggest specific receptor interactions but these have not been identified. It is also clear that the mechanism of action for tetrahydrocannabinol is different than that for other hallucinogens such as LSD. No cross-tolerance occurs between tetrahydrocannabinol and LSD or related compounds. The subjective effects for tetrahydrocannabinol in man differ from those of LSD. Also, rats trained to discriminate saline from Δ^9-THC do not generalize to LSD.[126]

SARs

The cannabinoids can be named either as monoterpenoid derivatives or as dibenzopyrans by the more formal IUPAC system (Structure 9) as shown below. The former method is more convenient for the naming of noncyclized compounds such as cannabidiol. The IUPAC nomenclature appears to be more widespread in the current literature and will be used in this monograph. The most active natural cannabinoid is the Δ^9 isomer. The isomeric Δ^8-THC is a minor component of cannabis and possesses only about 75% of the biological activity of the Δ^9 isomer. The SARs have been recently reviewed and can be summarized as follows:[127]

1. The complete tricyclic ring system appears necessary for activity.
2. Activity changes with the length of the 3-alkyl. Branching adjacent to the attachment site may increase activity.
3. A 4-alkyl substituent reduces activity.
4. The phenolic hydroxyl at C-1 is required. A labile ester retains activity but the 1-methoxy derivative is inactive.
5. The 6a,10a-trans ring fusion is necessary.
6. Only the levorotatory isomer is active, both for the Δ^9 and the Δ^8 isomers.
7. The natural Δ^8, and Δ^9 and synthetic $\Delta^{6a,10a}$ isomers are active. The Δ^7 and $\Delta^{9(11)}$ isomers are inactive.

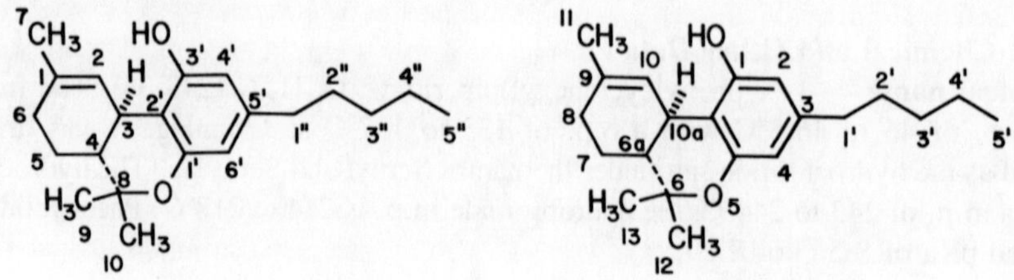

Monoterpenoid numbering
(−)-trans-Δ^1-tetra-
hydrocannabinol

IUPAC numbering
(−)-trans-Δ^9-tetra-)
hydrocannabinol

Structure 9

Pharmacokinetics

Tetrahydrocannabinol has complicated pharmacokinetic behavior. It apparently enters multiple body compartments and is converted into many metabolites, both active and inactive. It seems likely that the actual active species following THC administration may be the 11-hydroxy metabolite.[128,129] Tetrahydrocannabinol and its metabolites also bind tightly to blood proteins and may remain in fatty tissues for long periods. THC can be detected in user's blood for up to 6 days after the last marijuana use. After a single administration of THC, about 20% remains stored in the tissues and 20%, as metabolites, remains in the blood. Complete elimination of a single dose from the body may take 1 month.[130]

Tolerance, the diminished response to repeated administration of marijuana, has been observed.[131] The degree of tolerance is related to the frequency of use and dose of THC. It has been recently shown that pharmacokinetic and metabolic changes cannot account for the development of tolerance to the cardiovascular, psychological, and hypothermic action of THC.[132]

Uses and Dosage

The potency of marijuana available for ''street'' consumption has increased markedly over the past several years. Material confiscated in 1975 normally was found to contain 1% THC or less; 4 years later samples containing 5% THC were common. Recently, extracts of cannabis have been available as ''hash oil'' which contain 15 to 20% THC. Typical dosage with marijuana would consist of a 0.5 to 1-g marijuana cigarette, or ''joint'', containing 1 to 2 % THC. Because of the wide variation in smoking technique it is difficult to quantitate dosages. However, a cigarette containing 2% THC is thought to deliver to the lungs on the order of 10 mg of THC when smoked.[130] An i.v. dose of 1 to 2 mg would produce comparable effects. Approximately 30 mg is required to produce easily measured physiological and subjective effects following oral administration. Most measurable effects disappear within 3 to 6 hr following administration.[133]

Although the therapeutic potential of tetrahydrocannabinol or marijuana, as well as a number of synthetic tetrahydrocannabinol analogues, is being examined, only two medical indications have thus far emerged. The first of these is the treatment of glaucoma. In a recent study it was found that smoked marijuana significantly decreased intraocular pressure in hypertensive patients.[134] The results of several studies indicate that marijuana or THC-related drugs are able to reduce intraocular pressure in glaucoma patients both alone or in combination with more conventional antiglaucoma medications. Topical administration appears ineffective. However, there are presently no data from long-term studies which would indicate that these drugs can actually preserve vision in such individuals.

A second area of therapeutic potential is in the treatment of nausea resulting from cancer chemotherapy. Based on a number of studies, it appears that THC is superior to the phenothiazine antiemetics which are currently used. It has been concluded by one group of workers that oral THC represents a major advance in antiemetic therapy.[135]

Toxicity

Acute

Tetrahydrocannabinol is a relatively nontoxic material. There are no documented deaths which can be attributed to the direct effects of the drug. The most common adverse reaction on acute exposure is panic or anxiety, especially when the potency of the smoked marijuana is much higher than expected. THC also has pronounced dose-related cardiovascular effects. Following THC administration heart rate may increase up to 160 beats/min or more. Orthostatic hypotension is also observed. In persons with coronary artery disease the increased oxygen demand by the heart, and decreased oxygen delivery to the heart, may present problems.

As is true for any intoxicant, driving an automobile following marijuana use is very unwise. Several studies have indicated impairment of driving performance following cannabis use, suggesting that an increasing number of traffic accidents and fatalities may be attributable to THC intoxication.

Chronic

The chronic toxicity of marijuana is a subject which is at the heart of most of the controversy regarding social acceptability of this substance. Personal views and bias come into play and seem to prejudice many otherwise objective scientists. It ought to be obvious from the outset, however, that the repeated inhalation of any type of plant combustion products cannot promote good lung health, as has been demonstrated in the case of tobacco smoking. Furthermore, cannabis produces more tar than a comparable weight of tobacco and is smoked in such a way as to maximize tar deposition in the lungs.

In an experimental situation, significant worsening of pulmonary function was observed after 6 to 8 weeks of smoking a few marijuana cigarettes daily.[136,137] Chronic exposure may impair the lung's defense mechanisms, consistent with the clinical observation that heavy cannabis users often have bronchitis, coughs, hoarseness, or asthma-like conditions.

The findings regarding the endocrine effects of THC, and particularly with respect to the effect on sexual function in men, must still be considered controversial. Although supression of pituitary-testicular function, decreased sperm count, and sexual dysfunction have been reported,[138] other workers found no association between chronic marijuana use and decreased plasma testosterone.[139] In mice, THC has a biphasic effect on testosterone and LH levels.[140]

There have also been studies which indicate that marijuana use intereferes with short-term memory and may affect a number of intellectual functions which would be expected to impair classroom learning among student users. Yet, a recent study of a group of very heavy users of ganja, a particularly potent form of marijuana, failed to identify any intellectual or cognitive deficits.[141]

The most worrisome aspect of marijuana use has been pointed out by Cohen as being related to the chronic, heavy juvenile user.[142] The ages of users have consistently dropped over the last decade and there is an increasing incidence of classroom intoxication. The three elements noted by Cohen of (1) high density use over time, (2) inhalation of smoke products, and (3) frequent intermittent intoxication are most disturbing when marijuana use begins at an early age. Although there are no clear scientific data to indicate that neurological alterations persist or become cumulative over time, this is still an unknown which some scientists fear may ultimately be answered in a tragic way.

Physical-Chemical and Other Data

Chemical name — (IUPAC, Δ^9-THC) 3-*n*-pentyl-6a,7,8,10a-tetrahydro-6,6,9-trimethyl-6H-dibenzo[bd]pyran-1-ol; $C_{21}H_{23}{}_0O_2$ (314.45). THC is a viscous oil with b.p. 200°C at 0.02 mmHg. The absolute configuration was unequivocally established in 1967.[143] The pure (−) enantiomer has $[\alpha]_D = -150.5°$ (c = 0.53, $CHCl_3$). UV, IR, NMR, and mass spectral data have been published.[144]

REFERENCES

1. **Lewin, L.,** *Phantastica-Narcotic and Stimulating Drugs-Their Use and Abuse,* Routledge and Kegan Paul, London, 1964.
2. **Hollister, L. E.,** *Chemical Psychoses: LSD and Related Drugs,* Charles C Thomas, Springfield, Ill., 1968.
3. **Hollister, L. E.,** Psychotomimetic drugs in man, in *Handbook of Psychopharmacology,* Vol 11, Iversen, L. L., Iversen, S. D., and Snyder, S. H., Eds., Plenum Press, New York, 1978, 389.
4. **Schultes, R. E. and Hofmann, A.,** *The Botany and Chemistry of Hallucinogens,* 2nd ed., Charles C Thomas, Springfield, Ill, 1980.
5. **Jacobs, B. L.,** Dreams and hallucinations: a common neurochemical mechanism mediating the phenomenological similarity, *Neurosci. Biobehav. Rev.,* 2, 59, 1978.
6. **Corne, S. J. and Pickering, R. W.,** A possible correlation between drug-induced hallucinations in man and a behavioral response in mice, *Psychopharmacology,* 11, 65, 1967.
7. **Martin, W. R., Vaupel, D. B., Sloan, J. W., Bell, J. A., Nozaki, M., and Bright, L. D.,** The mode of action of LSD-like hallucinogens and their identification, in *The Psychopharmacology of Hallucinogens,* Stillman, R. C. and Willette, R. E., Eds., Pergamon Press, New York, 1978, 118.
8. **Aldous, F. A. B., Barrass, B. C., Brewster, K., Buxton, D. A., Green, D. M., Pinder, R. M., Rich, P., Skeels, M., and Tutt, K. J.,** Structure-activity relationships in psychotomimetic phenylalkylamines, *J. Med. Chem.,* 17, 1100, 1974.
9. **Calil, H. M.,** Screening hallucinogenic drugs, *Psychopharmacology,* 56, 87, 1978.
10. **Jacobs, B. L., Trulson, M. E., and Stern, W. C.,** Behavioral effects of LSD in the cat: proposal of an animal behavior model for studying the actions of hallucinogenic drugs, *Brain Res.,* 132, 301, 1977.
11. **Yim, G. K. W., Prah, T. E., Pfister, W. R., and Nichols, D. E.,** An economical screen for phenethylamine-type hallucinogens: mouse ear scratching, *Commun. Psychopharmacol.,* 3, 173, 1979.
12. **Winter, J. C.,** Hallucinogens as discriminative stimuli, *Fed. Proc.,* 33, 1825, 1974.
13. **Hoch, P. J., Cattell, J. P., and Pennes, H. H.,** Effects of mescaline and lysergic acid (d-LSD-25), *Am. J. Psychiatr.,* 108, 579, 1952.
14. **Isbell, H.,** Comparison of the reactions induced by psilocybin and LSD-25 in man, *Psychopharmacology,* 1, 29, 1959.
15. **Isbell, H., Wolbach, A. B., Wikler, A., and Miner, E. J.,** Cross tolerance between LSD and psilocybin, *Psychopharmacology,* 2, 147, 1961.
16. **Wolbach, A. B., Isbell, H., and Miner, E. J.,** Cross tolerance between mescaline and LSD-25 with a comparison of the mescaline and LSD reactions, *Psychopharmacology,* 3, 1, 1962.

17. **Appel, J. B. and Freedman, D. X.,** Tolerance and cross-tolerance among psychotomimetic drugs, *Psychopharmacology,* 13, 267, 1968.
18. **Winter, J. C.,** Tolerance to a behavioral effect of lysergic acid diethylamide and cross tolerance to mescaline in the rat: absence of a metabolic component, *J. Pharmacol. Exp. Ther.*, 178, 625, 1971.
19. **Hirschhorn, I. D. and Winter, J. C.,** Mescaline and lysergic acid diethylamide (LSD) as discriminative stimuli, *Psychopharmacology,* 22, 64, 1971.
20. **Schechter, M. D. and Rosecrans, J. A.,** Lysergic acid diethylamide (LSD) as a discriminative cue: drugs with stimulus properties, *Psychopharmacology,* 26, 313, 1972.
21. **Gaddum, J. H. and Hameed, K. A.,** Drugs which antagonize 5-hydroxytryptamine, *Br. J. Pharmacol.*, 9, 240, 1954.
22. **Wooley, D. W. and Shaw, E.,** A biochemical and pharmacological suggestion about certain mental disorders, *Proc. Natl. Acad. Sci. U.S.A.*, 40, 228, 1954.
23. **Cerletti, A. and Rothlin, E.,** Role of 5-hydroxytryptamine in mental disease and its antagonism to lysergic acid derivatives, *Nature (London),* 176, 785, 1955.
24. **Freedman, D. X.,** Effects of LSD-25 on brain serotonin, *J. Pharmacol. Exp. Ther.*, 134, 160, 1961.
25. **Rosecrans, J. A., Lovell, R. A., and Freedman, D. X.,** Effects of lysergic acid diethylamide on the metabolism of brain 5-hydroxytryptamine, *Biochem. Pharmacol.*, 16, 2011, 1967.
26. **Jacobs, B. L. and Trulson, M. E.,** The role of serotonin in the actions of hallucinogenic drugs, in *Serotonin Neurotransmissionn and Behavior* Jacobs, B. L. and Gelperin, A., Eds., MIT Press, Cambridge, Mass., 1981, 366.
27. **Haigler, H. J. and Aghajanian, G. K.,** Lysergic acid diethylamide and serotonin: a comparison of effects on serotoninergic neurons and neurons receiving a serotonergic input, *J. Pharmacol. Exp. Ther.,* 188, 688, 1974.
28. **Aghajanian, G. K., Foote, W. E., and Sheard, M. H.,** Action of psychotogenic drugs on midbrain raphe neurons, *J. Pharmacol. Exp. Ther.*, 171, 178, 1970.
29. **Rogawski, M. A. and Aghajanian, G. K.,** Serotonin autoreceptors on dorsal raphe neurons: structure-activity relationships of tryptamine analogs, *J. Neurosci.*, 1, 1148, 1981.
30. **Aghajanian, G. K., Haigler, H. J., and Bennett, J. L.,** Amine receptors in the CNS. III. 5-Hydroxytryptamine in the brain, in *Handbook of Psychopharmacology,* Vol 6, Iversen, L. L., Iversen, S. D., and Snyder, S. H., Eds., Plenum, Press, New York, 1975.
31. **Jacobs, B. L. and Trulson, M. E.,** Mechanisms of action of LSD, *Am. Sci.,* 67, 396, 1979.
32. **Trulson, M. E., Stark, A. D., and Jacobs, B. L.,** Comparative effects of hallucinogenic drugs on rotational behavior in rats with unilateral 6-hydroxydopamine lesions, *Eur. J. Pharmacol.,* 44, 113, 1977.
33. **Nichols, D. E.,** Structural correlation between apomorphine and LSD: involvement of dopamine as well as serotonin in the actions of hallucinogens, *J. Theor. Biol.*, 59, 167, 1976.
34. **Shulgin, A. T.,** Hallucinogens, in *Burger's Medicinal Chemistry,* Part III, 4th ed., Wolff, M. E., Ed., Interscience, New York, 1980, 1109.
35. **Brawley, P. and Duffield, J. C.,** The pharmacology of hallucinogens, *Pharmacol. Rev.*, 24, 31, 1972.
36. **Nichols, D. E.,** Structure-activity relationships of phenethylamine hallucinogens, *J. Pharm. Sci.,* 70, 839, 1981.
37. **Brimblecombe, R. W. and Pinder, R. M.,** *Hallucinogenic Agents,* Wright-Scientechnica, Bristol, 1975.
38. **Pieri, L., Pieri, M., and Haefely, W.,** LSD as an agonist of dopamine receptors in the striatum, *Nature (London),* 252, 586, 1974.
39. **Bennett, J. P., Jr. and Snyder, S. H.,** Serotonin and lysergic acid diethylamide binding in rat brain membranes: relationships to postsynaptic serotonin receptors, *Mol. Pharmacol.,* 12, 373, 1976.
40. **Burt, D. R., Creese, I., and Snyder, S. H.,** Binding interactions of lysergic acid diethylamide and related agents with dopamine receptors in the brain, *Mol. Pharmacol.*, 12, 631, 1976.
41. **Hofmann, A. and Tscherter, H.,** Isolierung von lysergsaure-alkaloiden aus der mexicanischen Zauberdroge ololiuqui (Rivea corymbosa (L.) Hall. f.), *Experientia,* 16, 414, 1960.
42. **Bach, N. J., Hall, D. A., and Kornfeld, E. C.,** Descarboxylysergic acid (9,10-didehydro-6-methylergoline), *J. Med. Chem.*, 17, 312, 1974.
43. **Hashimoto, H., Hayashi, M., Nakahara, Y., Niwaguchi, T., and Ishii, H.,** Hyperthermic effects of d-lysergic acid diethylamide (LSD) and its derivatives in rabbits and rats, *Arch. Int. Pharmacodyn.*, 228, 314, 1977.
44. **Hashimoto, H., Hayashi, M., Nakahara, Y., Niwaguchi, T., and Ishii, H.,** Actions of d-lysergic acid diethylamide (LSD) and its derivatives on 5-hydroxytryptamine receptors in the isolated uterine smooth muscle of the rat, *Eur. J. Pharmacol.*, 45, 341, 1977.
45. **Wagner, J. G., Aghajanian, G. K., and Bing, O. H. L.,** Correlation of performance test scores with "tissue concentration" of lysergic acid diethylamide in human subjects, *Clin. Pharm. Ther.,* 9, 635, 1968.
46. **Axelrod, J., Brady, R. O., Witkop, B., and Evarts, E. V.,** The distribution and metabolism of lysergic acid diethylamide, *Ann. N.Y. Acad. Sci.*, 66, 435, 1957.

47. **Siddik, Z. H., Barnes, R. D., Dring, L. G., Smith,R. L., and Williams, R. T.,** The fate of lysergic acid di [^{14}C]ethylamide ([^{14}C]LSD) in the rat, *Biochem. Soc. (London) Trans.*, 3, 290, 1975.
48. **Siddik, Z. H., Barnes, R. D., Dring, L. G., Smith, R. L., and Williams, R. T.,** The fate of lysergic acid di [^{14}C]ethylamide ([^{14}C]LSD) in the rat, guinea pig, and rhesus monkey and [^{14}C]iso-LSD in rat, *Biochem. Pharmacol.*, 28, 3093, 1979.
49. **Grinspoon, L., and Bakalar, J. B.,** *Psychedelic Drugs Reconsidered,* Basic Books Inc., New York, 1979, 192.
50. **Grof, S.,** *Ralms of the Human Unconscious,* Viking Press, New York, 1975.
51. **Grof, S.,** *LSD Psychotherapy,* Hunter House, Pomona, Calif., 1980.
52. **McCabe, O. L., Savage, C., Kurland, A., and Unger, S.,** Psychedelic (LSD) therapy of neurotic disorders: short term effects, *J. Psychedelic Drugs,* 5, 18, 1972.
53. **Brandrup, E. and Vanggaard, T.,** LSD treatment in a severe case of compulsive neurosis, *Acta Psychiatr. Scand,* 55, 127, 1977.
54. **Pahnke, W. N., Kurland, A. A., Unger, S., Savage, C., and Grof, S.,** The experimental use of psychedelic (LSD) psychotherapy, *JAMA,* 212, 1856, 1970.
55. **Grof, S. and Haifax, J.,** *The Human Encounter with Death,* E. P. Sutton, New York, 1977.
56. **Cohen, S.,** The psychotomimetic agents, *Prog. Drug Res.*, 15, 68, 1971.
57. **Dishotsky, N. I., Loughman, W. D., Mogar, R. E., and Lipscomb, W. R.,** LSD and genetic damage, *Science,* 172, 431, 1971.
58. **Stoll, A. and Hofmann, A.,** Amide der stereoisomeren lysergsauren und dihydro-lysergsauren, *Helv. Chim. Acta,* 38, 421, 1955.
59. **Nigman, I. C. and Holmes, J. L.,** Mass spectrometry of lysergic acid diethylamide, *J. Pharm. Sci.*, 58, 506, 1969.
60. **Perrin, D. D.,** *Dissociation Constants Of Organic Bases in Aqueous Solution,* Butterworths, London, 1965.
61. **Kapadia, G. J. and Fayez, M. B. E.,** The chemistry of peyote alkaloids, *Lloydia,* 36, 9, 1973.
62. **Shulgin, A. T.,** Mescaline: the chemistry and pharmacology of its analogs, *Lloydia,* 36, 46, 1973.
63. **Heffter, A.,** Ueber Cacteenalkaloide, *Chem. Ber.*, 29, 216, 1896.
64. **Späth, E.,** Uber die Anhalonium-Alkaloide. I. Anhalin and Mezcalin, *Monatash. Chem.*, 40, 129, 1919.
65. **Shulgin, A. T.,** Psychotomimetic drugs: structure-activity relationships, in *Handbook of Psychopharmacology,* Vol 11, Iversen, L. L., Iversen, S. D., and Snyder, S. H., Eds., Plenum Press, New York, 1978, 243.
66. **Shulgin, A. T.,** Stereospecific requirements for hallucinogenesis, *J. Pharm. Pharmacol.*, 25, 271, 1973.
67. **Patel, A. R.,** Mescaline and related compounds, *Prog. Drug Res.*, 11, 11, 1968.
68. **Charalampous, K. D., Walker, K. E., and Kinross-Wright, J.,** Metabolic fate of mescaline in man, *Psychopharmacology,* 9, 48, 1966.
69. **Friedhoff, A. J. and Hollister, L. E.,** Comparison of the metabolism of 3,4-dimethoxyphenylethylamine and mescaline in humans, *Biochem. Pharmacol.*, 15, 269, 1966.
70. **Zeller, E. A., Barsky, J., Berma, E. R., Charkas, M. S., and Fouts, J. R.,** Degradation of mescaline by amine oxidases, *J. Pharmacol. Exp. Ther.*, 124, 282, 1958.
71. **Schultes, R. E.,** The plant kingdom and hallucinogens. III, *Bull. Narcotics,* 22(1), 25, 1970.
72. **Blum, K., Futterman, S. L., and Pascarosa, P.,** Peyote, a potential ethnopharmacologic agent for alcoholism and other drug dependencies: possible biochemical rationale, *Clin. Toxicol.*, 11, 459, 1977.
73. **Huxley, A.,** *The Doors of Perception. Heaven and Hell,* Harper & Row, New York, 1963.
74. **Underhill, R.,** Peyote, in *Proc. 30th Int. Congr. of Americanists,* London, 1952, 148.
75. **Wolbach, A., Jr., Miner, E. J., and Isbell, H.,** Comparison of psilocin with psilocybin, mescaline and LSD-25, *Psychopharmacology,* 3, 219, 1962.
76. **Dorrance, D. L., Janiger, O., and Teplitz, R. L.,** Effect of peyote on human chromosomes, *JAMA,* 234, 299, 1975.
77. **Taska, K. J. and Schoolar, J. C.,** Peripheral tissue distribution, brain distribution and metabolism of mescaline-^{14}C in monkeys, *Arch. Int. Pharmacodyn. Ther.*, 202, 66, 1973.
78. **Nichols, D. E., Shulgin, A. T., and Dyer, D. C.,** Directional lipophilic character in a series of psychotomimetic phenethylamine derivatives, *Life Sci.*, 21, 569, 1977.
79. **Bellman, S. W.,** Mass spectral identification of some hallucinogenic drugs, *J. Assoc. Off. Anal. Chem.*, 51, 164, 1968.
80. **Heim, R. and Wasson, R. G.,** *Les Champignons Hallucinogenes du Mexique,* Museum National d'Histoire Naturelle, Paris, 1958.
81. **Hofmann, A.,** Teonanacatl and ololiuqui, two ancient magic drugs of Mexico, *Bull. Narcotics.*, 23 (1), 3, 1971.
82. **Hofmann, A. and Troxler, F.,** Identifizierung von psilocin, *Experientia,* 15, 101, 1959.
83. **Hofmann, A., Heim, R., Brack, A., Kobel, H., Frey, A., Ott, H., Petrzilka, Th., and Troxler, F.,** Konstitutions-aufklarung und synthese von Psilocybin, *Helv. Chim. Acta,* 42, 1557, 1959.

84. **Repke, D. B., Ferguson, W. J. and Bates, D. K.,** Psilocin analogs. I. Synthesis of 3-[2-(Dialkylamino)ethyl]- and 3-[2-(Cycloalkylamino)ethyl]-indol-4-ols, *J. Hetero. Chem.*, 14, 71, 1977.
85. **Repke, D. B., Ferguson, W. J., and Bates, D. K.,** Psilocin analogs, II. Synthesis of 3-[2-(Dialkylamino)ethyl]-3-[2-N-methyl-N-alkylamino)ethyl]-, and 3-[2-(Cycloalkylamino)-ethyl]indol-4-ols, *J. Hetero. Chem.*, 18, 175, 1981.
86. **Weidmann, H. and Cerletti, A.,** Studies on psilocybin and related compounds, *Helv. Physiol. Acta.*, 18, 174, 1960.
87. **Cerletti, A., Taeschler, M., and Weidmann, H.,** Pharmacologic studies on the structure-activity relationship of hydroxyindole alkylamines, *Adv. Pharmacol.*, 6 (Part B), 233, 1968.
88. **Leuner, H. and Baer, G.,** Two new short-acting hallucinogens of the psilocybin group, *Neuropsychopharmacology*, 4, 471, 1965.
89. **Kantor, R. E., Dudlettes, S. D., and Shulgin, A. T.,** 5-Methoxy-α-methyltryptamine (α,O-dimethylserotonin), a hallucinogenic homolog of serotonin, *Biol. Psychiatr.*, 15, 349, 1980.
90. **Migliaccio, G. P., Shieh, T-L. N., Byrn, S. R., Hathaway, B. A., and Nichols, D. E.,** Comparison of solution conformational preferences for the hallucinogens bufotenin and psilocin using 360 MHz proton nmr spectroscopy, *J. Med. Chem.*, 24, 206, 1981.
91. **Shulgin, A. T. and Carter, M. F.,** N,N-Diisopropyltryptamine (DIPT) and 5-methoxy-N,N-diisopropyltryptamine (5-MEO-DIPT). Two orally active tryptamine analogs with CNS activity, *Commun. Psychopharmacol.*, 4, 363, 1980.
92. **Shulgin, A. T. and Nichols, D. E.,** Characterization of three new psychotomimetics, in *The Psychopharmacology of Hallucinogens*, Stillman, R. C. and Willette, R. E., Eds., Pergamon, Elmsford, N.Y., 1978, chap. 6.
93. **Leung, A. Y. and Paul, A. G.,** Baeocystin and norbaeocystin: new analogs of psilocybin from Psilocybe baeocystis, *J. Pharm. Sci.*, 57, 1667, 1968.
94. **Horita, A. and Weber, L. J.,** The enzymatic dephosphorylation and oxidation of psilocybin and psilocin by mammalian tissue homogenates, *Biochem. Pharmacol.*, 7, 47, 1961.
95. **Horita, A.,** Some biochemical studies on psilocybin and psilocin, *J. Neuropsychiatr.*, 4, 270, 1963.
96. **Balestrieri, A.,** Studies on cross tolerance with LSD-25, UML-491 and JB-336, *Psychopharmacology*, 1, 257, 1960.
97. **Kalberer, F., Kreis, W., and Rutschmann, J.,** The fate of psilocin in the rat, *Biochem. Pharmacol.*, 11, 261, 1962.
98. **Sanders, E. and Bush, M. T.,** Distribution, metabolism and excretion of bufotenine in the rat with preliminary studies of its O-methyl derivative, *J. Pharmacol. Exp. Ther.*, 158, 340, 1967.
99. **Sanders-Bush, E., Oates, J. A., and Bush, M. T.,** Metabolism of bufotenine-2′-^{14}C in human volunteers, *Life Sci.*, 19, 1407, 1976.
100. **Petcher, T. J. and Weber, H. P.,** Crystal structure of the teonanacatl hallucinogens. II. Psilocin $C_{12}H_{15}N_2O$, *J. Chem. Soc. Perkin Trans. II*, p. 946, 1974.
101. **Weber, H. P. and Petcher, T. J.,** Crystal structures of the teonanacatl hallucinogens. I. Psilocybin, $C_{12}H_{17}N_2O_4P$, *J. Chem. Soc. Perkin Trans. II*, p. 942, 1974.
102. **Kuhnert-Brandstatter, M. and Heindl, W.,** Polymorphe modifikationene und solvate von psilocin und psilocybin, *Arch. Pharm.*, 309, 625, 1976.
103. **Johnstone, M., Evans, V., and Baigel, S.,** Sernyl (CI-395) in clinical anaesthesia, *Br. J. Anaesth.*, 31, 433, 1959.
104. **Collins, V. J., Gorospe, C. A., and Rovenstine, E. A.,** Intravenous nonbarbiturate, nonnarcotic analgesics: preliminary studies. I. Cyclohexylamines, *Anesth. Analg. Curr. Res.*, 39, 302, 1960.
105. **Lerner, S. E. and Burns, R. S.,** Phencyclidine use among youth: history, epidemiology, and acute and chronic intoxication, in National Institute on Drug Abuse, Research Monograph Series, 21, 66, 1978.
106. **Shannon, H. E.,** Evaluation of phencyclidine analogs on the basis of their discriminative stimulus properties in the rat, *J. Pharmacol. Exp. Ther.*, 216, 543, 1981.
107. **Fauman, M. and Fauman, B.,** The differential diagnosis of organic based psychiatric disturbance in the emergency department, *J. Am. Coll. Emergency Phys.*, 6(7), 315, 1977.
108. National Institute on Drug Abuse Research Monograph Series 21, Peterson, R. C. and Stillman, R. C., Eds., U.S. Government Printing Office, Washington, D.C., 1978.
109. **Aronow, R. and Done, A.,** Phencyclidine overdose: an emergency concept of management, *J. Am. Coll. Emergency Phys.*, 7 (2), 56, 1978.
110. **Kalir, A., Sadeh, S., Karoly, H., Shirin, E., Balderma, D., Edery, H., and Porath, G.,** 1-Phenylcycloalkylamine derivatives. III, *Isr. J. Chem.*, 13, 125, 1975.
111. **Kalir, A., Maayani, S., Rehavi, M., Elkavets, R., Pri-Bar, I., Buchman, O., and Sokolovsky, M.,** Structure-activity relationship of some phencyclidine derivatives: in vivo studies in mice, *Eur. J. Med. Chem.*, 13, 17, 1978.

112. **Vincent, J. P., Cavey, D., Kamenka, J. M., Geneste, P., and Lazdunski, M.,** Interaction of phencyclidines with the muscarinic and opiate receptors in the central nervous system, *Brain Res.*, 152, 176, 1978.
113. **Vincent, J. P., Kartalovski, B., Geneste, P., Kamenka, J. M., and Lazdunski, M.,** Interaction of phencyclidine (''angel dust'') with a specific receptor in rat brain membranes, *Proc. Natl. Acad. Sci. U.S.A.*, 76, 4678, 1979.
114. **Geneste, P., Kamenka, J. M., Ung, M. S. N., Herrmann, P., Goudal, R., and Trouiller, G.,** Determination conformationelle de derives de la phencyclidine en vue d'une correlation structure-activite, *Eur. J. Med. Chem.*, 14, 301, 1979.
115. **Geneste, P. and Kamenka, J. M.,** Etude par resonance magnetique nuclaire du 13C des chlorhydrates de-N-(phenyl-1-cyclohexyl)piperidines, *Org. Mag. Resonance,* 2, 579, 1975.
116. **Siegel, R. K.,** Phencyclidine and ketamine intoxication: a study of four populations of recreational users, in *NIDA Research Monograph Series,* 21, 119, 1978.
117. **Ober, R. E., Gwynn, G. W., Chang, T., McCarthy, D. A., and Glazko, A. J.,** Metabolism of 1-(1-phenylcyclohexyl)piperidine (Sernyl), *Fed. Proc.*, 22, 2265, 1963.
118. **Lin, D. C., Fentiman, A. F., Foltz, R. L., Forney, R. D., and Sunshine, I.,** Quantification of phencyclidine in body fluids by gas chromatography chemical ionization mass spectrometry and identification of two metabolites, *Biomed. Mass Spectrom.*, 2, 206, 1975.
119. **Wong, L. K. and Biemann, K.,** Metabolites of phencyclidine, *Clin. Toxicol.*, 9, 583, 1976.
120. **Corssen, G. and Domino, E. F.,** Dissociative anesthesia: further pharmacologic studies and first clinical experience with the phencyclidine derivative CI-581, *Anesth. Analg. (Cleveland),* 45, 29, 1966.
121. **James, S. H. and Schnoll, S. H.,** Phencyclidine: tissue distribution in the rat, *Clin. Toxicol.*, 2, 573, 1976.
122. **Reynolds, P. C.,** Clinical and forensic experiences with phencyclidine, *Clin. Toxicol.*, 9, 547, 1976.
123. **MacLeod, W. D., Jr., Green, D. E., and Seet, E.,** Automated analysis of phencyclidine in urine by probability based matching gc/ms, *Clin. Toxicol.*, 9, 561, 1976.
124. **Golden N. L., Sokal, R. J., and Rubin, I. L.,** Angel dust: possible effects on the fetus, *Pediatrics,* 65, 18, 1980.
125. **Done, A. K., Aronow, R., Miceli, J. N., and Lin, D. C.,** Pharmacokinetic observations in the treatment of phencyclidine poisoning. A preliminary report, in *Management of the Poisoned Patient,* Rumack, B. H. and Temple, A. R., Eds., Science Press, Princeton, 1977, 79.
126. **Barry, H. and Kubena, R. K.,** Discriminative stimulus characteristics of alcohol, marihuana, and atropine, in *Drug Addiction,* Vol 1, Singh, J., Miller, L., and Lal, H., Eds., Futura Publishing, Mt. Kisco, N.Y., 1972.
127. **Kettenes-Van den Bosch, J. J., Salemik, C. A., Van Noordwijk, J., and Khan, I.,** Biological activity of the tetrahydrocannabinols, *J. Ethnopharmacol.*, 2, 197, 1980.
128. **Lemberger, L., Crabtree, R. E., and Rowe, H. M.,** 11-Hydroxy-delta-9-tetrahydrocannabinol: pharmacology, disposition and metabolism of a major metabolite of marijuana in man, *Science,* 177, 62, 1972.
129. **Lemberger, L., Martz, R., Rodda, B., Forney, R., and Row, H.,** Comparative pharmacology of Δ^9-tetrahydrocannabinol and its metabolite, 11-hydroxy-Δ^9-tetrahydrocannabinol, *J. Clin. Invest.*, 52, 2411, 1973.
130. **Jones, R. T.,** in National Institute on Drug Abuse, Research Monograph Series 31, Marijuana Research Findings: 1980, U.S. Government Printing Office, Washington, D.C., 1980, 54.
131. **Fried, P. A.,** Behavioral and electroencephalographic correlates of the chronic use of marijuana. A review, *Behav. Biol.*, 21, 163, 1977.
132. **Hunt, C. A. and Jones, R. T.,** Tolerance and disposition of tetrahydrocannabinol in man, *J. Pharmacol. Exp. Ther.*, 215, 35, 1980.
133. **Harris, L. S.,** Cannabis: a review of progress, in *Psychopharmacology: A Generation of Progress,* Lipton, M. A., Dimascio, A., and Killam, K. F., Eds., Raven Press, New York, 1978, 1565.
134. **Crawford, W. J. and Merritt, J. C.,** Effects of tetrahydrocannabinol on arterial and intraocular hypertension, *Int. J. Clin. Pharmacol. Biopharm.*, 17, 191, 1979.
135. **Poster, D. S., Penta, J. S., Bruno, S., and MacDonald, J. S.,** Δ^9-tetrahydrocannabinol in clinical oncology, *JAMA,* 245, 2047, 1981.
136. **Tashkin, D. P., Shapiro, B. J., Lee, Y. E., and Harper, C. E.,** Subacute effects of heavy marijuana smoking on pulmonary function in healthy men, *N. Engl. J. Med.*, 294, 125, 1976.
137. **Tashkin, D. P., Shapiro, B. J., Ramanna, L., Taplin, G. V., Lee, Y. E., and Harper, C. E.,** Chronic effects of heavy marijuana smoking on pulmonary function in healthy young males, in *Pharmacology of Marihuana,* Vol. 1, Braude, M. C. and Szara, S., Eds., Raven Press, New York, 1976, 291.
138. **Kolodny, R. C., Masters, W. H., Kolodner, R. M., and Toro, G.,** Depression of plasma testosterone levels after chronic intensive marihuana use, *N. Engl. J. Med.*, 290, 872, 1974.
139. **Menddelson, J. H., Keuhnle, J., Ellingboe, J., and Babor, T. F.,** Plasma testosterone levels before, during and after chronic marihuana smoking, *N. Engl. J. Med.*, 291, 1051, 1974.

140. **Dalterio, S., Bartke, A., and Mayfield, D.,** Δ^9-Tetraydrocannabinol increases plasma testosterone concentrations in mice, *Science,* 213, 581, 1981.
141. **Schaeffer, J., Andrysiak, T., and Ungerleider, J. T.,** Cognition and long term use of ganja (cannabis), *Science,* 213, 465, 1981.
142. **Cohen, S.,** Adverse effects of marijuana: selected issues, *Ann. N.Y. Acad. Sci.,* 362, 19, 1981.
143. **Mechoulam, R. and Gaoni, Y.,** The absolute configuration of Δ^1-tetrahydrocannabinol, the major active constituent of hashish, *Tetrahed. Lett.,* 1109, 1967.
144. **Petrzilka, T. and Sikemeier, C.,** Uber inhaltsstoffe des haschisch. Umwandlung von (−)-$\Delta^{6,1}$-3,4-trans-tetrahydrocannabinol in (−)-$\Delta^{1,2}$-3,4-trans-tetrahydrocannabinol, *Helv. Chim. Acta,* 50, 2111, 1967.

LOCAL ANESTHETIC AGENTS

H. Jack Adams, Robert A. Ronfeld, and Bertil H. Takman

INTRODUCTION

The introduction of long-acting agents such as etidocaine and bupivacaine, since they permit prolonged surgical procedures, has stimulated the increased use of regional anesthetic procedures. Consequently, the study of local anesthesia and the search for new local anesthetic agents have received fresh impetus from the continuing increase in the numbers of regional anesthetic procedures being performed. Where suitable for the surgical procedure involved, regional anesthesia may cause less stress upon the patient than does general anesthesia and may be used when general anesthesia is contraindicated in a patient.

It is important to keep in mind that, as in other areas of medicine, there is a variety of agents available and that, in general, the agents fall into groups that cover specific areas of use. Although some agents such as lidocaine may be used for a variety of applications, there are no universal agents. If, for example, an agent is inherently short-acting or of intermediate duration, its duration of action can be increased by increasing the dosage only up to a point. Beyond that point there is little further gain in duration but an increased risk of local and systemic toxicity. For blocks of long duration the anesthesiologist has the choice of giving repeated doses or a continuous infusion of shorter-acting agents or of using agents that are inherently long-acting. A compound that has the physical-chemical properties desirable in a topical anesthetic agent may be completely unsuitable for infiltration and for central and peripheral nerve blocks, and an agent that is potent and long-acting when administered for spinal anesthesia may be too toxic for epidural anesthesia or intercostal nerve blocks in which considerably larger total dosages are required. It is not surprising, then, that it may take years for a newly-introduced local anesthetic agent to find its proper place in the anesthesiologist's armamentarium.

Definitions

The phrase "local anesthesia" is used to describe anesthesia, i.e., loss of sensation, in a more or less discrete area of the body. Unlike general anesthesia, it is not intended to produce unconsciousness. The loss of sensation may be localized to an area as small as a skin wheal or may include the entire lower half of the trunk and the lower extremities. Although local anesthesia produces local analgesia, i.e., block of pain fibers, it also blocks autonomic and motor fibers, and therefore the phrase local anesthesia is more appropriate for describing the condition produced by the application or injection of a local anesthetic agent.

"Regional anesthesia", rather than local anesthesia, is the term commonly employed by anesthesiologists to describe the production of anesthesia in discrete areas or regions of the body. The techniques used to produce regional anesthesia include local infiltration, field block, block of various nerve trunks and plexi, lumbar and caudal epidural anesthesia, and spinal (subarachnoid) anesthesia. The interested reader is referred to Covino and Vassallo for descriptions of these techniques.[1]

"Topical" or "surface anesthesia" describes the production of anesthesia by application of a local anesthetic agent, in a suitable formulation, to the skin, cornea, or mucous membranes.

All local anesthetic agents are nerve or conduction blocking agents, but all agents capable of blocking conduction in nerve fibers are not local anesthetic agents. The phrase "local anesthetic agent" is, by common consent, usually reserved for agents that have, or are investigated with the aim of ascertaining whether or not they have, the criteria necessary for being clinically useful. Agents such as tetrodotoxin, batrachotoxin, phenol, and ethanol, although they will at appropriate concentrations produce reversible nerve block, are not normally regarded as local anesthetic agents.

Historical Information

The history of local anesthesia in modern medicine begins with the isolation of cocaine from the leaves of *Erythroxylon coca* by Niemann in 1860 and subsequent studies by von Anrep and Koller,[2,3] among others, who demonstrated the clinical usefulness of this substance as a local anesthetic agent. A need had been filled, but during the ensuing years the toxicity and the strong addicting liabilities of cocaine became evident. This spurred the search for other compounds that would fill the need for a safe and effective local anesthetic agent. Procaine, synthesized by Einhorn in 1905,[4] was the first manmade local anesthetic agent to become widely used. Its toxicity is low, but it is not very potent, is unstable to autoclaving, and its duration of action, even when used with a vasoconstrictor agent, is too short for many surgical procedures. Consequently, the search for more potent agents with a longer duration of action and an acceptable toxicity continued. Because, however, procaine was the first local anesthetic agent to become widely used, its synonym, novocain, is to this day incorrectly used by both laymen and professionals as a generic term for any local anesthetic agent.

The continuing search for a better local anesthetic agent yielded tetracaine in 1931.[5] This is a potent agent with a long duration of action. It is too toxic for many types of block, and is used almost exclusively for topical and spinal anesthesia. In 1943 Löfgren synthesized lidocaine, and this agent, because it possesses an excellent combination of adequate potency, physical stability, moderate toxicity, and a duration of action sufficient for many surgical procedures, became in a short time the most widely used local anesthetic agent available.[6]

As emphasized in the Introduction, there are no universal agents. Although some, like lidocaine, are extremely versatile, they cannot adequately fill all the needs in the field of regional anesthesia. Thus, the search for new local anesthetic agents has continued almost unabated to the present day. Of the numerous compounds synthesized and tested each year, very few ever reach clinical trials and, of those that do, only a very small percentage are found to be sufficiently superior in one or more respects to currently available agents to become established. An example of such an agent is bupivacaine, which was synthesized in 1957 by af Ekenstam and co-workers.[7] Bupivacaine is a congener of mepivacaine, which is an agent similar to lidocaine in potency, toxicity, and duration of action. Bupivacaine, however, is more potent and more toxic than either of those compounds and also has a much longer duration of action. Such an agent is extremely useful for operative procedures of long durations and for reducing the need for postoperative analgesic agents. Etidocaine, first reported by Adams et al. in 1973, is also a potent agent with a long duration of action.[8] It has been reported to be especially useful in surgical procedures requiring good muscle relaxation.

The search for agents with an ultralong duration of action is prompted by the need for pain control in chronic pain. In spite of the unfortunate consequences that resulted from mixtures such as Efocaine, the possibility of obtaining a long duration of analgesia by means of some sort of depot or ''sustained-release'' preparation cannot be dismissed off hand.

Among the other approaches that have been pursued are cyclizing compounds, mixtures of tertiary and quaternary compounds, mixtures of conventional local anesthetic agents and nerve-blocking toxins such as tetrodotoxin and saxitoxin, and tetraethylammonium derivatives. Although at the time this chapter is being written none of these approaches has yielded a clinically useful preparation, the search continues along these and other lines. Even the failures yield information that may be useful in attaining the goal.

Screening

As is true for most classes of pharmacological agents, there are many test methods for screening local anesthetic agents and for evaluating efficacy and relative toxicity. In vitro studies using isolated amphibian or mammalian nerves will establish whether or not an agent will block conduction and, if so, its relative potency compared with other agents screened in the same test system. A more pragmatic approach, however, is to do initial screening in the whole animal. In this test system the agent has to contend with the barriers it must penetrate, the tissue circulation that is tending to carry it away before it can reach the nerve trunks, and, in some cases, enzyme systems that are bringing about its biotransformation. For these reasons, screening in the mouse, rat, or guinea pig is more comparable to the clinical situation than the in vitro test system is. A further advantage of using the whole animal is that it is possible to make preliminary assessments of systemic toxicity during screening procedures.

A typical primary screen should include assessments of local anesthetic activity (latency, frequency of block, duration of block) at various concentrations (dosages) and of acute toxicity. Early in the testing program there should be an assessment of tissue toxicity, especially of neurotoxicity, of compounds that warrant intensive study as potentially useful agents in man. It is also advantageous to evaluate test compounds in larger animals as soon as feasible, since comparisons of, for example, durations of action of compounds in smaller animals such as the mouse, rat, and guinea pig may be misleading.[9]

A series of screens may be designed to select the best compounds either for a specific use such a topical or spinal anesthesia or for a broad range of applications such as infiltration and peripheral and central neural blocks.

Screens for selecting a topical anesthetic agent may be designed to select the best agent for mucous membranes or intact skin or burned or abraded skin or for a broad range of uses.

It is essential that the chemist and the pharmacologist decide *a priori* what ''best'' means to them, and this decision should take into account such variables as latency, duration of action, sensory and motor components of the block, acute systemic toxicity, tissue toxicity, stability in solution, and compatibility with vasoconstrictor agents. Furthermore, initial batches of test compounds are often small, and for this reason the investigators should design the screening program to yield the greatest amount of information from the limited amount of substance available. See Tables 1 and 2 for descriptions of selected screening procedures.

References to Review Articles

For the convenience of the reader, these have been categorized according to the area or discipline emphasized in the review.

General Interest

Takman, Boyes and Vassallo: Local anesthetics.[10]

Chemistry

Takman and Adams: Local anesthetics.[11]

Pharmacokinetics and Metabolism

Mather and Tucker: Pharmacokinetics and biotransformation of local anesthetics.[12]

Tucker and Mather: Clinical pharmacokinetics of local anesthetics.[13]

Clinical Applications

Cousins and Mather: Clinical pharmacology of local anesthetics.[14]

Mather and Cousins: Local anesthetics and their current clinical use.[15]

Table 1
SCREENING FOR LOCAL ANESTHETIC AGENTS

Screening procedure	Principal area(s) of usefulness	Comments	Ref.
Peripheral nerve trunk			
In vitro: isolated amphibian and mammalian nerve trunks	Studying the effect of pH on nerve blocking activity; evaluating the use or frequency dependence of test compounds; studying blocking activity on sheathed and desheathed nerve trunks and on myelinated and nonmyelinated nerves; comparing conduction blocking agents under controlled conditions	Although in vitro systems have their place in the testing of local anesthetic agents and can yield information of value to the chemist and pharmacologist, they should not be used as the primary screen in testing programs designed to discover agents with potential for use in regional anesthesia in veterinary or human medicine	71, 71
In vivo: Mouse, rat and guinea pig	Comparing reference and test compounds on nerve trunks *in situ;* determining latency, frequency, and duration of block at various concentrations; evaluating the effects of vasoconstrictor agents on latency, frequency, and duration of block	These test methods require a minimum of equipment, are simple, and the data are reproducible. When used in conjunction with acute toxicity studies, for example, they provide sufficient information for rejection or acceptance of test compounds in a primary screen	
Mouse tail			73
Rat and guinea pig sciatic nerve block			72, 74
Rat infraorbital nerve block			75
Infiltration			
Guinea pig intradermal wheals	Comparing reference and test compounds *in situ;* determining latency, frequency, and duration of block at various concentrations Evaluating the effects of vasoconstrictor agents on latency, frequency, and duration of block	A simple and rapid test procedure; except with compounds possessing poor local anesthetic activity or with low concentrations of active agents the onset of block is almost instantaneous and the frequency of block is 100%; dose-duration curves with little or no slope are not uncommon, and the addition of vasoconstrictors to agents of short or intermediate duration may result in blocks of unexpectedly long duration; see text for further comments on this phenomenon	72
Topical			
Cornea	Primary screen for topical anesthetic activity; screening agents for use in ophthalmology	This test procedure is widely used as a primary screen for topical anesthetic agents, but the results obtained cannot be extrapolated to dermal application, which presents unique problems, nor to the mucosa of the respiratory and digestive tracts	
Guinea pig			76
Rabbit			72
Skin Guinea pig	Screen for agents that produce anesthesia in: burned skin; abraded skin; uninjured skin	The intact skin presents a formidable barrier to the penetration of many substances, including local anesthetic agents; injury such as burning or abrasion disrupts the barrier; screening of local anesthetics intended for dermal application should include testing on various preparations in order to select the agents best suited for general use or for a specific use such as burns	77

Table 1 (continued)
SCREENING FOR LOCAL ANESTHETIC AGENTS

Screening procedure	Principal area(s) of usefulness	Comments	Ref.
Mucous membranes of the respiratory tract	Screens for agents to be used on the upper respiratory tract, e.g., topical anesthesia for intubation	Unless the aim is to select an agent specifically for use in this area, a screening program for topical anesthetic agents should include these test procedures, or similar ones, in a battery of screens chosen to characterize the topical anesthetic properties of the test compounds	
Guinea pig			78
Rabbit			79

Table 2
PHARMACOKINETICS

Drug	T 1/2 (min)	Clearance (ℓ/min)	Volume of distribution (ℓ)	Plasma protein binding: % bound	Plasma protein binding: Plasma conc (μg/mℓ)	Ref.
Lidocaine	95	0.95	132	—	—	13
	95	0.88	122	—	—	56
	—	—	—	51	0.4	57
				75	5.0	
	—	—	—	61—80	1.8	31
Prilocaine	95	2.8	326	—	—	47
	—	—	—	55	—	58
Mepivacaine	114	0.78	128	—	—	13
	192	0.41	114	—	—	38
	126	0.76	129	—	—	47
	—	—	—	85	0.4	57
				66	5.0	
Bupivacaine	162	0.58	136	—	—	13
	210	0.47	142	—	—	29
	—	—	—	96	0.4	57
				85	5.0	
Etidocaine	162	1.11	259	—	—	13
	—	—	—	95	1.0	57
Procaine	0.75 (in vitro)	—	—	—	—	43
	1.4[a] (in vitro)					
	0.67 (in vitro)	—	—	—	—	44
	2.3[b] (in vitro)					
	7.7 (in vivo)	5.6	56	—	—	24
Chloroprocaine	0.38 (in vitro)	—	—	—	—	25
	5.0 (in vivo)	—	—	—	—	26
Cocaine	16-87 (in vivo)	—	—	—	—	27
	75 (intranasal)	—	—	—	—	28
	47 (oral)	—	—	—	—	

[a] Chronic obstructive pulmonary disease or heart failure.
[b] Liver disease.
[c] Subjects with atypical plasmacholinesterase.

CLASSIFICATION OF BLOCKING AGENTS

Nerve blocking agents, like other kinds of pharmacological agents, may be classified in many ways, e.g., by chemical structure, by the area of clinical application, or by physical properties. Dividing them into broad categories such as "ester vs. amide" or "topical vs. injectable" may serve a purpose, albeit a limited one. Takman has suggested the following classification based on the site and mechanism of action.[16]

1. Agents that penetrate the neuronal membrane with difficulty and either:
   - A. Act at the exterior opening of the sodium channel or
   - B. Act at the interior (axoplasmic) end of the sodium channel
2. Agents that penetrate the neuronal membrane relatively easily and either:
   - C. Act by means of a physicochemical mechanism or
   - BC. Act both at a receptor at the axoplasmic end of the sodium channel and by means of a physicochemical mechanism

Group A — The principal representatives of this group are the two marine toxins, tetrodotoxin and saxitoxin. Although they differ in structure from each other, they both act by blocking the sodium channels in the membranes of excitable cells.[17]

Group B — Principal representatives of this group are certain quaternary ammonium compounds that have been shown to block sodium channels readily when applied to the axoplasmic side of the membrane but with difficulty or not at all when applied outside the membrane.[18,19]

Group C — The common structural feature of compounds in this group is a lack of a positive charge. Some are uncharged molecules, some are weak acids, and some contain amine groups that are too weak to ionize to any significant degree in the physiological pH range. Examples of such agents are benzyl alcohol, phenol, and benzocaine.

Group BC — Most of the compounds in this group are amines, and those that have been or are in clinical use are either secondary or tertiary amines with pKa values ranging from about 7.7 to about 9.5. In the physiological pH range these agents exist as a mixture of the uncharged base and the charged ionic form or cation. The ratio of the percent present as the base to the percent present as the cation depends upon the pKa of the agent and the pH of the milieu. Both forms are capable of blocking sodium channels, but their mechanisms of action are different from each other.

See Section 3 for a discussion of the mechanism of action of the agents in these several groups.

MECHANISM OF ACTION

The mechanism of action of all the conventional local anesthetic agents discussed here is identical in that they block the initiation and conduction of nerve impulses by inhibiting ionic fluxes essential to these processes. Current evidence indicates that the inhibition or blocking of these fluxes may be achieved in several ways and that, in fact, most or all of the agents in common use have more than one site or mechanism of action.

The essential feature of membrane depolarization is a transient influx of sodium ions through the sodium channels. This can be blocked by plugging either the exterior or interior (axoplasmic) end of the channel and also by causing conformational changes in the channel. The end result is the same: conduction of the propagated impulse fails if a sufficient number of sodium channels are blocked per unit area of axonal membrane. There is strong evidence that the ionizable local anesthetic agents act in the cationic form by occupying a receptor site in the axoplasmic end of the sodium channel and that the cation has access to this receptor site only when the h-gate is open. In order to reach the receptor the local anesthetic molecule must penetrate the axonal membrane, and because of the lipid nature of this membrane, it penetrates more readily as the unionized base. There is also evidence that the lipophilic base form may contribute to the conduction block by effecting conformational changes in the sodium channel. Agents such as ethyl aminobenzoate (benzocaine), which are completely unionized at physiological pH values, are effective nerve blocking agents and probably act by this mechanism.

It should be clear from the foregoing discussion that an essential feature of a local anesthetic agent to be used for infiltration and for central and peripheral nerve blocks is a pKa that assures a fraction of the local anesthetic molecules will exist as the unionized base at physiological pH values, since the cationic (hydrophilic) form penetrates the axonal membrane with difficulty. In the clinical situation, an agent present exclusively as the cation might either not produce the desired nerve block or block only after a long latency period.

It has been known for some time that the degree of block obtained when an isolated nerve trunk is bathed in a solution containing a local anesthetic agent may depend not only upon the concentration of that agent in the solution but also upon the rate at which the nerve trunk is being stimulated. With many agents a partial block can be deepened by increasing the fequency of stimulation of the nerve trunk. This phenomenon, referred to in the literature as "frequency dependence" or "use dependence" is explained by the observation that the cationic form of the local anesthetic agent can gain access to the receptor only when the gate at the inner, i.e., axoplasmic, end of the sodium channel is open. The faster the rate of stimulation, the greater the statistical probability of a gate's being open when the charged form of the local anesthetic molecule is in a position to enter the channel. Local anesthetic agents, as a class, exhibit different degrees of frequency dependence and different rates of onset of frequency-dependent block. Even molecules that are uncharged at physiological pH values exhibit this phenomenon, and studies have shown that lipid solubility and size of the molecule may determine the frequency dependence exhibited by agents that block nerve conduction.[20]

A schematic diagram of the sodium channel is shown in Figure 1. This diagram and the preceding text on the mechanism of action of local anesthetic agents are much-simplified presentations of currently available information. Great advances in instrumentation and technology have provided new insights into the probable mechanisms of action of local anesthetic agents, but many questions remain to be answered. For example, we do not know the precise nature of the interaction between the local anesthetic molecule and the nerve membrane, although it is reasonably certain that the interaction involves both membrane proteins and acidic phospholipids. It is generally accepted that there is a true interaction between calcium and local anesthetic agents, but more recent evidence indicates that the observed results may be manifestations of an indirect antagonism.[2] We speak and write about a receptor site for the local anesthetic molecule in the sodium channel, but, again, more recent evidence points to the probable existence of two receptor sites.[3] Thus, attempting to present extremely complex phenomena in simple terms and in keeping with what is generally accepted by the scientific community is always a perilous proceeding.

STRUCTURE-ACTIVITY RELATIONSHIP

An extensive review of the structure-activity relationships (SARs) of local anesthetic agents is beyond the scope of this chapter. The published literature is voluminous, and we will of necessity limit our discussion to some general principles. The interested investigator must observe one *caveat:* for each general principle given there will almost certainly be one or more exceptions, chiefly because the observations made on the SARs within a homologous series of compounds may or may not apply to any other series of compounds. Those who

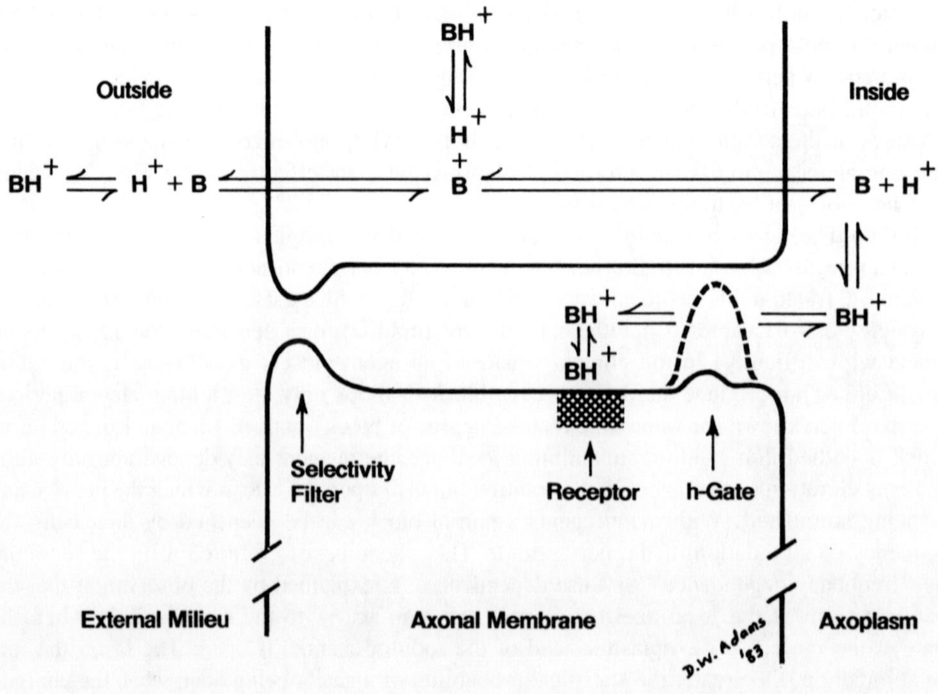

FIGURE 1. Schematic of a sodium channel. B represents the neutral local anesthetic base, and BH^+ represents its cationic form. The h-gate is shown in the open (solid line) and closed (broken line) states. Studies of use- or frequency-dependence indicate that quaternary compounds and cationic forms of tertiary compounds gain access to the receptor from the axoplasmic medium more readily when the h-gate is in the open state.

wish a detailed discussion of structure-activity should see, for example, the excellent review by Büchi and Perlia.[21]

To further narrow the discussion, we will exclude such widely disparate agents as, for example, benzyl alcohol and tetrodotoxin. The former is a relatively simple molecule for which no discrete receptor site has been proposed, and the latter is a complex molecule with a well-defined receptor but which does not fit within our definition of a local anesthetic agent. The discussion will include only those compounds that fit the general scheme:

lipophilic portion — intermediate chain — hydrophilic portion

since, of the compounds used in the practice of local anesthesia, these are the most numerous and have the widest range of uses. The lipophilic portion normally consists of at least one phenyl radical, the intermediate chain shows considerable variation but commonly has an ester or amide linkage, and the hydrophilic portion is most often a secondary or tertiary amine. The aim of the synthetic chemist is to obtain, by modifying the various portions of the molecule, a compound with the desired potency, duration of action, pKa, stability, and local and systemic toxicity. The chemist must strike a proper balance between lipophilicity and hydrophilicity in this hypothetical "ideal local anesthetic agent", since, as empasized by Wiedling and Tegnér, this is a *sine qua non* for a successful agent.[22] If the compound is to be used as an injectable agent, salts of it (normally the HCl) should have sufficient aqueous solubility to permit solutions of the appropriate concentrations to be made, and it may be desirable to have a compound that is compatible with a vasconstrictor agent such as epinephrine. It should be obvious from this that the burden on the synthetic chemist is a heavy one.

Modification of any part of a local anesthetic molecule may cause changes in lipophilicity, hydrophilicity, and charge distribution on the molecule. The following three examples have been cited by Covino and Vassallo and by Åström.[1,23]

1) 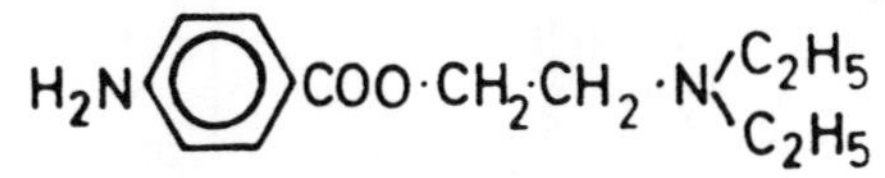 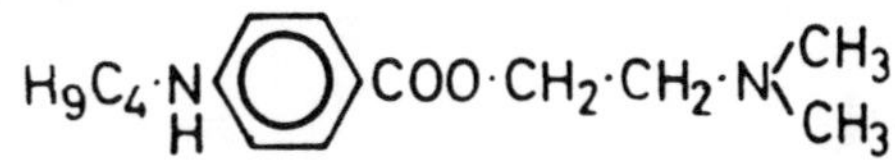

procaine tetracaine

2) 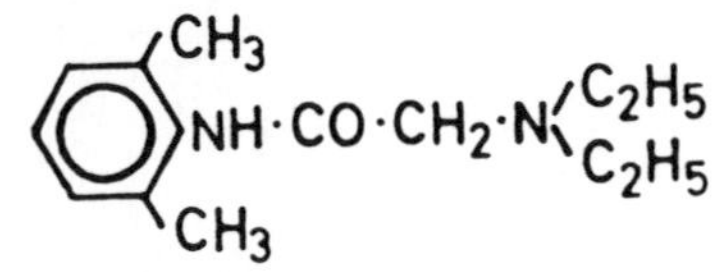 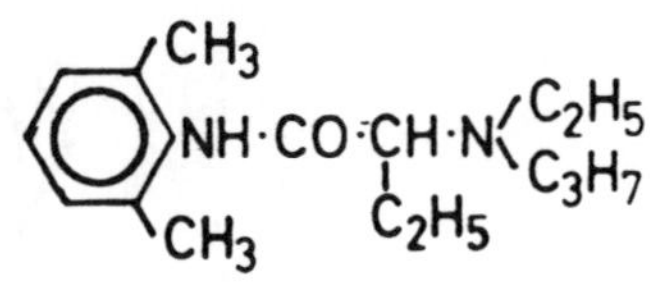

lidocaine etidocaine

3) 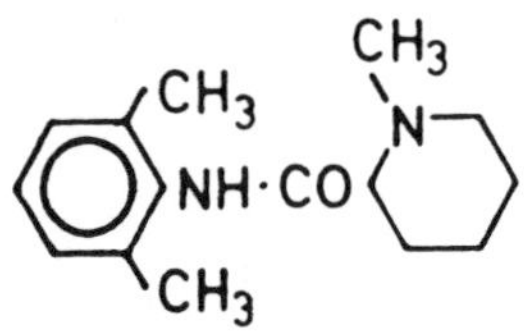 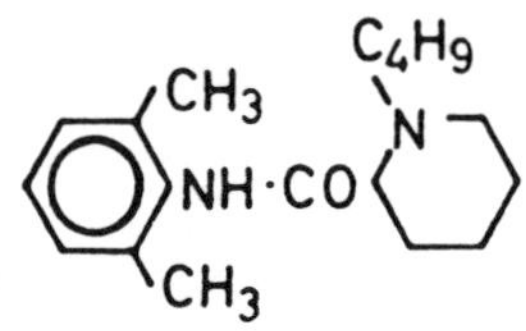

mepivacaine bupivacaine

In the first example both the lipophilic and hydrophilic portions have been modified. In the second example the intermediate chain and the hydrophilic portion have been modified. In both cases a marked increase in potency, toxicity, and duration of action have been effected. In the third example only the hydrophilic portion has been modified, but this is sufficient in itself to effect a highly significant increase in potency, toxicity, and duration of action.

Modification of the Lipophilic Portion

The phenyl radical, with or without substituents, is present in many local anesthetic agents, but it can be replaced by other radicals capable of contributing the required lipophilicity to the molecule. Substitutions on the phenyl radical play an extremely important role in conferring good local anesthetic activity on a molecule. Especially favorable are substitutions in the *ortho* and *para* positions. In addition to contributing to the lipophilicity of the molecule, substitutions in the *ortho* position lend stability, possibly because of steric hindrance, to linkages in the intermediate chain that are susceptible to enzymatic cleavage. For example,

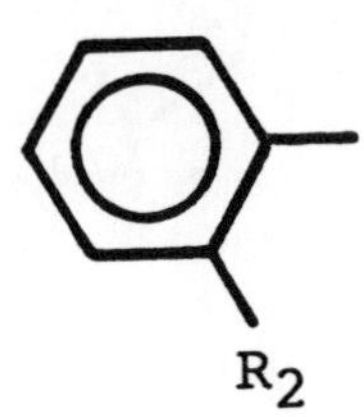

is more stable than

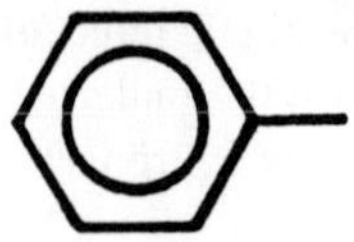

and

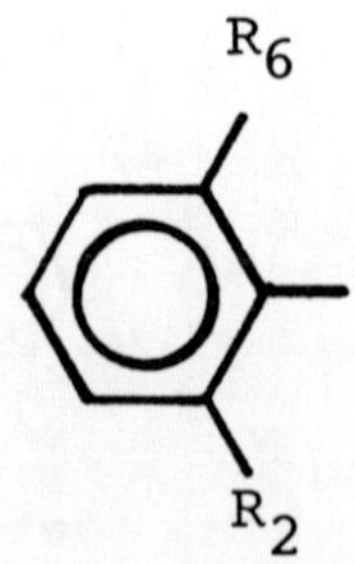

is much more stable than

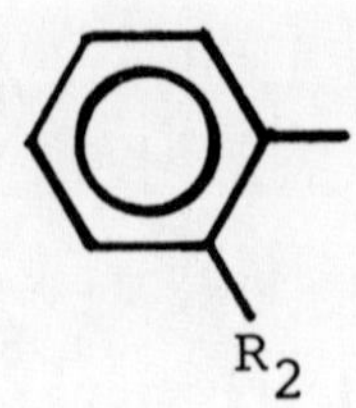

Modification of the Intermediate Chain

Like the other portions of the local anesthetic molecule, this is widely variable in structure. Marked effects on local anesthetic activity may be brought about by increasing or decreasing chain length and by introducing branching. The amide linkage is inherently more stable than the ester linkage.

Modification of the Hydrophilic Portion

Profound changes in local anesthetic activity can be effected by increasing and decreasing the size of the alkyl substituents on the nitrogen and by branching of these substituents. The kinds of alkyl groups may be more important than just the total number of carbons on the nitrogen in determining the properties of the compound. For example, in the lidocaine series

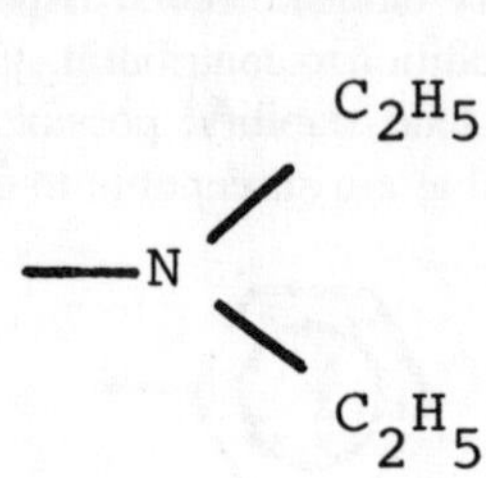

is more desirable than

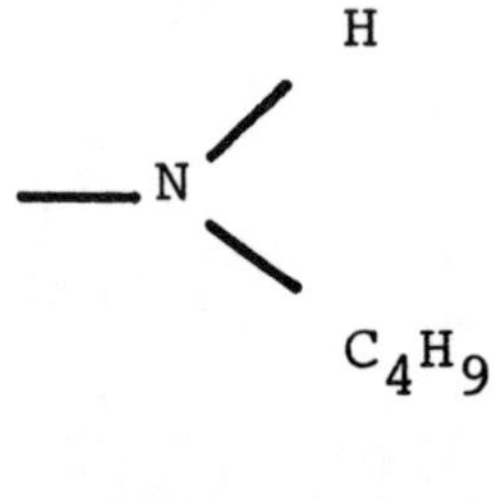

or

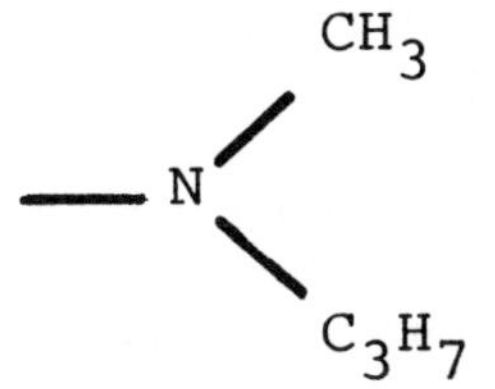

In summary, an examination of the great variety of structures capable of fitting into the BC category of nerve blocking agents (see section on Classification) makes it apparent that it is not so much specific structural requirements (e.g., interatomic distances, spatial relationships of various pharmacophore groups), such as are necessary for ganglionic or neuromuscular blocking agents, as a proper balance of physical-chemical properties that make a successful local anesthetic agent. There are, obviously, several basic requirements, at the molecular level, for agents of this type. They must have sufficient lipid solubility in the unionized state to penetrate the axonal membrane with ease. Agents that penetrate the membrane with difficulty may have an undesirably long latency period and a poor frequency of successful blocks. They must also be capable of existing as the cation at the pH of the axoplasm in order to attach, albeit reversibly, to the cationic receptor site in the sodium channel. And, finally, they must have sufficient hydrophilicity to remain in solution in the aqueous phases of the body tissues. A compound whose base form has an extremely low aqueous solubility may be limited to use as a topical agent.

METABOLISM AND PHARMACOKINETICS

Local anesthetics are somewhat unique, with respect to drug disposition, simply because of their intended use and mode of administration. For example, bioavailability cannot be measured in the typical manner. This is not only because the drugs are not usually administered orally but also because blood levels will probably not be a reflection of the availability of the drug at the site of action. Because of the unique mode of administration of local anesthetic agents, kinetics and metabolism do not have the typical influence on the drug's activity. That is, the undesirable effects are influenced more by the pharmacokinetic properties of the drug than are the desired therapeutic effects. Consequently, the optimal pharmacokinetic properties for local anesthetics run counter to those of many systemically administered drugs. In general, the shorter the half-life and greater the total body clearance, the safer and the more desirable the agent. Since toxicity is often the result of effects on the CNS, the degree of plasma protein binding could, at least theoretically, be important. However, the highly protein-bound agents are in general also the most potent in centrally mediated effects and local anesthetic activity. In contrast to other therapeutic classes, delaying or retarding systemic absorption increases the degree and duration of activity. Consequently, blood concentrations of drug have more relevance to the toxic effects than as a measure of efficacy.

One striking difference between the ester and amide agents is their rates of metabolism. The esters are hydrolyzed in both plasma and liver. Their rates of metabolism by plasma cholinesterase are remarkably rapid.

The in vitro half-lives of procaine and chloroprocaine are less than 1 min, and the half-life for tetracaine is only slightly longer. Consequently, it is almost impossible to measure these agents in the plasma after normal extravascular doses. Following an i.v. dose of 4300 mg of procaine, Seifen determined a terminal half-life of 7.7 min and a plasma clearance of 80 mℓ/kg/min.[24] However, it was not possible to detect chloroprocaine in plasma following a 250-mg i.v. dose administered over 30 min.[25] The implication of these data is that any toxic plasma level, such as that attained by an accidental i.v. injection, will only be momentarily sustained. Patients with atypical plasma cholinesterase might be expected to metabolize these drugs at a slower rate. Zsigmond et al. found the in vitro hydrolysis rate for subjects with atypical plasma cholinesterase to be six times slower for procaine and three times slower for chloroprocaine than with normal plasma cholinesterase.[26] They also administered chloroprocaine at doses of 3 and 5 mg/kg to the atypical subjects and recorded plasma half-lives of 5 min, with no observable toxicity. The rate of hydrolysis of cocaine is much slower. Plasma half-lives have varied from 16 to 80 min after an i.v. dose, 47 min after an oral dose, and 75 min after intranasal administration.[27,28] The maximum plasma concentrations with intranasal administration were dose dependent, and the time of the peak was earlier (35 min) when administered as crystals than as a solution (47 min).[28]

The amide agents are cleared at a much lower rate in the systemic circulation than are the esters. Also in contrast to the esters, they are stable in plasma. However, in all cases their elimination is dependent on metabolism, presumably in the liver, rather than renal elimination. Relative to most systemically administered drugs of any class, the amide local anesthetics have a rapid rate of metabolism and short half-lives.

Table 2 is a tabulation of published pharmacokinetic parameters for a number of local anesthetics. Tucker and Mather have published extensively on the pharmacokinetics of the amide agents, and their work is responsible for much of the data on these agents.[29] As pointed out by Tucker and Mather, total body clearance of these agents increases in the order bupivacaine — mepivacaine — lidocaine — etidocaine — prilocaine and bears no relationship to anesthetic potency, lipid solubility, or protein binding characteristics.[13] Extraction ratios of drug across the liver have been calculated and measured directly for some of these agents and vary from 0.38 for bupivacaine to 0.74 for etidocaine.[13,29]

As stated earlier, plasma protein binding is potentially an important consideration with these agents. Degree of binding, anesthetic potency, and lipophilicity roughly correlate with one another. At subtoxic concentrations the alpha$_1$-acid glycoprotein has been identified as the protein primarily responsible for the binding of these agents.[30,31] However, the concentration of this protein in plasma is only 23 μM. Consequently, the binding of the amide agents to this protein has been described as high affinity and low capacity. The free fraction increases dramatically for all the agents in the 0.5 to 10 μg/mℓ range (see Table 2). However, the free fraction changes very little for total plasma concentrations below the toxicity level. The volume of distribution is largest for the agents with the greatest degree of plasma binding, indicating extensive tissue binding for these agents as well.

Tucker and Mather list the following as the important determinants for the absorption into the peripheral circulation after administration for regional anesthesia: (1) the physicochemical and pharmacological properties of the specific agent; (2) the site of injection; (3) the dose, concentration, and volume of injection solution; (4) the presence or absence of a vasoconstrictor in the injection solution.[13] There are data that indicate that the more lipophilic agents such as bupivacaine and etidocaine are absorbed more slowly and hence may have a greater margin of safety.[13] Tucker et al. observed an increase in the absorption rate of mepivacaine when the dose was kept constant and the concentration of the injection solution was increased.[32] Epinephrine, generally used at dilutions of 1:100,000 and 1:200,000, reduces the rate of absorption and the peak plasma concentration for all regional block procedures.

It is now well established that local anesthetics can easily cross the placenta during maternal regional anesthesia. The degree to which this occurs depends more on maternal plasma concentrations and protein binding than on the physicochemical properties of the specific agent. The fetal and neonatal effects of regional anesthesia in obstetrics has been reviewed extensively by Ralston and Shnider.[33] There are data on the ability of the neonate to metabolize the amide local anesthetics;[34-39] however, this is not one of the important safety factors in the choice of an agent for obstetrical anesthesia. The clearance of lidocaine on a kilogram basis is just as high in the neonate as in the adult.[37] This is not true for the other

Table 3
PHARMACOKINETICS — NEONATE

Drug	T 1/2 (min)	Clearance mℓ/min/kg	Volume of distribution (ℓ/kg)	Plasma protein binding (% bound) Maternal	Neonate	Ref.
Lidocaine	192	10.2	2.75	—	—	37
	—	—	—	56	24	42
Mepivacaine	520	2.0	1.75	—	—	38
Bupivacaine	485	—	—	—	—	59
	1080	—	—	—	—	35
	—	—	—	95	65	42
	—	—	—	85	70	40
	—	—	—	91	51	41
Etidocaine	385	—	—	—	—	39
Procaine	1.4	—	—	—	—	44
Chloroprocaine	0.72	—	—	—	—	25

agents, and in all cases the half-lives are longer in the neonate (Table 3).[34-39] The plasma protein binding of the amides is significantly less in the neonate (see Table 3).[39-42] This is probably due to a lower concentration of the alpha$_1$-acid glycoprotein in the neonate. Consequently, the umbilical-to-maternal ratios of total drug in plasma, at birth, are greater than unity. The umbilical-to-maternal ratio is proportional to the degree of binding in maternal plasma, and, in general, the ratio of nonbound or free drug at birth is equal to one.[41,42]

There are general patterns in the metabolism of these agents. As stated above, the esters are easily hydrolyzed in the plasma.[25,43,44]

Procaine — This compound is hydrolyzed to para-aminobenzoic acid and diethylaminoethanol by plasma pseudocholinesterase. Diethylaminoethanol may undergo further metabolism, and the para-aminobenzoic acid may be excreted unchanged or as a conjugation product.

Tetracaine — This compound is also hydrolyzed, albeit more slowly than procaine. Cleavage of the ester linkage yields parabutylaminobenzoic acid and diethylaminoethanol.

The amides are hydrolyzed in the liver but at a much slower rate. The degree of amide hydrolysis appears to be structurally dependent, occurring most extensively with lidocaine and prilocaine. Other pathways of metabolism for the amides are N-dealkylation and ring hydroxylation.

Prilocaine — The metabolism of this agent has not been studied extensively. The primary metabolite that has been identified is *o*-toluidine.[45] Presumably *N*-hydroxyl-toluidine is responsible for the methemoglobinemia observed with prilocaine. There are animal data that indicate that the D(−) isomer is metabolized faster that the L(+) isomer.[46] In rats there are amidases in the kidney and lung as well as the liver that are capable of hydrolyzing prilocaine. There is indirect evidence to indicate that this also occurs in man, since the total body clearance of prilocaine is greater than liver blood flow.[47]

Lidocaine* — This agent undergoes N-dealkylation, amide hydrolysis, and ring hydroxylation.[48,49] The metabolic pathways have been identified in a number of species.[48] In man the primary drug-related moiety excreted in urine is the conjugate of 4-hydroxy-2,6-xylidine.[48,49] Amide hydrolysis occurs directly with lidocaine as well as with the deethylated metabolite.[49] The neonate appears to have a substantial ability to dealkylate lidocaine but may be deficient, relative to the adult, in amide hydrolysis and ring hydroxylation.[37]

Mepivacaine — Urine data indicates that aromatic ring hydroxylation followed by con-

* A second monograph on Lidocaine may be found in the "Antiarrhythmic Agents" chapter of the *CRC Handbook of Anti-Inflammatory and Cardiovascular Agents.*

jugation accounts for 25% of the dose in man and 60% of the dose in the rat.[51,50] N-dealkylation appears to be a minor pathway. In the neonate, the most outstanding difference is the lack of ring hydroxylation, a deficiency similiar to that observed with lidocaine.[36,38]

Bupivacaine — Only a limited amount of data is available for this agent. In man, intact bupivacaine and aromatic ring hydroxylated and N-dealkylated metabolites account for less than 5% of the dose in urine.[40,52] In the rat, only 50% of a radioactive dose was excreted in the urine, of which conjugated 3′-hydroxy-bupivacaine was the major component.[53] Very possibly the monkey is unique in its ability to hydrolyze the amide bond of this agent, since pipecolic acid accounts for 52% of the dose.[53]

Etidocaine — The metabolism of this agent follows the aforementioned pathways: N-dealkylation, aromatic ring hydroxylation, and amide hydrolysis. Only 31% of a dose was accounted for in urine, with 2-amino-2′-butyroxylidine and 4-hydroxy-xylidine accounting for 18% of the dose.[54] Eight different ring-hydroxylated metabolites of etidocaine, the two possible secondary amines, and the primary amine accounted for only 10% of an oral dose.[55] All in all, it appears that etidocaine is subjected to extensive metabolism, without any single metabolite accounting for a significant portion of the dose.

TOXICITY

Toxicology of pharmacological agents is an extremely important subject and, unfortunately, one that is much misunderstood. One must keep in mind the sage comment of Paracelsus: "All things are poisonous, for there is nothing without poisonous qualities. It is only the dose that makes a thing a poison." The spectrum of toxicities is such that at one end there are substances lethal at minute dosages and at the other end substances that are lethal only if massive quantities are administered. That latter case is responsible for the old joke among toxicologists and pharmacologists about adding the test animal to the drug rather than vice versa.

Lethality, however, is only one measure of toxicity. In general, it is the upper end of the dose-response curve of the pharmacological activity of a substance. If we begin with a low enough dosage, we will produce no measurable effect. As the dosage is increased, measurable effects will be elicited and will become more pronounced until, finally, a dosage level is reached that is lethal to 100% of the animals to which it is administered.

Toxicity may result from a single dose administered statim, from continuous infusion (e.g., an i.v. infusion), or from repeated doses. It may be manifested, for example, as convulsions, loss of consciousness, death, or as morphological and functional changes in tissues and organs, which may, in turn, be reversible or may ultimately be lethal.

The toxicity of a substance may be either localized (local tissue toxicity) or generalized (systemic toxicity). As with many agents that are administered parenterally by injection or applied topically, local tissue toxicity of local anesthetic agents is an important issue. There are certainly a number of local anesthetic agents that have acceptable systemic toxic potential but that cause an unacceptable degree of tissue damage when applied topically or administered around nerve trunks. Such agents may be particularly undesirable as spinal anesthetic agents because of the chances of producing arachnoiditis. For these reasons, tissue irritation studies, which will be discussed under "Testing Procedures", are an important part of local anesthetic screening programs. Local tissue toxicity and systemic toxicity do not necessarily go hand in hand and must, therefore, be evaluated separately.

Local (tissue) toxicity may be evaluated grossly (e.g., the rabbit or guinea pig intradermal wheal test) or microscopically. In the former, an arbitrary system of numbers is often used to express the degree of tissue irritation or damage. In skin irritation studies, manifestations of tissue toxicity may range from mild reddening to frank necrosis. Neurotoxicity may be evaluated by subjecting central or peripheral nerve tissue to graded concentrations of an agent and examining the tissues microscopically for morphological changes. This is an extremely sensitive method, and morphological changes may be observed in nervous tissue of animals that did not exhibit signs of motor or sensory deficit.[60,61]

Systemic toxicity resulting from acute or chronic administration of a substance may be evaluated by endpoints such as ataxia, convulsions, or death. Using groups of animals at each dosage level and graded dosages, the investigator may obtain data that will permit the calculation of the dose that statistically causes convulsions (CD_{50}) or death (LD_{50}) in 50% of the animals to which it is administered. These values are useful in establishing the toxicities of compounds relative to one another. What is too often forgotten, if it is learned at all, is that an LD_{50} is *not* a physical constant such as a melting point or a pK_a: it is a number based on a biological endpoint and, as such, may, even under the most well-controlled conditions, show considerable variability.[62-65] Assuming that an investigator takes adequate precautions to keep constant such variables as species, strain, age and sex of the animals, the number of animals per cage, food and water, site and rate of administration and form (concentration, vehicle, etc.) in which the test agent is administered, the resulting LD_{50} value and confidence limits may be influenced by still other variables such as time of day the substance is administered, the season, ambient temperature, relative humidity, and the estrus cycle of female animals.

The myotoxic effects of local anesthetic agents are well documented. Benoit and Belt demonstrated the rapidity with which morphological changes occur following injection of solutions of local anesthetic agents into rat skeletal muscle.[66] All local anesthetic agents tested to date exhibit myotoxicity in this animal model, and the presence of a vasocontrictor such as epinephrine increases the severity of morphological changes induced. Although most local anesthetic agents produce myonecrosis without detectable effects on other tissues, certain agents, such as piperocaine, cause necrosis of nerves and blood vessels as well.

Yagiela et al. have studied the effects of lidocaine with epinephrine on rat and human skeletal muscle and conclude that the rat is a good model for the study of myonecrosis induced by local anesthetic agents.[67] The authors add a note of caution about the injection of local anesthetic agents into traumatized muscles for the relief of pain, since, depending upon the agent selected and the dosage administered, the damage may outweigh the benefits obtained.

The possible adverse effects of local anesthetic agents upon the fetus and the neonate are, understandably, a matter of concern. Recently Moroshima et al. showed, in studies with sheep, that the mean doses required to produce convulsions, hypotension, respiratory arrest, or circulatory collapse were significantly higher for the fetus (41.9 $\pm$ 6 mg/kg) than for the neonate (18.4 $\pm$ 2.2 mg/kg) and significantly higher for the neonate than for the adult (5.8 $\pm$ 1.8 mg/kg).[68] The authors suggest that the dose required for the neonate may be higher than for the adult because the neonate exhibits a larger volume of distribution. The higher doses required by the fetus may be the result of placental clearance of the drug into the mother's circulation and better maintenance of normal levels of blood gases in the fetus.

The subject of toxicity encompasses drug interactions, but drug interactions may involve either the toxic or the therapeutic actions. A classical example of the latter is the interaction of sulfonamides and agents such as procaine, benzocaine, and tetracaine. These compounds yield para-aminobenzoic acid during biotransformation and consequently may reduce the effectiveness of therapy with sulfonamide drugs. Local anesthetic agents have, at sufficiently high tissue levels, depressant effects on the CNS and on various components of the cardiovascular system. Because of this, they must be used with caution in the presence of other drugs that have a depressant effect on the CNS or cardiovascular system. A different type of interaction occurs between local anesthetics and depolarizing neuromuscular blocking agents: local anesthetics may displace agents such as succinylcholine and decamethonium from their plasma protein binding sites and, as a consequence, prolong the apnea produced by the neuromuscular blocking agents. The toxicity of mixtures of local anesthetic agents has stimulated both rhetoric and research, but the question of whether the toxicities of two different local anesthetics exhibit additive effects, antagonism, or true potentiation has not been completely answered satisfactorily.[93,94] The clinician who administers mixtures of local anesthetic agents or administers, for example, one agent topically and the other parenterally to a patient should at least be aware of the possibility of additive toxic effects. Vasoconstrictors have, in general, beneficial effects when added to solutions of local anesthetic agents. Latency is shortened, depth and duration of block are increased, and absorption into the systemic circulation is slowed. Although the net result depends upon the local anesthetic used and the site of administration, the presence of a vasoconstrictor usually decreases the potential for systemic toxicity of parenterally administered local anesthetic agents. On the other hand, the presence of a vasoconstrictor may enhance the myotoxic and neurotoxic effects of a local anesthetic.[67,94]

DRUG MONOGRAPHS

A. Chemical names and structures — See Table 4.
B. Physical-chemical properties: dissociation constants — See Table 5.
C. Toxicities — See Table 6.
D. Dosage and uses — See Table 7.
E. Synonyms — See Table 8.

OTHER CLASSES OF BLOCKING AGENTS NOT CURRENTLY IN USE

Certain marine and amphibian toxins have received considerable attention because of their highly specific modes of action. The marine toxins tetrodotoxin and saxitoxin, for example, inhibit impulse formation and conduction by physically blocking the external end of the sodium channel.[80,81] Because these toxins and conventional local anesthetic agents act at different sites to block nerve conduction, they exhibit true potentiation when appropriate concentrations of a toxin and a local anesthetic agent are used concomitantly to produce nerve block.[82] Adams et al. have demonstrated that in laboratory animals extremely long blocks can be effected by the use of such mixtures.[83,84] Ultralong-acting agents would indeed be highly desirable in certain types of pain control, but to date neither saxitoxin nor tetrodotoxin, alone or with a local anesthetic agent, has been demonstrated to be clinically acceptable.[85,86] There is still the intriguing possibility that even if the naturally occurring toxins are not found to be clinically useful, their structures may inspire chemists to synthesize an entirely new class of agents with ultralong durations of block.[86,87]

Batrachotoxin, obtained from certain South American frogs, produces nerve block by activating sodium channels and causing depolarization of the neuronal membrane.[81] Neurophysiologists have used it as a tool to elucidate the structure and functions of the sodium channel. Minute quantities, e.g., total doses of 2 to 4 μg, are capable of producing spinal anesthesia up to 12 hr in duration in the sheep, but these doses may also induce histopathological changes in the spinal cord and roots.[88]

A novel approach in the search for ultralong-acting local anesthetic agents was pursued by Ross and Åkerman.[89,90] They reasoned that since the unionized lipophilic form of the lcoal anesthetic molecule penetrates the neuronal membrane relatively easily and since the ionized hydrophilic form, which is also the form that blocks by fitting onto a cationic receptor site in the axoplasmic end of the sodium channel, penetrates the membrane with difficulty, it might be possible to produce very long blocks by means of an agent that penetrated as the unionized tertiary amine form and then cyclized to yield the charged quaternary form. They conducted studies in laboratory animals with certain haloalkylamines and demonstrated the feasibility of this approach. No clinically acceptable compounds have been identified to date, but this concept continues to be an intriguing one.

Table 4
CHEMICAL NAMES AND STRUCTURES

Generic name	Chemical name	Structure
Benoxinate	4-amino-3-butoxybenzoic acid, 2-(diethylamino)ethyl ester	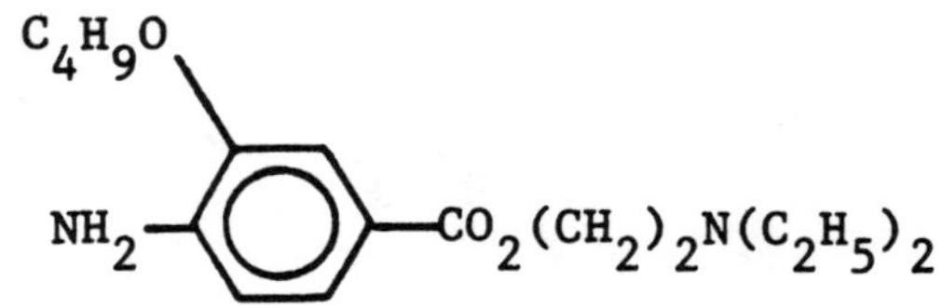
Benzocaine	4-aminobenzoic acid, ethyl ester	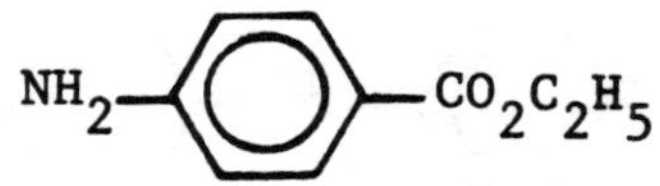
Bupivacaine	1-butyl-*N*-(2,6-dimethylphenyl)-2-piperidinecarboxamide	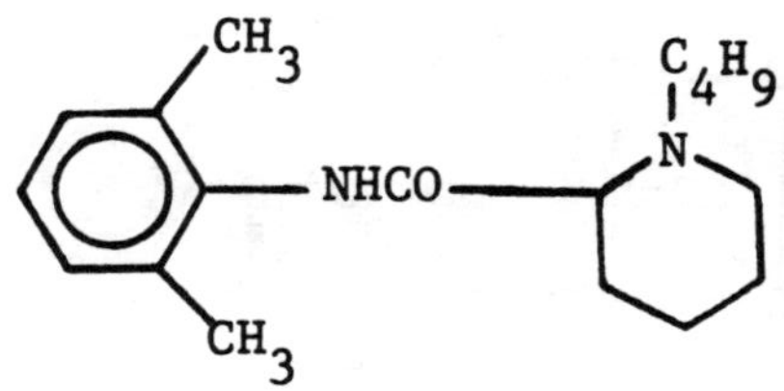
Butacaine	3-(dibutylamino)-1-propanol, 4-aminobenzoate (ester)	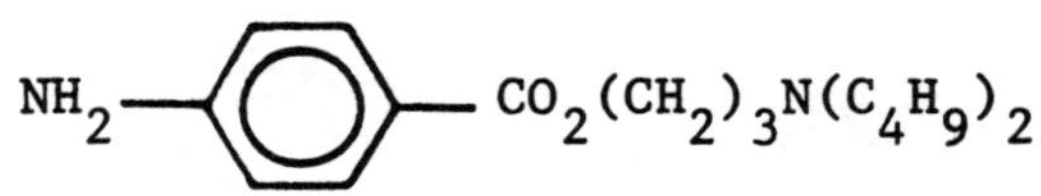
Butamben	4-aminobenzoic acid, butyl ester	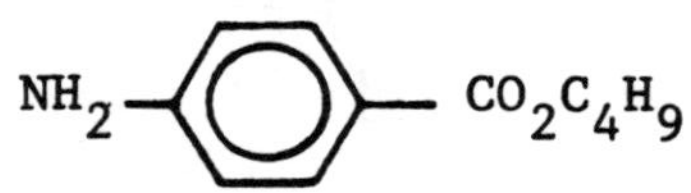

Table 4 (continued)
CHEMICAL NAMES AND STRUCTURES

Generic name	Chemical name	Structure
2-Chloroprocaine	4-amino-2-chlorobenzoic acid, 2-diethylaminoethyl ester	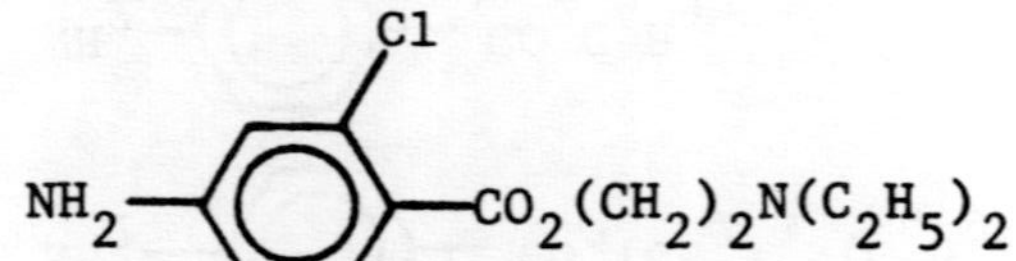
Cocaine	[1R-(exo, exo)]-3-(benzoyloxy)-8-methyl-8-azabicyclo[3.2.1]octane-2-carboxylic acid, methyl ester	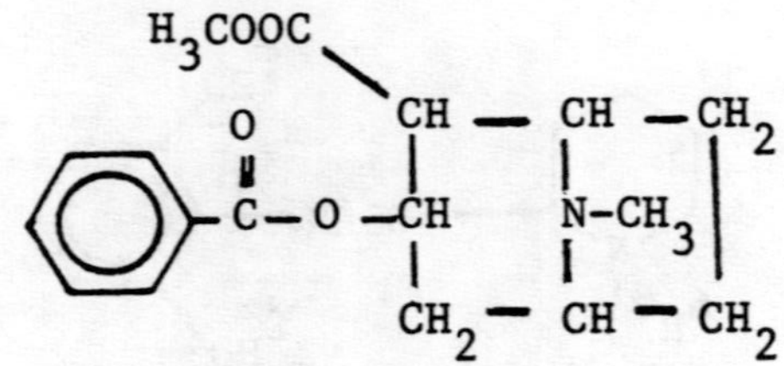
Cyclomethycaine	4-(cyclohexyloxy)benzoic acid, 3-(2-methyl-1-piperidinyl)propyl ester	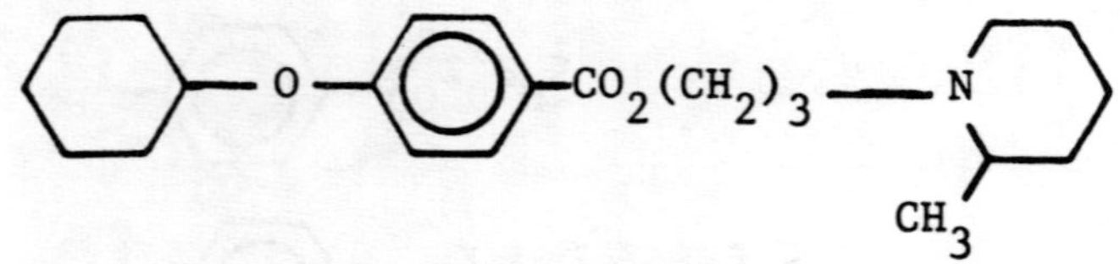
Dibucaine	2-butoxy-*N*-[2-(diethylamino)-ethyl]-4-quinolinecarboxamide	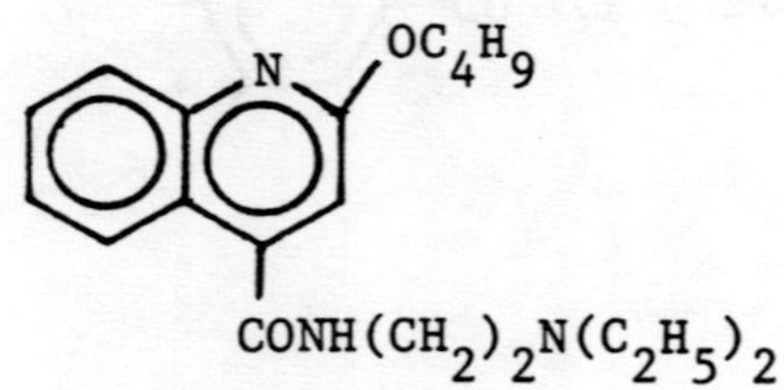

Dimethisoquin	2-[(3-butyl-1-isoquinolinyl)oxy]-*N*,*N*-dimethylethanamine	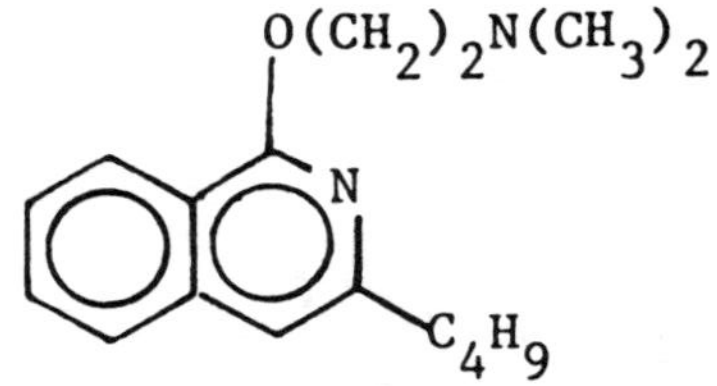
Diperodon	3-(1-piperidinyl)-1,2-propanediol, bis(phenylcarbamate) (ester)	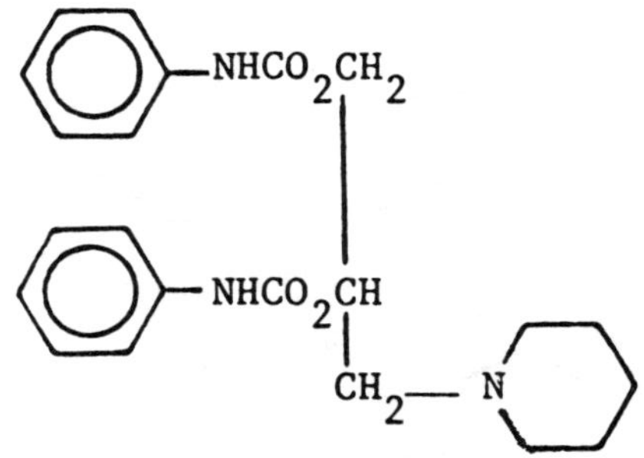
Dyclonine	1-(4-butoxyphenyl)-3-(1-piperidinyl)-1-propanone	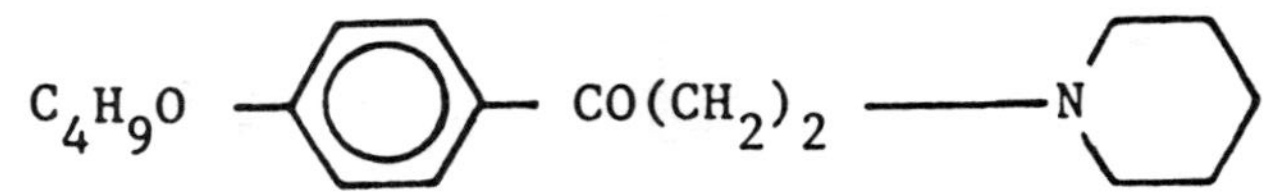
Etidocaine	*N*-(2,6-dimethylphenyl)-2-(ethylpropylamino)butanamide	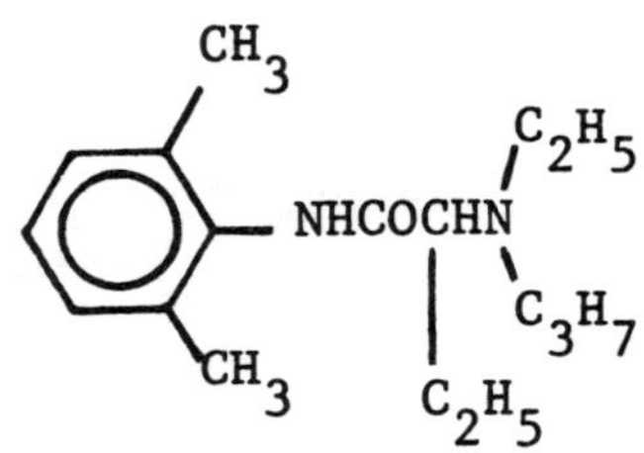
Hexylcaine	1-cyclohexylamino-2-propanol, benzoate (ester)	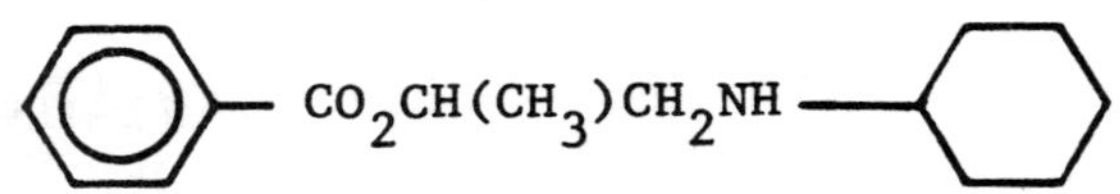

Table 4 (continued)
CHEMICAL NAMES AND STRUCTURES

Generic name	Chemical name	Structure
Lidocaine	2-(diethylamino)-*N*-(2,6-dimethylphenyl)acetamide	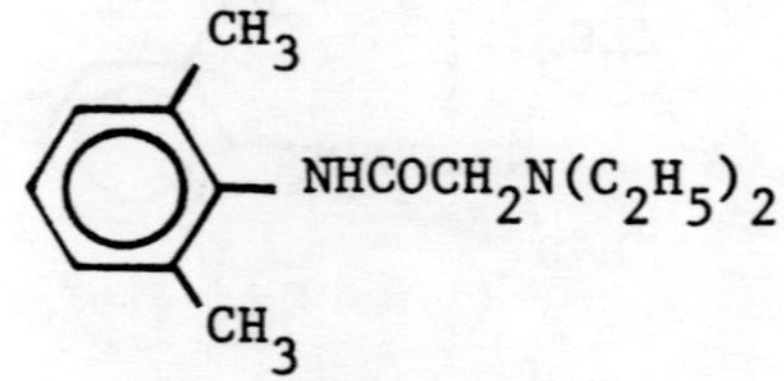
Mepivacaine	*N*-(2,6-dimethylphenyl)-1-methyl-2-piperidinecarboxamide	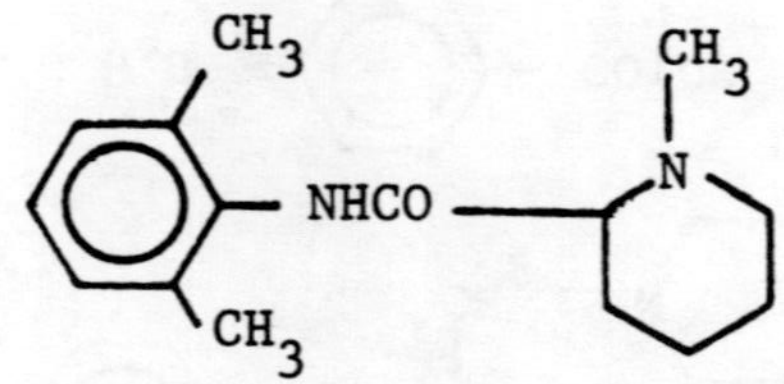
Piperocaine	3-(2-methyl-1-piperidinyl)-propanol, benzoate	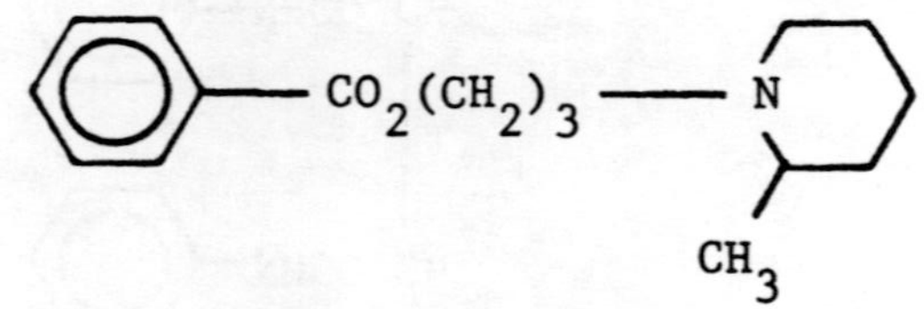
Pramoxine	4-[3-(4-butoxyphenoxy)propyl]-morpholine	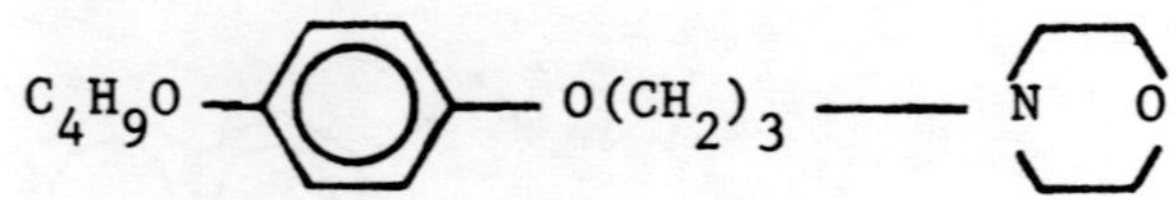

Prilocaine	*N*-(2-methylphenyl)-2-(propylamino)-propanamide	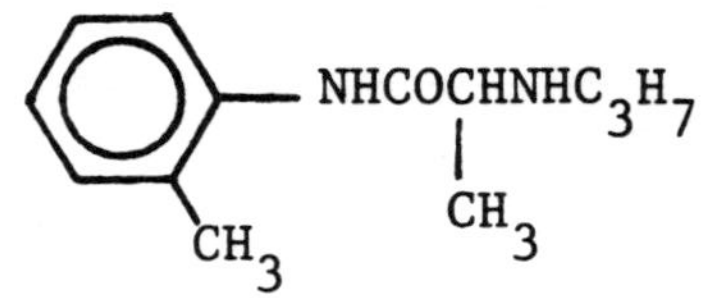
Procaine	4-aminobenzoic acid, 2-(diethylamino)ethyl ester	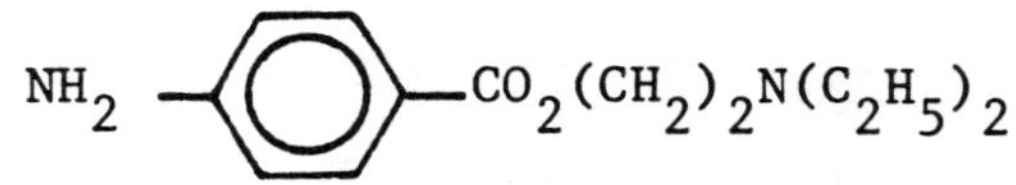
Proparacaine	3-amino-4-propoxybenzoic acid, 2-(diethylamino)ethyl ester	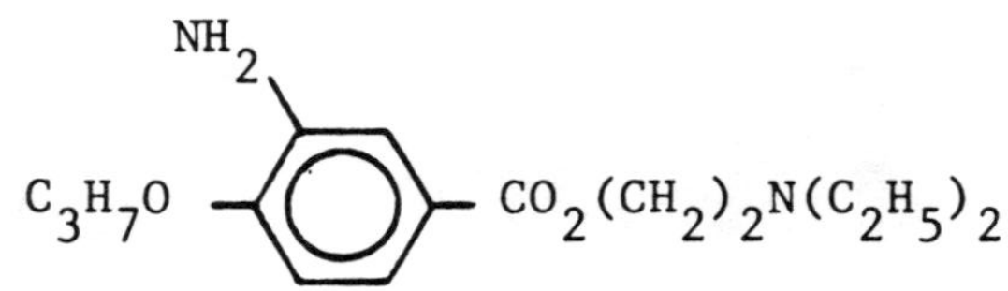
Propoxycaine	4-amino-2-propoxybenzoic acid, 2-(diethylamino)ethyl ester	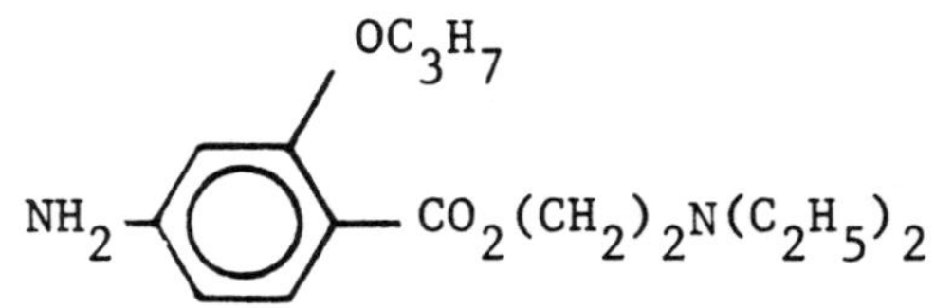
Tetracaine	4-(butylamino)benzoic acid, 2-(dimethylamino)ethyl ester	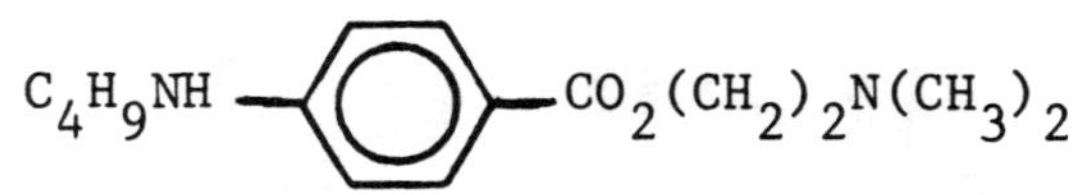

Table 5
PHYSICAL-CHEMICAL PROPERTIES: DISSOCIATION CONSTANTS

Generic name	pKa Water	pKa 50% Ethanol	Ref.
Benoxinate	8.94	—	21
Benzocaine	3.5	—	69
Bupivacaine	8.1	—	69
Butacaine	—	8.25	21
Chloroprocaine	9.0	—	69
	8.7		13
Cocaine	8.7	—	69
Cyclomethycaine	—	8.56	21
Dibucaine	8.8	—	69
Dimethisoquin	—	7.71	21
Dyclonine	—	8.21	21
Etidocaine	7.7	—	69
Hexylcaine	9.3	—	69
Lidocaine	7.9	—	69
Mepivacaine	7.6	—	69
Piperocaine	9.8	—	69
Pramoxine	7.08	—	21
Prilocaine	7.9	—	69
Procaine	8.9	—	69
	9.02		21
Proparacaine	8.96	—	21
Propoxycaine	9.05	—	21
Tetracaine	8.2	—	69
	8.5		13

More recently, Scurlock and Curtis have examined the possibility of producing ultralong nerve blocks by means of certain derivatives of tetraethylammonium.[91,92] Reversible blocks of 2 to 3 weeks in duration have been produced in laboratory animals, but further studies are required in order to determine whether or not compounds of this type will be clinically useful.

Table 6
TOXICITIES

Acute Toxicities in Female Mice: I.V. and S.C. Administration

Compound and salt form	LD_{50} and 95% confidence limits mg as base/kg of body weight	
	I.V.	S.C.
Bupivacaine HCl	6.4 (5.5—7.3)[a]	45 (38—54)[a]
Dibucaine HCl	5 (4—6.3)	27 (20—41)
Etidocaine HCl	6.7 (5.8—7.5)[a]	99 (85—147)[a]
Lidocaine HCl	26 (23—33)[a]	211 (183—256)[a]
Mepivacaine HCl	—	278 (227—333)
Prilocaine HCl	—	550 (359—905)
Procaine HCl	—	287 (197—480)
Tetracaine HCl	4.1 (2.9—5.3)[a]	32 (25—42)[a]

Acute Toxicities in Female Mice: Intraperitoneal Administration

Compound and salt form	Mol Wt[b] base	Mol Wt salt	LD_{50} and 95% confidence limits mg of base/kg of body weight
Benoxinate HCl	308	345	41 (35—47)
Bupivacaine HCl	288	343	40 (28—56)
Butacaine Sulfate	306	711	33 (30—77)
Cocaine HCl	303	340	63 (53—76)
Cyclomethycaine sulfate	360	458	109 (87—120)
Dibucaine HCl	343	380	23 (15—31)
Diperodon HCl	397	434	197 (169—229)
Etidocaine HCl	276	312	62 (53—82)
Falicain HCl	275	312	46 (39—53)
Hexylcaine HCl	261	298	165 (108—221)
Lidocaine HCl	234	271	133 (121—150)[c]
Mepivacaine HCl	246	283	95 (82—108)
Oxethazaine HCl	468	504	42 (30—61)
Phenacaine HCl	298	353	51 (44—55)
Proparacaine HCl	294	331	45 (20—66)
Propoxycaine HCl	294	331	68 (62—76)
Pyrrocaine HCl	232	269	150 (130—179)
Tetracaine HCl	264	301	55 (44—69)[c]

[a] LD_{50}s for these compounds are from Reference 8. LD_{50}s for the remaining compounds are from unpublished data of Adams and Takman.

[b] Molecular weights have been rounded off to first whole number.

[c] LD_{50}s for lidocaine and tetracaine are from Reference 70; the other LD_{50}s are from unpublished data of Adams and Takman.

Table 7
DOSAGE AND USES: INJECTION AND TOPICAL APPLICATION

Generic name	Principal recommended clinical uses	Conc available (%)	Maximum recommended single adult dose	Principal site(s) of biotransformation	Comments
Bupivacaine	Local infiltration Peripheral nerve block Retrobular anesthesia Sympathetic nerve block Lumbar and caudal epidural anesthesia	0.25, 0.5, 0.75	175 mg without epinephrine 225 mg with epinephrine 1:200,000	Liver	A long-acting agent that is widely used for regional anesthesia
Chloroprocaine	Infiltration Peripheral nerve block Lumbar and caudal epidural anesthesia	1.0, 2.0, 3.0	800 mg without epinephrine 1000 mg with epinephrine 1:200,000	Plasma	A short-acting agent; it is hydrolyzed about four times faster than procaine in human plasma
Dibucaine	Topical Spinal anesthesia	Topical, 0.25, 0.5, 1.0 Parenteral, 0.0667, 0.2, 0.25, 0.5	50 mg	Liver	One of the most potent, most toxic, and longest acting of the agents in current use
Etidocaine	Percutaneous infiltration Peripheral nerve block Lumbar and caudal epidural anesthesia	0.5, 1.0, 1.5	300 mg without epinephrine 400 mg with epinephrine, 1:200,000	Liver	A long-acting agent; appears to be especially useful in those procedures requiring good motor block
Lidocaine	Topical Percutaneous infiltration I.V. regional anesthesia Peripheral nerve block Lumbar and caudal epidural anesthesia Spinal anesthesia	Topical, 2.0, 2.5, 4.0, 5.0, 10 Parenteral, 0.5, 1.0, 1.5, 2.0, 4.0, 5.0	300 mg without epinephrine 500 mg with epinephrine, 1:200,000	Liver	One of the most versatile and widely used local anesthetic agents

Mepivacaine	Infiltration Peripheral nerve block Lumbar and caudal epidural anesthesia	1.0, 1.5, 2.0	400 mg	Liver	Similar in potency, toxicity, and duration of action to lidocaine
Prilocaine	Infiltration Peripheral nerve block Lumbar and caudal epidural anesthesia	1.0, 2.0, 3.0	600 mg	Liver Kidney (?) Lung (?)	One of the metabolites of prilocaine, *o*-toluidine, is responsible for producing methemoglobinemia
Procaine	Infiltration Peripheral nerve block Spinal anesthesia	1.0, 2.0, 10	1000 mg	Plasma	The first synthetic local anesthetic agent to gain wide acceptance
Propoxycaine	Infiltration Peripheral nerve block Sympathetic nerve block	0.4% (with 2% procaine) 0.5	100 mg	Plasma	Not used for epidural or spinal anesthesia
Tetracaine	Topical Spinal anesthesia	Topical 2% Spinal anesthesia 1% solution and as soluble salt form, 20 mg/ampule	15—20 mg (in spinal anesthetic procedures)	Plasma	Like dibucaine, it is too toxic for routine use in many regional anesthetic procedures but is an excellent spinal anesthetic agent with a long duration of action
Dimethisoquin	Surface anesthesia of the skin and mucocutaneous junctions	0.5%			Use in the eyes and on the eyelids is proscribed
Diperodon	Surface anesthesia of the skin and mucocutaneous junctions	1%			Reported to be as potent as cocaine and to have a longer duration of action.
Dyclonine	Surface anesthesia of the skin and mucous membranes of the mouth, esophagus, upper respiratory tract, rectum, urethra, and vagina	0.5, 1			Contraindicated in cystoscopic procedures following i.v. pyelography: iodine-containing contrast media can cause a precipitate, which interferes with visualization
Hexylcaine	Surface anesthesia of intact mucous membranes of the respiratory, upper GI, and urinary tracts	Up to 5			Was formerly used for regional block procedures, but this use has been discontinued

Table 7 (continued)
DOSAGE AND USES: INJECTION AND TOPICAL APPLICATION

Generic name	Principal recommended clinical uses	Conc available (%)	Maximum recommended single adult dose	Principal site(s) of biotransformation	Comments
Piperocaine	Used almost exclusively for surface anesthesia although it may also be used for infiltration and peripheral nerve block anesthesia	Powder (for solution) for surface anesthesia; 2% solution for infiltration and peripheral nerve block			Use of this agent appears to have declined, perhaps because it offered no marked advantage over other currently available agents
Pramoxine	Surface anesthesia of the skin and mucous membranes	Up to 1			Contraindicated in bronchoscopy and gastroscopy. Use in the eye and on nasal mucosa not recommended
Proparacaine	Surface anesthesia in ophthalmological procedures	0.5			About as potent as tetracaine but reported to cause little or no initial irritation upon application
Benoxinate	In ophthalmology	0.4			Relatively short duration of action makes it suitable for removal of foreign bodies in the eye and for short operative procedures on the cornea or conjunctiva
Benzocaine	Surface anesthesia of skin and mucous membranes	0.5—20			One of the most widely used agents for the relief of sunburn, pruritus, and minor burns; present in numerous over-the-counter preparations

Butacaine	Surface anesthesia of the eye, nose, and throat	Up to 2	Onset faster and duration longer than cocaine when applied to mucous membranes; in general, does not affect accommodation or intraocular pressure
Butamben	Surface anesthesia of the skin	Up to 1	Sustained anesthetic effect and low incidence of systemic toxic reactions are attributed to low aqueous solubility and sequent poor absorption
Cocaine	Surface anesthesia of the mucosa of the upper respiratory tract		It is the only local anesthetic agent in use that has intrinsic vasoconstrictor activity and addiction liabilities; the vasoconstrictor activity is a consequence of its ability to block uptake of norepinephrine by adrenergic nerve terminals
Cyclomethycaine	Surface anesthesia of the skin and mucous membranes of the rectum, vagina, urethra, and urinary bladder	Up to 1	Reported to be relatively ineffective when applied to the mucous membranes of the mouth, nose, trachea, bronchi, and eye

Table 8
SYNONYMS: GENERIC AND COMMON PROPRIETARY NAMES

Generic	Proprietary
Benoxinate	Dorsacaine
Oxybuprocaine[a]	
Benzocaine	Numerous proprietary names
Ethylaminobenzoate	
Bupivacaine	Marcadaine, Sensorcaine
Butacaine	Butyn Sulfate
Butamben	Butesin
Carticaine	Ulfracain
Chloroprocaine	Nesacaine
Cyclomethycaine	Surfacaine
Dibucaine	Nupercaine
Cinchocaine	
Dimethisoquin	Quotane
Diperodon	Diothane
Diperocaine	
Dyclonine	Dyclone
Etidocaine	Duranest
Hexylcaine	Cyclaine
Lidocaine	Xylocaine[b]
Lignocaine[a]	
Mepivacaine	Carbocaine, Polocaine
Oxethazaine	Oxaine
Oxetacaine	
Phenacaine	Holocaine
Piperocaine	Metycaine
Pramoxine	Tronothane
Prilocaine	Citanest
Propitocaine	
Procaine	Novocaine[b]
Proparacaine	Ophthaine, Alcaine, Ophthetic
Proxymetycaine[a]	
Propoxycaine	Blockain
Pyrrocaine	Dynacaine
Tetracaine	Pontocaine[b]
Amethocaine[a]	

[a] U.K.
[b] Numerous other proprietary names.

REFERENCES

1. **Covino, B. G. and Vassallo, H. G.,** *Local Anesthetics: Mechanisms of Action and Clinical Use,* Grune & Stratton, New York, 1976.
2. **von Anrep, B.,** Ueber die physiologische Wirkung des Cocain, *Pflügers Arch. Ges. Physiol.*, 21, 38, 1880.
3. **Koller, K.,** Ueber die Verwendung des Cocain zur Anästhesierung am Auge, *Wien. Med. Wochenschr.*, 34, 1276, 1884.
4. **Brann, H.,** Ueber einige neue örtliche Anaesthetica, *Dtsch. Med. Wochenschr.*, 31, 1667, 1905.
5. **Fussgänger, R. and Schaumann, O.,** A new local anesthetic of the novocain series (Pantocain), *Arch. Exp. Pathol. Pharmakol.*, 160, 53, 1931.
6. **Löfgren, N.,** *Studies in Local Anesthetics. Xylocaine, A New Synthetic Drug,* Haeggström, Stockholm, 1948; reprinted, Worcester, Mass., 1958.

7. **af Ekenstam, B., Egnér, B., and Pettersson, G.,** Local anesthetics, I. N-alkyl pyrrolidine and N-alkyl piperidine carboxylic acid amides, *Acta Chem. Scand.*, 11, 1183, 1957.
8. **Adams, H. J., Kronberg, G. H., and Takman, B. H.,** Local anesthetic activity and acute toxicity of (±)-2-(N-ethylpropylamino) - 2′,6′-butyroxylidide, a new long-acting agent, *J. Pharm. Sci.*, 61, 1829, 1972.
9. **Adams, H. J. and Takman, B. H.,** unpublished observations.
10. **Takman, B. H., Boyes, R. N., and Vassallo, H. G.,** Local anesthetics, in *Principles of Medicinal Chemistry,* Foye, W. O., Ed., Lea & Febiger, Philadelphia, 1981, chap. 14.
11. **Takman, B. H. and Adams, H. J.,** Local anesthetics, in *Burger's Medicinal Chemistry,* Part III, Wolff, M. E., Ed., John Wiley & Sons, New York, 1981, chap. 51.
12. **Mather, L. E., and Tucker, G. T.,** Pharmacokinetics and biotransformation of local anesthetics, in *Int. Anesthesiol. Clin.*, 16, 23, 1978.
13. **Tucker, G. T. and Mather, L. E.,** Clinical pharmacokinetics of local anaesthetics, *Clin. Pharmacokinet.*, 4, 241, 1979.
14. **Cousins, M. J. and Mather, L. E.,** Clinical pharmacology of local anesthetics, *Anaesth. Intens. Care,* 8, 257, 1980.
15. **Mather, L. E. and Cousins, M. J.,** Local anaesthetics and their current clinical use, *Drugs,* 18, 185, 1979.
16. **Takman, B. H.,** The chemistry of local anaesthetic agents: classification of blocking agents., *Br. J. Anaesth.*, 47, 183, 1975.
17. **Narahashi, T., Haas, H. G., and Therien, E. F.,** Saxitoxin and tetrodotoxin: comparison of nerve blocking mechanism, *Science,* 157, 1441, 1967.
18. **Frazier, D. T., Narahashi, T., and Yamada, M.,** The site of action and active form of local anesthetics. II. Experiments with quaternary compounds, *J. Pharmacol. Exp. Ther.*, 171, 45, 1970.
19. **Narahashi, T. and Frazier, D. T.,** Site of action and active form of local anesthetics, in *Neurosciences Research,* Vol. 4, Ehrenpreis, S. and Solnitzky, O. C., Eds., Academic Press, New York, 1971, 65.
20. **Strichartz, G. R.,** Current concepts of the mechanism of action of local anesthetics, *J. Dent. Res.*, 60, 1460, 1981.
21. **Büchi, J. and Perlia, X.,** Structure-activity relations and physico-chemical properties of local anesthetics, *International Encyclopedia of Pharmacology and Therapeutics,* Vol. 1, Sect. 8, Lechat, P., Ed., Pergamon Press, Oxford, 1971, chap. 2.
22. **Wiedling, S. and Tegnér, C.,** Local anaesthetics, in *Progress in Medicinal Chemistry,* Ellis, G. P. and West, G. B., Butterworth, Washington, 1963, chap. 7.
23. **Åström, A.,** Classification and structure-activity relationship of local anesthetic agents, in *Anesthésiques Locaux en Anesthésia et Réanimation,* Librairie Arnette, Paris, 1974.
24. **Seifen, A. B., Ferrari, A. A., Seifen, E. E., Thompson, D. S., and Chapman, J.,** Pharmacokinetics of intravenous procaine infusion in humans, *Anesth. Analg.*, 58, 382, 1979.
25. **O'Brien, J. E., Abbey, V., Hinsvark, O., Perel, J., and Finster, M.,** Metabolism and measurement of chloroprocaine, an ester-type local anesthetic, *J. Pharm. Sci.*, 68, 75, 1979.
26. **Zsigmond, E. K., Hinsvark, O. W., Chandra, P., and Kothary, S. P.,** Pharmacokinetics of 2-chloroprocaine in E_1^a E_1^a *Anesthesiology,* 51, S222, 1979.
27. **Javaid, J. J., Fischman, M. W., Schuster, C. R., Dekirmenjian, H., and Davis, J. M.,** Cocaine plasma concentrations: relation to physiological and subjective effects in humans, *Science,* 202, 227, 1978.
28. **Wilkinson, P., Van Dyke, C., Jatlow, P., Barash, P., and Byck, R.,** Intranasal and oral cocaine kinetics, *Clin. Pharmacol. Ther.*, 27, 386, 1980.
29. **Tucker, G. T., and Mather, L. E.,** Pharmacokinetics of local anesthetic agents, *Br. J. Anaesth.*, 47, 213, 1975.
30. **Piafsky, K. M. and Knoppert, D.,** Binding of local anesthetics to $alpha_1$ acid glycoprotein, *Clin. Res.*, 26, 836A, 1978.
31. **Routledge, P. A., Barchowsky, A., Bjornsson, T. D., Kitchell, B. B., and Shand, D. G.,** Lidocaine plasma protein binding, *Clin. Pharmacol. Ther.*, 27, 347, 1980.
32. **Tucker, G. T., Moore, D. C., Bridenbaugh, P. O., Bridenbaugh, L. D., and Thompson, G. E.,** Systemic absorption of mepivacaine in commonly used regional block procedures, *Anesthesiology,* 37, 277, 1972.
33. **Ralston, D. H. and Shnider, S. M.,** The fetal and neonatal effects of regional anesthesia in obstetrics, *Anesthesiology,* 48, 34, 1978.
34. **Brown, W. U., Bell, G. C., Lurie, A. O., Weiss, J. B., Scanlon, J. W., and Alper, M. H.,** Newborn blood levels of lidocaine and mepivacaine in the first postnatal day following maternal epidural anesthesia, *Anesthesiology,* 42, 698, 1975.
35. **Caldwell, J., Moffatt, J. R., Smith, R. L., Lieberman, B. A., Cawston, M. O., and Beard, R. W.,** Pharmacokinetics of bupivacaine administered epidurally during childbirth, *Br. J. Clin. Pharm.*, 3, 956P, 1976.

36. **Meffin, P., Long, G. L., and Thomas, J.,** Clearance and metabolism of mepivacaine in the human neonate, *Clin. Pharmacol. Ther.*, 14, 218, 1973.
37. **Mihaly, G. W., Moore, R. G., Thomas, J., Triggs, E. J., Thomas, D., and Shanks, C. A.,** The pharmacokinetics and metabolism of the anilide local anesthetics in neonates. I. Lignocaine, *Eur. J. Clin. Pharmacol.*, 13, 143, 1978.
38. **Moore, R. G., Thomas, J., Triggs, E. J., Thomas, D. B., Burnard, E. D., and Shanks, C. A.,** The pharmacokinetics and metabolism of the anilide local anesthetics in neonates. III. Mepivacaine, *Eur. J. Clin. Pharmacol.*, 14, 203, 1978.
39. **Morgan, D., McQuillan, D., and Thomas, J.,** Pharmacokinetics and metabolism of the anilide local anesthetics in neonates. II. Etidocaine, *Eur. J. Clin. Pharmacol.*, 13, 365, 1978.
40. **Mather, L. E., Long, G. L., and Thomas, J.,** The binding of bupivacaine to maternal and foetal plasma proteins, *J. Pharm. Pharmacol.*, 23, 359, 1971.
41. **Thomas, J., Long, G., Moore, G., and Morgan, D.,** Plasma protein binding and placental transfer of bupivacaine, *Clin. Pharmacol. Ther.*, 19, 426, 1975.
42. **Tucker, G. T., Boyes, R. N., Bridenbaugh, P. O., and Moore, D. C.,** Binding of anilide-type local anesthetics in human plasma. II. Implications *in vivo* with special reference to transplacental distribution, *Anesthesiology*, 33, 304, 1970.
43. **DuSouich, P. and Erill, P.,** Altered metabolism of procainamide and procaine in patients with pulmonary and cardiac diseases, *Clin. Pharmacol. Ther.*, 21, 101, 1977.
44. **Reidenberg, M. M., James, M., and Dring, L. G.,** The rate of procaine hydrolysis in serum of normal subjects and diseased patients, *Clin. Pharmacol. Ther.*, 13, 279, 1972.
45. **Åkerman, B., Åström, A., Ross, S., and Telč, A.,** Studies on the absorption, distribution and metabolism of labeled prilocaine and lidocaine in some animal species, *Acta Pharmacol. Toxicol.*, 24, 389, 1966.
46. **Åkerman, B. and Ross, S.,** Stereospecificity of enzymatic biotransformation of the enantiomers of prilocaine, *Acta Pharmacol. Toxicol.*, 28, 445, 1970.
47. **Arthur, G. R., Scott, D. H. T., Boyes, R. N., and Scott, D. B.,** Pharmacokinetic and clinical pharmacological studies with mepivacaine and prilocaine, *Br. J. Anaesth.*, 51, 481, 1979.
48. **Keenaghan, J. B. and Boyes, N. R.,** The tissue distribution, metabolism and excretion of lidocaine in rats, guinea pigs, dogs and man, *J. Pharmacol. Exp. Ther.*, 180, 459, 1972.
49. **Nelson, S. D., Garland, W. A., Breck, G. D., and Trager, W. F.,** Quantitation of lidocaine and several metabolites utilizing chemical ionization, mass spectrometry and stable isotope labeling, *J. Pharm. Sci.*, 66, 1180, 1977.
50. **Meffin, P. J., and Thomas, J.,** The relative rates of formation of the phenolic metabolites of mepivacaine in man, *Xenobiotica*, 3, 625, 1973.
51. **Thomas, J. and Meffin, P. J.,** Aromatic hydroxylation of lidocaine and mepivacaine in rats and humans, *J. Med. Chem.*, 15, 1046, 1972.
52. **Reynolds, F.,** Metabolism and excretion of bupivacaine in man: a comparison with mepivacaine, *Br. J. Anaesth.*, 43, 33, 1971.
53. **Goehl, T. J., Davenport, J. B., and Stanley, M. J.,** Distribution, biotransformation and excretion of bupivacaine in the rat and monkey, *Xenobiotica*, 3, 761, 1973.
54. **Thomas, J., Morgan, D., and Vine, J.,** Metabolism of etidocaine in man, *Xenobiotica*, 6, 39, 1976.
55. **Vine, J., Morgan, D., and Thomas, J.,** The identification of eight hydroxylated metabolites of etidocaine by chemical ionization mass spectrometry, *Xenobiotica*, 8, 509, 1978.
56. **Rowland, M., Thomson, P. D., Guichard, A., and Melmon, K. L.,** Disposition kinetics of lidocaine in normal subjects, *Ann. N.Y. Acad. Sci.*, 179, 383, 1971.
57. **Tucker, G. T., Boyes, R. N., Bridenbaugh, P. O., Moore, D. C.,** Binding of anilide type local anesthetics in human plasma. I. Relationships between binding, physicochemical properties, and anesthetic activity, *Anesthesiology*, 33, 287, 1970.
58. **Eriksson, E.,** Prilocaine, an experimental study in man of a new local anesthetic with special regards to efficacy toxicity and excretion, *Acta Chir. Scand. Suppl.*, 358, 1966.
59. **Magno, R., Berlín, A., Karlsson, K., and Kjellmer, I.,** Anesthesia for Cesarian section. IV. Placental transfer and neonatal elimination of bupivacaine following epidural analgesia for elective Cesarian section, *Acta Anaesthesiol. Scand.*, 20, 141, 1976.
60. **Adams, H. J., Mastri, A. R., Doherty, D., and Charron, D.,** Spinal anesthesia with batrachotoxin in sheep and microsopic examination of spinal cords and roots, *Pharmacol. Res. Commun.*, 10, 719, 1978.
61. **Adams, H. J., Mastri, A. R., Eicholzer, A. W., and Kilpatrick, G.,** Morphological effects of intrathecal etidocaine and tetracaine on the rabbit spinal cord, *Anesth. Analg. (Cleveland)*, 53, 904, 1974.
62. **Griffith, J. F.,** Interlaboratory variations in the determination of acute oral LD_{50}, *Toxicol. Appl. Pharmacol.*, 6, 726, 1964.
63. **Lamanna, C. and Hart, E. R.,** Relationship of lethal toxic dose to body weight of the mouse, *Toxicol. Appl. Pharmacol.*, 13, 307, 1968.

64. **Rowan, A. N.,** The LD_{50} test: a critique and suggestions for alternatives, *Pharm. Technol.*, p. 65, April 1981.
65. **Brown, V. K. H.,** Acute toxicity testing — a critique, in *Testing For Toxicity*, Gorrod, J. W., Ed., Taylor & Francis, London, 1981, chap. 3.
66. **Benoit, P. W. and Belt, W. D.,** Destruction and regeneration of skeletal muscle after treatment with a local anesthetic, bupivacaine, *J. Anat.*, 107, 547, 1970.
67. **Yagiela, J. A., Benoit, P. W., Buoncristiani, R. D., Peters, M. P., and Fort, N. F.,** Comparison of myotoxic effects of lidocaine with epinephrine in rats and humans, *Anesth. Analg. (Cleveland)*, 60, 471, 1981.
68. **Moroshima, H. O., Pedersen, H., Finster, M., Sakuma, K., Bruce, S. L., Gutche, B. B., Stark, R. I., and Covino, B. G.,** Toxicity of lidocaine in adult, newborn and fetal sheep, *Anesthesiology*, 55, 57, 1981.
69. **de Jong, R. H.,** *Local Anesthetics*, Charles C Thomas, Springfield, Ill., 1977, 281.
70. **Adams, H. J., Kronberg, G. H., and Takman, B. H.,** Primary local anesthetic testing and acute toxicity of N,N'-disubstituted 2,3-diamino-2',6'-propionoxylidides, *J. Pharm. Sci.*, 62, 1677, 1973.
71. **Paterson, J. G. and Hamilton, J. T.,** A nerve chamber bioassay of local anesthetic activity, *Arch. Int. Pharmacodyn. Ther.*, 183, 360, 1970.
72. **Camougis, G. and Takman, B. H.,** Nerve and muscle preparations (as applied to local anesthetics), in *Methods in Pharmacology*, Vol. 1, Schwartz, A., Ed., Appleton-Century-Crofts, New York, 1971, chap. 1.
73. **Bianchi, C.,** A simple quantitative method for testing local anesthetics, *Br. J. Pharmacol.*, 11, 104, 1956.
74. **Luduena, F. P. and Hoppe, J. O.,** Local anesthetic activity, toxicity and irritancy of 2-alkoxy analogs of procaine and tetracaine, *J. Pharmacol. Exp. Ther.*, 104, 40, 1952.
75. **Ready, L. B. and Fink, B. R.,** Experimental evaluation of local anesthetic solutions using rat infraorbital nerve block, *Can. Anaesth. Soc. J.*, 27, 58, 1980.
76. **Chance, M. R. A. and Lobstein, H.,** The value of the guinea-pig corneal reflex for tests of surface anaesthesia, *J. Pharmacol. Exp. Ther.*, 82, 203, 1944.
77. **Campbell, A. H., Stasse, J. A., Lord, G. H., and Willson, J. E.,** *In vivo* evaluation of local anesthetics applied topically, *J. Pharm. Sci.*, 57, 2045, 1968.
78. **Åkerman, S. B. A.,** Surface anesthetic properties of the new local anesthetic amide etidocaine (Duranest): a laboratory evaluation, *Br. J. Anaesth.*, 47, 923, 1975.
79. **Åström, A. and Persson, N. H.,** The toxicity of some local anesthetics after application on different mucous membranes and its relation to anesthetic action on the nasal mucosa of the rabbit, *J. Pharmacol. Exp. Ther.*, 132, 87, 1961.
80. **Evans, M. H.,** Tetrodotoxin, saxitoxin and related substances: their applications in neurobiology, *International Review Neurobiology*, Vol. 15, Pfeiffer, C. C. and Smythies, J. R., Eds., Academic Press, New York, 1971, 83.
81. **Narahashi, T.,** Chemicals as tools in the study of excitable membranes, *Physiol. Rev.*, 54, 518, 1974.
82. **Staiman, A. and Seeman, P.,** Different sites of membrane action for tetrodotoxin and lipid-soluble anaesthetics, *Can. J. Physiol. Pharmacol.*, 53, 513, 1975.
83. **Adams, H. J., Blair, M. R., Jr., and Takman, B. H.,** The local anesthetic activity of tetrodotoxin alone and in combination with vasoconstrictors and local anesthetics, *Anesth. Analg. (Cleveland)*, 55, 568, 1976.
84. **Adams, H. J., Blair, M. R., Jr., and Takman, B. H.,** The local anesthetic activity of saxitoxin alone and with vasoconstrictor and local anesthetic agents, *Arch. Int. Pharmacodyn. Ther.*, 224, 275, 1976.
85. **Scurlock, J. E. and Curtis, B.,** The clinical potential of tetrodotoxin, *Anesthesiology*, 51, S233, 1979.
86. **Covino, B. G.,** Ultralong-acting local anesthetic agents, *Anesthesiology*, 54, 263, 1981.
87. **Ritchie, J. M. and Greene, N. M.,** Local anesthetics in *The Pharmacological Basis of Therapeutics*, Gilman, A. G., Goodman, L. S., and Gilman, A., Eds., Macmillan, New York, 1980, chap. 15.
88. **Adams, H. J., Mastri, A. R., Doherty, D., and Charron, D.,** Spinal anesthesia with batrachotoxin in sheep and microscopic examination of spinal cords and roots, *Pharmacol. Res. Commun.*, 10, 719, 1978.
89. **Ross, S. B. and Åkerman, S. B. A.,** Formation of a piperidinium derivative from N-(5'-chloropentyl)-N-methylaminoaceto-2,6-xylidide in relation to the sustained local anesthetic action on the sciatic nerve of the guinea-pig *in vivo*, *Nature New Biol.*, 230, 274, 1971.
90. **Ross, S. B. and Åkerman, S. B. A.,** Cyclization of N-ω-haloalkyl-N-methylaminoaceto-2,6-xylidide derivatives in relation to their local anesthetic effect *in vitro* and *in vivo*, *J. Pharmacol. Exp. Ther.*, 182, 351, 1972.
91. **Scurlock, J. E. and Curtis, B. M.,** Tetraethylammonium derivatives: ultralong-acting local anesthetics?, *Anesthesiology*, 54, 265, 1981.
92. **Curtis, B. M. and Scurlock, J. E.,** The mechanism of action of local anesthesia by tetraethylammonium derivatives, *Anesthesiology*, 54, 270, 1981.
93. **de Jong, R. H.,** *Local Anesthetics*, Charles C. Thomas, Springfield, Ill., 1977, 257.

94. **Brecher, M. J., Greenberg, R. A., and Greene, N. M.,** Drug interactions (toxicity) associated with drug combinations, *Acta Anaesthesiol. Scand.*, 16, 22, 1972.
95. **Selander, D., Brattsand, R., Lundborg, G., Nordberg, C., and Olsson, Y.,** Local anesthetics: importance of mode of application, concentration and adrenaline for appearance of nerve lesions, *Acta Anaesthesiol. Scand.*, 23, 127, 1979.

INDEX

A

B

C

D

E

F

G

H

I

J

K

L

M

N

Q

R

S

T

U

V

W

X

Z